Advances in
INORGANIC CHEMISTRY
AND
RADIOCHEMISTRY

Volume 23

CONTRIBUTORS TO THIS VOLUME

R. P. Burns
Alwyn G. Davies
Lawrence B. Ebert
J. Fenner
Richard J. Lagow
C. A. McAuliffe
F. P. McCullough
John A. Morrison
Geoffrey A. Ozin
William J. Power
A. Rabenau
Henry Selig
Peter J. Smith
G. Trageser

Advances in
INORGANIC CHEMISTRY
AND
RADIOCHEMISTRY

EDITORS

H. J. EMELÉUS

A. G. SHARPE

*University Chemical Laboratory
Cambridge, England*

VOLUME 23

1980

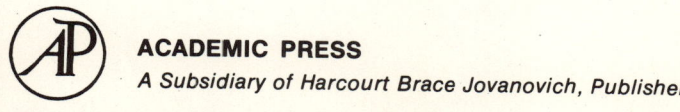

ACADEMIC PRESS
A Subsidiary of Harcourt Brace Jovanovich, Publishers
New York   London   Toronto   Sydney   San Francisco

COPYRIGHT © 1980, BY ACADEMIC PRESS, INC.
ALL RIGHTS RESERVED.
NO PART OF THIS PUBLICATION MAY BE REPRODUCED OR
TRANSMITTED IN ANY FORM OR BY ANY MEANS, ELECTRONIC
OR MECHANICAL, INCLUDING PHOTOCOPY, RECORDING, OR ANY
INFORMATION STORAGE AND RETRIEVAL SYSTEM, WITHOUT
PERMISSION IN WRITING FROM THE PUBLISHER.

ACADEMIC PRESS, INC.
111 Fifth Avenue, New York, New York 10003

United Kingdom Edition published by
ACADEMIC PRESS, INC. (LONDON) LTD.
24/28 Oval Road, London NW1 7DX

LIBRARY OF CONGRESS CATALOG CARD NUMBER: 59-7692

ISBN 0-12-023623-0

PRINTED IN THE UNITED STATES OF AMERICA

80 81 82 83   9 8 7 6 5 4 3 2 1

CONTENTS

LIST OF CONTRIBUTORS ix

Recent Advances in Organotin Chemistry

ALWYN G. DAVIES AND PETER J. SMITH

I. Introduction	1
II. Preparation and Reactions of Organotin Compounds	2
III. Structural Organotin Chemistry	28
IV. Biological and Environmental Aspects	41
V. Industrial Applications	51
References	62

Transition Metal Vapor Cryochemistry

WILLIAM J. POWER AND GEOFFREY A. OZIN

I. Introduction	80
II. Atomic, Diatomic, and Cluster Species	81
III. New Cluster Techniques	101
IV. Cluster Complexes	114
V. Classical Inorganic Ligands	130
VI. Organometallic Complexes	145
VII. Conclusion	166
Addendum	166
Addendum References	168
References	169

New Methods for the Synthesis of Trifluoromethyl Organometallic Compounds

RICHARD J. LAGOW AND JOHN A. MORRISON

I. Introduction	178
II. Plasma-Generated Trifluoromethyl Radicals as a Synthetic Reagent .	181
III. Bis(trifluoromethyl)mercury as a Synthetic Reagent	192
IV. Synthesis of Trifluoromethyl Organometallic Compounds by Direct Fluorination	197

V. A New General Synthesis for Trifluoromethyl Organometallic Compounds and Other Sigma-Bonded Metal Compounds Based on Metal Vapor as a Reagent 203
References 208

1,1-Dithiolato Complexes of the Transition Elements

R. P. Burns, F. P. McCullough, and C. A. McAuliffe

I. Introduction 211
II. The Ligands 212
III. Transition-Metal Complexes 215
References 269

Graphite Intercalation Compounds

Henry Selig and Lawrence B. Ebert

I. Introduction 281
II. Covalent Compounds of Graphite 283
III. Lamellar Compounds 285
IV. Residue Compounds 314
V. Applications of Intercalation Compounds 315
References 319

Solid-State Chemistry of Thio-, Seleno-, and Tellurohalides of Representative and Transition Elements

J. Fenner, A. Rabenau, and G. Trageser

I. Introduction 330
II. Group IB 332
III. Group IIB 351
IV. Group IIIB and Lanthanides 357
V. Group IVB 364
VI. Group VB 364
VII. Group VIB 370
VIII. Group VIIB 379
IX. Group VIIIB 381
X. Group IIIA 382
XI. Group IVA 389

CONTENTS

XII. Group VA	400
Appendix	412
References	413
Appendix References	425
SUBJECT INDEX	427
CONTENTS OF PREVIOUS VOLUMES	436

LIST OF CONTRIBUTORS

Numbers in parentheses indicate the pages on which the authors' contributions begin.

R. P. BURNS (211), *Department of Chemistry, University of Manchester, Institute of Science and Technology, Manchester M60 1QD, England*

ALWYN G. DAVIES (1), *Chemistry Department, University College London, London WC1H 0AJ, England*

LAWRENCE B. EBERT (281), *Exxon Research and Engineering Company, P.O. Box 45, Linden, New Jersey 07036*

J. FENNER* (329), *Max-Planck-Institut für Festkörperforschung, Heisenbergstrasse 1, D-7000 Stuttgart 80, West Germany*

RICHARD J. LAGOW (177), *Department of Chemistry, University of Texas at Austin, Austin, Texas 78712*

C. A. MCAULIFFE (211), *Department of Chemistry, University of Manchester, Institute of Science and Technology, Manchester M60 1QD, England*

F. P. MCCULLOUGH† (211), *Department of Chemistry, University of Manchester, Institute of Science and Technology, Manchester M60 1QD, England*

JOHN A. MORRISON (177), *Department of Chemistry, University of Illinois, Chicago Circle, Chicago, Illinois 60680*

GEOFFREY A. OZIN (79), *Lash Miller Chemical Laboratories and Erindale College, University of Toronto, Toronto, Ontario M5S 1A1, Canada*

WILLIAM J. POWER‡ (79), *Lash Miller Chemical Laboratories and Erindale College, University of Toronto, Toronto, Ontario M5S 1A1, Canada*

A. RABENAU (329), *Max-Planck-Institut für Festkörperforschung, Heisenbergstrasse 1, D-7000 Stuttgart 80, West Germany*

HENRY SELIG (281), *Department of Inorganic and Analytical Chemistry, Hebrew University of Jerusalem, Jerusalem, Israel*

PETER J. SMITH (1), *International Tin Research Institute, Greenford, Middlesex UB6 7AQ, England*

G. TRAGESER (329), *Max-Planck-Institut für Festkörperforschung, Heisenbergstrasse 1, D-7000 Stuttgart 80, West Germany*

---

* Present address: Th. Goldschmidt AG, Goldschmidtstrasse 100, D-4300 Essen 1, West Germany.
† Present address: BOC Limited, TechSep, London N18 3BW, England.
‡ Present address: Imperial Oil Limited, Sarnia, Ontario N7T 7M1, Canada.

Advances in
INORGANIC CHEMISTRY
AND
RADIOCHEMISTRY

Volume 23

RECENT ADVANCES IN ORGANOTIN CHEMISTRY

ALWYN G. DAVIES

Chemistry Department, University College London, London, England

and

PETER J. SMITH

International Tin Research Institute, Greenford, Middlesex, England

I. Introduction	1
II. Preparation and Reactions of Organotin Compounds	2
A. The Sn–C Bond	2
B. The Sn–H Bond	15
C. The Sn–O, Sn–N, and Sn–S Bonds	16
D. Compounds Having Sn–Sn Bonds	21
E. Tin–Metal Bonds, Sn–M	22
F. Functionally Substituted Compounds	24
G. $R_3Sn \cdot$ Radicals	25
H. R_2Sn: Stannylenes	26
III. Structural Organotin Chemistry	28
A. X-Ray Investigations of Organotin(IV) Compounds	29
B. ^{119}Sn Mössbauer Spectroscopy	40
IV. Biological and Environmental Aspects	41
A. Biological Activity and Mode of Toxic Action	41
B. Environmental Degradation	48
V. Industrial Applications	51
A. Biological Uses (R_3SnX Compounds)	51
B. Nonbiological Uses (R_2SnX_2 and $RSnX_3$ Compounds)	58
References	62

I. Introduction

The literature on organotin chemistry up to 1970 is summarized in the excellent monographs by Neumann (1) and Poller (2), and in the volumes edited by Sawyer (3). Since that date, periodic reviews of advances in the field, and accounts of some selected aspects have been published, but there has been no attempt to bring the general survey up to date.

In this article, we have attempted to review the progress that has been made since 1970, and to give an account of the present status of the field, briefly sketching in the earlier background.

About 1000 papers are published annually on organotin chemistry, and we have been able to include only 5% of these. We have deliberately avoided treating in depth those aspects of the subject that have been thoroughly reviewed recently, and, in particular, we have avoided duplicating the excellent surveys of the use of organotin compounds in organic synthesis (4), of ^{119m}Sn Mössbauer spectroscopy (5–9), and of ^{119}Sn NMR spectroscopy (10–12) that are available.

The older literature on organotin chemistry is reviewed in references (13) to (16), and the periodical surveys are listed in references (17) to (30). Recent reviews of specific aspects of the subject are referred to at the appropriate place in the text, or are listed in references (31) to (34).

II. Preparation and Reactions of Organotin Compounds

A. The Sn–C Bond

The synthesis and properties of the tin–carbon bond were reviewed in reference (35). Three recent volumes of Gmelin, written by H. Schumann and I. Schumann, comprehensively cover the literature, up to the end of 1973, on tetraalkyltin compounds R_4Sn (36), R_3SnR' (37), and $R_2SnR'_2$, $R_2SnR'R''$, and $RR'SnR''R'''$ (38), and are invaluable sources of reference.

Four principal methods have been developed for forming a bond between tin and carbon.

(i) "Direct" synthesis from metallic tin, or by the analogous, oxidative addition-reactions of tin(II) compounds.

$$Sn° \xrightarrow{RX} RSn^{II}X \xrightarrow{RX} R_2Sn^{IV}X_2$$

(ii) From organic derivatives of more electropositive metals.

$$SnX_4 + 4\,RM \rightarrow R_4Sn + 4\,MX$$

(iii) From tin–lithium and tin–sodium compounds.

$$R_3SnM + R'X \rightarrow R_3SnR' + MX$$

(iv) By hydrostannation of an alkene or alkyne.

$$R_3SnH + R'_2C{=}CR'_2 \rightarrow R_3SnCR'_2CR'_2H$$

Advances since 1970 have been made in the improvement and extension of these established, general methods, rather than in the development of fundamentally new processes.

1. "Direct" Synthesis

The "direct" synthesis (39) has obvious attractions as an industrial process, but, in the absence of a catalyst, it proceeds readily only for allyl and benzyl halides, and much attention has been directed towards finding suitable promoters for the reactions.

One approach (40) has been to conduct the reaction in the presence of a more electropositive metal, often as an alloy. In the presence of magnesium, tin reacts with ethyl bromide to give tetraethyltin, and various additives promote the reaction, the sequence of effectiveness being carbitols $\sim I^- >$ tetrahydrofuran, tetrahydrothiophene $>$ ether $\sim$ triethylamine $\sim Br^-$; the ions ClO_4^-, PF_6^-, BF_4^-, and BPh_4^- are without effect. It is suggested that this reflects the coordination of the additive (L) to the Grignard reagent that is first formed, making it more reactive towards metallic tin.

$$EtBr + Mg + 2 L \rightarrow EtMgBrL_2 \rightleftharpoons Et^{-\ +}MgBrL_2$$
$$4\ Et^{-\ +}MgBrL_2 + Sn \rightarrow SnEt_4 + 2\ MgX_2 + 2\ Mg$$

For the reaction of butyl bromide with tin, to give Bu_3SnBr and Bu_2SnBr_2 (approximately equimolar), tetrabutylammonium iodide was found to be the best catalyst, and the mechanism was proposed (41) to be as follows.

$$EtBr + Mg + 2 L \rightarrow EtMgBrL_2 \rightleftharpoons Et^{-+}MgBrL_2$$

$$4\ Et^{-+}MgBrL_2 + Sn \rightarrow SnEt_4 + 2\ MgBr_2 + 2\ Mg$$

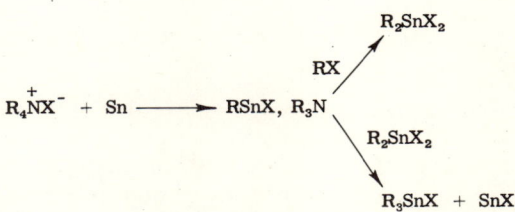

Metallic tin reacts with methyl halides and 2-halogenopropanoates at 135°, catalyzed by magnesium and butyl iodide in tetrahydrofuran, to give compounds $MeSnX_2CH_2CH_2CO_2R$, from which various other derivatives were prepared (42).

Cuprous iodide catalyzes the reaction of various alkyl chlorides, bromides, and iodides in hexamethylphosphoric triamide (HMPT), to give the complexed product R_2SnX_2, which can then be further alkylated with a Grignard reagent, or can be hydrolyzed to the oxide and converted into various other compounds, R_2SnY_2 (43). This promises to be a useful laboratory method, e.g.,

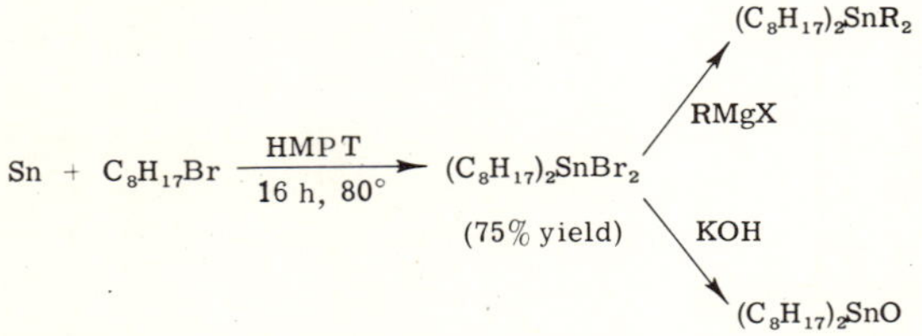

$$Sn + C_8H_{17}Br \xrightarrow[16\ h,\ 80°]{HMPT} (C_8H_{17})_2SnBr_2 \text{ (75\% yield)}$$

$$(C_8H_{17})_2SnBr_2 \xrightarrow{RMgX} (C_8H_{17})_2SnR_2$$

$$(C_8H_{17})_2SnBr_2 \xrightarrow{KOH} (C_8H_{17})_2SnO$$

Monoalkyltin(IV) compounds can be prepared under suitable conditions from tin(II) compounds (44–46). Tin(II) bis(acetylacetonate) (acac) and related compounds react readily with a variety of alkyl bromides or iodides, to give the product $RSn(acac)_2X$, e.g.,

$$Sn^{II}(acac)_2 + CH_2I_2 \xrightarrow{24\ h} ICH_2Sn(acac)_2I\ (90\%)$$

The reaction is catalyzed by light, suggesting a homolytic mechanism.

$$R\cdot + SnX_2 \rightarrow R\dot{S}nX_2$$
$$R\dot{S}nX_2 + RY \rightarrow RSnX_2Y + R\cdot$$

Similarly, dicyclopentadienyltin(II) reacts with methyl iodide to give $Me(C_5H_5)_2SnI$, which is alkylated by more dicyclopentadienyltin, to give $MeSn(C_5H_5)_3$ (45).

A very useful route to monoalkyltin trihalides involves the reaction of stannous bromide with alkyl bromides in the presence of 5 mol % of a trialkylantimony compound (47). A typical example is shown in the following equation.

$$C_{18}H_{37}Br + SnBr_2 \xrightarrow[150°,\ 10\ h]{5\ mol\ \%\ Et_3Sb} C_{18}H_{37}SnBr_3\ (100\%)$$

2. Formation from Organic Compounds of Other Metals

The alkylation of tin tetrachloride with organolithium compounds, Grignard reagents, or organoaluminum compounds remains the most common route to tetraalkyltins, and thence, by the Kocheshkov disproportionation, to the various organotin halides.

$$SnCl_4 + 4\ RM \rightarrow R_4Sn + 4\ MCl$$
$$n\ R_4Sn + (4-n)SnCl_4 \rightarrow 4\ R_nSnCl_{4-n}$$

These two methods, together with the Wurtz modification of co-reacting an alkyl halide and tin tetrachloride with metallic sodium, are used industrially (34).

Some typical recent examples are shown in the following equations.

$$Bu^t_2SnClF + Bu^tLi \longrightarrow Bu^t_3SnCl \qquad (48)$$

$$SnCl_4 + 4\ CH_2=CHMgBr \longrightarrow (CH_2=CH)_4Sn\ (88\%) \qquad (49)$$

$$Me_2SnCl_2 + 2\ HC\equiv CMgBr \longrightarrow Me_2Sn(C\equiv CH)_2 \xrightarrow{BEt_3} Me_2Sn\text{-(vinyl-Et, BEt}_2) \qquad (50)$$

$$Ph_2SnCl_2 + \text{(2,2'-dilithiodiphenylmethane)} \longrightarrow \text{dibenzostannepine (SnPh}_2) \quad (23\%) \qquad (51)$$

$$Me_2SnCl_2 + \text{(2,2'-dilithiodiphenyl ether)} \longrightarrow \text{phenoxastannine (SnMe}_2) \quad (31\%) + \text{dimer} \quad (4\%) \qquad (52)$$

By the same principle, starting from the appropriate, metal-substituted, organolithium compound, products containing both tin and another metal have been synthesized, e.g.,

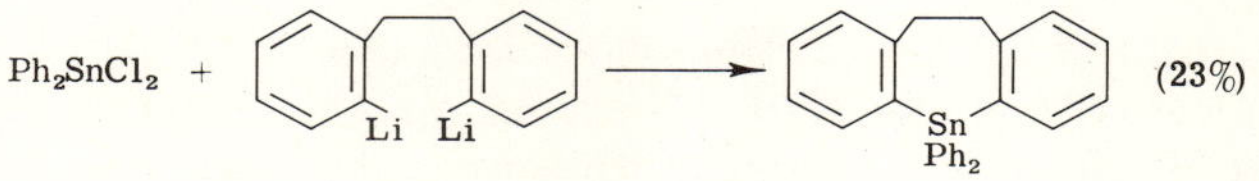

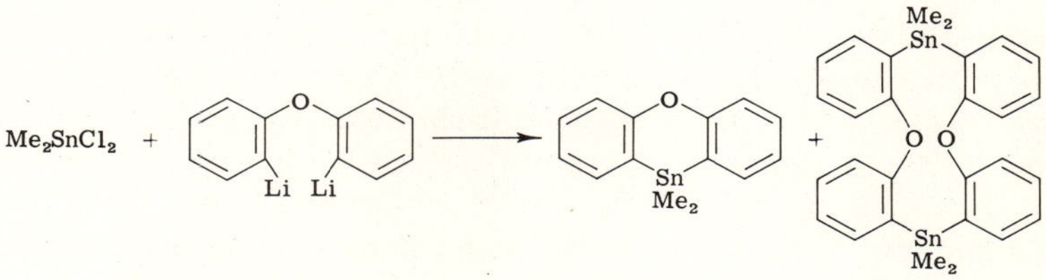

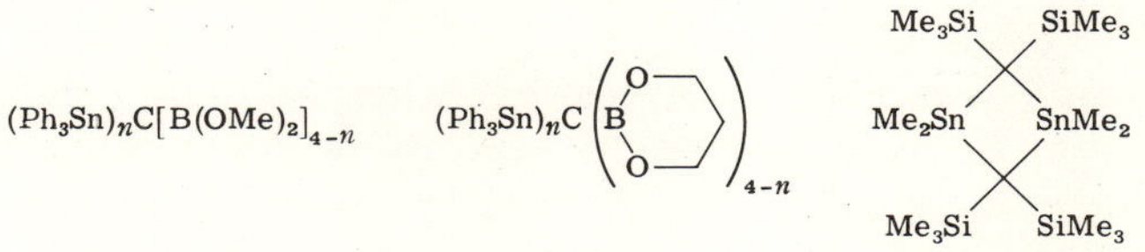

$$(Ph_3Sn)_nC[B(OMe)_2]_{4-n} \qquad (Ph_3Sn)_nC\!\left(B\!\begin{array}{c}O\\O\end{array}\!\right)_{\!4-n} \qquad \begin{array}{cc}Me_3Si & SiMe_3\\ Me_2Sn\!\!\times\!\!SnMe_2\\ Me_3Si & SiMe_3\end{array}$$

$$(53) \qquad\qquad (54) \qquad\qquad (55)$$

By a similar process, stannylsilylthiomethanes have been prepared, and then converted by the Petersen reaction into stannylthiyl alkenes (56).

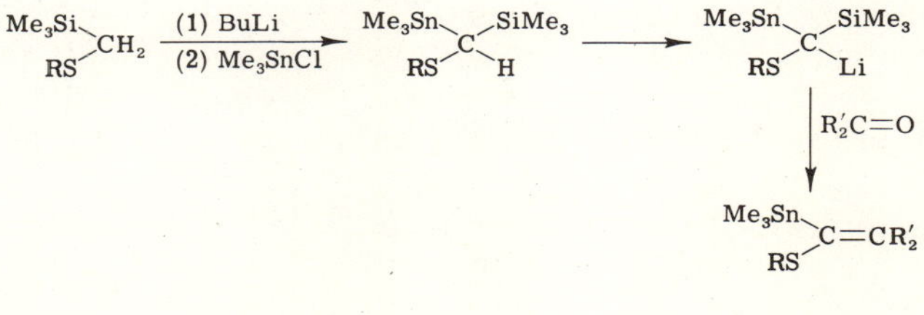

$$\begin{array}{c}Me_3Si\\ RS\end{array}\!\!CH_2 \xrightarrow[(2)\ Me_3SnCl]{(1)\ BuLi} \begin{array}{c}Me_3Sn\\ RS\end{array}\!\!C\!\!\begin{array}{c}SiMe_3\\ H\end{array} \longrightarrow \begin{array}{c}Me_3Sn\\ RS\end{array}\!\!C\!\!\begin{array}{c}SiMe_3\\ Li\end{array}$$

$$\downarrow R'_2C=O$$

$$\begin{array}{c}Me_3Sn\\ RS\end{array}\!\!C=CR'_2$$

Seyferth prepared α-halogenoalkyl-lithium and -magnesium compounds by treating the appropriate *gem*-dihalides with butyllithium or with Grignard reagents at low temperature, and then used the products to prepare acyclic and cyclic α-halogenoalkyltin compounds (*57–60*). Typical examples are shown in the following equations.

$$Me_3SnNEt_2 + HCI_3 \rightarrow Me_3SnCI_3 \xrightarrow[-100°]{Pr^iMgCl} Me_3SnCI_2MgCl \xrightarrow{Me_3SnCl} (Me_3Sn)_2CI_2$$

Iodomethylzinc iodide or bromomethylzinc bromide have likewise been used for preparing the compounds Me_3SnCH_2I, Me_3SnCH_2Br, $Me_2Sn(CH_2I)_2$, $Me_2Sn(CH_2Br)_2$, $Me_2PhSnCH_2I$, Ph_3SnCH_2I, and $Sn(CH_2I)_4$ (*61*). It is probable that the formation of tetraethyltin in 52% yield when diethyl sulfate is electrolyzed with a zinc cathode and a tin anode similarly involves an organozinc intermediate (*62*).

The availability of arylcopper(I) compounds has provided a useful, one-step route to triorganotin halides (*63*): the introduction of the final, aryl group is relatively slow, and the need to revert to the Kocheshkov disproportionation to remove the fourth organic group is avoided. By reaction between the appropriate arylcopper compounds and tin halides in ether or benzene at 0° or room temperature, such compounds as 2-$(Me_2NCH_2)C_6H_4SnPh_2Br$ and 2,6-$(MeO)_2C_6H_3SnMe_2Br$ can be prepared in high yield.

Ylids can also serve as the alkylating nucleophiles; an example is as follows (*64*).

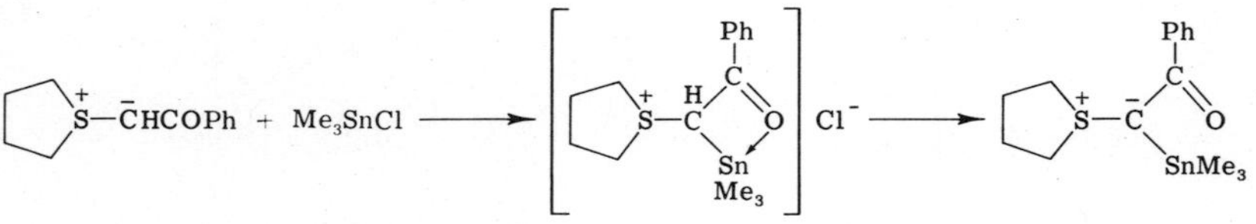

The disproportionation reactions of organotin compounds may also be regarded as alkylations by organometallic compounds, as they involve transfer of an alkyl group from one tin atom to another. An ingenious application of this has been described in which α,ω-distannanes are caused to disproportionate into the corresponding tetraalkyltins and 1,1-dialkylstannacycloalkanes (*65*).

$$BrMg(CH_2)_nMgBr + 2\ R_3SnCl \rightarrow R_3Sn(CH_2)_nSnR_3 \xrightarrow{270-300°} R_4Sn + R_2Sn(CH_2)_n$$

The reaction is best suited to the preparation of the stannacyclopentanes ($n = 4$) and stannacyclohexanes ($n = 5$), which are formed in >85% yield. In contrast, the direct reaction between a di-Grignard or dilithium reagent and a dialkyltin dichloride gives the stannacycloalkane in only 10–30% yield, together with a substantial proportion of polymer. Attempts to make stannacyclobutanes by the disproportionation reaction were unsuccessful, although they may, apparently, be prepared by the direct reaction (66).

3. Formation by Hydrostannation

The hydrostannation reaction can proceed either by a free-radical mechanism, or, with polar-substituted alkenes or alkynes, by a polar mechanism, respectively resulting in anti-Markownikoff or Markownikoff orientation. Both types of reaction are particularly suitable for preparing functionally substituted, organotin compounds.

By the addition of organotin hydrides to norbornene and norbornadiene, and subsequent reactions of the products, a variety of norbornyl-, norbornenyl-, and nortricyclyl-tin compounds has been isolated, and identified (67–69).

1-Buten-3-ynes react by 1,4-addition to give allenic tin compounds; o-diethynylbenzene reacts to afford the benzostannepin (70),

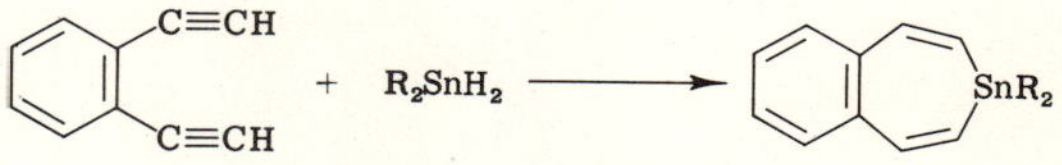

($R_2 = Me_2$, Et_2, or EtPh)

and o-divinylbenzene gives the tetrahydro compound, together with the cyclic dimer.

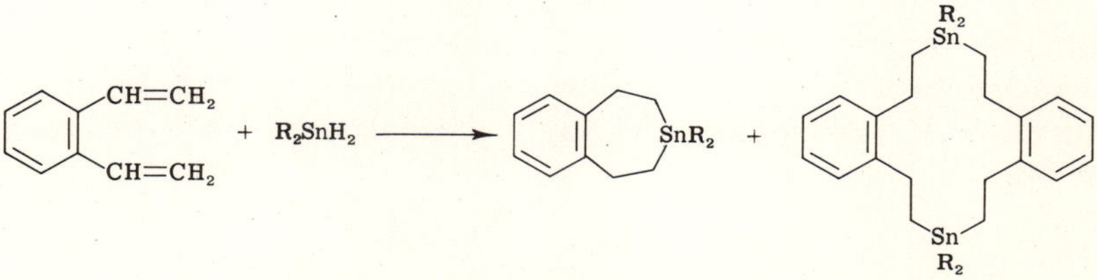

($R_2 = Et_2$ or Ph_2)   ($R_2 = Ph_2$)

The stanna-2,5-cyclohexadienes can be transformed into other metallocycloalkenes, as shown in the following equations.

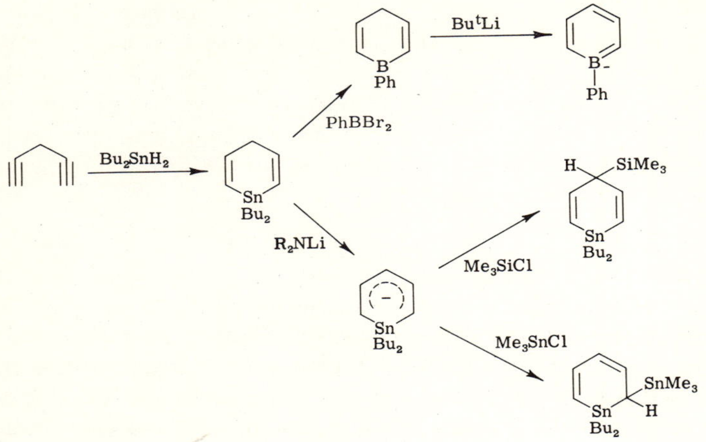

Alkenes carrying C=N, OH, or COCH$_3$ substituents give the corresponding, functionally substituted, organotin compounds, and stannaoxacyclopentanes and stannaoxacyclopentenes have been prepared from dialkyltin dihydrides and allylic alcohols or propargylic acetates, e.g. (75),

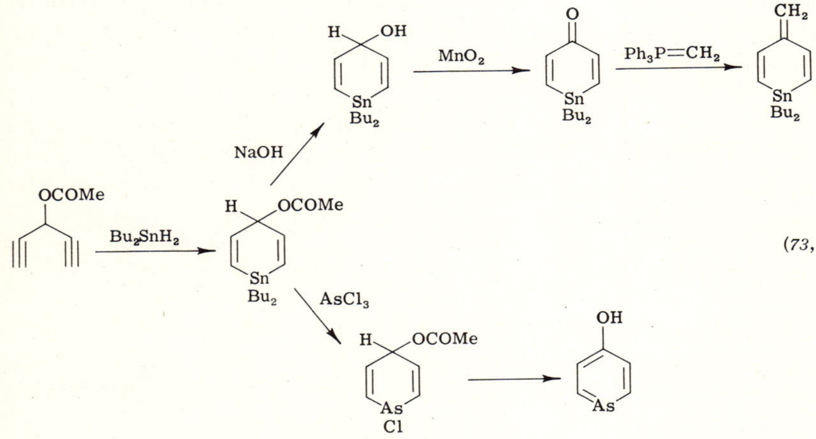

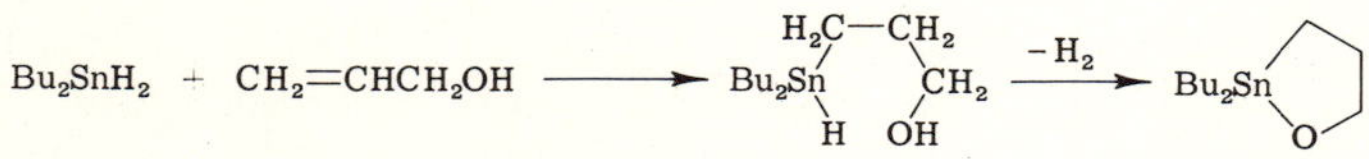

Hydrostannation of chiral menthyl esters of substituted acrylic acids proceeds stereoselectively, providing a route to optically active alkyltin compounds having an optical purity of 10–20% (76).

$$R_3SnH + CH_3CH=CHCO_2(-)Men \xrightarrow{u.v.} CH_3\overset{*}{C}HCH_2CO_2(-)Men$$
$$(Men = menthyl) \qquad\qquad\qquad\qquad |$$
$$SnR_3$$

$$\downarrow LiAlH_4$$

$$CH_3\overset{*}{C}HCH_2CH_3 \xleftarrow[\text{(b) LiAlH}_4]{\text{(a) TosCl}} CH_3\overset{*}{C}HCH_2CH_2OH$$
$$|\qquad\qquad\qquad\qquad\qquad |$$
$$SnR_3 \qquad\qquad\qquad\qquad\qquad SnR_3$$

From the industrial point of view, the most important advance in hydrostannation is the reaction developed by the AKZO company for preparing organotin di- or trichlorides carrying a β-aldehyde, ketone, ester, or acid chloride group (77, 78). Hydrogen chloride is passed into an ethereal solution of the unsaturated compound in the presence of powdered tin or tin(II) chloride. The reactions apparently proceed through the chlorotin hydrides, and, with $SnCl_2$, the hydride dietherate, $HSnCl_3(OEt_2)_2$, may be preformed, and then caused to add to the unsaturated compound in a second step. Two examples are shown in the equations.

$$Sn + 2\ HCl + 2\ CH_2=CMeCO_2Me \rightarrow Cl_2Sn(CH_2CHMeCO_2Me)_2\ (85\%)$$
$$SnCl_2 + HCl + Me_2C=CHCOMe \rightarrow Cl_3SnCMe_2CH_2COMe\ (80\%)$$

The products are used as intermediates for PVC stabilizers. Further applications of the halogenotin hydrides can readily be envisaged.

4. *Formation from Sn–Li and Sn–Na Compounds*

Synthetic and mechanistic aspects of the reactions of the alkalimetal derivatives of organotin compounds, R_3SnM ("organostannylanionoids") have been reviewed (79, 80). They may be prepared by reactions of the types shown in the following equations.

$$Me_4Sn \xrightarrow{Na\ in\ liq.\ NH_3} Me_3SnNa \qquad (81)$$

$$Me_6Sn_2 \xrightarrow[\text{or Li in THF}]{Na\ in\ tetraglyme} 2\ Me_3SnM \qquad (82)$$

$$Pr_3SnCl \xrightarrow[\text{or Li in THF}]{Na\ in\ liq.\ NH_3} Pr_3SnM \qquad (83)$$

$$RCl + SnCl_4 \xrightarrow{Na\ in\ C_6H_{12}} R_3SnNa \qquad (81)$$

$$Bu_3SnCl \xrightarrow{LiNPr_2^i} Bu_3SnLi \qquad (81a)$$

Primary and secondary alkyl halides react with the reagents R_3SnM largely by substitution, but *tert*-alkyl halides, if they can, undergo elimination. The stereochemistry of the reaction is dependent on the structure of R and X, the solvent, and the nature of M, suggesting three possible mechanisms: S_N2 at C, S_N2 at X, and a radical-pair process (84–86).

Recent examples of these reactions are the preparation of a series of norbornyl-, norbornenyl-, and norbornadienyl-tin compounds (87); 1-AdSnMe$_3$, (1-Ad)$_4$Sn; and 2-AdSnMe$_3$ (Ad = adamantyl) (81, 83); Pr$_3$SnCH=CHCH=CH$_2$ (83); and (Me$_3$Sn)$_4$C (82).

By reaction with the appropropriate aryl halides can be prepared a variety of aryltin compounds that are not accessible from the reactions involving arylmagnesium halides and organotin halides (88, 89); there is evidence that an aryne intermediate may be involved (90). However, for some purposes, such as the addition to carbonyl compounds, oxiranes, and oxetanes, to give hydroxyalkyltin compounds, the Sn–Mg reagents may have advantages (see Section II,E) (91–93).

5. Cleavage of the Sn–C Bond

Heterolytic cleavage of the tin–carbon bond is reviewed in references (94–96). Cleavage by electrophiles (e.g., HgX$_2$ or halogen) is dominated by electrophilic attack at carbon, and cleavage by nucleophiles principally involves nucleophilic attack at tin. Much of the interest in these processes centers on the intermediate mechanisms that may exist between these extremes, in which electrophilic attack is accompanied by some nucleophilic assistance, and vice versa. Allylic, allenic, and propargylic compounds show a special reactivity by a special (S_E2' or $S_E2\gamma$) mechanism.

The earlier work on acidolysis of the aryl–tin bond is reviewed in reference (97). Attachment of the proton to the aryl ring is rate-determining, and the Hammett ρ-factor for the reaction has been shown to be solvent-dependent (98).

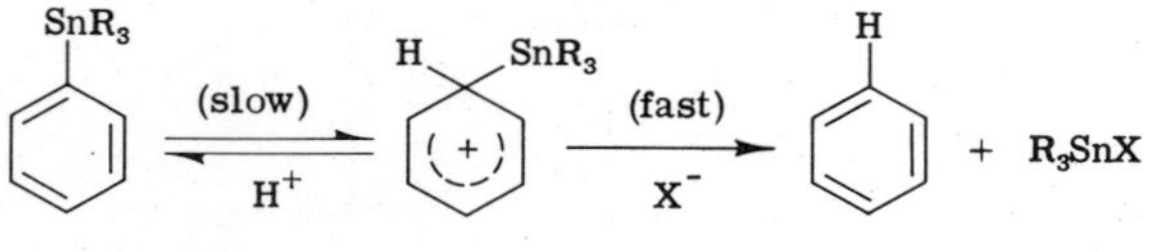

Benzyl and heteroaryl (e.g., furanyl or thienyl) groups can be cleaved from the tin under basic conditions also, and nucleophilic attack on the tin is now assisted by attack of the solvent (e.g., water) on the organic group, through a transition state of the type HO---H---R---$\bar{\text{Sn}}$R$_3$OH (99–101).

Acidolysis of the alkyl–tin bond provides a useful route from tetraalkyltins to alkyltin carboxylates, and is discussed in Section II,C.

With functionally substituted, alkyltin compounds, the functional substituent may become involved in the cleavage process, resulting in an intramolecular reaction, e.g.,

$$Me_3SnCH_2CH_2CH_2OTs \xrightarrow{125°} Me_3SnOTs + \text{cyclo-}C_3H_6 \,(82\%) \quad (102)$$

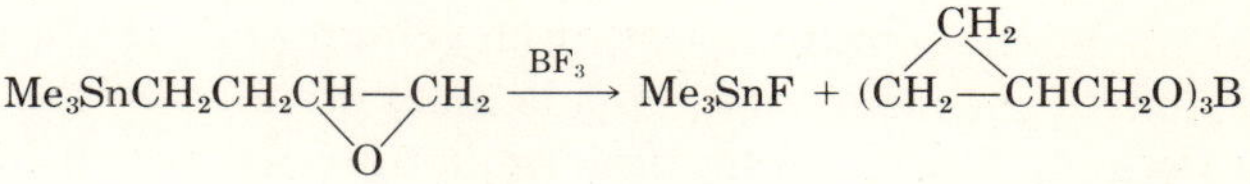

$$Me_3SnCH_2CH_2CH\underset{O}{-}CH_2 \xrightarrow{BF_3} Me_3SnF + (\overset{CH_2}{\underset{}{CH_2}}\!\!-\!\!CHCH_2O)_3B \quad (103)$$

The mechanism of the cleavage of the alkyl–tin bond by mercuric halides and carboxylates has been thoroughly investigated, and the evidence is in favor of an open S$_E$2 transition state (*104–110*).

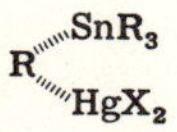

$$R\underset{HgX_2}{\overset{SnR_3}{\rightleftharpoons}}$$

The reactions of tetraorganotins with ICl, IBr, and ClCN have been investigated as synthetic routes to aryl cyanides (*111*) and alkyltin bromides and chlorides (*112, 113*).

The reaction of bromine with optically active *sec*-butyltin compounds Bu^{s}SnR$_3$, to give *sec*-butyl bromide, can give retention or inversion in the *sec*-butyl group, depending on the nature of the group R (*114*), and the inversion that is observed with *sec*-butyltrineopentyltin (*115*) appears to be the exception rather than the rule.

A lot of attention has also been paid to the reaction of organotin compounds with sulfur dioxide to give organotin sulfinates (*116–119*).

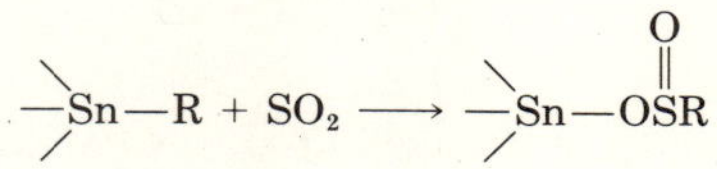

$$>\!\!Sn\!-\!R + SO_2 \longrightarrow\; >\!\!Sn\!-\!\overset{O}{\underset{}{\overset{\|}{O}}}SR$$

The topic has been reviewed (*120*).

The reactivity of various groups (R) follows the sequence allyl, benzyl > aryl > alkyl, and usually proceeds readily to the stage of R$_2$Sn(OSOR)$_2$ (*121, 122*), but pentafluorophenyl- and trifluorovinyl-tin bonds are usually unreactive. The reactivity is enhanced by such ligands as bipyridyl (*123*).

Allyl- and vinyl-tin compounds react with retention in the structure of the organic group (*124*), and a kinetic study of the reaction of aryl and benzyl compounds suggested that the reaction is best represented as proceeding through an S$_E$i transition state (*125–127*).

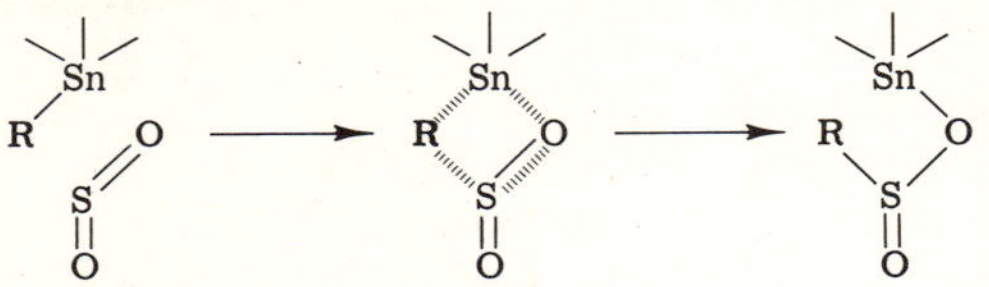

$$\underset{R}{\overset{|}{\text{Sn}}}\!\!\diagdown\!\!\underset{\underset{\text{O}}{\overset{\|}{\text{S}}}}{\text{O}} \longrightarrow \underset{R}{\overset{|}{\text{Sn}}}\!\!\cdots\!\!\underset{\underset{\text{O}}{\overset{\|}{\text{S}}}}{\text{O}} \longrightarrow \underset{R}{\overset{|}{\text{Sn}}}\!\!\diagdown\!\!\underset{\underset{\text{O}}{\overset{\|}{\text{S}}}}{\text{O}}$$

Angle-strain in the small-ring stannacycloalkanes confers on them an anomalously high reactivity. This is most obvious in the stannacyclopentanes which, for example, undergo ionic polymerization in polar solvents such as methanol, react readily with acetic acid (*128*), and react additively with diarylsulfurdiimides (*129*).

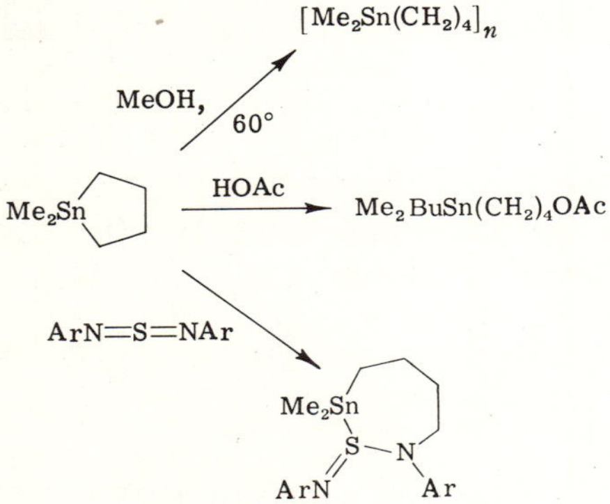

$$\text{Me}_2\text{Sn}\!\!\diagup\!\!\diagdown \quad \xrightarrow{\text{MeOH, } 60°} [\text{Me}_2\text{Sn}(\text{CH}_2)_4]_n$$
$$\xrightarrow{\text{HOAc}} \text{Me}_2\text{BuSn}(\text{CH}_2)_4\text{OAc}$$
$$\xrightarrow{\text{ArN=S=NAr}} \underset{\underset{\text{ArN}}{}}{\text{Me}_2\text{Sn}}\!\!\diagdown\!\!\underset{\text{Ar}}{\text{N}}$$

Cleavage of the tin–carbon bond can also be achieved by bimolecular, homolytic substitution (S$_H$2) at the tin center (*130–132*).

$$\text{X}\cdot + -\!\!\overset{\diagdown}{\underset{\diagup}{\text{Sn}}}\!\!-\text{R} \longrightarrow \text{X}-\!\!\overset{\diagup}{\underset{\diagdown}{\text{Sn}}}\!\!- + \text{R}\cdot$$

The alternative of an S$_H$2 process at the α-carbon center has not yet been identified. Examples of the unimolecular thermolysis or photolysis of the tin–carbon bond are known (see Section II,G), but have not yet been investigated extensively.

Whereas bromine radicals (*133*) and succinimidyl radicals (*134*) react by the S$_H$2 mechanism at the tin center in tetraalkyltins, but not in alkyltin halides, alkoxyl radicals (*135*) and ketone triplets (*136*) react with alkyltin halides, but not tetraalkyltins; this may reflect the conflicting, electronic demands of the radical reagents which, as electrophilic species, should be more reactive towards tetraalkyltins than alkyltin halides, but which would also tend to make use of a 5d orbital on tin to establish a 5-coordinate transition state or intermediate, and these orbitals are more accessible in the alkyltin halides. There is some inconclusive evidence that at least some of these reactions do involve a transient Sn(V) radical intermediate (132, 137).

The relative reactivities of alkyltin compounds towards *tert*-butoxyl radicals, ketone triplets, and succinimidyl radicals are dominated by the steric effect of the alkyl ligands ($R^p > R^s$), but that towards bromine atoms follows the reverse sequence ($R^p < R^s$).

The stannacyclopentanes are again particularly reactive (138). Alkoxyl radicals will now react at the tin center in the fully alkylated compounds, with opening of the ring, and benzoyloxyl and alkylthiyl radicals will also induce ring cleavage. For example, 1,1-dibutylstannacyclopentane reacts homolytically with benzenethiol to give tributyl(phenylthio)tin.

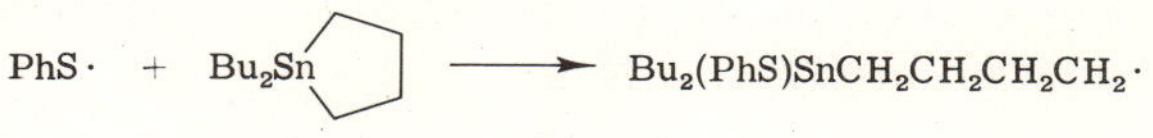

$$PhS\cdot + Bu_2Sn\diagup\!\!\!\diagdown \longrightarrow Bu_2(PhS)SnCH_2CH_2CH_2CH_2\cdot$$

$$Bu_2(PhS)SnCH_2CH_2CH_2CH_2\cdot + PhSH \longrightarrow Bu_3SnSPh + PhS\cdot$$

6. Allylic and Related Tin Compounds

The allylic, allenic, propargylic, 2,4-dienylic, cyclopentadienylic, and related tin compounds present special, structural features and show special reactivity by both heterolytic and homolytic mechanisms.

Allyltin compounds can be prepared by simple modifications of the usual reaction involving allyl Grignard reagents (139), by the 1,4-addition of trialkyltin hydrides to 1,3-dienes (140, 141), or by the reaction of an aldehyde or ketone with the appropriate, tin-carrying, Wittig reagents (142).

$$CH_2\!\!=\!\!CHCH_2MgCl + SnCl_4 \longrightarrow (CH_2\!\!=\!\!CHCH_2)_4Sn \xrightarrow{SnX_4} (CH_2\!\!=\!\!CHCH_2)_nSnX_{4-n}$$

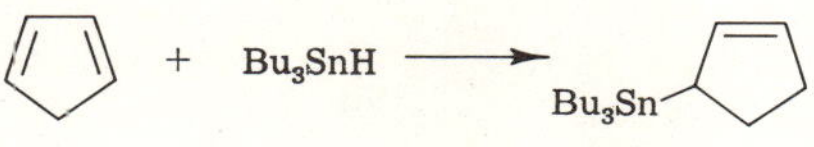

$$\text{cyclopentadiene} + Bu_3SnH \longrightarrow Bu_3Sn\text{-cyclopentenyl}$$

$$\overset{+}{CH_2}\!\!=\!\!CHPPh_3 + Me_3SnLi \longrightarrow Me_3SnCH_2CH\!\!=\!\!PPh_3 \xrightarrow{R_2C=O} Me_3SnCH_2CH\!\!=\!\!CR_2$$

The allyltin halides can then be obtained by a disproportionation reaction between tetraallyltin and tin tetrachloride or tetrabromide, a reaction that is exothermic (143, 144).

Allyltin compounds can also be formed by elimination from the tin derivatives of allyldialkylcarbinols (*145*).

$$Bu_3SnOCR_2CH_2CH=CH_2 \rightleftharpoons Bu_3SnCH_2CH=CH_2 + O=CR_2$$

Photoelectron spectroscopy shows that the carbon–tin bond prefers that orientation in which it lies parallel to the $p\pi$ orbitals of the double bond, to permit carbon–metal hyperconjugation (*146*).

Cleavage of the Sn–C bond is dominated by electrophilic or homolytic attack at the γ-carbon of the allylic group, leading to allylic rearrangement, and these reactions [e.g., acidolysis (*147–149*)] usually occur more readily than with the corresponding alkyltin compounds. The reaction with thiocyanogen is considered to be an S$_E$2γ process, but that with iodine in a polar medium may be accompanied by an S$_E$2 component (*150, 151*). Complete allylic rearrangement is involved in the addition to carbonyl compounds (*152–154*) and, presumably, also in the carbonyl elimination (*145*).

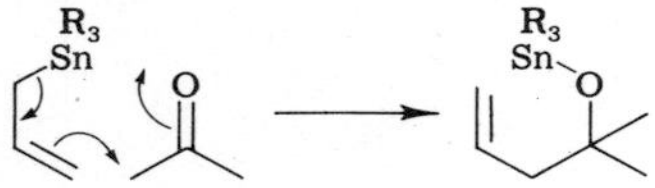

The S$_H$2γ mechanism is most clearly shown by the exchange of R$_3$Sn· (*155, 156*), and by the reaction with alkyl halides (*157, 158*).

$$R'\cdot CH_2=CH-CH_2SnR_3 \rightarrow R'CH_2CH=CH_2 + \cdot SnR_3$$
$$R_3Sn\cdot + XR' \rightarrow R_3SnX + R'\cdot$$

The equilibrium between propargyl- and allenyl-tin compounds is not spontaneous, but it occurs in the presence of Lewis acids or coordinating solvents, and an ion-pair mechanism has been proposed (*159*). Substitution by iodine, or addition to chloral, occurs with propargyl/allenyl rearrangement (*160, 161*), analogous to the allylic rearrangement already mentioned.

$$Sn-\overset{1}{C}-\overset{2}{C}\equiv\overset{3}{C} \xrightarrow{Cl_3CCHO} Cl_3CCH-\overset{3}{C}=\overset{2}{C}=\overset{1}{C}$$
$$\underset{OSn}{|}$$

$$\overset{1}{C}=\overset{2}{C}=\overset{3}{C}-Sn \xrightarrow{Cl_3CCHO} Cl_3CCH-\overset{1}{C}-\overset{2}{C}\equiv\overset{3}{C}$$
$$\underset{OSn}{|}$$

Compounds in which the R$_3$Sn group is attached to a 2,4-dienyl group, such as cyclopentadiene, cycloheptadiene, cycloheptatriene, and cyclononatetraene, whose formulas are shown, are fluxional.

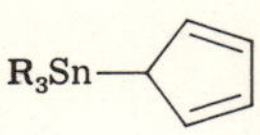

R_3Sn—[cyclopentadienyl] (162,163) R_3Sn—[cycloheptatrienyl] (164)

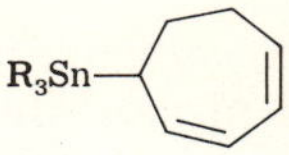

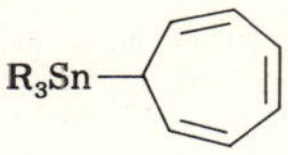

R_3Sn—[cycloheptatrienyl] (165) R_3Sn—[cyclooctatetraenyl] (166)

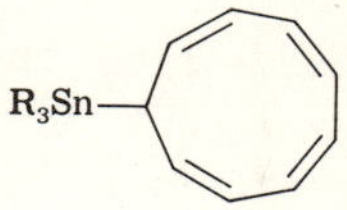

The chemistry of cyclopentadienyltin compounds is reviewed in references (167) and (168). The ready disproportionation of $(C_5H_5)_4Sn$ with $SnCl_4$ (169), and the sensitivity of the cyclopentadienyl–tin bond to acidolysis and to photolysis (170) suggests that these compounds may find application in synthesis.

B. The Sn–H Bond

A comprehensive review of the literature on the organotin hydrides up to the end of 1974 is available in a recent volume of Gmelin (170a).

The hydrides are usually prepared by reducing an organotin chloride, alkoxide, or oxide with lithium aluminum hydride or with poly(methylsiloxane), $[MeSiHO]_n$.

The reduction of tributyltin methoxide with optically active methylphenyl-1-naphthylsilane involves retention of configuration at the silicon atom and follows second-order kinetics (171). The reaction between tributyltin methoxide and ring-substituted dimethylphenylsilanes shows a Hammett ρ-value of $+0.903$, and that between dimethylphenylsilane and ring-substituted tributyltin phenoxides shows a ρ-value of -1.319; this is compatible with the reactions proceeding through a 4-centered (SNi-Si) transition state (172, 173).

$$\begin{array}{c} R_3Sn\text{--}OR \\ | \quad\quad | \\ H\text{---}SiR_3 \end{array}$$

The mixed halide-hydrides Ph_2SnHCl and Ph_2SnHBr have been prepared from the reaction of diphenyltin dihydride with the appropriate diphenyltin dihalides (174).

The tin hydrides find important applications as reducing agents. Many of their reactions (particularly the reduction of alkyl halides and the hydrostannation of simple alkenes and alkynes) are known to proceed through $R_3Sn\cdot$ intermediates, and this aspect of their chemistry is referred to in Section II,G.

The synthetic applications have been reviewed (175). General developments have included the generation of the tin hydride *in situ* from sodium borohydride and a catalytic amount of trialkyltin chloride (*176, 177*). Small proportions of dialkyltin dihalides or dialkoxides have been shown to catalyze the hydrostannation of aldehydes and ketones by dialkyltin dihydrides, probably through the formation of intermediates R_2SnXH; under these conditions, many aldehydes and ketones can be reduced (*178*) to alcohols at 30°.

Promising experiments have also been made at immobilizing the tin hydride on a polymer, but, as yet, regeneration of the hydride has been incomplete (*179*).

Trimethyltin hydride has been shown to add to trimethylvinyltin, and triethyltin hydride to triethylvinyltin, to give both the 1,1- and the 1,2-distannylethanes, whereas triphenyltin hydride reacts with triphenylvinyltin to give only the 1,2-adduct (*180*).

$$R_3SnH + R_3SnCH{=}CH_2 \rightarrow (R_3Sn)_2CHCH_3 + R_3SnCH_2CH_2SnR_3$$

C. The Sn–O, Sn–N, and Sn–S Bonds

Organotin chlorides, R_nSnCl_{4-n}, are usually obtained from the Kocheshkov disproportionation between tetraalkyltins and tin tetrachloride, and other organotin derivatives, R_nSnX_{4-n}, are then prepared by substitution reactions of the chlorides. The chemistry of the chlorides is reviewed in reference (*181*).

For preparing the acetates, thallous acetate has been used as the reagent as an alternative to sodium acetate (*182*). Alternatively, the carboxylates can be prepared directly by acidolysis of the tetraalkyltins (*183*).

$$R_4Sn + n\ R'CO_2H \rightarrow R_{4-n}Sn(OCOR')_n + n\ RH$$

Successive groups are replaced with increasing difficulty. Vinyl groups are replaced more readily than saturated alkyl groups, and, with trifluoroacetic acid, two vinyl groups are displaced exothermically at room temperature, and a third after several hours of heating (*184*). With trifluoromethanesulfonic acid, all four vinyl groups are displaced at $-78°$, to give a compound that was shown by Mössbauer spectroscopy to contain both Sn(II) and Sn(IV), and that was assigned the formula $Sn^{II}[Sn^{IV}(SO_3CF_3)_6]$. Trivinyltin carboxylates have also been prepared from the reaction between tetravinyltin and mercury(I) carboxylates, which may be generated electrochemically *in situ* (*185*).

Tetraallyltin is more reactive than tetravinyltin, but, with methanol as the solvent, acidolysis can be restricted to the stage of the formation of the triallyltin or diallyltin carboxylates (*186*).

Trimethyltin chloride reacts with carboxylic acids at 100° to give the corresponding chloride carboxylates Me$_2$Sn(Cl)OCOR (*187, 188*), and diethyltin dihydride, triethyltin hydride, hexaethylditin, and bis(triethyltin) oxide have been shown to react with lead tetraacetate to give diethyltin diacetate or triethyltin acetate, as appropriate (*189*).

The organotin alkoxides R$_3$SnOR′ and R$_2$Sn(OR′)$_2$ can be prepared by treating the appropriate organotin chlorides with sodium alkoxides, and this procedure has been extended to the preparation of the monoalkyltin trialkoxides, RSn(OR′)$_3$ (*190*), which serve as useful reagents for the synthesis of other monoalkyltin derivatives. Alternatively, the trialkoxides can be prepared by alcoholysis of the tris(amino) compounds RSn(NR$_2'$)$_3$ (*191*).

The trialkyltin alkoxides can often be prepared more conveniently by azeotropic dehydration of the appropriate bis(trialkyltin) oxide and alcohols, or by heating together the bis(trialkyltin) oxide and dialkyl carbonate (*192*). The latter reaction involves formation, and then decarboxylation, of the alkyl trialkyltin carbonate, e.g.,

$$(Bu_3Sn)_2O + (MeO)_2CO \rightarrow Bu_3SnOMe + Bu_3SnOCO_2Me \rightarrow 2\ Bu_3SnOMe + CO_2.$$

There is a growing interest in the tin enolates that can be prepared by treating enol acetates with trialkyltin methoxides, e.g. (*193*),

$$Bu_3SnOMe + CH_2=CHOCOCH_3 \rightarrow Bu_3SnOCH=CH_2 + MeOCOCH_3.$$

The products are in metallotropic equilibrium between the *O*-bonded (enol) and *C*-bonded (keto) isomers, and the topic has been reviewed (*194*).

Many organotin derivatives of functionally substituted alcohols have been prepared, partly for their structural interest, and partly for their use as reaction intermediates. In particular, alkylstannatranes, RSn(OCH$_2$CH$_2$)$_3$N (*195–202*) and the halogenoalkoxytin compounds R$_3$SnO(CH$_2$)$_n$X (n = 2–5) (*203–206*) have attracted much attention. A variety of dialkyltin and trialkyltin derivatives of carbohydrates has also been prepared, in order to modify the reactivity of specific hydroxyl groups (*207–211b*). For example, azeotropic dehydration of a mixture of D-glucose and bis(tributyltin) oxide effects stannylation of the 1-, 4-, and 6-hydroxyl groups.

The lower trialkyltin hydroxides and oxides, which are usually readily interconverted, have been characterized by IR and Mössbauer spectroscopy (*212*). The dimer of di-*n*-butyltin oxide (Bu$_2$SnO)$_2$ has been reported to be formed as a crystalline solid when dibutyltin dichloride is hydrolyzed with ammonium hydroxide (*213*).

Trialkyltin methoxides react with anhydrous hydrogen peroxide in ether to give the rather unstable bis(trialkyltin) peroxides, $R_3SnOOSnR_3$ (*214*). Under the same conditions, dialkyltin dimethoxides give polymeric peroxides, $(R_2SnOO)_n$, but, if an aldehyde is present, monomeric peroxides of the following structure are obtained (*215*).

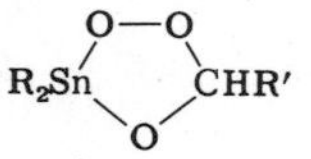

$$R_2Sn\underset{O}{\overset{O-O}{\diagup\diagdown}}CHR'$$

The chemistry of the peroxides of the Group IV metals has been reviewed (*216*).

The aminotin compounds are less readily prepared, and are more reactive both in substitution and addition processes, than the alkoxides. The established routes to, and reactions of, these compounds are exemplified by recent work on the aziridine derivatives (*217*).

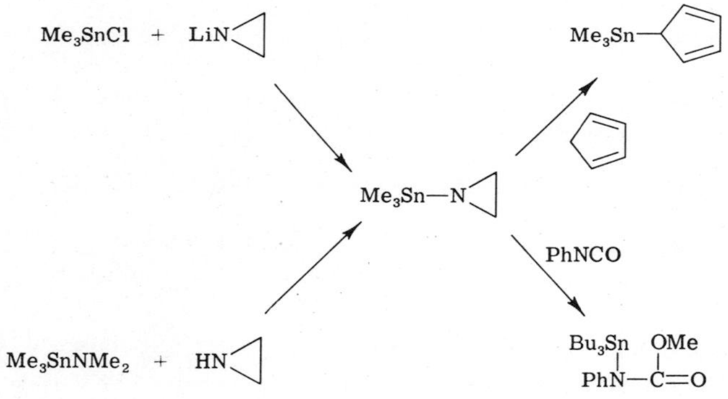

$$Me_3SnCl + LiN{\triangleleft} \quad\quad Me_3Sn-{\triangleleft}$$

$$Me_3Sn-N{\triangleleft}$$

$$Me_3SnNMe_2 + HN{\triangleleft} \quad\quad \underset{PhN-C=O}{\overset{Bu_3Sn\ \ OMe}{|\ \ \ \ \ \ |}}$$

(with PhNCO)

The aminotin compounds can also be prepared from the reaction between aminosilanes and alkoxytin compounds, and this reaction has been extended to the preparation of the first sulfinylaminotin compounds (*218*).

$$Bu_3SnOMe + Me_3SiNMe_2 \rightarrow Bu_3SnNMe_2 + Me_3SiOMe$$
$$Bu_3SnOMe + Me_3SiNSO \rightarrow Bu_3SnNSO + Me_3SiOMe$$

The aminotin compounds react with aldehydes by addition and elimination, to give enamines (*219*), but some ketones, by acidolysis, give tin enolates, e.g. (*220*),

$$CH_3COCHMe_2 + Bu_3SnNEt_2 \longrightarrow Bu_3SnCH_2COCHMe_2 + CH_2=C\underset{OSnBu_3}{\overset{CHMe_2}{\diagup\diagdown}}$$

80%   20%

N,N-Dialkylamides are much less reactive than ketones or esters, and acidolysis gives the C-bonded product (221).

$$Bu_3SnNEt_2 + CH_3CONR_2 \rightarrow Bu_3SnCH_2CONR_2 + HNEt_2$$

Organotin enamines can be prepared by treating organotin halides with the lithium or magnesium derivatives of enamines, and also by treating distannazanes with tin enolates (222, 223).

$$Sn-N-Sn + C=C-O-Sn \longrightarrow \underset{\underset{Sn}{|}}{C=C-N} + \underset{\underset{Sn}{|}}{C-C=N} + Sn-O-Sn$$

Like the tin enolates, the tin enamines are metallotropic.

Cyanamide can be stannylated under various conditions to give bis(stannyl)carbodiimides (224–227), and bis(stannyl)sulfurdiimides have been prepared from the reaction between S_4N_4 and trimethyldimethylaminotin (228).

$$2 R_3SnX + H_2NC\equiv N \rightarrow R_3SnN=C=NSnR_3$$
$$Me_3SnNMe_2 + N_4S_4 \rightarrow Me_3SnN=S=NSnMe_3$$

The organotin thiolates are more readily prepared, and are less reactive, than either the alkoxy or the amino compounds, and the alkynylthiyltin compounds $Me_3SnSC\equiv CPh$, $Me_3SiC\equiv CSSnMe_3$, and $Me_3CC\equiv CSSnBu_3$ have recently been prepared by extension of the established, general methods (229).

Dithiastannacyclopentanes and dithiastannacyclohexanes have been obtained by treating diphenyltin dichloride with the appropriate lead dithiolates (230), e.g.,

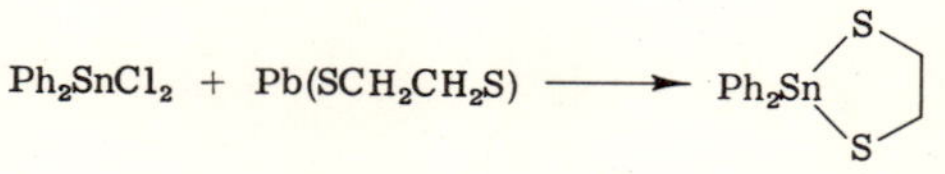

$$Ph_2SnCl_2 + Pb(SCH_2CH_2S) \longrightarrow Ph_2Sn\begin{matrix}S\\ \diagup \\ \diagdown \\ S\end{matrix}$$

Sodium dicyanoethylenedithiolate reacts with trimethyl- or triphenyl-tin chloride to give anionic trialkylstannadithiacyclopentenes, but dialkyltin dichlorides undergo dealkylation (231).

Dialkyltin compounds R_2SnXY, where X and Y are dissimilar ligands, are readily accessible, often merely by disproportionation between the compounds R_2SnX_2 and R_2SnY_2. Typical examples of such compounds that have been characterized are as follows: $Me_2SnClBr$, Et_2SnBrI, Bu_2SnClI (232) $MeSnFCl_2$ (233), $MeSnCl(SO_3F)_2$ (234), $Me_2SnCl(OMe)$ (235), $Me_2SnCl(NMe_2)$ (236), $Me_2SnCl(P^tBu_2)$ (237, 238), $Me_2SnCl(OCOR)$ (239–241), $BuSnCl(O^iPr)_2$, $BuSn(O^iPr)(OCOCH_3)_2$ (242), $Bu_2Sn(OPh)OCOPh$ (243), $Bu_2Sn(OEt)(OCH_2CH_2NH_2)$ (244), and $^tBu_2SnCl(OH)$ (245).

By a similar disproportionation involving stannoxanes, $(R_2SnO)_n$, stannathianes, $(R_2SnS)_n$, and stannazanes, $(R_2SnNR)_n$, or by a controlled solvolysis reaction, oligomeric, functionally substituted stannoxanes, stannathianes, and stannazanes, can be prepared. The most familiar of these are the tetraalkydistannoxanes, $XR_2SnOSnR_2X$.

In recent years, the stannoxanes $N_3Me_2SnOSnMe_2OMe$, $N_3SnR_2OSnR_2OSnR_2N_3$, $N_3SnR_2OSnR_2OSnR_2OSnR_2N_3$ (246), $R_3SnOSnR_2'X$ (e.g., $Me_3SnOSnBu_2Cl$), $ClBu_2Sn(OSnBu_2)_nCl$ ($n = 2$–12), and $BuSn[(OSnBu_2)_nCl]_3$ ($n = 1$–4) (247), the stannathianes $ClMe_2SnSSnMe_2Cl$, $FBu_2SnSSnBu_2F$, $Cl_2PhSn(SSnBu_2)_nCl$ ($n = 1$ and 2), and $Cl_2BuSn(SSnBu_2)_2Cl$ (248), and the stannazanes $ClMe_2SnNEtSnMe_2Cl$ and $Cl_2MeSnNEtSnMe_2Cl$ (249) have been characterized.

A number of mixed metalloxanes, $R_n'MOSnR_rX$ (M = Hg, Tl, Si, Ge, or Pb), have similarly been synthesized from the stannoxanes $(R_2SnO)_n$ and the metal compounds $R_n'MX$ (250), and the borostannoxanes $B(OSnR_3)_3$ and $(RO)_2BOSnR_2OSnR_2OB(OR)_2$ have also been characterized (251, 252).

Whereas most dialkylbis(dialkylamino)tin compounds react with primary amines to give cyclotristannazanes di-*tert*-butylbis(dimethylamino)tin reacts to give a cyclodistannazane, from which the cyclodistannathiane and cyclodistannaphosphazane can be prepared (253).

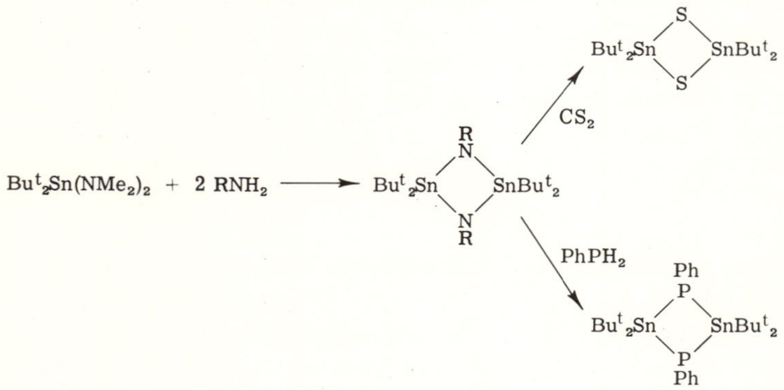

NMR spectroscopy shows that, in solution, $(Me_2SnS)_3$ and $(Me_2SnNEt)_3$ are in equilibrium with the corresponding cyclotristannaazadithiane and cyclotristannadiazathiane (254).

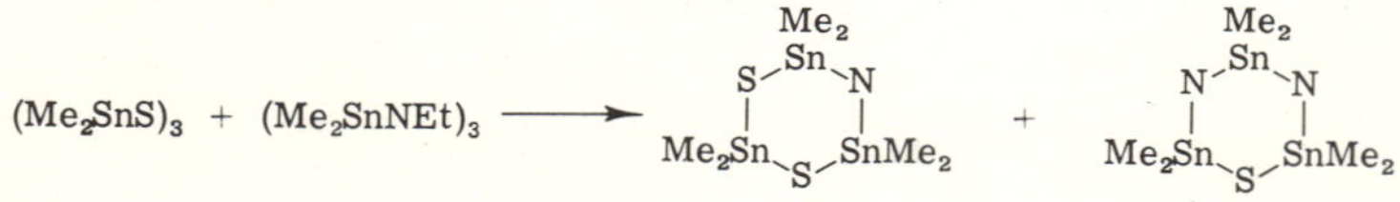

In a rather different type of process, oligomeric dimethyltin and tris(dimethyltin sulfide) react together to give the dithiatristannacyclopentane (255).

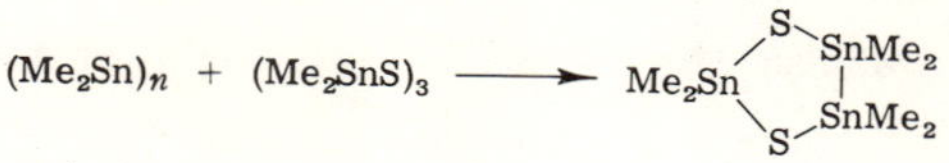

$$(Me_2Sn)_n + (Me_2SnS)_3 \longrightarrow Me_2Sn\underset{S\diagdown SnMe_2}{\overset{S\diagup SnMe_2}{\diagup}}$$

D. Compounds Having Sn–Sn Bonds

Tin–tin bonds are usually best prepared by reducing an Sn–O or Sn–N bonded compound with a tin hydride. For example, trimethyl(diethylamino)tin is reduced by alkyltin trihydrides to give decaorganotetratins (256).

$$3\ Me_3SnNEt_2 + RSnH_3 \rightarrow RSn(SnMe_3)_3 + 3\ Et_2NH$$

(R = e.g., Me, Pe, or Ph)

In a modification of this method, the Sn–O bonded compound can be generated *in situ* by partial acidolysis of a tin hydride, and, from the reaction between diphenyltin dihydride and carboxylic acids, a number of 1,2-bis(acyloxy)-1,1,2,2-tetraphenylditins, $(RCO_2)Ph_2SnSnPh_2(O_2CR)$ (e.g., R = CH_3, CF_3, Ph_3Si, or Ph_3Ge), have been prepared (257, 258).

The simple hexaalkylditins, R_3SnSnR_3', do not disproportionate on heating, but, in oxolane (tetrahydrofuran) or acetonitrile in the presence of a base such as a Grignard reagent, or in the more strongly basic solvent hexamethylphosphoric triamide (HMPT), disproportionation readily occurs at room temperature, and, in HMPT, addition occurs to such alkynes as phenylacetylene and diphenylbutadiyne. The disproportionation is considered to proceed by nucleophilic attack upon tin (259, 260), e.g.,

$$BrMgMe + R_3SnSnR_3' \rightarrow MeSnR_3 + BrMgSnR_3'$$
$$BrMgSnR_3' + R_3'SnSnR_3 \rightarrow R_3'SnSnR_3' + BrMgSnR_3,\ etc.$$

A similar mechanism probably applies to the reaction of the reagents $LiMMe_4$ (M = B, Al, Ga, or Tl), (261), e.g.,

$$Li^+\ Me_3\overline{Tl}\text{—}Me + Me_3Sn\text{—}SnMe_3 \rightarrow Me_4Sn + Li^+\ Me_3\overline{Tl}SnMe_3$$

The tin–tin bond is also cleaved by alkylmercuric halides, triethyltin halides, trimethyllead chloride, and, catalytically, dicobalt octacarbonyl (266).

$$RHgX + Me_3SnSnMe_3 \rightarrow Me_3SnX + RHgSnMe_3 \rightarrow RSnMe_3 + Hg^\circ \ (262, 263)$$

$$R_3SnX + Me_3SnSnMe_3 \rightarrow Me_3SnX + RSnMe_3 + (Me_2Sn^{II})_n \ (264)$$

$$Me_3PbCl + Me_3SnSnMe_3 \rightarrow Me_3SnCl + Me_4Sn + Pb^{II}Cl_2 \ (265)$$

$$Me_3SnSnMe_3 \xrightarrow{Co_2(CO)_8} Me_4Sn + (Me_2Sn^{II})_n$$

A cyclic mechanism has been tentatively proposed for the reaction involving trialkyltin halides.

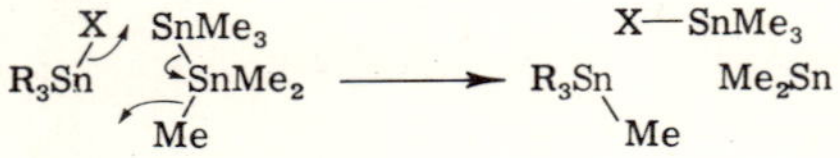

$$R_3Sn \overset{X}{\underset{Me}{\diagdown}} \overset{SnMe_3}{\underset{}{\diagup}} \ \longrightarrow \ R_3Sn \overset{X—SnMe_3}{\underset{Me}{\diagdown}} Me_2Sn$$

If the groups R in R_3SnSnR_3 are very bulky, the Sn–Sn bond is weakened by steric strain, and dissociation can now occur on heating. Hexakis(2,4,6-trimethylphenyl)- and hexakis(2,4,6-triethylphenyl)-ditin can be prepared by heating the corresponding triaryltin hydrides with azoisobutyronitrile (AIBN) (267).

$$2 \ Ar_3SnH \xrightarrow[100^\circ]{AIBN} Ar_3SnSnAr_3 \rightleftharpoons 2 \ Ar_3Sn\cdot$$

In addition, the former compound shows the esr spectrum of $Ar_3Sn\cdot$ at 180°, and the latter does so at 100°. The Sn–Sn bond-dissociation energies are 190 ± 8 kJ · mol^{-1} and 125 ± 5 kJ · mol^{-1}, respectively, that is, they are considerably less than that for $Me_3SnSnMe_3$ (210–240 kJ · mol^{-1}).

E. TIN–METAL BONDS, Sn–M

The preparation of trialkyltin compounds of lithium, R_3SnLi, and their use for preparing the organotin compounds, R_3SnR', has been discussed in a previous section.

Much interest has also been shown in compounds in which tin is bonded to a metal of Group II.

Grignard reagents having bulky alkyl groups react with trialkyltin hydrides to give compounds having a Sn–Mg bond, and are synthetically useful as a source of nucleophilic R_3Sn; in particular, they react with carbonyl compounds, oxiranes, and oxetanes to give the α-, β-, or γ-hydroxyalkyltin compounds in good yield (268–270).

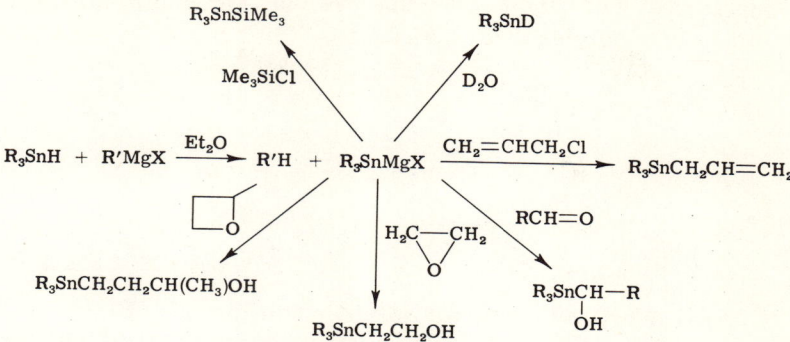

Similarly, triphenyltin hydride reacts with diethylzinc or diethylcadmium in a strongly solvating solvent, such as oxolane (tetrahydrofuran) or 1,2-dimethoxyethane, to give the solvated, metal–metal-bonded products (271).

$$Ph_3SnH + Et_2Zn \rightarrow (Ph_3Sn)_2Zn$$

$$Ph_3SnH + Et_2Cd \rightarrow (Ph_3Sn)_2Cd$$

Likewise, triphenyltin hydride reacts with ethylzinc chloride, or triphenyltin chloride with metallic zinc, to give the compound $Ph_3SnZnCl$, which is stable in the presence of a strongly coordinating ligand, but, in its absence, apparently undergoes an intermetallic shift of the organic group, so that protic acids react to liberate benzene (272).

$$Ph_3SnH + EtZnCl \searrow$$
$$Ph_3SnZnCl \longrightarrow Ph_2Sn, PhZnCl \xrightarrow{MeOH} PhH$$
$$Ph_3SnCl + Zn \nearrow$$

The Mössbauer spectrum of the rearranged compound corresponds with that of a Sn(IV) compound, and the most probable structure appears to be that of a chlorine-bridged dimer.

$$\begin{array}{c} Ph_2Sn \overset{Cl}{\diagdown} ZnPh \\ | \qquad \quad | \\ PhZn \underset{Cl}{\diagup} SnPh_2 \end{array}$$

The bis(trialkyltin) compounds of mercury are formed when trialkyltin hydrides react with dialkylmercury compounds, bis(trialkylsilyl)mercuries with trialkyltin halides, or hexamethylditin with methylmercuric halides (*273–276*).

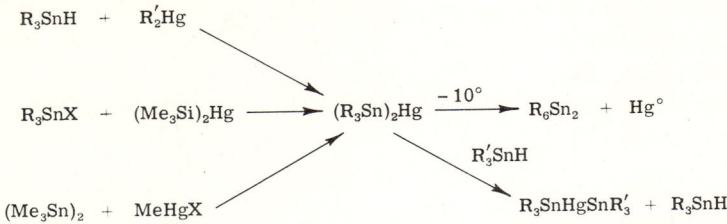

The products are yellow or red solids; when R = Me, Et, Pr, or Bu, they decompose below $-10°$, but when R = Ph, or, particularly, when R = Me_3SiCH_2, the products are more stable. They are oxidized immediately in air to the corresponding distannoxanes, readily exchange the trialkyltin group with trialkyltin hydrides, and add across polar-substituted alkynes or azo compounds.

The compounds $R_3SnHgCR_3$ can be prepared by transmetallation with the corresponding silicon compound.

$$(Me_3Si)_2Hg + Bu^tHgX \rightarrow Me_3SiHgBu^t + Me_3SiX$$

$$Me_3SiHgBu^t + Me_3SnOMe \rightarrow Me_3SnHgBu^t + Me_3SiOMe$$

The product is a yellow oil that reacts with benzylidenemalonodinitrile by 1,4-addition of the Bu^t and $SnMe_3$ fragments, and decomposes at 37°, giving a CIDNP effect (*277, 278*).

Relatively little work has been carried out on the derivatives of the metals of Group III, but (trimethyltin)lithium has been shown to react with the trialkylmetallic compounds R_3M (M = B, Al, Ga, In, or Tl) to give the products $Me_3Sn\overline{M}R_3$ Li^+, which decompose (B immediately, Ga and Zn in two days) to give $(Me_3Sn)_3SnLi$ (*279*).

F. Functionally Substituted Compounds

An improved route to iodomethylzinc iodide has made the iodomethyltin compounds, R_3SnCH_2I, more accessible (*280*). Dimetallic compounds of the types Me_3SnCBr_2MgCl (*281*), $Me_3SnCClBrLi$ (*282*), and $Me_3SnCHIZnI$ (*280, 283*), have been prepared in solution at low temperature, and thence, the ditin compounds $(Me_3Sn)_2CClBr$, $(Me_3Sn)_2CBr_2$, $(Me_3Sn)_2CHI$ (*284*).

These compounds can then be used as a source of a tin-substituted carbene, e.g. (*284*),

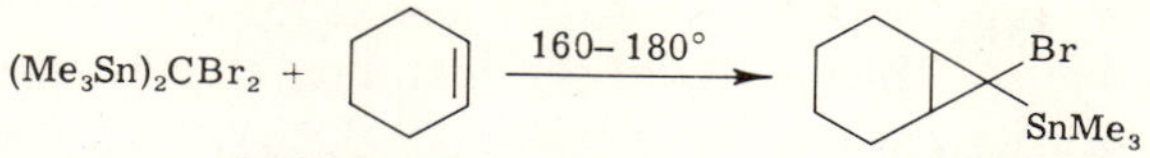

$$(Me_3Sn)_2CBr_2 + \text{[cyclohexene]} \xrightarrow{160-180°} \text{[bicyclic]} \begin{matrix} Br \\ SnMe_3 \end{matrix}$$

The iodomethyltin compounds react with amines to give aminomethyltin compounds, $R_3SnCH_2NR_2$, and with phosphines to give the phosphonium ions, $Me_3SnCH_2P^+R_3$, and tris(tributylstannylmethyl)amine has been prepared from (tributyltin)lithium and tris(phenylthiomethyl)amine (285).

$$Bu_3SnLi + (PhSCH_2)_3N \rightarrow (Bu_3SnCH_2)_3N$$

The classic route to halogenomethyltin compounds is the methylenation of tin halides with diazomethane, and this reaction has been used as the basis for the preparation of a series of thiomethyltin compounds (286).

$$SnBr_4 \xrightarrow{CH_2N_2} Sn(CH_2Br)_4 \xrightarrow{RSNa} Sn(CH_2SR)_4$$
$$\downarrow Br_2 \quad (R = Bu \text{ or } Ph)$$
$$Br_2Sn(CH_2Br)_2 \xrightarrow{RSNa} (RS)_2Sn(CH_2SR)_2$$

G. $R_3Sn\cdot$ Radicals

Trialkyltin radicals are important intermediates in the reduction of alkyl halides, and in the hydrostannation of alkenes (1, 287).

$$R_3Sn\cdot + R'X \rightarrow R_3SnX + R'\cdot$$
$$R'\cdot + R_3SnH \rightarrow R'H + R_3Sn\cdot$$
$$R_3Sn\cdot + C=C \rightarrow R_3SnCC\cdot$$
$$R_3SnCC\cdot + R_3SnH \rightarrow R_3SnCCH + R_3Sn\cdot$$

These reactions are well established, have been reviewed elsewhere (288, 289), and will not be considered in detail here.

New sources of $R_3Sn\cdot$ radicals that have been developed include the reversible thermal dissociation of bis(trialkylstannyl)pinacols (290–292), the β-scission of β-stannylalkyl radicals (293), and the photolysis of cyclopentadienyltin compounds (294).

$$\begin{matrix} Ph_2COSnMe_3 \\ | \\ Ph_2COSnMe_3 \end{matrix} \rightleftharpoons 2\, Ph_2\dot{C}OSnMe_3 \rightleftharpoons 2\, Ph_2CO + 2\, Me_3Sn\cdot$$

$$Me_3SnCH_2CHMe_2 \xrightarrow{Bu^tO\cdot} Me_3SnCH_2\dot{C}Me_2 \rightarrow Me_3Sn\cdot + H_2C=CMe_2$$

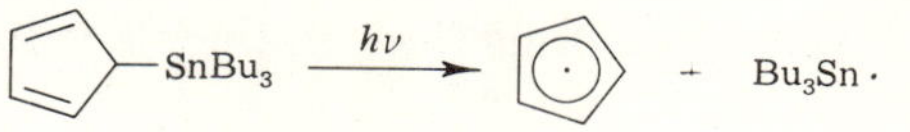

$$\text{[Cp]}-SnBu_3 \xrightarrow{h\nu} \text{[Cp}\cdot\text{]} + Bu_3Sn\cdot$$

The last two reactions are useful for esr studies involving free radicals. Until recently, the only trialkyltin radical that had been observed directly, in solution, by esr was Me$_3$Sn· (*295*), but many more have now been reported (e.g., Et$_3$Sn·, Pr$_3$Sn·, and Bu$_3$Sn·) (*296*). Bulky ligands [e.g., (PhCMe$_2$CH$_2$)$_3$Sn·] increase the persistence of the radicals, so that esr observation is easier (*297*), and tris(2,3,5-trimethylphenyl)tin and tris(2,3,5-triethylphenyl)tin radicals, at 180° and 100°, respectively, are in thermal equilibrium with the corresponding hexaarylditins (*298*).

$$Ar_3SnH \xrightarrow[100°]{AIBN} 2\ Ar_3Sn· \rightleftharpoons Ar_3SnSnAr_3$$
$$(Ar = 2,3,5\text{-Me}_3C_6H_2 \text{ or } 2,3,5\text{-Et}_3C_6H_2)$$

Very much more persistent radicals, with a half-life of up to a year, and with the structures [(Me$_3$Si)$_2$CH]$_3$Sn· and [(Me$_3$Si)$_2$CH]$_2$RSn· (R = Pr^i, Bu^t, Me, Et, Bu, or cyclopentadienyl) have been prepared by the photolysis of [(Me$_3$Si)$_2$CH]$_2$Sn(II), or by photolyzing a mixture of the appropriate halide [(Me$_3$Si)$_2$CH]$_2$RSnX with an "electron rich" olefin (*299, 300*).

Homolytic substitution (S$_H$2) by tin radicals at halogen centers (X) has been investigated extensively, the reactivity of the alkyl halides RX increasing in the sequence R = R^p < R^s < R^t, and X = F < Cl < Br < I (*301, 302*).

The reaction of tributyltin hydride with ring-substituted benzyl chlorides gives a Hammett ρ-factor of +0.81, confirming the "nucleophilic" character of the Bu$_3$Sn· radical (*303*).

Other centers at which the S$_H$2 reaction by R$_3$Sn· radicals has been established include H (in R$_3$SnH) (*304*), N (in tetraalkyltetrazenes) (*305*), (but not P in diphosphines) (*306*), O (in peroxides) (*307–309*), S (in disulfides) (*310, 311*), and Se and Te (in diselenides and ditellurides) (*312*).

Apart from the familiar addition of R$_3$Sn· radicals to alkenes and alkynes, addition has also been shown to occur at O in C=O (in aldehydes, ketones, 1,2-diketones, and esters) (*313, 314*), S in C=S (in thioesters) (*315, 316*) and in R$_2$P(S)P(S)R$_2$ (*317*), and Se in C=Se (in selenoketones) (*318*).

H. R$_2$Sn: STANNYLENES

Like the Sn(III) radicals, the Sn(II) stannylenes are familiar both as persistent species of unlimited life and as transient, highly reactive intermediates. Many of the compounds that, in the older literature, were referred to as monomeric R_2Sn species, are now recognized to be cyclic oligomers.

Dicyclopentadienyltin(II) is readily prepared from cyclopentadienyllithium or -sodium and tin(II) chloride, and is the best known, monomeric, R_2Sn compound. The planes of the $^5\eta$-cyclopentadienyl groups subtend an angle of $\sim 55°$, and the unshared pair of electrons can act as a ligand towards such Lewis acids as BF_3 and $AlCl_3$ (*319*).

The cyclopentadienyl groups are readily displaced by protic acids HX (e.g., alcohols, phenols, thiols, and oximes), providing a convenient route to other Sn(II) compounds (*320–323*).

$$(C_5H_5)_2Sn + 2\ HX \rightarrow SnX_2 + 2\ C_5H_6$$

Dicyclopentadienyltin also takes part in oxidative addition reactions with such reagents as iodomethane, diiodomethane, ethyl bromoacetate, and diphenyl disulfide, and there is evidence that the reactions involve a radical chain-mechanism (*324, 325*).

$$X\cdot + (C_5H_5)_2Sn \rightarrow (C_5H_5)_2\dot{S}nX$$
$$(C_5H_5)_2\dot{S}nX + XY \rightarrow (C_5H_5)_2SnXY + X\cdot$$

Another interesting organotin(II) compound is the highly sterically hindered $[(Me_3Si)_2CH]_2Sn$, which can be prepared by the reaction of $(Me_3Si)_2CHLi$ with $SnCl_2$ or $Sn[N(SiMe_3)_2]_2$ (*326–329*).

Like dicyclopentadienyltin, it undergoes oxidative addition-reactions with alkyl halides, and, again, there is evidence for a homolytic chain-mechanism (*330, 331*).

A single-crystal, X-ray diffraction analysis of the structure has recently been performed that shows that the compound is, in fact, a tin–tin bonded dimer, having an Sn–Sn bond length of 276 pm, similar to that in hexaphenylditin; this was interpreted in terms of overlap of a filled sp_zp_y orbital with the vacant p_z orbitals on the other tin atom resulting in a "bent," weak, Sn–Sn double bond (*332*).

Transient dialkylstannylene intermediates R_2Sn: (reviewed in ref. *333*) can be prepared by the thermolysis of distannanes ClR_2SnSnR_2Cl, R_3SnSnR_2Cl, or HR_2SnSnR_2H, e.g., (*334, 335*)

$$ClBu_2SnSnBu_2Cl \xrightarrow{120-130°} Bu_2SnCl_2 + [Bu_2Sn:]$$
$$Me_3SnSnMe_2Br \xrightarrow{20°} Me_3SnBr + [Me_2Sn:]$$

Alternatively, the oligomeric dialkylstannanes can be photolyzed (*336*).

$$(Bu_2Sn)_n \xrightarrow{h\nu} Bu_2\dot{S}n(SnBu_2)_{n-2}\dot{S}nBu_2 \rightarrow n\ Bu_2Sn:$$

The stannylenes from either source will insert into the Sn–Sn, Sn–R, or Sn–H bonds of organotin compounds, and react with alkyl halides, disulfides, or peroxides as shown in the reaction scheme below, but only the stannylenes that are generated photolytically will react with carbonyl compounds, and it appears that the stannylenes may exist in two forms, perhaps related as singlet and triplet, or a complexed and uncomplexed species.

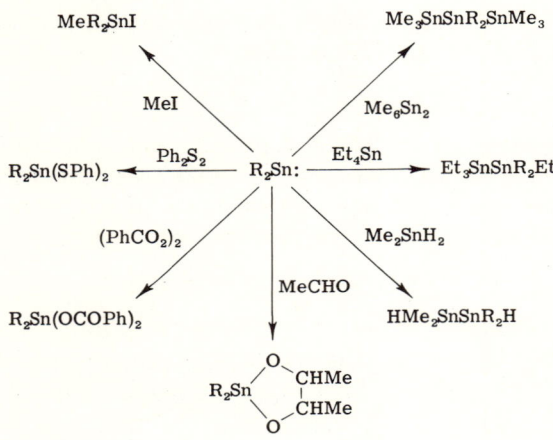

III. Structural Organotin Chemistry

Before 1963, when Hulme (*337*) used X-ray crystallography to demonstrate that the monopyridine adduct of trimethyltin chloride contains 5-coordinate tin, it had been generally assumed that most organotin(IV) compounds are simple, tetrahedral molecules containing 4-coordinate tin atoms. In recent years, however, the well established techniques of proton NMR and infrared spectroscopy have been supplemented by ^{119m}Mössbauer (*5–9*) and ^{119}Sn NMR spectroscopy (*10–12, 338*), and these have stimulated X-ray investigations of a large number of organotin compounds (*339, 340*). Many derivatives are now known that contain not only 5- and 6-, but even 7-, coordinate tin atoms, and a selection of these will be discussed in this section. This increased knowledge of the structural chemistry of organotin compounds is of considerable importance in understanding the mode of action operative in their many applications (see Section V) and explaining the variations that are observed in their toxicity (see Section IV).

A. X-Ray Investigations of Organotin(IV) Compounds

1. 4-Coordinate Compounds

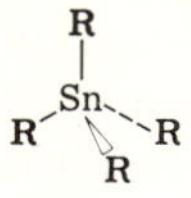

R_4Sn type 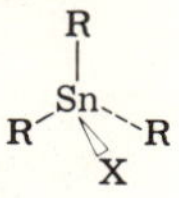 R_3SnX type 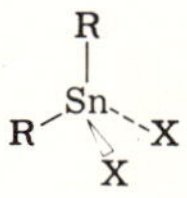 R_2SnX_2 type 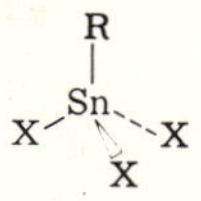 $RSnX_3$ type

The simplest bonding situation that can occur in organotin(IV) compounds consists of hybridization of the $5s$- and three $5p$- orbitals of the tin atom, to produce four tetrahedrally oriented bonds, and a coordination number of 4. Thus, the R_4Sn structural type is expected to be a tetrahedral molecule, as has been demonstrated for $Sn(C_6F_5)_4$ (341), $Sn(C_6H_4Me\text{-}3)_4$ (342), $Sn(C_6H_4Me\text{-}4)_4$ (343), $Sn(2\text{-thienyl})_4$ (344), and $Sn(C_5H_5)_4$ (345). The absolute configuration of optically active (+)-2-triphenylstannylbutane has also been determined crystallographically (346).

If one (or more) alkyl or aryl groups in a tetraorganotin compound, R_4Sn, is (are) replaced by an electronegative substituent, X, the tin atom in the resulting compounds, R_nSnX_{4-n}, has a marked tendency to increase its coordination number above 4, as described in the following subsections. However, if the R or X groups are fairly bulky, the coordination number may remain 4, as in $[(Me_3Si)_2CH]_3SnCl$ (347), and in the triphenyltin thiophenoxides, $Ph_3SnSC_6H_4{}^tBu\text{-}4$ (348) and $Ph_3SnSC_6H_4Me\text{-}2$ (349), all of which have a regular, tetrahedral R_3SnX type of structure. Similarly, the cyclic dimethyltin sulfide trimer, $(Me_2SnS)_3$, is (350, 351) a six-membered ring in a boat conformation, with near-tetrahedral tin atoms of the R_2SnX_2 type. Structural data on 4-coordinate, mono-organotin compounds, $RSnX_3$, are very scarce, but an X-ray study of methyltin sesquisulfide, $(MeSnS_{1.5})_4$, in which the tin atom is in a tetrahedral $RSnX_3$ type of geometry involving one alkyl group and three sulfur atoms, showed the presence of an adamantane-like cage (352).

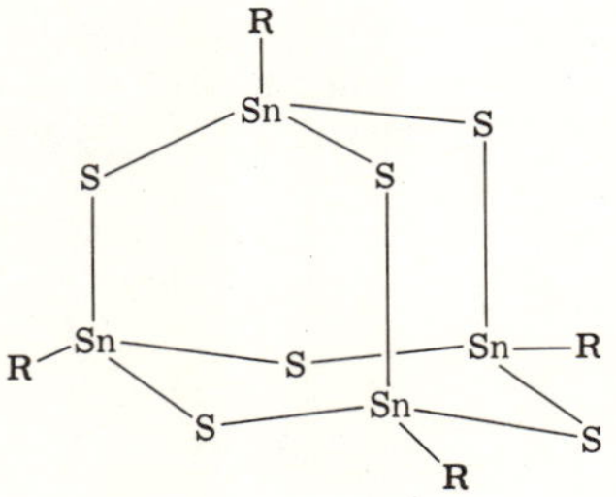

2. 5-Coordinate Compounds

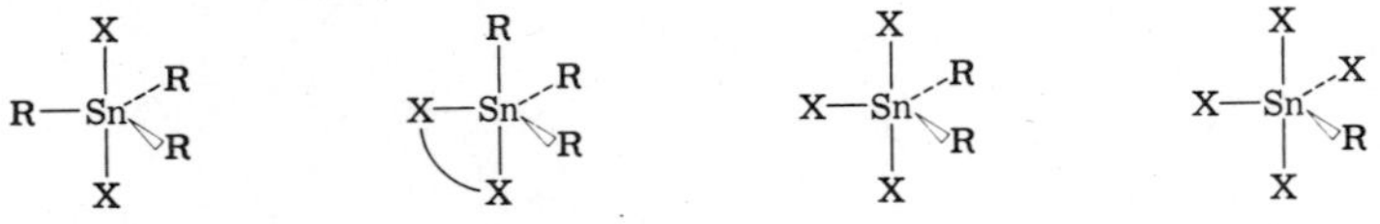

R_3SnX_2 type

cis-R_3SnX_2 type

cis-R_2SnX_3 type

$RSnX_4$ type

The tetraorganotin compounds, R_4Sn, show no tendency to increase their coordination number, owing to their weak, Lewis acidity conferred by the four electron-releasing alkyl groups. It has, however, been claimed (*353*) that trimethyl(trifluoromethyl)tin forms a 1:1 adduct with hexamethylphosphoric triamide, and that this may be isolated in the solid state.

Replacement of one of the organic groups by an electronegative radical, X, decreases the electron density at the tin atom, and thus increases its ability to act as an electron acceptor. Compounds of the type R_3SnX readily form 5-coordinate, trigonal, bipyramidal complexes of the R_3SnX_2 type, in which the three organic groups are situated in an equatorial plane at right angles to the linear X—Sn—X axis. The 1:1 adduct of trimethyltin chloride with triphenylphosphine-acetylmethylene, $Me_3SnCl, Ph_3P:CHCOMe$, has been shown to be of this type (*354*). Trimethyltin acetate also has a tin atom that is pentacoordinate with an R_3SnX_2 geometry, but the structure is polymeric, with planar, Me_3Sn units linked by bridging, bidentate, acetate groups (*355*).

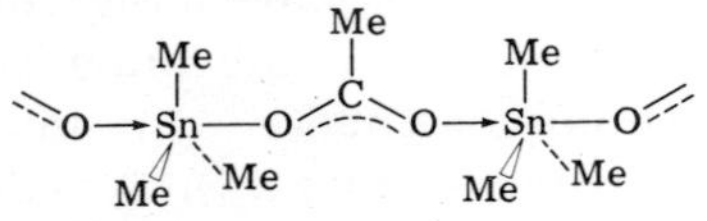

Many other R_3SnX compounds adopt this self-associated, infinite-chain structure in the solid state, and examples of common bridging atoms or groups include X = F, NCS, NCO, OMe, NO_3, or OH. Rather more unusual, however, are triorganotin compounds that have the same R_3SnX_2 geometry, but contain a tin atom that is pentacoordinate through intramolecular coordination via one of the R groups. Two compounds that are known to have a structure of this type are the mixed triorganotin bromides, dimethyl(4-bromo-1,2,3,4-tetraphenyl-*cis,cis*-1,-3-butadienyl)tin bromide (*356*)

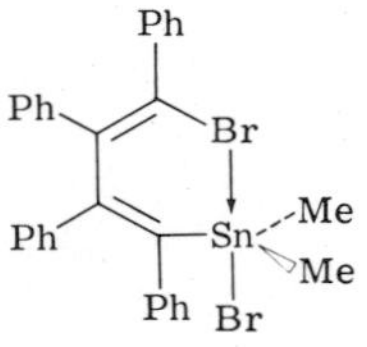

and C,N-{2-[(dimethylamino)methyl]phenyl} diphenyltin bromide (*357*).

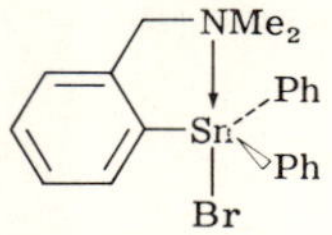

If the ligand X_2 in a pentacoordinate triorganotin compound is potentially bidentate, such as the anion of 1,3-diphenyl-1,3-propanedione or of *N*-benzoyl-*N*-phenylhydroxylamine, the tin atom is constrained to a *cis*-R_3SnX_2 type of geometry, e.g., the triphenylstannyl derivatives of

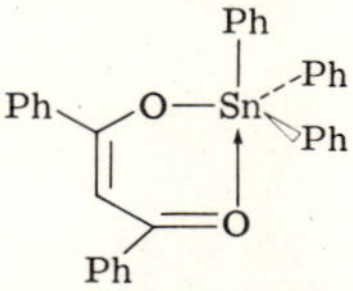

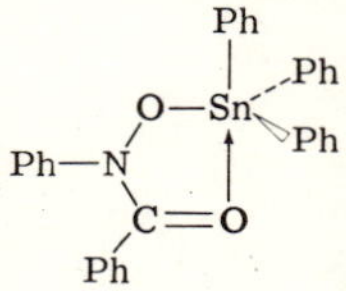

these anions (*358, 359*). No examples of triorganotin compounds that have a *meridional* -R_3SnX_2 type of geometry have yet been demonstrated crystallographically.

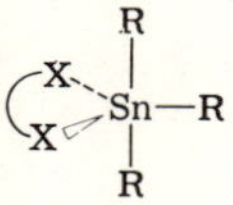

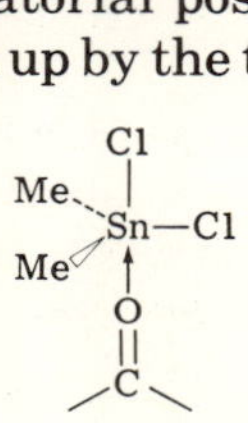

The diorganotin compounds, R_2SnX_2, are stronger Lewis acids than the triorganotins, R_3SnX; and the 5-coordinate complexes, R_2SnX_3, which result from the addition of a donor atom or group to the R_2SnX_2 compound, are not very common, mainly because of the tendency of the tin atom in these compounds to increase its coordination number to 6 by accepting *two* donor molecules (and, in some cases, even up to 7, as will be described later). However, the existence of a number of pentacoordinate complexes having a *cis*-R_2SnX_3 geometry has been demonstrated by X-ray crystallography. The anion in quinolinium dimethyltrichlorostannate, $(C_9H_8N)^+$ $(Me_2SnCl_3)^-$, is a distorted, trigonal bipyramid having the two methyl groups occupying equatorial positions (*360*). Salicylaldehyde forms a 1:1 complex with dimethyltin dichloride, and X-ray studies revealed (*361*) a similar tinatom geometry, with two methyl groups occupying equatorial positions of a trigonal bipyramid and the other three sites taken up by the two chlorine atoms and a donor carbonyl oxygen atom from the aldehyde. Dimeric tetrabutyl-1,3-bis(trichloroacetoxy)distannoxane, [(Cl$_3$CO·O·Bu$_2$Sn)^{2}O]$_2$, has a ladder structure involving both monodentate and bidentate carboxylate groups, with the tin atoms occupying a *cis* R$_2$SnX$_3$ geometry (*362*).

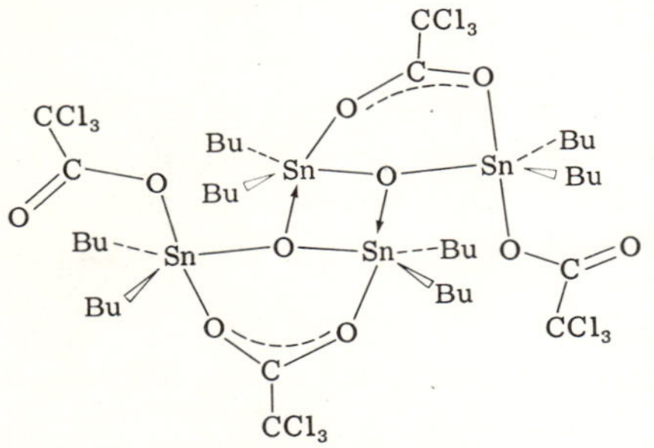

Monoorganotin compounds, RSnX$_3$, also show a marked tendency to increase their coordination number from 4 to 6, or 7, and there are few examples of compounds of the type RSnX$_4$ (which contain pentacoordinate tin). The anion in tetraphenylarsonium tetrachloromonomethylstannate, (Ph$_4$As)$^+$(MeSnCl$_4$)$^-$, is one such example, and consists of a trigonal bipyramid with the methyl group occupying an equatorial site (*363*). A similar RSnX$_4$ type of geometry, with the organic group again occupying an equatorial position, was found (*364*) in the intramolecularly pentacoordinate, ketiminotin trichloride.

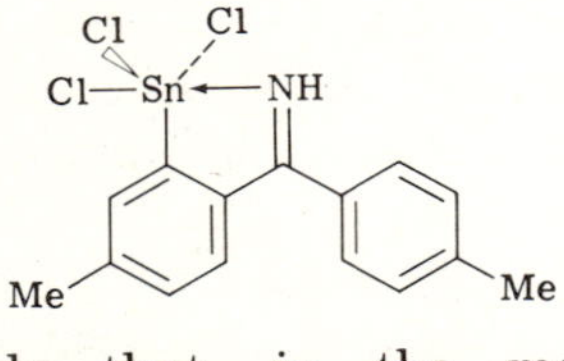

It is quite probable that, in the monoorganostannatranes, RSn(OCH$_2$CH$_2$)$_3$N, the tin atom also occupies a trigonal, bipyramidal geometry, but with the organic group forced into an axial site (*195,*

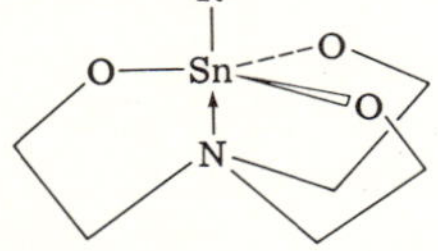

198). This configuration has not yet been demonstrated by X-ray studies, perhaps due to the difficulty in obtaining good crystalline samples. Recently, Tzschach reviewed the whole field of intramolecularly pentacoordinate, organotin compounds (*198*), including the stannatranes.

3. 6-Coordinate Compounds

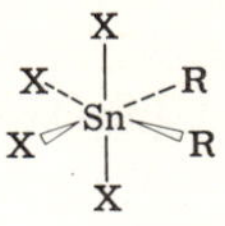

cis-R_2SnX_4 type

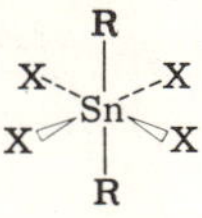

trans-R_2SnX_4 type

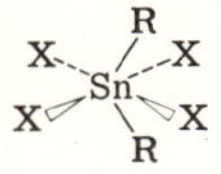

distorted trans-R_2SnX_4 type

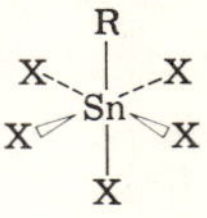

$RSnX_5$ type

In triorganotin compounds of the type R_3SnX, as mentioned previously, the tin atom is only a weak acceptor, and these compounds tend to increase their coordination number only to 5. Consequently, examples of 6-coordinate, triorganotin complexes containing three inorganic groups or ligands, R_3SnX_3, have not yet been demonstrated crystallographically. The 1:1 adducts of trimethyltin chloride with 2,2-bipyridyl, and of trimethyltin isothiocyanate with 1,10-phenanthroline are believed to contain 6-coordinate tin on the basis of the infrared-spectral properties (*365*).

In contrast, the diorganotin compounds, R_2SnX_2, show a strong tendency to increase their coordination number to 6 by accepting two donor molecules, thus leading to octahedral complexes of the type R_2SnX_4, which can exist as the cis or trans isomers, as already illustrated. The cis-R_2SnX_4 orientation is often found to be sterically favorable when two bidentate ligands are present, as in Me_2Snox_2 (*366*), $Me_2Sn(O\cdot NH\cdot CO\cdot Me)_2$ (*367*), and $Ph_2Sn(S\cdot CS\cdot NEt_2)_2$(*368*). Dimethyltin bis(acetylacetonate), however, has a trans-R_2SnX_4 geometry (*369*), as have the two diphenyltin dichloride adducts Ph_2SnCl_2,bipy (*370*) and $Ph_2SnCl_2,2Me_2SO$ (*371*). Dimethyltin difluoride consists of an infinite, two-dimensional network of tin and fluorine atoms, with each tin linearly bridged to its four neighbors by one, symmetrically disposed, fluorine atom, and the methyl groups situated above and below this plane, thus completing a regular, octahedral trans-R_2SnX_4 type of coordination of the tin atom (*372*).

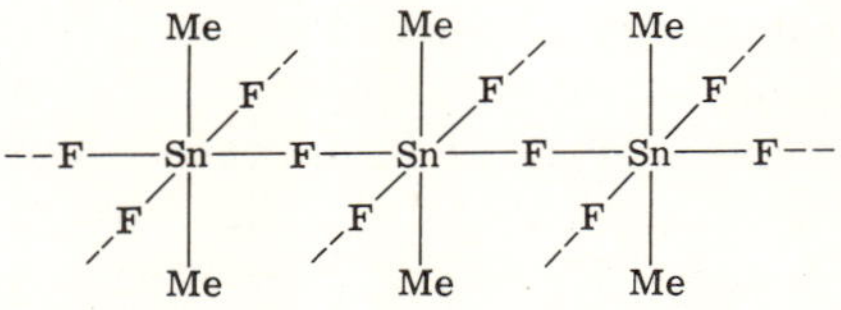

Dimethyltin bis(fluorosulfonate), $Me_2Sn(SO_3F)_2$, has a similar, polymeric structure (*373*).

Interesting variations of this trans-R_2SnX_4 structure exist; they usually take the form of a distorted geometry in which the C–Sn–C bond angle is lessened below 180°, and the four inorganic groups remain in a plane about the tin atom, but two pairs of groups are at different distances from the central tin atom. (This was illustrated at the beginning of this subsection.) This type of structure is exemplified by diethyltin dichloride and dibromide (*374*), which consist of chains of molecules, with each tin atom in a distorted, *trans*-R_2SnX_4 environment of two ethyl groups and coplanar chlorine, and bridging chlorine atom pairs.

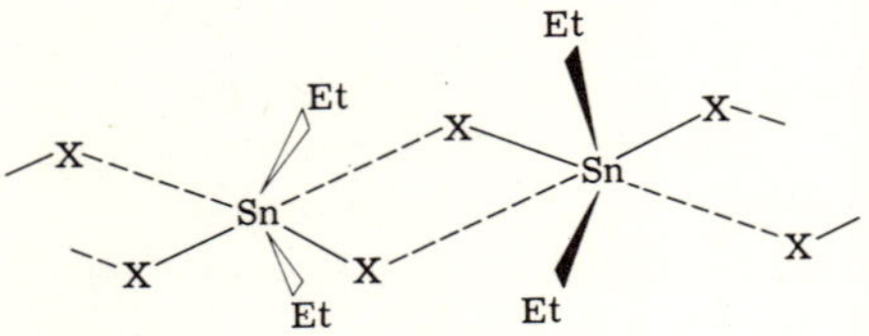

(X = Cl or Br)

Dimethyltin dichloride has a similar chain structure (*375*). In diethyltin diiodide (*374*), dimethyltin diisothiocyanate (*376, 377*), and dichloro bis(chloromethyl)stannane (*378*), however, the distorted, *trans*-R_2SnX_4 geometry of each tin atom is completed by two bridging bonds involving the halogen or pseudohalogen atoms on the same, neighboring molecule.

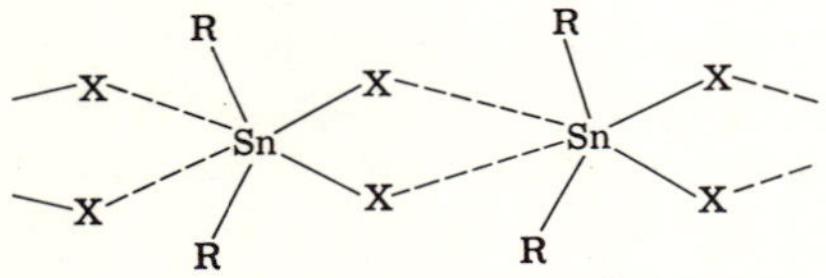

(R = Me, X = NCS; R = Et, X = I;
R = CH_2Cl, X = Cl)

The same, distorted, octahedral geometry is also found in a number of monomeric diorganotin complexes having two bidentate ligands, such as $Me_2Sn(O \cdot NMe \cdot CO \cdot Me)_2$ (*379*) and $Me_2Sn(S \cdot CS \cdot NMe_2)_2$ (*380*), or one tetradentate group, such as $Me_2Sn(salen)$ (*381*).

Monoorganotin compounds, $RSnX_3$, like the diorganotin derivatives, have a strong tendency to increase their coordination number up to 6 by accepting two donor molecules, leading to octahedral complexes of the $RSnX_5$ type. Two examples of molecules having this geometry are chloromonophenyltin bis(*N,N*-diethyldithiocarbamate) (*382*)

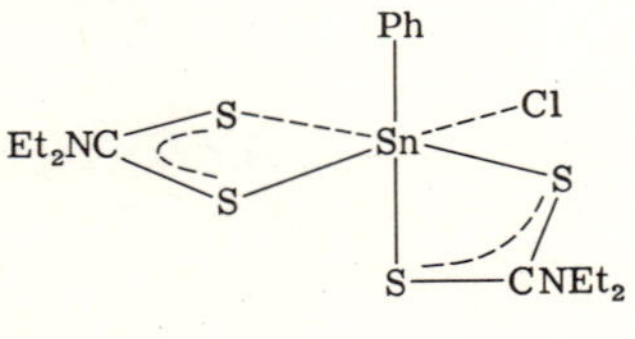

and the dimeric, monoethyltin hydroxide dichloride hydrate, $EtSn(OH)Cl_2·H_2O$ (383).

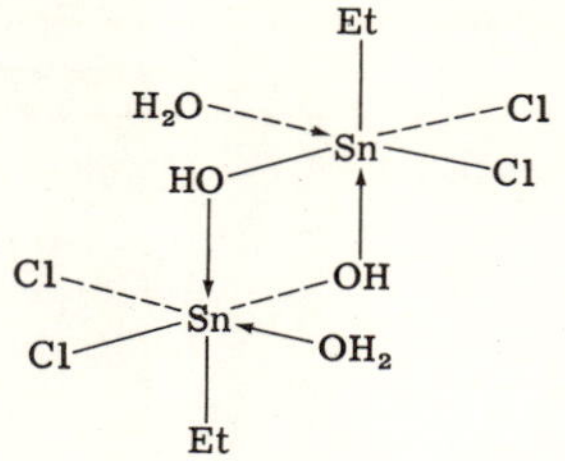

$$\begin{array}{c}
\text{Et} \\
H_2O\cdots\overset{|}{\underset{|}{Sn}}\cdots Cl \\
HO \quad\quad Cl \\
Cl\cdots\overset{|}{\underset{|}{Sn}}\cdots OH \\
Cl \quad\quad OH_2 \\
\text{Et}
\end{array}$$

4. 7-Coordinate Compounds

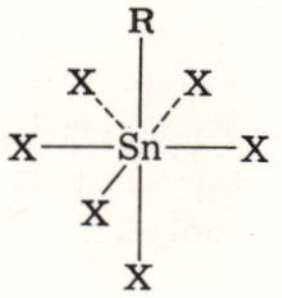

$RSnX_6$ type 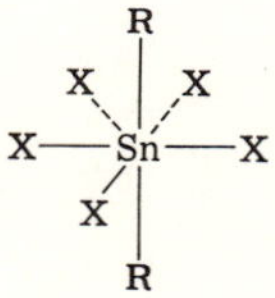 R_2SnX_5 type

Monoorganotin compounds having three bidentate, donor ligands, such as $MeSn(NO_3)_3$ (384) and $MeSn(SCSNEt_2)_3$ (385), contain a tin atom occupying a pentagonal, bipyramidal geometry, as just illustrated. Certain diorganotin complexes have also been shown to possess a tin atom in a pentagonal, bipyramidal structure of the R_2SnX_5 type, namely, $Me_2Sn(NCS)_2$, terpy (386), $[Ph_2Sn(NO_3), 3Me_2SO]^+ NO_3^-$ (387), $(^nPr_2SO)Ph_2Sn(NO_3)·O·CO·CO·O·SnPh_2(NO_3)$ $(^nPr_2SO)$ (388), and $Ph_2Sn(NO_3)_2, Ph_3PO$ (461).

5. Survey of Organotin(IV) Crystal Structures

The X-ray crystal structures of organotin compounds that have been described since Ho and Zuckerman's earlier compilation (339) are listed systematically in Table I.[1]

The following abbreviations are used: $salenH_2$ = bis(salicylaldehyde)ethylenediimine, salH = salicylaldehyde, SAB = 2-hydroxy-N-(2-hydroxybenzylidine)aniline dianion, SAT = 2-(o-hydroxyphenyl)benzothiazoline dianion, SAP = N-(2-hydroxyphenyl)salicyladldimine dianion, and HMPT = hexamethylphosphoric triamide, and structures marked with an asterisk (*) are polymeric, usually by intermolecular association.

---

[1] Since this review was written, Zuckerman has up-dated his earlier compilation (339) to 1978 [Zubieta, J. A., and Zuckerman, J. J., *Progr. Inorg. Chem.*, **24**, 251 (1978)].

TABLE I

THE STEREOCHEMISTRIES OF ORGANOTIN(IV) COMPOUNDS THAT HAVE BEEN DETERMINED BY X-RAY CRYSTALLOGRAPHY

Compound	Geometry of Sn Atom	Ref.
Mono-, Di-, and Tri-methyltin Compounds, Me_nSnX_{4-n}		
$MeSn(SCSNEt_2)_3$	Distorted $RSnX_6$	385
$Me_2Sn(acac)_2$	trans-R_2SnX_4	369
$Me_2Sn(salen)$	Distorted trans-R_2SnX_4	381
$Me_2\overline{SnSCH_2CH_2S}$	Distorted R_2SnX_2	389
$Me_2Sn(O \cdot NMe \cdot CO \cdot Me)_2$	Distorted trans-R_2SnX_4	379
$Me_2Sn(O \cdot NH \cdot CO \cdot Me)_2$	cis-$R_2SnX_4^*$	367
$Me_2Sn(O \cdot NH \cdot CO \cdot Me)_2 \cdot H_2O$	Distorted trans-$R_2SnX_4^*$	367
$Me_2Sn(SCSNMe_2)_2$	Distorted trans-R_2SnX_4	380
$Me_2Sn(SAB)$	Distorted cis-R_2SnX_3 (dimeric)	390
$Me_2Sn(SAP)$	Distorted cis-R_2SnX_3	391
$Me_2Sn(OH)NO_3$	Distorted cis-R_2SnX_3 (dimeric)	392
$(Me_2Sn)_3(PO_4)_2 \cdot 8 H_2O$		393
inner Sn atoms	trans-$R_2SnX_4^*$	
outer Sn atoms	Distorted trans-$R_2SnX_4^*$	
Me_3SnOMe	$R_3SnX_2^*$	394
$Me_3SnONC_6H_{10}$	$R_3SnX_2^*$	395
$Me_3SnN(Me)NO_2$	$R_3SnX_2^*$	396
Me_3SnNCO, Me_3SnOH	$R_3SnX_2^*$	397
$Me_3SnOCHO$	$R_3SnX_2^*$	398
$Me_3SnOCOMe$	$R_3SnX_2^*$	355
$Me_3SnOCOCF_3$	$R_3SnX_2^*$	355
$Me_3SnOCOCH_2NH_2$	$R_3SnX_2^*$	399
$Me_3SnOCOPy-2 \cdot H_2O$	$R_3SnX_2^*$	400
Me_3SnNO_3	$R_3SnX_2^*$	401
Me_3SnSO_2Me	$R_3SnX_2^*$	402,403
$Me_3SnSO_2CH_2C\vdots CH$	$R_3SnX_2^*$	404
$Me_3SnSO_3Ph \cdot H_2O$	$R_3SnX_2^*$	405
$(Me_3Sn)_3(OH)CrO_4$	$R_3SnX_2^*$	406
Mono- and Di-ethyltin Compounds, $EtSnX_3$ and Et_2SnX_2		
$EtSn(OH)Cl_2 \cdot H_2O$	$RSnX_5$ (dimeric)	383
Et_2SnCl_2	Distorted trans-$R_2SnX_4^*$	374
Et_2SnBr_2	Distorted trans-$R_2SnX_4^*$	374
Et_2SnI_2	Distorted trans-$R_2SnX_4^*$	374
Di- and Tri-n-butyltin Compounds, Bu_2SnX_2 and Bu_3SnX		
$Bu_2\overline{SnO(CH_2)_3O}$	Distorted trans-$R_2SnX_4^*$	407
$Bu_2\overline{SnSCH_2CH_2S}$	Space-group and unit-cell data only	408

TABLE I (*Continued*)

Compound	Geometry of Sn Atom	Ref.
Bu$_3$SnF	Incomplete determination	409
(Bu$_3$Sn)$_2$SO$_4$	Incomplete determination	450

Mono-, Di- and Tri-phenyltin Compounds, Ph$_n$SnX$_{4-n}$

Compound	Geometry of Sn Atom	Ref.
PhSn(SCSNEt$_2$)$_2$Cl	Distorted RSnX$_5$	382
(^{n}Pr$_2$SO)Ph$_2$Sn(NO$_3$)O·CO·CO·OSnPh$_2$(NO$_3$)(^{n}Pr$_2$SO)	R$_2$SnX$_5$	388
Ph$_2$Sn(SCSNEt$_2$)$_2$	Distorted *cis*-R$_2$SnX$_4$	368
Ph$_2$Sn[SPS(OEt)$_2$]$_2$	Distorted *trans*-R$_2$SnX$_4$	465
Ph$_2$Sn(glygly)	Distorted *cis*-R$_2$SnX$_3$	410
Ph$_2$Sn(SAT)	Distorted *cis*-R$_2$SnX$_3$	411
Ph$_2$Sn(SAB)	Distorted *cis*-R$_2$SnX$_3$	412
Ph$_3$SnOH	R$_3$SnX$_2^*$	413
Ph$_3$SnO·CPh·CH·CO·Ph	*cis*-R$_3$SnX$_2$	358
Ph$_3$SnO·NPh·CO·Ph	*cis*-R$_3$SnX$_2$	359
Ph$_3$SnSPy-4	R$_3$SnX$_2^*$	414
Ph$_3$SnNCS	R$_3$SnX$_2^*$	415
Ph$_3$SnNCO	R$_3$SnX$_2^*$	416
Ph$_3$SnSC$_6$H$_4^t$Bu-4	R$_3$SnX	348
Ph$_3$SnSC$_6$H$_2$Me$_3$-2,4,6	R$_3$SnX	349
Ph$_3$SnSC$_6$H$_4$Me-2	R$_3$SnX	349
Ph$_3$SnSC$_6$H$_2$F-4,Br$_2$-2,6	R$_3$SnX	349

Other Mono-, Di-, and Tri-organotin Compounds, R$_n$SnX$_{4-n}$

Compound	Geometry of Sn Atom	Ref.
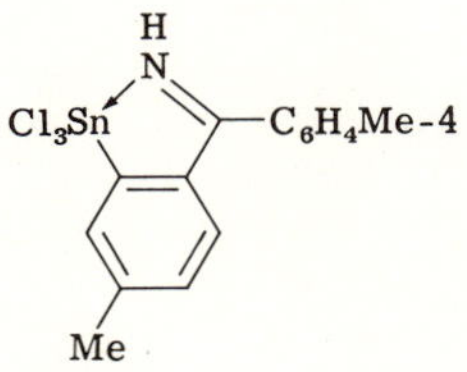 Cl$_3$Sn-indazolyl(NH)-C$_6$H$_4$Me-4, 6-Me (structure shown)	RSnX$_4$	364
(ClCH$_2$)$_2$SnCl$_2$	Distorted *trans*-R$_2$SnX$_4^*$	378
(Ferrocenyl)$_2$SnCl$_2$	R$_2$SnX$_2$	417
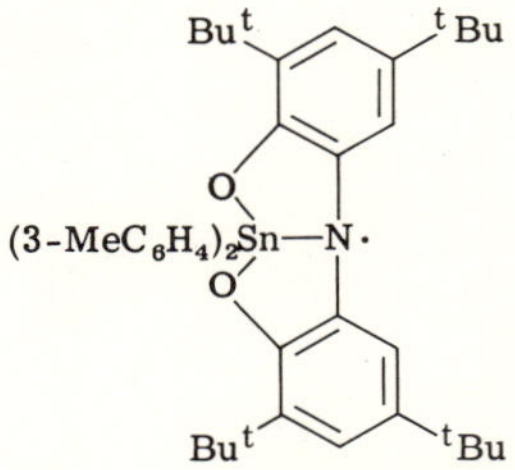 (3-MeC$_6$H$_4$)$_2$Sn—N(catecholate with ^{t}Bu groups)	Distorted *cis*-R$_2$SnX$_3$	418

(*continued*)

TABLE I (Continued)

Compound	Geometry of Sn Atom	Ref.
2-(Me$_2$NCH$_2$)C$_6$H$_4$SnPh$_2$Br	R$_3$SnX$_2$	357
2,6-(Me$_2$NCH$_2$)$_2$C$_6$H$_3$SnMe$_2^+$Br$^-$	R$_3$SnX$_2$	463
2-(Me$_2$NCHMe)C$_6$H$_4$SnMePhBr	R$_3$SnX$_2$	464
[(Me$_3$Si)$_2$CH]$_3$SnCl	R$_3$SnX	347
Symmetrical and Unsymmetrical Tetraorganotins, R$_4$Sn and R$_3$SnR^1		
Sn(C$_6$F$_5$)$_4$	R$_4$Sn	341
Sn(C$_6$H$_4$Me-3)$_4$	R$_4$Sn	342
Sn(2-thienyl)$_4$	R$_4$Sn	344
Sn(C$_5$H$_5$)$_4$	R$_4$Sn(−60°)	345
1,1-(Me$_3$Sn)$_2$C$_5$H$_4$	R$_3$SnR^1(−60°)	419
Ph$_3$SnC$_7$H$_7$	R$_3$SnR^1	420
Ph$_3$SnCHMeEt	R$_3$SnR^1	346
Ph$_3$SnCH$_2$I	R$_3$SnR^1	346a
Compounds Containing Tin–Metal Bonds		
Ph$_3$SnSnPh$_3$	R$_3$SnX	422
Ph$_3$Sn^{a}SnNO$_3$(AsPh$_3$) Tin a	R$_3$SnX (dimeric)	421
(Ph$_3$Sn^a)$^b_3$SnNO$_3$ Tin a	Distorted R$_3$SnX	423
Tin b	See ref.	423
Ph$^a_3$SnSnNO$_3$ Tin a	Distorted R$_3$SnX	423
[Me$_3$SnRu$_4$(CO$_4$)]$_2$	R$_3$SnX	424
[Me$_2$SnFe(CO$_4$)]$_2$	Distorted R$_2$SnX$_2$	425
{Me$_2$Sn Fe(CO$_4$)}$_2$Sn{Fe(CO$_4$)$_2$SnMe$_2$	R$_2$SnX$_2$	426
{(σ-C$_5$H$_5$)$_2$SnFe(CO)$_4$}$_2$	Distorted R$_2$SnX$_2$	427
Ph$_3$Sn{Fe(CO)(π-C$_5$H$_5$)(π-PhC⋮CPh)}	R$_3$SnX	428
cis-(Ph$_3$Sn)$_2$Fe(CO)$_4$	R$_3$SnX	429
{Me$_2$SnCo(CO)(π-C$_5$H$_5$)}$_2$	Distorted R$_2$SnX$_2$	430
Ph$_2$Sn{Co(CO)$_2$(π-C$_7$H$_8$)}$_2$	Distorted R$_2$SnX$_2$	431
trans-(Me$_3$Sn)$_2$Ni(PPh$_3$)$_2$Cl$_2$	R$_3$SnX	432
trans-(Ph$_3$Sn)$_2$Os(CO)$_4$	R$_3$SnX	433
PhCl$_2$SnMo(CO)$_2$(C$_7$H$_7$)	Distorted RSnX$_3$	434
Ph$_2$ClSnMo(CO)$_2$(C$_7$H$_7$)	Distorted R$_2$SnX$_2$	434
(CH$_2$:CH)$_2$Sn{Mn(CO)$_5$}$_2$	R$_2$SnX$_2$	435
^{t}Bu$_2$(Py)SnCr(CO)$_5$	Flattened tetrahedron	436
[Ph$_3$SnNi{N(CH$_2$CH$_2$PPh$_2$)$_3$}]$^+$BPh$_4^-$	R$_3$SnX	451
Organotin Oxides and Sulfides		
Ph$_3$SnOSiPh$_3$	Unit-cell data only	437
(Bu$_2$SnOCOCCl$_3$)$_2$O	Distorted cis-R$_2$SnX$_3$ (dimeric)	362
Ph$_3$SnSSnPh$_3$	R$_3$SnX	462

TABLE I (Continued)

Compound	Geometry of Sn Atom	Ref.
$Ph_3SnSPbPh_3$	R_3SnX	438
$(Me_2SnS)_3$		
tetragonal form	R_2SnX_2	350
monoclinic form	R_2SnX_2	351
Ionic Organotin Halide Complexes		
$(Ph_4As)^+ (MeSnCl_4)^-$	$RSnX_4$	363
$(C_9H_8N)^+ (Me_2SnCl_3)^-$	Distorted cis-R_2SnX_3	360
$(PyH)_2^+(Me_2SnCl_4)^{2-}$	trans-R_2SnX_4	439
$[Ph_2Sn(NO_3), 3\ DMSO]^+\ NO_3^-$	R_2SnX_5	389
$(BzPPh_3)^+(Bu_3SnCl_2)^-$	R_3SnX_2	440
$[Me_3Sn(HMPT)_2]^+ (Me_3SnBr_2)^-$	R_3SnX_2	440a
Covalent Organotin Halide, Pseudohalide, Carboxylate, and Nitrate Complexes		
$Me_2SnCl_2,\ 2\ DMF$	trans-R_2SnX_4	441
$Me_2SnCl_2,\ 2\ HMPT$	trans-R_2SnX_4	441
$Me_2SnBr_2,\ 2\ HMPT$	trans-R_2SnX_4	441
$Me_2SnCl_2,\ salenH_2$	trans-$R_2SnX_4^*$	442
$Me_2SnCl_2, salH$	cis-R_2SnX_3	361
$Me_2SnCl_2,\ 2\ DMSO$	trans-R_2SnX_4	441,443
$Me_2SnBr_2,\ 2\ DMSO$	trans-R_2SnX_4	441
$Me_2Sn(NCS)_2,\ terpy$	R_2SnX_5	386
$Me_2SnCl_2,\ Ni(salen)$	Distorted trans-R_2SnX_4	444
$Me_3SnCl,\ Ph_3P{:}CHCOMe$	R_3SnX_2	354
$(CH_2{:}CH)_2Sn(OCOCF_3)_2,\ bipy$	Distorted trans-R_2SnX_4	445
$Ph_2Sn(NO_3)_2,\ Ph_3PO$	R_2SnX_5	461
$Ph_2SnCl_2,\ bipy$	trans-R_2SnX_4	370
$Ph_2SnCl_2,\ 2\ DMSO$	trans-R_2SnX_4	371
$Ph_3SnNO_3,\ Ph_3PO$	R_3SnX_2	446
$Ph_3SnNO_3,\ Ph_3AsO$	R_3SnX_2	447
$Ph_3SnNO_3,\ PyO$		
monoclinic form	Distorted R_3SnX_2	448
triclinic form	Distorted R_3SnX_2	449
$MeSnCl_3,\ 2\ HMPT$	$RSnX_5$	441
$MeSnBr_3,\ 2\ HMPT$	$RSnX_5$	441
$MeSnCl_3,\ 2\ DMF$	$RSnX_5$	441
$Me_3SnCl,\ HMPT$	R_3SnX_2	440a
$Ph_2SnCl_2,\ benzothiazole$	cis-R_2SnX_3	440b

B. ^{119}Sn Mössbauer Spectroscopy

The various, solid-state stereochemistries just described may often be distinguished fairly readily by ^{119}Sn Mössbauer spectroscopy (5–9, 452), particularly from the value of the quadrupole splitting parameter, ΔE_Q (see Table II).

TABLE II

^{119}Sn Mössbauer Quadrupole Splittings for Organotin Compounds of Known Stereochemistry

Stereochemistry	Example	ΔE_Q (mm·sec^{-1})	Ref.
RSnX$_3$	(MeSnS$_{1.5}$)$_4$ (352)	1.40	195
R$_2$SnX$_2$	(Me$_2$SnS)$_3$ (350,351)	1.82	453
R$_3$SnX	[(Me$_3$Si)$_2$CH]$_3$SnCl (347)	2.18	454
R$_4$Sn	(C$_6$F$_5$)$_4$Sn (341)	0.00	455
cis-R$_3$SnX$_2$	Ph$_3$SnO·NPh·CO·Ph (359)	1.94	400
R$_3$SnX$_2$	Me$_3$SnOCOMe (355)	3.00–4.00^a	
R$_2$SnX$_3$	Me$_2$SnCl$_2$, salH (361)	3.32	456
RSnX$_4$	(364) [Cl$_3$Sn–C$_6$H$_4$Me-4 structure with NH and Me substituents]	1.59	364
cis-R$_2$SnX$_4$	Ph$_2$Sn(SCSNEt$_2$)$_2$ (368)	1.74	457
trans-R$_2$SnX$_4$	Me$_2$Sn(acac)$_2$ (369)	4.02	400
distorted trans-R$_2$SnX$_4$	Me$_2$Sn(salen) (381)	3.10–3.70^b	
RSnX$_5$	PhSn(SCSNEt$_2$)$_2$Cl (382)	1.66	457
RSnX$_6$	MeSn(SCSNEt$_2$)$_3$ (385)	1.97	458
R$_2$SnX$_5$	Me$_2$Sn(NCS)$_2$, terpy (386)	4.29	459

^{a} Lower end of ΔE_Q range for symmetrical X—Sn · · · X unit.
^{b} ΔE_Q increases as CSnC opens.

For example, octahedral trans-R$_2$SnX$_4$ complexes give approximately twice the quadrupole splitting observed for the cis-octahedral analogs (7, 8). More recently, temperature-dependent Mössbauer measurements have been used in conjunction with Raman spectroscopy to determine molecular weights (453) and lattice rigidity (460) of various organotin compounds.

IV. Biological and Environmental Aspects

A. Biological Activity and Mode of Toxic Action

1. R_4Sn Compounds

The biological effects of tetraorganotins, R_4Sn, in mammals appear to be caused principally by the R_3SnX compound, which is produced by their *in vivo* (*466*) and *in vitro* (*466, 467*) conversion, particularly in the liver. Studies in mice and dogs (*468*) showed that tetraethyltin was the most toxic, followed by tetramethyltin, and, thereafter, the toxicity decreased with increasing alkyl chain-length. A similar pattern of toxicity was found for the R_3SnX, R_2SnX_2, and $RSnX_3$ compounds (*469–471*). Following administration of the R_4Sn compound, delayed development of toxic symptoms is usually observed (*468*), due to the *in vivo* formation of the more toxic R_3SnX derivative.

2. R_3SnX Compounds

Progressive introduction of organic groups at the tin atom in any member of the R_nSnX_{4-n} series produces a maximum biological activity when $n = 3$, i.e., for the triorganotin compounds, R_3SnX (*469–471*). If the chain length of the *n*-alkyl group is increased within any trialkyltin series, R_3SnX, the highest mammalian toxicity is attained when R = Et (*471*). For insects, however, the trimethyltins are usually the most toxic (*472*); for Gram-negative bacteria, the tri-*n*-propyltins (*473*), and for Gram-positive bacteria and fungi, the tributyltins show the highest activity (*473, 474*) (see Fig. 1). Further increase in the *n*-alkyl chain-length produces a sharp drop in the biological activity, and the tri-*n*-octyltin compounds are essentially nontoxic to all living species.

Variation of the inorganic radical, X, within any particular series of R_3SnX compounds usually has very little effect on the biological activity, and it is the nature of the R_3Sn moiety that is of prime importance. The triphenyltin compounds also show a high fungicidal activity (*473, 474*), and tricyclohexyl- (*475*) and trineophyl-tin (*476*) derivatives are very active against mites. In addition, bulky R groups, such as those in bis(trineophyltin) oxide, appear to lessen the mammalian toxicity (*476*).

The lower trialkyltin compounds are able to inhibit mitochondrial, oxidative phosphorylation (*471, 477*) and, therefore, disrupt the funda-

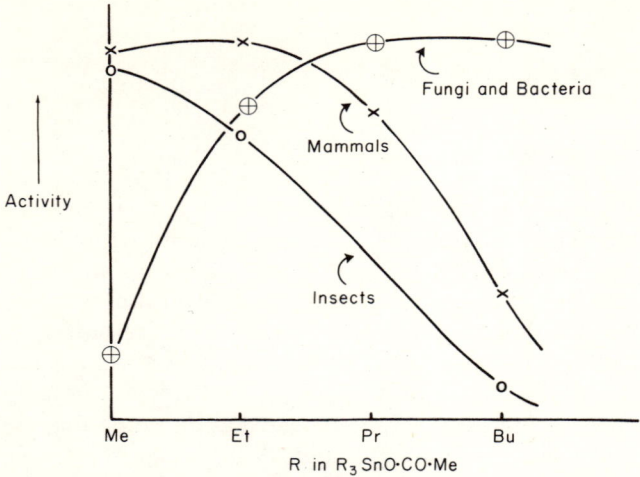

FIG. 1. Dependence of the biological activity of tri-n-alkyltin acetates on the nature of the alkyl group for different species.

mental energy processes in living systems. Their remarkable biological activity pattern may be dependent upon the effectiveness of their interaction at an active site (or sites) which involves coordination to an amino acid (478). The exact, chemical nature of these high-affinity binding-sites has recently been the focus of much attention (477, 479, 480).

In the case of simple amino acids and dipeptides, esterification of the carboxyl groups occurs on heating in toluene with the appropriate bis(triorganotin) oxide or triorganotin hydroxide (481, 482), the water being removed azeotropically.

$$2\ H_2NCHR^1CO_2H + (R_3Sn)_2O \rightarrow 2\ H_2NCHR^1CO_2SnR_3 + H_2O$$

The resulting derivatives, unlike most organotin carboxylates, are hydrolyzed relatively readily in air. An X-ray crystallographic study (399) showed that trimethyltin glycinate has an intermolecularly associated, polymeric structure, with bridging amino groups.

In trimethyltin acetate, however, bridging occurs through the carboxylate group (355), as described in Section III. An X-ray study of the diphenyltin derivative of the dipeptide glycylglycine was also reported recently (*410*).

Limouzin and Maire (*66*) used ESCA spectroscopy to study complex-formation between tributyltin chloride and a number of amino acids. They concluded that, although the $-NH_2$ groups are not basic enough to coordinate to tin, complexation does occur at the $-SH$ group of cysteine. It has also been found that the $Sn-S-C$ linkage in S-(tributylstannyl)cysteine, $Bu_3SnSCH_2CH(NH_3^+)CO_2^-$, is quite stable to hydrolysis (*483*). Another possible point of attachment of trialkyltins to amino acids is via N-1 of the imidazole ring in histidine (*484, 485*).

On a macromolecular scale, trialkyltins are known to bind to a number of proteins (see Table III).

Studies *in vitro* with bis(tributyltin) oxide in snail protein indicated a high order of activity with a number of amino acids (as well as lipids), and it has been suggested that mortality arises from direct reaction of the organotin with proteins (*489*). Elliott and Aldridge (*486*) found that two molecules of triethyltin chloride bind to one molecule of either rat or cat hemoglobin, and that at least one of these binding sites involves histidine residues. Recent work by Maire (*66*), Griffiths (*490*), and Tan (*490a*) and their co-workers suggested that $-SH$ groups may also be involved in binding the trialkyltins.

The ^{119m}Sn Mössbauer spectra of triethyltin compounds bound to high-affinity, protein sites indicate that the quadrupole splitting, ΔE_Q, is less than 2.3 mm·sec^{-1} (*479, 491*), which would be consistent with tetrahedral, R_3SnX, tin-atom geometry, or a 5-coordinate, *cis*-R_3SnX_2 chelated structure. However, the trialkyltin–histidine complex, having a planar, R_3Sn unit, suggested earlier by Rose and Lock (*485,*

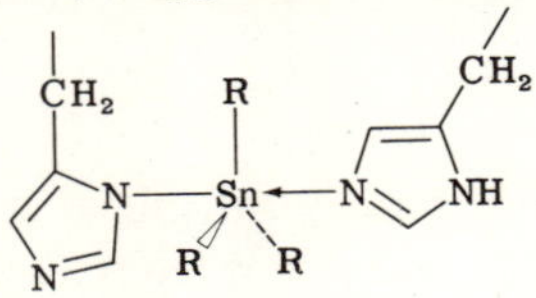

TABLE III

SOME PROTEINS THAT BIND TRIALKYLTIN COMPOUNDS

Protein	R_3SnX Compound	Ref.
1. Cat hemoglobin	$(Et_3Sn)_2SO_4/Et_3SnCl$	*486*
2. Rat hemoglobin	$(Et_3Sn)_2SO_4/Et_3SnCl$	*486*
3. Rat-brain myelin	Et_3SnCl	*487*
4. Protein fraction from guinea pig liver	Et_3SnCl	*488*
5. Snail-tissue protein	$(Bu_3Sn)_2O$	*489*
6. Yeast mitochondrial membrane	Et_3SnCl	*480*

488), may be excluded (479), as this R_3SnX_2 tin-atom geometry would be expected to give a $\Delta E_Q \geqslant 3$ mm·sec^{-1} (400).

Farrow and Selwyn (479) found that the intramolecularly 5-coordinate, mixed triorganotin bromide, prepared by Noltes and his co-workers (63, 492), namely,

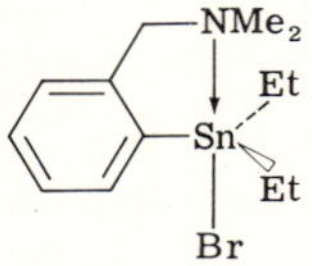

is more effective as an inhibitor of the mitochondrial ATPase than the simple triorganotin compounds, but that 5-coordinate, triorganotin compounds having two intramolecular Sn ← N bonds (63, 463, 492)

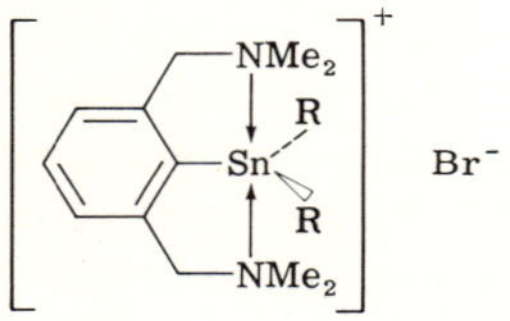

are feeble inhibitors. Tzschach and co-workers (493) observed that the tributyltin derivative

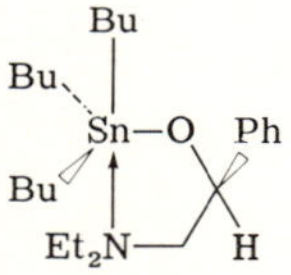

which contains a chelating diethylamino alcohol group, was approximately one-eighth as toxic orally to mice as the 4-coordinate bis(tributyltin) oxide, and this was ascribed to the lack of affinity of the chelated tributylstannyl complex for the active sites on the protein. 4-Hydroxybutyldibutyltin acetate, in which the hydroxyl group may also be in-

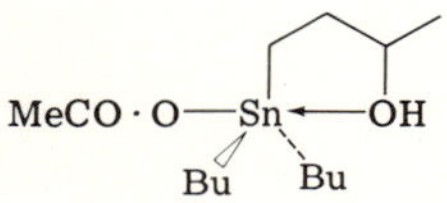

tramolecularly coordinated to tin, was found by Aldridge and co-workers (*494*) to be about half as toxic to mice (treated intraperitoneally) as the 4-coordinate, tributyltin chloride.

As the intramolecularly 5-coordinate, mixed triorganotin bromides have no tendency to react with monodentate, donor ligands, such as pyridine (*492*), these observations would appear to indicate that the halogen of the triorganotin bromide undergoes a chemical exchange-reaction at the active sites on the protein; this is in line with the toxicity, relatively independent of X, of trialkyltin compounds that contain simple, nonchelating X groups. An exchange with thiol groups of the type

$$R_3SnX + HS— \rightarrow R_3SnS— + HX$$

would give tetrahedral, R_3SnS— groups having a ΔE_Q of the observed magnitude.

Ascher and Nemny (*495*) found that residues of triphenyltin acetate on glass, resulting from the evaporation of acetone solutions thereof, were, on contact to houseflies, less toxic with rising concentration. As triphenyltin acetate is likely to be a self-associated polymer in the solid state [similar to trimethyltin acetate (*355*)] and in concentrated solutions, it was suggested (*495*) that the monomer, which exists in dilute solutions, is toxic to insects, and the polymer, nontoxic. Interestingly, in this connection, a triphenyltin methacrylate copolymer has (*470*) a very low mammalian toxicity (acute, oral LD_{50} for mice >2000 mg/kg).

3. R_2SnX_2 Compounds

The dialkyltin compounds show a similar trend of decreasing toxicity with increasing length of the alkyl chain, and certain di-*n*-octyltin derivatives have been used for many years in food-contact applications, as described in Section V.

The toxic action of the lower di-*n*-alkyltins in mammals, which is quite different from that of the tri-*n*-alkyltin analogs, is due to their ability to combine with enzymes (such as lipoic acid or lipoyl dehydrogenase) possessing two thiol groups in the correct positions, and thereby to interfere with α-keto acid oxidation (*477*). Two possible reactions may be envisaged. Here, the dialkyltin compound (a) combines with the dithiol groups to form a tetrahedral, dialkylstannadithiaheterocyclic derivative, with the elimination of HCl, or (b) forms an octahedral complex in which the dithiol group acts as a neutral, bidentate ligand. The observed toxicological data for the dialkyltin derivatives may be most readily rationalized in terms of reaction pathway (b).

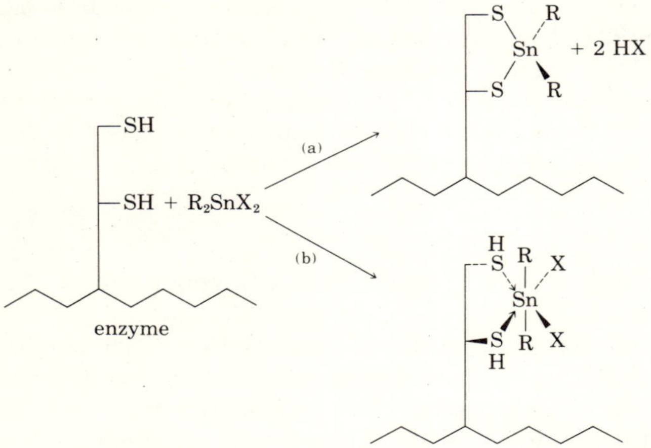

For example, the decrease in toxicity of the dialkyltin dichlorides with increasing alkyl chain-length (see Table IV) is paralleled by a drop in their Lewis acidity and, hence, a lowered tendency to complex the dithiol ligand. Although diethyltin dichloride has a relatively high mammalian toxicity, introduction of a 2-methoxycarbonyl substituent renders the resulting compound, $(MeOCOCH_2CH_2)_2SnCl_2$, essentially nontoxic (see Table II). In all probability, the latter has an octahedral structure with intramolecular, chelating, carboxylate groups (77), a structure found for other carbonyl-substituted, alkyltin compounds

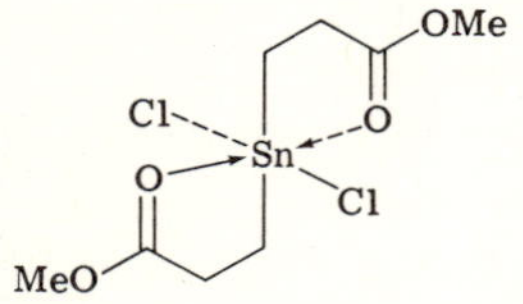

(496–498), and, therefore, it has no tendency to complex the dithiol groups. Further substitution of the two chlorine atoms by isooctylthioglycolate groups produces the compound, $(MeOCOCH_2CH_2)_2Sn(SCH_2COO^iOct)_2$, in which the carbonyl groups are now likely to be free (cf. 496, 498) and, in common with other dialkyltin bis(isooctylthioglycolates), the mammalian toxicity is still very low

TABLE IV
ACUTE, ORAL TOXICITIES OF DIALKYLTIN DICHLORIDES (470)

R in R_2SnCl_2	LD_{50} (rats) (mg/kg)
Me	74–237
Et	66–94
^{n}Bu	112–219
^{n}Oct	4000–7000
$C_{16}H_{33}$	10,000
$MeOCOCH_2CH_2-$	2350

(470). In this case, the diminution in Lewis acidity caused by the introduction of the two, less-electronegative, sulfur ligands prevents any reaction with the dithiol enzyme, regardless of the nature of the alkyl group.

It may, therefore, be seen that the mammalian toxicity of the lower dialkyltin compounds, unlike that of their trialkyltin counterparts, is markedly dependent upon the nature of the X groups; this is probably true for species other than mammals (e.g., fungi) if the mode of action is similar.

4. $RSnX_3$ Compounds

The monoorganotins do not appear to have any important toxic action in mammals (469, 470, 471), but they show the familiar pattern of decreasing toxicity with increasing alkyl chain-length, with the maximum again falling at the monoethyltin derivative (see Table V).

TABLE V
VARIATION OF TOXICITY OF MONOALKYLTIN TRICHLORIDES WITH ALKYL CHAIN LENGTH (470, 499)

R in $RSnCl_3$	Acute, oral LD_{50} (rats) (mg/kg)
Me	1370
Et	200 (LD_{100})
^{n}Bu	2200
^{n}Oct	3800

B. Environmental Degradation

1. Metabolic Breakdown

In view of the increasing number of industrial applications of organotins, which are described in the next section, a knowledge of their metabolic fate in mammals is obviously of considerable importance.

The earliest work in this field, by Cremer (466), showed that tetraethyltin is metabolized *in vitro* and *in vivo* (rats) to a triethyltin derivative and, later, it was further demonstrated that triethyltin compounds are converted *in vitro* into diethyltin derivatives (500). The latter are broken down *in vivo* to monoethyltins, which are eliminated from the body within a short time (501). Other trialkyltins appear to behave similarly (500).

However, very recent studies by Fish and his co-workers (467) with butyltin compounds showed that the primary, metabolic reaction is not Sn—C bond-cleavage but carbon hydroxylation of the *n*-butyl group. Using [1-^{14}C]tetrabutyltin in an *in vitro* study, the major, primary metabolite was identified as a 2-hydroxybutyltributyltin derivative that underwent a rapid β-elimination reaction to afford 1-butene and a tributyltin compound (467).

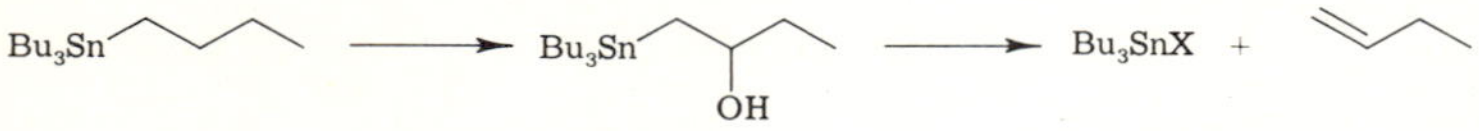

Bu$_3$Sn⁀⁀⁀ ⟶ Bu$_3$Sn⁀(OH)⁀ ⟶ Bu$_3$SnX + ⁀⁀

A similar approach with [1-^{14}C]tributyltin acetate showed that carbon hydroxylation occurred at the α-, as well as at the β-, carbon atoms, followed by Sn—C bond-cleavage, to afford dibutyltin derivatives.

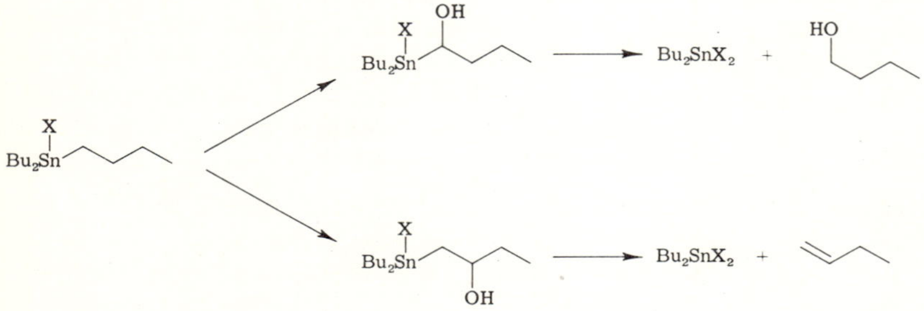

Studies with the same compound in mice showed an essentially identical pattern of breakdown (467).

It was suggested that the greater susceptibility of the α- and β-carbon–hydrogen bonds to hydroxylation is consistent with a free-radical pathway for this reaction. In line with this concept, it is known, for example, that an α- or β-trialkylstannyl substituent confers an enhanced reactivity on a CH group towards the abstraction of hydrogen by *tert*-butoxy radicals (*132*).

Tricyclohexyltin hydroxide is metabolized *in vivo* to inorganic tin via di- and monocyclohexyltin derivatives (*502*), and *in vitro* studies suggested that the major, metabolic reaction is carbon-hydroxylation of the cyclohexyl group (*503*). Studies *in vivo* using either triphenyl[^{113}Sn]tin acetate (*467*) or triphenyl[^{113}Sn]tin chloride (*504*) in rats showed that these compounds are metabolized to yield substantial amounts of di- and monophenyltin derivatives, although no significant quantities of hydroxylated metabolites have been identified (*503*) in this case.

2. Photolytic and Microbiological Breakdown

It has been well established that triphenyltin compounds are broken down photochemically to inorganic tin via the di- and monophenyltin derivatives both under laboratory (*505, 506*) and natural (*507*) conditions. In soil, triphenyltin acetate is converted microbiologically (*505*) into inorganic tin, as is tricyclohexyltin hydroxide (*502*). The latter compound is also photochemically broken down to inorganic tin (*502, 508*).

Bis(tributyltin) oxide is known to break down to inorganic tin under UV irradiation in laboratory conditions (*509, 510*), and the decomposition may be accelerated by absorbing the organotin compound on a cellulosic matrix (*511*). As bis(tributyltin) oxide is known to react rapidly with carbon dioxide (atmospheric, or trapped in various cellulosic materials, such as cotton or wood) (*512*), to form bis(tributyltin) carbonate, $(Bu_3SnO)_2CO$, the observed UV degradation pattern may be rationalized in terms of more-ready breakdown of the carbonate than of the oxide, due to the presence of the carbonyl chromophore. The half-life of bis(tributyltin) oxide in pond water has recently been given as 16 days (*513*). Diorganotin compounds have also been shown to decompose to inorganic tin under UV irradiation (*514, 515*).

It may, therefore, be concluded that, within a generally consistent pattern of behavior, organotins are degraded in the environment to afford nontoxic, inorganic tin species. A generalized, environmental-breakdown scheme for the commercially used tributyl- and triphenyltin derivatives (*516*), that is probably applicable to other triorganotins, is illustrated in Scheme 1.

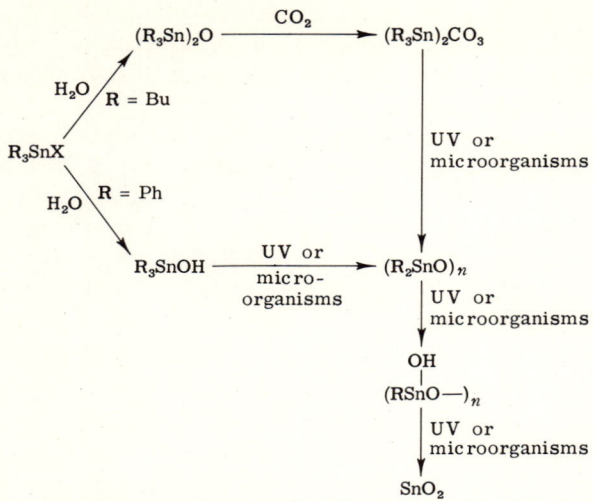

SCHEME 1. Environmental-degradation scheme for tributyl- and triphenyltin compounds (516).

The possibility of biomethylation of inorganic tin residues in the environment has recently been discussed by Wood and co-workers (517–519), who found that methylation of certain tin(II) salts (e.g., $SnCl_3^-$) by methylcobalamin in aqueous solution at pH 1 requires the presence of an oxidizing agent, such as Co(III) or Fe(III), and that monomethyltin(IV) species are formed. No reaction was observed between Sn(II) and methylcobalamin in the absence of an oxidizing agent (518), or between Sn(IV) and methylcobalamin under a variety of conditions of pH and complexing ligands (519). It was suggested that the $SnCl_3^-$ species is oxidized to a trichlorostannyl radical, $\dot{S}nCl_3$, which could then cleave the Co–C bond homolytically to produce $MeSnCl_3$. The following biological cycle for tin was proposed (517).

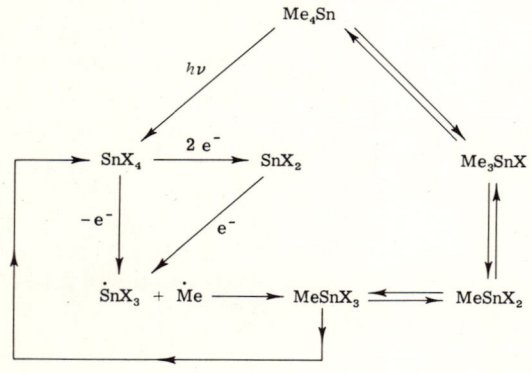

Brinckman and his co-workers, using a strain of *Pseudomonas* bacteria isolated from the Chesapeake Bay (*520*), obtained evidence of a biomethylation of $SnCl_4$ to a dimethyltin species under laboratory conditions. At the present time, however, very little is known about (a) the distribution of methyltin compounds in the environment,[2] (b) whether they bioaccumulate in food chains (*521*), and (c) the chemical nature and oxidation state of inorganic tin residues. The toxicity pattern of the Me_nSnX_{4-n} series is obviously very important in this context. All monomethyltins, $MeSnX_3$, so far studied have a low mammalian toxicity (*470*), whereas, with dimethyltin compounds, Me_2SnX_2, the nature of the X group is of considerable importance, as mentioned previously. For example, dimethyltin bis(isooctylthioglycolate) has a low mammalian toxicity (*470*), and is approved in a number of countries for use as a stabilizer in PVC food-packaging (see Section V,B,1). The trimethyltins, Me_3SnX, are the most toxic members of the series, regardless of the nature of X, and show acute, oral, LD_{50} values (rats) in the range of 10 to 20 mg/kg (*470*).

V. Industrial Applications

Although the first organotin compound was prepared by Frankland as long ago as 1849 (*522*), it is only during the last decade that their use in industry has risen dramatically—from ~5000 tons in 1965 to at least 30,000 tons at the present time. Some two-thirds of the current total of organotins are used for nontoxic types of applications (R_2SnX_2 and $RSnX_3$ derivatives), and the remaining 7000–10,000 tons are used as biocides (R_3SnX derivatives) (see Table VI). This rapid increase in the industrial consumption of organotins may be attributed principally to the remarkable diversity of their applications, coupled with their favorable toxicological and environmental properties, as described in the preceding section.

The tetraorganotins, R_4Sn, are used primarily as intermediates in the industrial synthesis of organotins from $SnCl_4$ (see Section II), and have no large industrial application.

A. Biological Uses (R_3SnX Compounds)

The pioneering work that paved the way for the widespread industrial use of organotins as pesticides was conducted in 1954 at the Insti-

---

[2] Recent analytical studies have indicated that mono-, di-, and trimethyltin compounds are present in environmental samples at nanogram levels [Braman, R. S., and Tompkins, M.A., *Analyt. Chem.* **51**, 12 (1979); Hodge, V. F., Seidel, S. L., and Goldberg, E. D., *ibid.* **51**, 1256 (1979)].

TABLE VI

APPROXIMATE WORLD PRODUCTION OF ORGANOTIN CHEMICALS

Compounds	Uses	Tons
A. R_2SnX_2 and $RSnX_3$ compounds	1. PVC stabilizers	
	United States	8000
	Japan	5500
	United Kingdom and Western Europe	6500
	2. Homogeneous Catalysts	1000
B. R_3SnX compounds	1. Triphenyltin agrochemicals	1000
	2. Others (see Table VII)	8000
	Total	30,000

tute of Organic Chemistry, TNO, Utrecht, Holland, by van der Kerk and Luijten (*474*), who demonstrated the high fungicidal and bactericidal activity of tributyl- and triphenyltin compounds (*473, 474*). Table VII summarizes the major biocidal applications of various triorganotin compounds in current use.

1. Agrochemicals

The first organotin compound to reach commercialization in agriculture, namely, triphenyltin acetate ("Brestan"), was introduced in the early 1960s by Farbwerke Hoechst A. G. in West Germany for control of the potato-blight fungus *Phytophthora infestans,* and the sugar-beet fungus *Cercospora beticola*. This was soon followed by another triphenyltin fungicide, triphenyltin hydroxide ("Duter"), developed by Philips–Duphar, N. V., in Holland and having a spectrum of activity similar to that of the acetate. It has recently been estimated (*523*) that the consumption of these two organotin agrochemicals alone in Europe and Japan is of the order of ~1000 tons per annum.

The next major discovery in this field, which was the result of a joint research effort in 1968 between M and T Chemicals, Inc., and the Dow Chemical Company in the United States (*475, 524*), was that tricyclohexyltin hydroxide ("Plictran") possesses a very high activity against certain types of mites, and this compound was subsequently introduced by Dow as an acaricide for use on apple, pear, and citrus-fruit trees. A second triorganotin acaricide, bis(trineophyltin) oxide ("Vendex" or "Torque"), has recently been introduced by Shell Chemical Company (*476*). Two other tricyclohexyltin compounds are currently under de-

TABLE VII
Biocidal Uses of R_3SnX Compounds

Compound	Application
Ph_3SnOAc	Agrochemical fungicides
Ph_3SnOH	
$(cyclo\text{-}C_6H_{11})_3SnOH$	Agrochemical miticides
$[(PhCMe_2CH_2)_3Sn]_2O$	
$Bu_3SnOCOPh$	Disinfectant
Bu_3SnF	Biocides in marine antifouling paints
Ph_3SnX (X = Cl, F, OH, or OAc)	
$(-CHMe\cdot CHCO_2SnBu_3-)_n$	
$Bu_3SnOSnBu_3$	Wood preservation
	Stone preservation
	Textile preservation
	Slimicide in paper industry
	Biocide in antifouling rubbers
	In-can fungicide for paints

velopment (524, 525) as acaricides. Tris(2-norbornyltin) hydroxide also shows a high acaricidal activity (524).

As only one compound in approximately 10,000 tested will actually reach commercialization as a plant-protection agent (526) it is quite remarkable for one metal to have four of its organic compounds in commercial use. The main advantages of the organotin agrochemicals are their low phytotoxicity, their generally low toxicity to nontarget organisms, and their relatively rapid breakdown in the environment to nontoxic, inorganic tin residues.

2. Tributyltin Disinfectants

The high antibacterial activity of tributyltin derivatives, particularly against Gram-positive bacteria (473), has led to their use as disinfectants, although a second chemical toxicant is usually added to extend the biological activity spectrum to cover Gram-negative bacteria. One such product ("Incidin"), containing tributyltin benzoate and formaldehyde, was developed by Henkel in West Germany (527). Various combinations of tributyltin benzoate and a quaternary ammonium salt have been evaluated as bactericides for use on hospital floors (528).

3. Preservation of Cellulosic Materials and Stonework

Bis(tri-n-butyltin) oxide, $Bu_3SnOSnBu_3$, is an organotin chemical very effective in, and widely used for, the protection of such cellulosic materials as cotton textiles, wood, and cellulose-based household-fillers against fungal attack (*529–531*). In the case of wood, for example, fungal protection may be achieved by dipping, brushing, or vacuum-impregnating the material with a 1% (w/w) solution of the organotin compound in an organic solvent, such as white spirit (*531*). An organic insecticide is usually added to the solution to prevent attack by wood-destroying insects (*530*).

The high retentive capacity of cellulosic materials for the organotin oxide was originally considered (*530, 532*) to be due to its chemical reaction with the terminal hydroxyl groups of the cellulose chains (see Scheme 2). Very recent work has, however, shown (*512*) that, in fact, bis(tributyltin) oxide reacts rapidly with the carbon dioxide trapped in wood, to form bis(tributyltin) carbonate, $(Bu_3SnO)_2CO$, which, unlike the trialkyltin oxide, is a polymeric, self-associated species containing 5-coordinate tin, probably having the following structure.

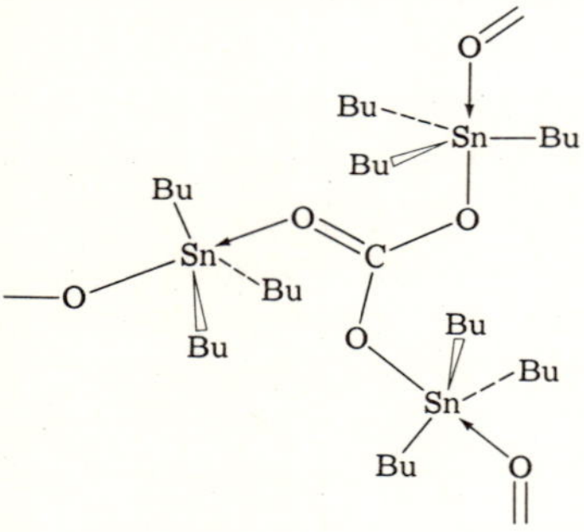

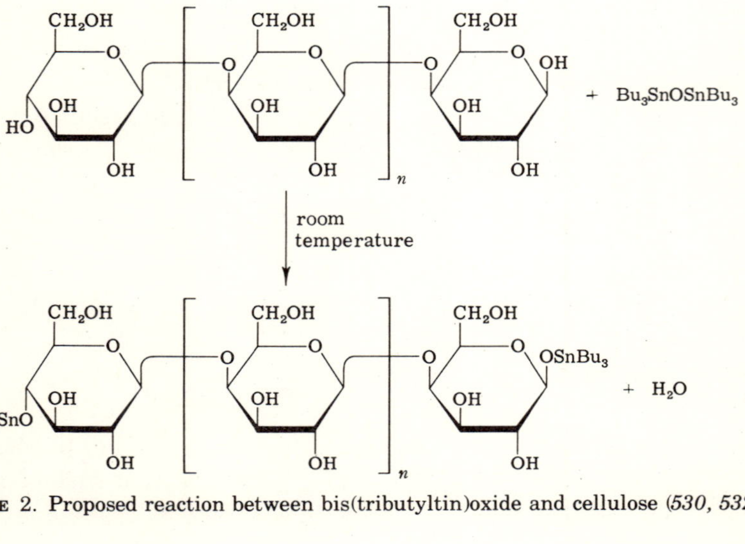

SCHEME 2. Proposed reaction between bis(tributyltin)oxide and cellulose (*530, 532*).

The polymeric organotin carbonate is less volatile than the oxide and, in this connection, tris(tributylstannyl) phosphate, $(Bu_3Sn)_3PO_4$, which is likely to have a similarly self-associated structure [cf., $(Me_2Sn)_3(PO_4)_2 \cdot 8\ H_2O$ (393)], is also being used as a wood preservative (533) in Holland.

Recently, there has been much interest in developing water-soluble tributyltin biocides to lessen the costs of application, and to prevent fire hazards when treating material in confined spaces. Bis(tributyltin) oxide itself has a very low aqueous solubility (~0.001%), but it may be made water-dispersible by the addition of certain (534, 535) quaternary ammonium salts. Formulations of this type, although currently under development[3] as wood preservatives (534), have been used extensively in the United Kingdom for the treatment of stonework to eradicate fungal growths, algae, mosses, and lichens (535).

In order to avoid the necessity of adding a quaternary ammonium compound to solubilize the organotin derivative, a number of new, anionic, tributyltin salts of the type $(R_3R^1P)^+$ $(Bu_3SnCl_2)^-$, where $R = R^1 = {}^nBu$ or Ph; $R = Ph, R^1 = Bz$, were prepared (440, 536). Although these contain discrete, trigonal, bipyramidal tributyldichlorostannate anions (440), their solubility in water was found to be very low.

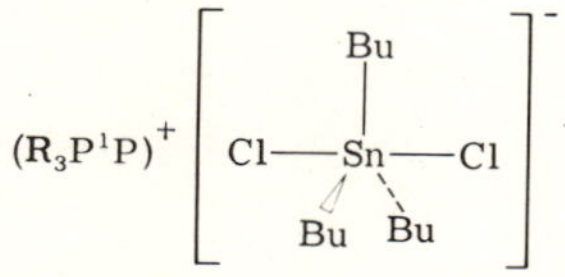

$$(R_3P^1P)^+ \left[Cl - \underset{\underset{Bu\ Bu}{|}}{\overset{Bu}{Sn}} - Cl \right]^-$$

Similarly, the cationic tributyltin complexes (537), $(Bu_3SnL_2)^+BPh_4^-$, where $L = Me_2SO, Ph_3PO, Ph_3AsO$, or PyO, are also insoluble in water. However, tributyltin alkanesulfonates, Bu_3SnSO_3R (particularly where $R = Me, Et, {}^nBu$, or ${}^tBu$), dissolve in water to the extent of 0.8 to 1.5% (538, 539), which is a remarkably high solubility for a tributyltin compound, and a concentration quite adequate for most biocidal applications (see the foregoing). The resulting, aqueous solutions appear to be quite stable on standing in normal daylight for 2 to 3 months.

4. Marine Antifouling Paints and Related Systems

Marine fouling, namely, the attachment of such marine organisms as barnacles, algae, tubeworms, hydroids, and sponges to a surface im-

---

[3] One such formulation has recently been approved for use in Sweden (Anonymous, "Wood Preservatives Approved by the Swedish Wood Preservation Institute," Svenska Traskyddinstitutet, Stockholm, 1 Jan., 1979).

mersed in sea water, can cause serious problems. In the case of large tankers and other ocean-going vessels, marine fouling increases hull friction as the ship moves through the water, fuel consumption goes up, and the maximum speed attainable is decreased. Fouling may also cause severe mechanical damage to marine instruments and outboard-motor engines.

The most realistic method of dealing with this fouling problem is to protect the ship's hull with an antifouling, coating system that functions by releasing chemicals toxic to marine organisms, and thus prevents their attachment to the hull surface. The two vital components of such an antifouling system are: (a) a toxicant that (i) is effective against a wide range of fouling species at low concentrations, (ii) will constitute only a minimum toxic hazard during application to a vessel, and (iii) will not contribute significantly to environmental pollution, and (b) a coating medium that will release the toxicant at a low, but steady, rate and preserve its film integrity on the hull surface.

Conventional, organotin-based, antifouling paints containing up to 20% (by weight) of a tributyl- or triphenyl-tin derivative incorporated in a standard paint-vehicle have been in use for many years (*540–542*). The most common triorganotin compounds employed are those that have a polymeric, intermolecularly anion-bridged, R_3SnX_2 type of structure, such as tributyltin fluoride, triphenyltin fluoride, and triphenyltin hydroxide (*541, 543*). Bis(tributyltin) oxide is also used, and, in some cases, is known (*543*) to react with carboxylic acids in the paint matrix to form the corresponding tributyltin carboxylate. The colorlessness of these triorganotin additives allows the use of any color in the paints, and, additionally, they are not subject to bimetallic corrosion problems when used on lightweight, aluminum hulls (*541*).

Poller (*498, 544*) prepared a number of tributyl- and triphenyl-stannyl esters of sucrose hydrogenphthalate and succinate, and found that, as potential antifoulants, these were at least three times as effective against the marine alga, *Enteromorpha,* as bis(tributyltin) oxide, even though they contain almost one third the tin (see Table VIII). A new antifouling paint that also contains tributyltin compounds has recently been developed in Norway (*545*).

As mentioned earlier, water-based, tributyltin-based biocides are attracting much interest at the present time, and one such formulation has been developed in the United Kingdom for the prevention of fouling in sea-water cooling-systems (*546*).

For the majority of merchant ships, which dry-dock at intervals of about 12 to 18 months, the conventional paint-formulations containing simple triorganotin additives are quite satisfactory, and provide adequate protection over this period. However, giant tankers involve capital investment that runs into millions of pounds, and their economical

TABLE VIII

ACTIVITY OF TRIBUTYL- AND TRIPHENYL-STANNYLSUCROSE COMPOUNDS AGAINST *Enteromorpha* IN SEA WATER MODIFIED WITH ALGAL NUTRIENTS (498, 544)

Compound	Concentration[a]	
	1 ppm	0.1 ppm
$Bu_3SnOCOC_6H_4COOsucrose$	+	+
$Ph_3SnOCOC_6H_4COOsucrose$	+	+
$Bu_3SnOCOCH_2CH_2COOsucrose$	+	+
$Ph_3SnOCOCH_2CH_2COOsucrose$	+	+

[a] Key: + = effective [minimum concentration at which $(Bu_3Sn)_2O$ is effective = 0.3 ppm].

operation requires that they spend a high proportion of their time at sea, with a period of at least 2 to 2.5 years (preferably 4 to 5 years) between dry-docking periods. A long-term, antifouling coating is even more essential for naval vessels, and many of the more recent developments in this field have originated in the laboratories of the United States and Australian Navies, whose ships spend considerable periods in tropical environments conducive to marine fouling.

The first approach in the search for longer-life, or "second generation," antifouling coatings has been to incorporate the triorganotin moiety in an organometallic polymer system. This has been achieved by co-polymerizing tributyltin acrylate (or methacrylate) with other co-monomers, such as vinyl chloride (547–549), to give linear poly-

$$n/2 \; \underset{\underset{Cl}{|}}{\overset{\overset{H}{|}}{C}}=\underset{\underset{H}{|}}{\overset{\overset{H}{|}}{C}} \; + \; n/2 \; \underset{\underset{H}{|}}{\overset{\overset{H}{|}}{C}}=\underset{\underset{CO_2SnBu_3}{|}}{\overset{\overset{R}{|}}{C}}$$

(R = H or Me)

$$\left[-\underset{\underset{H}{|}}{\overset{\overset{H}{|}}{C}}-\underset{\underset{Cl}{|}}{\overset{\overset{H}{|}}{C}}-\underset{\underset{H}{|}}{\overset{\overset{H}{|}}{C}}-\underset{\underset{C=O}{|}}{\overset{\overset{R}{|}}{C}}- \right]_n$$
$$ OSnBu_3$$

mers having trialkylstannyl groups chemically bound to the polymer backbone, cf., the simple R_3SnX additives, where the toxic R_3SnX species are weakly held together in the solid state by self-association. The resulting polymers are then formulated into marine coatings, and tested in immersion trials at marine sites. Foul-free periods of at least four years have been observed with these slow-release, polymeric systems (549).

A second, parallel approach, has been to incorporate the tributyltin toxicant into an elastomeric matrix to produce long-life, antifouling, rubber coatings. The precursor for these systems was the "Nofoul" antifouling rubber-sheet marketed by B. F. Goodrich in the United States and described in detail by Cardarelli (*550*). Although originally developed as a protective coating for ships' hulls, these controlled-release, bis(tributyltin) oxide-containing rubbers, in the form of pellets, have been extensively investigated as molluscicides (*489, 550*) for use against the vector snails that transmit the notorious, water-borne, tropical disease, schistosomiasis (*551*). The safe concentration level of bis(tributyltin) oxide wherein the snails may be destroyed in infected tropical waters with minimal damage to the fish life has recently been estimated (*552*) at 0.12–0.27 µg/liter.

B. Nonbiological Uses (R_2SnX_2 and $RSnX_3$ Compounds)

1. PVC Stabilizers

The largest single use for organotin compounds is the stabilization of PVC (*553*), some 20,000 tons of chemicals currently being used. PVC is degraded both by heat (to which it is subjected during processing at 180–200°) and by long-term exposure to sunlight, producing severe discoloration, a rapid deterioration in physical properties, and progressive embrittlement until the polymer completely disintegrates. This phenomenon is caused by the elimination of hydrogen chloride from the polymer, starting from the labile, allylic chlorine atoms, and resulting in the formation of a polyene. The degradation may be pre-

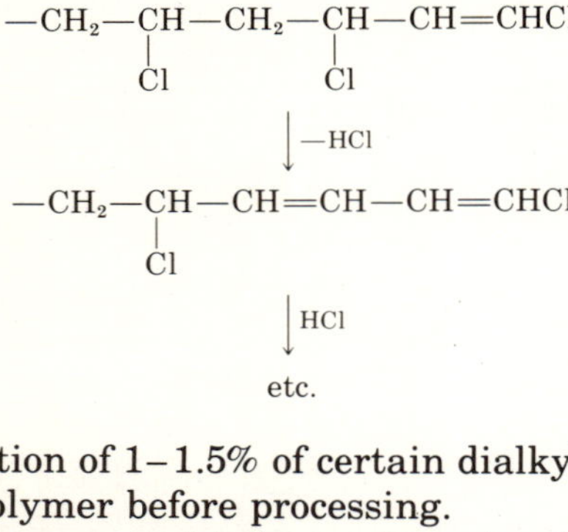

$$-CH_2-\underset{Cl}{CH}-CH_2-\underset{Cl}{CH}-CH=CHCl$$

$$\downarrow -HCl$$

$$-CH_2-\underset{Cl}{CH}-CH=CH-CH=CHCl$$

$$\downarrow HCl$$

etc.

vented by the addition of 1–1.5% of certain dialkyltin compounds (see Table IX) to the polymer before processing.

In general, the diorganotin bis(isooctylthioglycolates) are used in applications that require good stability to heat, e.g., PVC drink-bottles and food packaging, whereas the dialkyltin bis(carboxylates) are useful for providing long-term stability to light, e.g., in PVC roofing. Di-*n*-octyltin bis(isooctylthioglycolate) and maleate have a low mammalian toxicity and are used in many countries as stabilizers for PVC food-

TABLE IX
Some Organotin Stabilizers for PVC

R	X	Ref.
(a) $R_2Sn(SCH_2COO^iOct)_2$		
Me	SCH_2COO^iOct	554
$^nBuOCH_2CH_2$	SCH_2COO^iOct	78,555
^nBu	SCH_2COO^iOct	553
^nOct	SCH_2COO^iOct	556
$PhCH_2$	SCH_2COO^iOct	557
(b) $R_2Sn(OCOR^1)_2$		
^nBu	$(OCOCHCHCOO)_n$	553
^nBu	$OCOCHCHCOOR$	553
^nBu	$OCO^nC_{11}H_{23}$	553
^nOct	$(OCOCHCHCOO)_n$	556

packaging and drink containers; dimethyltin bis(isooctylthioglycolate) is also approved in some European countries, e.g., West Germany, for stabilizing food-contact, PVC packaging, and in the United States as a heat stabilizer for use in PVC, potable-water piping; di-n-butyl- and di-(2-butoxycarbonylethyl)-tin stabilizers are currently used in non-food-contact PVC. In most cases, up to 60% of the corresponding monoalkyltin compound, $RSnX_3$, is added to the dialkyltin stabilizer, R_2SnX_2, as it is found that this combination gives a synergistic improvement in the stabilizing activity (553). Monobutyltin sesquisulfide, $(BuSnS_{1.5})_4$, is used as a stabilizer in its own right for certain grades of PVC in West Germany (558), and dilauryltin compounds also show promise (559).

The di- and monoalkyltin compounds are considered to be effective as stabilizers because they (i) inhibit the onset of the dehydrochlorination reaction by exchanging their anionic groups, X, with the reactive, allylic chlorine atoms in the polymer; (ii) react with, and thereby scavenge, the hydrogen chloride that is produced and that would otherwise induce further elimination; (iii) produce the compound HX, which may also help to inhibit other undesirable side reactions; and (iv) prevent breakdown of the polymer initiated by atmospheric oxidation, i.e., by acting as antioxidants.

The dialkyltin dichloride formed[4] by reaction of the dialkyltin stabilizer with the polymer, or with the hydrogen chloride liberated, is itself

---

[4] A Mössbauer study of PVC stabilized with 4% di-n-butyltin bis(isooctylthioglycolate) indicates that the tin species formed in the polymer is $Bu_2Sn(Cl)SCH_2COO^iOct$ [Allen, D. W., Brooks, J. S., Clarkson, R. W., Mellor, M. T. J., and Williamson, A. G., *Chem. Ind. (London)* 663 (1979)].

a Lewis acid catalyst for further dehydrochlorination (*496, 498*). Poller (*496*) synthesized the stabilizer di-(4-oxopentyl)tin bis(isooctylthioglycolate), in which the Lewis acidity of the resulting diorganotin dichloride was suppressed because of intramolecular coordination of the two carbonyl groups to tin. Subsequent tests showed (*416*) that the

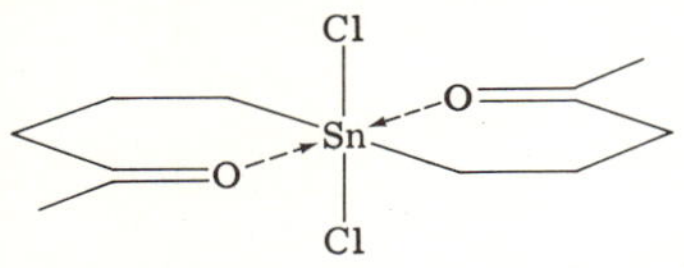

stabilizer was more than twice as effective as the corresponding butyltin derivatives. In this connection, it is interesting that di-(2-methoxycarbonyl)ethyltin dichloride, produced from the corresponding isooctylthioglycolate stabilizer, also contains chelated carbonyl groups (*77, 78*).

2. Homogeneous Catalysts

Dibutyltin diacetate, dilaurate, and di-(2-ethylhexanoate) are used as homogeneous catalysts for room-temperature-vulcanizing (RTV) silicones. The dialkyltin compounds bring about the cross-linking of the oligomeric siloxanes, to produce flexible, silicone rubbers having a host of different uses, such as electrical insulators and dental-impression molds. Recent work has also shown (*560*) that various dibutyltin dicarboxylates catalyze both the hydrolysis and gelation of ethyl silicate under neutral conditions.

The same dibutyltin compounds are used in the industrial manufacture of poly(urethane) foams, the first step in which involves the addition of a polyether glycol to 2,4-diisocyanotoluene, to produce the urethane prepolymer having isocyanate end-groups.

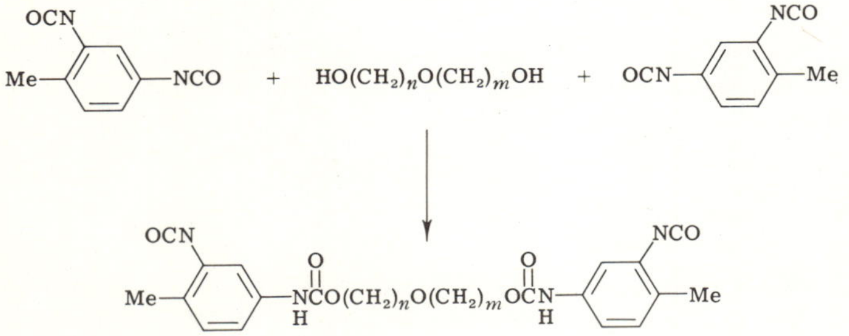

In the second, or "foaming," stage, water is added to the prepolymer to produce the polyurethane and carbon dioxide gas. The organotin compounds catalyze both reaction steps in the manufacturing process.

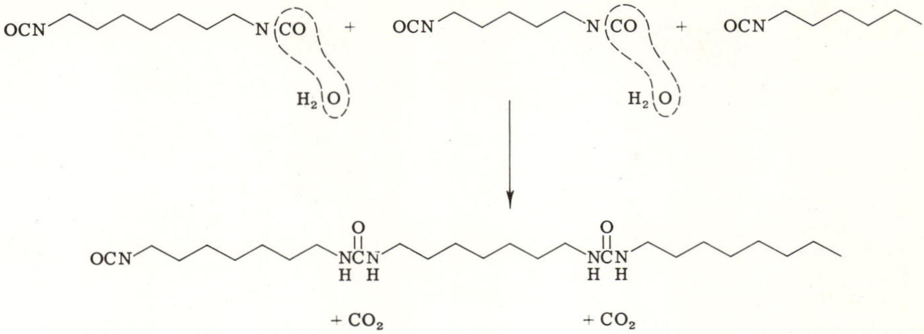

$$OCN\text{~~~~~~}N(CO) + H_2O \;\; + \;\; OCN\text{~~~~~}N(CO) + H_2O \;\; + \;\; OCN\text{~~~~~}$$

$$\downarrow$$

$$OCN\text{~~~~~~}\underset{H\;\;H}{NCN}\text{~~~~~~}\underset{H\;\;H}{NCN}\text{~~~~~~}$$

$$+ CO_2 \qquad\qquad + CO_2$$

Certain monobutyltin compounds have recently been introduced (561) as esterification catalysts, e.g., in the reaction of phthalic anhydride with octanol to produce dioctyl phthalate.

3. Treatment of Glass

Dimethyltin dichloride is used in the glass industry as an alternative to stannic chloride for coating glass with a thin film of stannic oxide (562). The dialkyltin compound vapor is brought into contact with the glass surface at temperatures of 500–600°C, where decomposition and oxidation occurs.

$$Me_2SnCl_2 + O_2 \rightarrow SnO_2 + 2\ MeCl$$

The thickness of the film of SnO_2 varies from ~10.0 nm to several μm, depending on the desired application (563).

At the low end of the thickness scale (<100.0 nm), films of SnO_2 are used in the glass industry for strengthening glasses, bottles, jars, and other items subjected to vigorous usage. If thickness of the film on the glass is of the same order as the wavelength of visible light (~100.0–1000.0 nm), thin-film interference occurs, to give the article an iridescent, decorative coating. Finally, very thick films of SnO_2 on glass are useful when electrical conductivity combined with optical transparency is required, e.g., for de-icing aircraft windscreens (563). Other organotin compounds have also been investigated recently for this application (564, 565, 566).

Acknowledgment

The permission of the International Tin Research Council, London, to publish this review is gratefully acknowledged (by P.J.S.).

References

1. Neumann, W. P., "The Organic Chemistry of Tin." Wiley, New York, 1970.
2. Poller, R. C., "The Chemistry of Organotin Compounds." Logos, London and Academic Press, New York, 1970.
3. A. K. Sawyer, ed., "Organotin Compounds," Vols. 1, 2, and 3. Dekker, New York, 1971, 1972.
4. Pereyre, M., and Pommier, J.-C., *J. Organomet. Chem. Library* **1,** 161 (1976).
5. Smith, P. J., *Organomet. Chem. Rev., Sect. A* **5,** 373 (1970).
6. Ruddick, J. N. R., *Rev. Silicon, Germanium, Tin, Lead Compd.* **2,** 115 (1976).
7. Parish, R. V., *Prog. Inorg. Chem.* **15,** 101 (1972).
8. Bancroft, G. M., and Platt, R. H., *Adv. Inorg. Chem. Radiochem.* **15,** 59 (1972).
9. Zuckerman, J. J., *Adv. Organomet. Chem.* **9,** 22 (1970).
10. Smith, L., and Smith, P. J., *Inorg. Chim. Acta Rev.* **7,** 11 (1973).
11. Kennedy, J. D., and McFarlane, W., *Rev. Silicon, Germanium, Tin, Lead Compd.* **1,** 235 (1974).
12. Smith, P. J., and Tupčiauskas, A. P., *Annu. Rep. NMR Spectrosc.* **8,** 291 (1978).
13. Ingham, R. K., Rosenberg, S. D., and Gilman, H. *Chem. Rev.* **60,** 459 (1960).
14. Kocheshkov, K. A., Zemlyanskii, N. N., Sheverdina, N. I., and Panov, E. M., "Metodi Elementoorganicheskoi Khimii. Germanii, Olovo, Svenets." Nauka, Moscow, 1968 (in Russian).
15. Coates, G. E., and Wade, K., "Organometallic Compounds," Vol. 1: The Main Group Elements. Methuen, London, 1967.
16. Coates, C. E., Aylett, B., and Wade, K., "Organometallic Compounds," Vol. 1: The Main Group Elements, Methuen, London, 1978.
17. Emeleus, H. J., and Aylett, eds., "Organometallic Derivatives of the Main Group Elements," MTP *Int. Rev. Sci., Ser. One* **4** (1972).
18. P. G. Harrison, MTP *Int. Rev. Sci., Ser. Two* **4** (1972), Chapter 3.
19. Bulten, E. J., "Tin: Annual Survey Covering the Year 1970." *Organomet. Chem. Rev.* **9,** 248 (1972).
20. Bulten, E. J., "Tin: Annual Survey Covering the Year 1971." *J. Organomet. Chem.* **53,** 1 (1973).
21. Harrison, P. G., "Tin: Annual Survey Covering the Year 1972." *J. Organomet. Chem.* **58,** 49 (1973).
22. Harrison, P. G., "Tin: Annual Survey Covering the Year 1973." *J. Organomet. Chem.* **79,** 17 (1974).
23. Harrison, P. G., "Tin: Annual Survey Covering the Year 1974." *J. Organomet. Chem.* **109,** 363 (1976).
24. Harrison, P. G., "Tin: Annual Survey Covering the Year 1975." *J. Organomet. Chem. Library* **4,** 367 (1977).
25. Armitage, D. A., *Chem. Soc. Spec. Period. Rep.: Organomet. Chem.* **1,** Chapter 5 (1972).
26. Armitage, D. A., *Chem. Soc. Spec. Period. Rep.: Organomet. Chem.* **2,** Chapter 6 (1973).
27. Armitage, D. A., *Chem. Soc. Spec. Period. Rep.: Organomet. Chem.* **3,** Chapter 6 (1974).
28. Armitage, D. A., *Chem. Soc. Spec. Period. Rep.: Organomet. Chem.* **4,** Chapter 6 (1975).
29. Armitage, D. A., *Chem. Soc. Spec. Period. Rep.: Organomet. Chem.* **5,** Chapter 6 (1976).

30. Armitage, D. A., *Chem. Soc. Spec. Period. Rep.: Organomet. Chem.* **6,** Chapter 6 (1977).
31. Zuckerman, J. J., ed., "Organotin Compounds: New Chemistry and Applications." *Adv. Chem. Ser.* **157,** 227 (1976).
32. Crompton, T. R., ed., "Chemical Analysis of Organometallic Compounds," Vol. 3: Elements of Group IV–B, Chapter 9. Academic Press, New York, 1974.
33. Price, J. W., and Smith, R., "Gmelin Handbuch der Analytischen Chemie" W. Fresenius, ed. Teil 3, Band 49: Tin Springer-Verlag, Berlin and New York, 1977.
34. Bokranz, A., and Plum, H., "Industrial Manufacture and Use of Organotin Compounds," Schering, A. G., D-4619, Bergkamen, Germany, 1975; *Fortschr. Chem. Forsch.* **16,** 365 (1971).
35. Luijten, J. G. A., and van der Kerk, G. J. M., Synthesis and properties of the tin–carbon bond, *in* "Organometallic Compounds of the Group IV Elements," (A. G. MacDiarmid, ed.), Vol. 1, Part 2. Arnold, New York, 1968.
36. Schumann, H., and Schumann, I., Zinn-organische Verbindungen SnR_4, *in* "Gmelin Handbuch der Anorganischen Chemie," Band 26, Teil 1. Springer and New York, Berlin, 1975.
37. Schumann, H., and Schumann, I., Zinn-organische Verbindungen, R_3SnR', *in* "Gmelin Handbuch der Anorganischen Chemie," Band 26, Teil 2. Springer-Verlag, Berlin and New York, 1975.
38. Schumann, H., and Schumann, I., Zinn-organische Verbindurgen, $R_2SnR'_2$, $R_2SnR'R''$, $RR'SnR''R'''$, *in* "Gmelin Handbuch der Anorganischen Chemie," Band 26, Teil 3. Springer-Verlag, Berlin and New York, 1975.
39. Poller, R. C., and Murphy, J., *J. Organomet. Chem. Library,* **9,** 189 (1979).
40. Galli, R., Giannaccari, B. M., and Cassar, L., *J. Organomet. Chem.* 25, 429 (1970).
41. Matschiner, H., Voigtländer, R., and Tzschach, A., *J. Organomet. Chem.,* **70,** 387 (1974).
42. Matsuda, S., and Nomura, M., *J. Organomet. Chem.,* **25,** 101 (1970).
43. Fostein, P., and Pommier, J.-C., *J. Organomet. Chem.,* **114,** C7 (1976).
44. Bos, K. D., Budding, H. A., Bulten, E. J., and Noltes, J. G., *Inorg. Nucl. Chem. Lett.* **9,** 963 (1973).
45. Bos, K. D., Bulten, E. J., and Noltes, J. G., *J. Organomet. Chem.* **99,** 397 (1975).
46. Wakeshima, I., and Kijima, I., *J. Organomet. Chem.* **76,** 37 (1974).
47. Bulten, E. J., *J. Organomet. Chem.* **97,** 167 (1975).
48. Kandil, S. A., and Allred, A. L., *J. Chem. Soc. A* p. 2987 (1970).
49. Glockling, F., Lyle, M. A., and Stobart, S. A., *J. Chem. Soc., Dalton Trans.* p. 2537 (1974).
50. Killian, L., and Wrackmeyer, B., *J. Organomet. Chem.* **132,** 213 (1977).
51. Corey, J. Y., Duebar, M., and Malaizde, M., *J. Organomet. Chem.* **36,** 49 (1972).
52. Meinema, H. A., and Noltes, J. G., *J. Organomet. Chem.* **63,** 243 (1973).
53. Matteson, D. S., and Larson, G. L., *J. Organomet. Chem.* **57,** 225 (1973).
54. Matteson, D. S., and Wilcsek, R. J., *J. Organomet. Chem.* **57,** 231 (1973).
55. Seyferth, D., and Lefferts, J. L., *J. Am. Chem. Soc.* **96,** 6237 (1974).
56. Gröbel, B. T., and Seebach, D., *Chem. Ber.* **110,** 852 (1977).
57. Seyferth, D., Armbrecht, F. M., Lambert, R. L., and Tronich, W., *J. Organomet. Chem.* **44,** 299 (1972).
58. Seyferth, D., and Lambert, R. L., *J. Organomet. Chem.* **54,** 123 (1973).
59. Seyferth, D., Lambert, R. L., and Massol, M., *J. Organomet. Chem.* **88,** 255 (1975).
60. Seyferth, D., and Lambert, R. L., *J. Organomet. Chem.* **88,** 287 (1975).
61. Seyferth, D., and Andrews, S. B., *J. Organomet. Chem.* **30,** 151 (1971).

62. Mengoli, G., and Daolio, S., *J. Chem. Soc., Chem. Commun.* p. 76 (1976).
63. van Koten, G., Schaap, C. A., and Noltes, J. G., *J. Organomet. Chem.* **99,** 157 (1975).
64. Itoh, K., Kato, S., and Ishii, Y., *J. Organomet. Chem.* **34,** 293 (1972).
65. Bulten, E. J., and Budding, H. A., *J. Organomet. Chem.* **110,** 167 (1976).
66. Limouzin, Y., and Maire, J. C., ref. *31,* p. 227.
67. Kuivila, H. G., Kennedy, J. D., Tien, R. Y., Tyminiski, I. J., Pelczar, F. L., and Khan, O. R., *J. Org. Chem.* **36,** 2083 (1971).
68. Kennedy, J. D., Kuivila, H. G., Pelczar, F. L., Tien, R. Y., and Considine, J. L., *J. Organomet. Chem.* **61,** 167 (1973).
69. Juenge, E. C., Hawkes, S. J., and Snider, T. E., *J. Organomet. Chem.* **51,** 189 (1973).
70. Leusink, A. J., Budding, H. A., and Noltes, J. G., *J. Organomet. Chem.* **24,** 375 (1970).
71. Ashe, A. J., and Shu, P., *J. Am. Chem. Soc.* **93,** 1804 (1971).
72. Jutzi, P., and Baumgärtner, J., *J. Organomet. Chem.* **148,** 247 (1978).
73. Märkl, G., Baier, H., and Heinrich, S., *Angew. Chem., Int. Ed. Engl.* **14,** 710 (1975).
74. Märkl, G., and Rampal, J. B., *Tetrahedron Lett.* p. 2325 (1977).
75. Massol, M., Barreau, J., Satgé, J., and Bonyssieres, B., *J. Organomet. Chem.* **80,** 47 (1974).
76. Rahm, A., and Pereyre, M., *J. Organomet. Chem.* **88,** 79 (1975).
77. Hutton, R. E., and Oakes, V., *J. Chem. Soc., Chem. Commun.* 803 (1976).
78. Hutton, R. E., and Oakes, V., ref. *31,* p. 227.
79. Kuivila, H. G., ref. *31,* p. 41.
80. Davis, D. D., and Gray, C. L., *Organomet. Chem. Rev.* **6,** 283 (1970).
81. Roberts, R. M. G., *J. Organomet. Chem.* **63,** 159 (1973).
81a. Still, W. C., *J. Am. Chem. Soc.* **100,** 1481 (1978).
82. Kuivila, H. G., and Di Stefano, F. V., *J. Organomet. Chem.* **122,** 171 (1976).
83. Juenge, E. C., Snider, T. E., and Lee, Y.-C., *J. Organomet. Chem.* **22,** 403 (1970).
84. Jensen, F. R., and Davis, D. D., *J. Am. Chem. Soc.* **93,** 4047 (1971).
85. Koermer, G. S., Hall, M. L., and Traylor, T. G., *J. Am. Chem. Soc.* **94,** 7205 (1972).
86. Kuivila, H. G., Considine, J. L., and Kennedy, J. D., *J. Am. Chem. Soc.* **94,** 7206 (1972).
87. Jones, C. H. W., Jones, R. C., Partington, P., and Roberts, R. M. G., *J. Organomet. Chem.* **32,** 201 (1971).
88. Kuivila, H. G., and Wursthorne, K. R., *Tetrahedron Lett.* p. 4357 (1975).
89. Kuivila, H. G., and Wursthorne, K. R., *J. Organomet. Chem.* **105,** C6 (1975).
90. Quintard, J.-P., Hauvette, S., and Pereyre, M., *J. Organomet. Chem.* **112,** C11 (1976).
91. Lahournère, J.-C., and Valade, J., *C. R. Hebd. Acad. Sci., Ser. C* **270,** 2080 (1970).
92. Lahournère, J.-C., and Valade, J., *J. Organomet. Chem.* **22,** C3 (1970).
93. Lahournère, J.-C., and Valade, J., *J. Organomet. Chem.,* **33,** C4 (1971).
94. Abraham, M. H., "Electrophilic Substitution at a Saturated Carbon Atom" (C. H. Bamford and C. F. H. Tipper, eds.). Elsevier, Amsterdam, 1973.
95. Eaborn, C., Cleavages of aryl–silicon and related bonds by electrophiles, *J. Organomet. Chem.* **100,** 43 (1975).
96. Reutov, O. A., The mechanisms of the substitution reactions of non-transition metal organometallic compounds, *J. Organomet. Chem.* **100,** 219 (1975).
97. Eaborn, C., *Pure Appl. Chem.,* **19,** 375 (1969).
98. Eaborn, C., Thompson, A. R., and Walton, D. R. M., *J. Organomet. Chem.* **29,** 257 (1971).

99. Eaborn, C., Jones, J. R., and Seconi, G., *J. Organomet. Chem.* **116,** 83 (1976).
100. Eaborn, C., and Seconi, G., *J. Chem. Soc., Perkin Trans. 2* p. 925 (1976).
101. Alexander, R., Eaborn, C., and Traylor, T. G., *J. Organomet. Chem.* **21,** P65 (1970).
102. Wardell, J. L., ref. *31,* p. 113.
103. Kuivila, H. J., and Scarpa, N. M., *J. Am. Chem. Soc.* **92,** 6990 (1970).
104. Abraham, M. H., Dadjour, D. F., Gielen, M., and de Poorter, B., *J. Organomet. Chem.* **84,** 317 (1975).
105. Abraham, M. H., and Dadjour, D. F., *J. Chem. Soc., Perkin Trans. 2* p. 233 (1974).
106. Abraham, M. H., and Dorrell, F. J., *J. Chem. Soc., Perkin Trans. 2* p. 444 (1973).
107. Abraham, M. H., Dadjour, D. F., and Holloway, C. J., *J. Organomet. Chem.* **52,** C27 (1973).
108. Abraham, M. H., and Hogarth, M. J., *J. Chem. Soc. A* p. 1474 (1971).
109. Abraham, M. H., *J. Chem. Soc. A* p. 1061 (1971).
110. Abraham, M. H., and Grellier, P. L., *J. Chem. Soc., Perkin Trans. 2* p. 623 (1975).
111. Bartlett, E. H., Eaborn, C., and Walton, D. R. M., *J. Organomet. Chem.* **46,** 267 (1972).
112. Folaranni, A., McLean, R. A. N., and Wadibia, N., *J. Organomet. Chem.* **73,** 59 (1974).
113. Bhattacharya, S. N., Raj, P., and Srivastava, R. C., *J. Organomet. Chem.* **105,** 45 (1976).
114. Rahm, A., and Pereyre, M., *J. Am. Chem. Soc.* **99,** 1672 (1977).
115. Jensen, F. R., and Davis, D. D., *J. Am. Chem. Soc.* **93,** 4049 (1971).
116. Kunze, U., Lindner, E., and Koola, J. D., *J. Organomet. Chem.* **57,** 319 (1973).
117. Lindner, E., and Frembs, D., *J. Organomet. Chem.* **49,** 425 (1973).
118. Lindner, E., Ritter, G., and Haag, A., *J. Organomet. Chem.* **24,** 119 (1970).
119. Vitzthum, G., Kunze, U., and Lindner, E., *J. Organomet. Chem.* **21,** P38 (1970).
120. Lindner, E., *Rev. Silicon, Germanium, Tin, Lead Compd.* **1,** 35 (1972).
121. Fong, C. W., and Kitching, W., *J. Organomet. Chem.* **22,** 95 (1970).
122. Moore, C. J., and Kitching, W., *J. Organomet. Chem.* **59,** 225 (1973).
123. Koola, J. D., and Kunze, U., *J. Organomet. Chem.* **77,** 325 (1974).
124. Fong, C. W., and Kitching, W., *J. Organomet. Chem.* **59,** 213 (1973).
125. Fong, C. W., and Kitching, W., *J. Am. Chem. Soc.* **93,** 3791 (1971).
126. Fong, C. W., and Kitching, W., *J. Organomet. Chem.* **59,** 213 (1973).
127. Moore, C. J., and Kitching, W., *J. Organomet. Chem.* **59,** 225 (1973).
128. Bulten, E. J., and Budding, H. A., *J. Organomet. Chem.* **137,** 165 (1977).
129. Hänssgen, D., and Odenhause, E., *J. Organomet. Chem.* **124,** 143 (1977).
130. Ingold, K. U., and Roberts, B. P., "Free Radical Substitution Reactions." Wiley-Interscience, New York, 1971.
131. Davies, A. G., and Roberts, B. P., *in* "Free Radicals" (J. Kochi, ed.), Vol. 1, p. 547. Wiley-Interscience, New York, 1973.
132. Davies, A. G., ref. *31,* Chap. 2.
133. Boué, S., Gielen, M., and Nasielski, J., *J. Organomet. Chem.* **9,** 461 (1967).
134. Davies, A. G., Roberts, B. P., and Smith, J. M., *J. Chem. Soc., Perkin Trans. 2* p. 2221 (1971).
135. Davies, A. G., Roberts, B. P., and Scaiano, J. C., *J. Chem. Soc., Perkin Trans. 2* p. 223 (1972).
136. Davies, A. G., and Scaiano, J. C., *J. Chem. Soc., Perkin Trans. 2* p. 1777 (1973).

137. Davies, A. G., Muggleton, B., Roberts, B. P., Tse, M.-W., and Winter, J., *J. Organomet. Chem.* **118,** 289 (1978).
138. Davies, A. G., Roberts, B. P., and Tse, M.-W., *J. Chem. Soc., Perkin Trans. 2* p. 1499 (1977).
139. Abel, E. W., and Rowley, R. I., *J. Organomet. Chem.* **84,** 199 (1975).
140. Albert, H.-J., Neumann, W. P., Kaiser, W., and Ritter, H.-P., *Chem. Ber.* **103,** 1372 (1970).
141. Schröer, U., and Neumann, W. P., *J. Organomet. Chem.* **105,** 183 (1976).
142. Hannon, S. J., and Traylor, T. G., *J. Chem. Soc., Chem. Commun.* p. 630 (1975).
143. Fishwick, M. F., and Wallbridge, M. G. H., *J. Organomet. Chem.* **25,** 69 (1973).
144. Fishwick, M. F., and Wallbridge, M. G. H., *J. Organomet. Chem.* **136,** C46 (1977).
145. Schweig, A., Weidner, U., and Manuel, G., *J. Organomet. Chem.* **54,** 145 (1973).
146. Schweig, A., Weidner, U., and Manuel, G., *J. Organomet. Chem.* **54,** 145 (1973).
147. Peruzzo, V., Plazzogna, G., and Tagliavini, G., *J. Organomet. Chem.* **40,** 121 (1972).
148. Puruzzo, V., and Tagliavini, A., *J. Organomet. Chem.* **66,** 437 (1974).
149. Magravite, J. A., Verdona, J. A., and Kuivila, H. G., *J. Organomet. Chem.* **104,** 303 (1976).
150. Bullpitt, M. L., and Kitching, W., *J. Organomet. Chem.* **84,** 321 (1972).
151. Roberts, R. M. G., *J. Organomet. Chem.* **24,** 675 (1970).
152. Servens, C., and Pereyre, M., *J. Organomet. Chem.* **35,** C20 (1970).
153. Servens, C., and Pereyre, M., *J. Organomet. Chem.* **26,** C4 (1971).
154. Abel, E. W., and Rowley, R. I., *J. Organomet. Chem.* **84,** 199 (1975).
155. Albert, H.-J., Neumann, W. P., and Ritter, H.-P., *Justus Liebigs Ann. Chem.* **737,** 152 (1970).
156. Schröer, U., and Neumann, W. P., *J. Organomet. Chem.* **105,** 183 (1976).
157. Kosigi, M., Kurimo, K., Takayama, K., and Migita, T., *J. Organomet. Chem.* **56,** C11 (1973).
158. Grignon, J., Servens, C., and Pereyre, M., *J. Organomet. Chem.* **96,** 225 (1975).
159. Guillerm, G., Maganem, F., Lequan, M., and Brower, K. R., *J. Organomet. Chem.* **67,** 43 (1974).
160. Simo, M. S., Jean, A., and Lequan, M., *J. Organomet. Chem.* **35,** C23 (1970).
161. Lequan, M., and Guillerm, G., *J. Organomet. Chem.* **54,** 153 (1973).
162. Kison, A. V., Korenevsky, V. A., Sergeyev, N. M., and Ustynyuk, Yu. A., *J. Organomet. Chem.* **34,** 93 (1972).
163. Grishin, Yu. K., Sergeyev, N. M., and Ustynyuk, Yu. A., *J. Organomet. Chem.* **34,** 105 (1972).
164. Curtis, M. D., and Fink, R., *J. Organomet. Chem.* **38,** 299 (1972).
165. Larrabee, R. B., *J. Am. Chem. Soc.* **93,** 1510 (1971).
166. Boche, G., and Haidenhain, F., *J. Organomet. Chem.* **121,** C49 (1976).
167. Abel, E. W., Dunster, M. O., and Waters, A., *J. Organomet. Chem.* **49,** 287 (1973).
168. Larrabee, R. B., *J. Organomet. Chem.* **74,** 313 (1974).
169. Schröer, U., Albert, H.-J., and Neumann, W. P., *J. Organomet. Chem.* **102,** 291 (1975).
170. Davies, A. G., and Tse, M.-W., *J. Chem. Soc., Chem. Commun.* p. 353 (1978).
170a. Schumann, H., and Schumann, I., "Organotin Hydrides," "Gmelin Handbuch der Anorganischen Chemie," Band 35, Teil 4. Springer-Verlag, Berlin and New York, 1976.
171. Pereyre, M., and Pijselman, J., *J. Organomet. Chem.* **25,** C27 (1970).
172. Pijselman, J., and Pereyre, M., *J. Organomet. Chem.* **32,** C72 (1971).
173. Pijselman, J., and Pereyre, M., *J. Organomet. Chem.* **63,** 139 (1973).

174. Brum, J. E., May, G. S., Sawyer, A. K., Schofield, R. E., and Sprague, W. E., *J. Organomet. Chem.* **124,** 13 (1977).
175. Pereyre, M., and Pommier, J.-C., *J. Organomet. Chem. Library* **1,** 161 (1976).
176. Corey, E. J., and Suggs, J. W., *J. Organomet. Chem.* **40,** 2554 (1975).
177. Lipowitz, J., and Bowman, S. A., *Aldrichimica Acta* **6,** 1 (1973).
178. Knocke, P., and Neumann, W. P., *Justus Liebigs Ann. Chem.* p. 1486 (1974).
179. Weinshenker, N. M., Crosby, G. A., and Wong, J. Y., *J. Org. Chem.* **40,** 1966 (1975).
180. Bulten, E. J., and Budding, H. A., *J. Organomet. Chem.* **111,** C33 (1976).
181. Clark, H. C., and Puddephatt, R. J., Synthesis and properties of the tin–halogen and tin–halogenoid bond, *in* "Organometallic Compounds of the Group IV Elements" (A. G. MacDiarmid, ed.), Vol. 2, Part 2, Chapter 3. Arnold, New York, 1968.
182. Knips, U., and Huber, F., *J. Organomet. Chem.* **107,** 9 (1976).
183. Garner, C. D., and Hughes, B., *J. Chem. Soc. D* p. 1306 (1974).
184. Batchelor, R. J., Riddick, J. N. R., Sams, J. R., and Aubke, F., *Inorg. Chem.* **16,** 1414 (1977).
185. Peruzzo, V., Plazzogna, G., and Tagliavini, G., *J. Organomet. Chem.* **24,** 347 (1970).
186. Peruzzo, V., Plazzogna, G., and Tagliavini, G., *J. Organomet. Chem.* **40,** 121 (1972).
187. Wang, C. S., and Shreeve, J. M., *J. Chem. Soc., Chem. Commun.* p. 151 (1970).
188. Wang, C. S., and Shreeve, J. M., *J. Organomet. Chem.* **38,** 287 (1972).
189. Christen, U., and Neumann, W. P., *J. Organomet. Chem.* **39,** C58 (1972).
190. Gaur, D. P., Srivastava, G., and Mehrotra, R. C., *J. Organomet. Chem.* **63,** 221 (1973).
191. Kennedy, J. D., McFarlane, W., Smith, P. J., White, R. F. M., and Smith, L., *J. Chem. Soc., Perkin Trans. 2* p. 1785 (1973).
192. Davies, A. G., Kleinschmidt, D. C., Palan, P. R., and Vasishtha, S. C., *J. Chem. Soc. C* p. 3972 (1971).
193. Lutzenko, I. F., Baukov, Y. I., and Belavin, I. Yu., *J. Organomet. Chem.* **24,** 359 (1970).
194. Baukov, Yu. I., and Lutzenko, I. F., *Organomet. Chem. Rev.* **6,** 355 (1970).
195. Davies, A. G., Smith, L., and Smith, P. J., *J. Organomet. Chem.* **39,** 279 (1972).
196. Zschunke, A., Tzschach, A., and Pönicke, K., *J. Organomet. Chem.* **51,** 197 (1973).
197. Tzschach, A., and Pönicke, K., Korecz, I., and Burger, K., *J. Organomet. Chem.* **59,** 199 (1973).
198. Tzschach, A., and Pönicke, K., *Kem. Kozl.* **41,** 141 (1974).
199. Tzschach, A., and Pönicke, K., *Z. Anorg. Allg. Chem.* **404,** 121 (1974).
200. Tzschach, A., and Pönicke, K., *Z. Anorg. Allg. Chem.* **413,** 136 (1975).
201. Zeldin, M., and Ochs, J., *J. Organomet. Chem.* **86,** 369 (1975).
202. Zschunke, A., Tzschach, A., and Jurkschat, K., *J. Organomet. Chem.* **112,** 273 (1976).
203. Delmond, B., and Pommier, J.-C., *J. Organomet. Chem.* **26,** C7 (1971).
204. Delmond, B., Pommier, J.-C., and Valade, J., *J. Organomet. Chem.* **34,** 91 (1972).
205. Delmond, B., Pommier, J.-C., and Valade, J., *J. Organomet. Chem.* **47,** 337 (1973).
206. Delmond, B., Pommier, J.-C., and Valade, J., *J. Organomet. Chem.* **50,** 121 (1973).
207. Wagner, D., Verheyden, J. P. H., and Moffatt, J. G., *J. Org. Chem.* **39,** 24 (1974).
208. Auge, C., David, S., and Veyrières, A., *J. Chem. Soc., Chem. Commun.* p. 375 (1976).
209. Nashed, M. A., and Anderson, L., *Tetrahedron Lett.* p. 3503 (1976).
210. Munavu, R. M., and Szmant, H. H., *J. Org. Chem.* **10,** 1832 (1976).
211. Crowe, A. J., and Smith, P. J., *J. Organomet. Chem.* **110,** C57 (1976).
211a. Ogawa, T., and Matsui, M., *Carbohydrate Res.* **62,** C1 (1978).

211b. Nashed, M. A., and Anderson, L. *Carbohydrate Res.* **56,** 325 (1977).
212. Brown, J. M., Chapman, A. C., Harper, R., Mowthorpe, D. J., Davies, A. G., and Smith, P. J., *J. Chem. Soc. D* p. 338 (1972).
213. Stapfer, C. H., Dworkin, R. D., and Weisfeld, L. B., *J. Organomet. Chem.* **24,** 355 (1970).
214. Salomon, M. F., and Salomon, R. G., *J. Am. Chem. Soc.* **99,** 3500 (1977).
215. Sannley, R. L., Aue, W. A., and Shubber, A. K., *J. Organomet. Chem.* **38,** 281 (1972).
216. Brandes, D., and Blaschette, A., *J. Organomet. Chem.* **78,** 1 (1974).
217. Bishop, M. E., and Zuckerman, J. J., *J. Am. Chem. Soc.* **16,** 1749 (1977).
218. Armitage, D. A., and Sinden, A. W., *J. Organomet. Chem.* **44,** C43 (1972).
219. Pommier, J.-C., and Roubineau, A., *J. Organomet. Chem.* **50,** 101 (1973).
220. Brocas, J.-M., and Pommier, J.-C., *J. Organomet. Chem.* **121,** 45 (1976).
221. Roubineau, A., and Pommier, J.-C., *J. Organomet. Chem.* **107,** 63 (1976).
222. Brocas, J.-M., de Jeso, B., and Pommier, J.-C., *J. Organomet. Chem.* **120,** 217 (1976).
223. de Jeso, B., and Pommier, J.-C., *J. Organomet. Chem.* **122,** C1 (1976).
224. Cardona, R. A., and Kupchik, E. J., *J. Organomet. Chem.* **34,** 129 (1972).
225. Feiccabrino, J. A., and Kupchik, E. J., *J. Organomet. Chem.* **56,** 167 (1973).
226. Feiccabrino, J. A., and Kupchik, E. J., *J. Organomet. Chem.* **73,** 319 (1974).
227. Feiccabrino, J. A., and Kupchik, E. J., *J. Organomet. Chem.* **93,** 325 (1975).
228. Hänssgen, D., and Roelle, W., *J. Organomet. Chem.* **56,** C14 (1973).
229. Harris, S. J., and Walton, D. R. M., *J. Organomet. Chem.* **127,** C1 (1977).
230. Cragg, R. H., and Taylor, A., *J. Organomet. Chem.* **99,** 391 (1975).
231. Breitschneider, E. S., and Allen, C. W., *J. Organomet. Chem.* **38,** 43 (1972).
232. Armitage, D. A., and Tarassoli, A., *Inorg. Chem.* **14,** 1210 (1975).
233. Levchuk, L. E., Sams, J. R., and Aubke, F., *Inorg. Chem.* **11,** 43 (1972).
234. Yeats, P. A., Sams, J. R., and Aubke, F., *Inorg. Chem.* **11,** 2634 (1972).
235. Chapman, A. C., Davies, A. G., Harrison, P. G., and McFarlane, W., *J. Chem. Soc. C* p. 821 (1970).
236. Davies, A. G., and Kennedy, J. D., *J. Chem. Soc. C* p. 759 (1970).
237. Schumann, H., du Mont, W. W., and Kooth, H.-J., *Chem. Ber.* **109,** 237 (1976).
238. Schumann, H., du Mont, W.-W., Corvan, P. J., and Zuckerman, J. J. *J. Organomet. Chem.* **128,** 187 (1977).
239. Wang, C. S., and Shreeve, J. M., *J. Chem. Soc., Chem. Commun.* p. 151 (1970).
240. Wang, C. S., and Shreeve, J. M., *J. Organomet. Chem.* **38,** 287 (1972).
241. Cohen, A. D., and Dillard, C. R., *J. Organomet. Chem.* **25,** 421 (1970).
242. Gaur, D. P., Srivastava, G., and Mehrotra, R. C., *J. Organomet. Chem.* **63,** 221 (1973).
243. Stapfer, C. H., Dworkin, R. D., and Weisfeld, L. B., *J. Organomet. Chem.* **24,** 355 (1970).
244. Gaur, D. P., Srivastava, G., and Mehrotra, R. C., *J. Organomet. Chem.* **63,** 213 (1973).
245. Chu, C. K., and Murray, J. D., *J. Chem. Soc. A* p. 360 (1971).
246. Matsuda, H., Mori, F., Kashiwa, A., Matsuda, S., Kasai, N., and Jitsumori, K., *J. Organomet. Chem.* **34,** 341 (1972).
247. Davies, A. G., Harrison, P. G., and Palan, P. R., *J. Chem. Soc. C* p. 2030 (1970).
248. Davies, A., and Harrison, P. G., *J. Chem. Soc. C* p. 2035 (1970).
249. Davies, A. G., and Kennedy, J. D., *J. Chem. Soc. C* p. 759 (1970).

250. Davies, A. G., and Harrison, P. G., *J. Chem. Soc. C* p. 1769 (1971).
251. Mehrotra, S. K., Srivastava, G., and Mehrotra, R. C., *J. Organomet. Chem.* **65,** 361 (1974).
252. Mehrotra, S. K., Srivastava, G., and Mehrotra, R. C., *J. Organomet. Chem.* **65,** 367 (1974).
253. Hännsgen, D., Kuna, J., and Ross, B., *Angew. Chem., Int. Ed. Engl.* **13,** 607 (1974).
254. Davies, A. G., and Kennedy, J. D., *J. Chem. Soc. D* p. 759 (1970).
255. Mathiasch, B., *J. Organomet. Chem.* **122,** 345 (1976).
256. Mitchell, T. N., and El-Bahairy, M., *J. Organomet. Chem.* **141,** 43 (1977).
257. Plazzogna, G., Puruzzo, V., and Tagliavinni, G., *J. Organomet. Chem.* **24,** 667 (1970).
258. Plazzogna, G., Puruzzo, V., and Tagliavinni, G., *J. Organomet. Chem.* **66,** 57 (1974).
259. Bulten, E. J., Budding, H. A., and Noltes, J. G., *J. Organomet. Chem.* **22,** C5 (1970).
260. Bulten, E. J., and Budding, H. A., *J. Organomet. Chem.* **78,** 385 (1974).
261. Weibel, A. T., and Oliver, J. P., *J. Organomet. Chem.* **57,** 313 (1978).
262. McWilliam, D. C., and Wells, P. R., *J. Organomet. Chem.* **85,** 335 (1975).
263. McWilliam, D. C., and Wells, P. R., *J. Organomet. Chem.* **85,** 347 (1975).
264. McWilliam, D. C., and Wells, P. R., *J. Organomet. Chem.* **85,** 165 (1975).
265. Arnold, D. P., and Wells, P. R., *J. Organomet. Chem.* **108,** 345 (1975).
266. Bulten, E. J., and Budding, H. A., *J. Organomet. Chem.* **82,** 121 (1974).
267. Buschhaus, H. U., and Neumann, W. P., *Angew. Chem., Int. Ed. Engl.* **17,** 60 (1978).
268. Lahournère, J.-C., and Valade, J., *C. R. Hebd. Acad. Sci. Ser. C* **270,** 1080 (1970).
269. Lahournère, J.-C., and Valade, J., *J. Organomet. Chem.* **22,** C3 (1970).
270. Lahournère, J.-C., and Valade, J., *J. Organomet. Chem.* **33,** C4 (1971).
271. des Tombe, F. J. A., van der Kerk, G. J. M., Creemers, H. M. J. C., Corey, N. A. D., and Noltes, J. G., *J. Organomet. Chem.* **44,** 247 (1972).
272. des Tombe, F. J. A., van der Kerk, G. J. M., and Noltes, J. G., *J. Organomet. Chem.* **43,** 323 (1977).
273. Kruglaya, O. A., Kalinina, G. S., Petrov, B. I., and Vyazankin, N. S., *J. Organomet. Chem.* **46,** 51 (1972).
274. Blaukat, U., and Neumann, W. P., *J. Organomet. Chem.* **63,** 27 (1973).
275. Mitchell, T. N., *J. Organomet. Chem.* **92,** 311 (1975).
276. McWilliam, D. C., and Wells, P. R., *J. Organomet. Chem.* **85,** 347 (1975).
277. Mitchell, T. N., *J. Organomet. Chem.* **71,** 29 (1974).
278. Mitchell, T. N., *J. Organomet. Chem.* **71,** 39 (1974).
279. Weibel, A. T., and Oliver, J. P., *J. Am. Chem. Soc.* **94,** 8590 (1972).
280. Seyferth, D., and Andrews, S. B., *J. Organomet. Chem.* **30,** 151 (1971).
281. Seyferth, D., Armbrecht, F. M., Lambert, R. L., and Tronich, W., *J. Organomet. Chem.* **44,** 299 (1972).
282. Seyferth, D., Lambert, R. L., and Hanson, E. M., *J. Organomet. Chem.* **24,** 647 (1970).
283. Seyferth, D., Andrews, S. B., and Lambert, R. L., *J. Organomet. Chem.* **37,** 69 (1972).
284. Seyferth, D., and Lambert, R. L., *J. Organomet. Chem.* **91,** 31 (1975).
285. Abel, E. W., and Rowley, R. J., *J. Organomet. Chem.* **97,** 159 (1975).
286. Brasington, R. D., and Poller, R. C., *J. Organomet. Chem.* **40,** 115 (1972).
287. Kuivila, H. G., *Acc. Chem. Res.* **1,** 299 (1968).
288. Davies, A. G., ref. *1*, p. 26.

289. Poller, R. C., *Rev. Silicon, Germanium, Tin, Lead Compd.* **3**, 243 (1978).
290. Hillgartner, H., Schroeder, B., and Neumann, W. P., *J. Organomet. Chem.* **42**, C83 (1972).
291. Hillgärtner, H., Neumann, W. P., and Schroeder, B., *Justus Liebigs Ann. Chem.* p. 586 (1975).
292. Neumann, W. P., Schroeder, B., and Ziebarth, M., *Justus Liebigs Ann. Chem.* p. 2279 (1975).
293. Davies, A. G., and Tse, M.-W., *J. Chem. Soc., Perkin Trans. 2* p. 145 (1978).
294. Davies, A. G., and Tse, M.-W., *J. Chem. Soc., Chem. Commun.* p. 353 (1978).
295. Watts, G. B., and Ingold, K. U., *J. Am. Chem. Soc.* **94**, 491 (1972).
296. Lehnig, M., *Tetrahedron Lett.* p. 3663 (1977).
297. Buschhaus, H. A., Lehnig, M., and Neumann, W. P., *J. Chem. Soc., Chem. Commun.* p. 129 (1977).
298. Buschhaus, H. A., and Neumann, W. P., *Angew. Chem., Int. Ed. Engl.* **17**, 59 (1978).
299. Davidson, P. J., Hudson, A., Lappert, M. F., and Lednor, P. W., *J. Chem. Soc., Chem. Commun.* p. 829 (1973).
300. Gynane, M. J. S., Harris, D. H., Lappert, M. F., Power, P. P., Riviere, P., and Riviere-Baudet, M., *J. Chem. Soc., Dalton Trans.* p. 2004 (1977).
301. Cooper, J., Hudson, A., and Jackson, R. A., *J. Chem. Soc., Perkin Trans. 2* p. 1056 (1973).
302. Coates, D. A., and Tedder, J. M., *J. Chem. Soc., Perkin Trans. 2* p. 1570 (1973).
303. Grady, G. L., Danyliw, T. J., and Rabidaux, P., *J. Organomet. Chem.* **142**, 67 (1977).
304. Lehnig, M., *Tetrahedron Lett.* p. 3663 (1977).
305. Hollaender, J., Neumann, W. P., and Lind, H., *Chem. Ber.* **106**, 2395 (1973).
306. Avar, G., and Neumann, W. P., *J. Organomet. Chem.* **131**, 207 (1977).
307. Albert, H. J., Neumann, W. P., and Schneider, K., *Chem. Ber.* **106**, 411 (1973).
308. Christen, U., and Neumann, W. P., *Chem. Ber.* **106**, 421 (1973).
309. Rotenberg, K., Neumann, W. P., and Avar, G., *Chem. Ber.* **110**, 1628 (1977).
310. Neumann, W. P., and Schwindt, J., *Chem. Ber.* **108**, 1339 (1975).
311. Lehnig, M., Schwindt, J., and Neumann, W. P., *Chem. Ber.* **108**, 1355 (1975).
312. Scaiano, J. C., Schmid, P., and Ingold, K. U., *J. Organomet. Chem.* **121**, C4 (1976).
313. Cooper, J., Hudson, A., and Jackson, R. A., *J. Chem. Soc., Perkin Trans. 2* p. 1933 (1973).
314. Schroeder, B., Neumann, W. P., and Hillgärtner, H., *Chem. Ber.* **107**, 3494 (1974).
315. Neumann, W. P., and Schwindt, J., *Chem. Ber.* **108**, 1346 (1975).
316. Barton, D. H. R., and McCombie, S. W., *J. Chem. Soc., Perkin Trans. 2* p. 1574 (1975).
317. Avar, G., and Neumann, W. P., *J. Organomet. Chem.* **131**, 215 (1977).
318. Scaiano, J. C., and Ingold, K. U., *J. Am. Chem. Soc.* **99**, 2079 (1977).
319. Harrison, P. G., and Richards, J. A., *J. Organomet. Chem.* **108**, 35 (1976).
320. Harrison, P. G., *J. Chem. Soc., Chem. Commun.* 544 (1972).
321. Harrison, P. G., and Stobart, S. R., *J. Chem. Soc., Dalton Trans.* p. 940 (1973).
322. Ewings, P. F. R., and Harrison, P. G., *J. Chem. Soc., Dalton Trans.* p. 1719 (1975).
323. Ewings, P. F. R., and Harrison, P. G., *J. Chem. Soc., Dalton Trans.* p. 2015 (1975).
324. Bos, K. D., Bulten, E. J., and Noltes, J. F., *J. Organomet. Chem.* **99**, 397 (1975).
325. Albert, H.-J., and Schröer, U., *J. Organomet. Chem.* **60**, C6 (1973).
326. Davidson, P. J., and Lappert, M. F., *J. Chem. Soc., Chem. Commun.* p. 317 (1973).
327. Davidson, P. J., Harris, D. H., and Lappert, M. F., *J. Chem. Soc., Dalton Trans.* p. 2269 (1976).

328. Cotton, J. D., Davidson, P. J., Lappert, M. F., Donaldson, J. D., and Ger, J. S., *J. Chem. Soc., Dalton Trans.* p. 2286 (1976).
329. Cotton, J. D., Davidson, P. J., and Lappert, M. F., *J. Chem. Soc., Dalton Trans.* p. 2275 (1976).
330. Gynane, M. J. S., Lappert, M. F., Miles, S. J., and Power, P. P., *J. Chem. Soc., Chem. Commun.* p. 256 (1973).
331. Gynane, M. J. S., Lappert, M. F., Miles, S. J., and Power, P. P., *J. Chem. Soc., Chem. Commun.* p. 192 (1978).
332. Goldberg, D. E., Harris, D. H., Lappert, M. F., and Thomas, K. M., *J. Chem. Soc., Chem. Commun.* p. 261 (1976).
333. Neumann, W. P., *Rev. Silicon, Germanium, Tin, Lead Compd.* In press.
334. Schröer, U., and Neumann, W. P., *Angew. Chem., Int. Ed. Engl.* **14**, 246 (1975).
335. Grugel, C., Neumann, W. P., and Seifert, P., *Tetrahedron Lett.* p. 2205 (1977).
336. Neumann, W. P., and Schwarz, A., *Angew. Chem., Int. Ed. Engl.* **14**, 812 (1975).
337. Hulme, R., *J. Chem. Soc.* p. 1524 (1963).
338. Petrosyan, V. S., *Prog. Nucl. Magn. Reson. Spectrosc.* **11**, 115 (1977).
339. Ho, B. Y. K., and Zuckerman, J. J., *J. Organomet. Chem.* **49**, 1 (1973).
340. Bokii, N. G., *Itogi Nauki Tekh. Kristallokhim.* **10**, 94 (1974).
341. Karipides, A., Forman, C., Thomas, R. H. P., and Reed, A. T., *Inorg. Chem.* **13**, 811 (1974).
342. Karipides, A., and Oertel, M., *Acta Crystallogr., Sect. B* **33**, 683 (1977).
343. Karipides, A., and Wolfe, K., *Acta Crystallogr., Sect. B* **31**, 605 (1975).
344. Karipides, A., Reed, A. T., Haller, D. A., and Hayes, F., *Acta Crystallogr., Sect. B* **33**, 950 (1977).
345. Kulishov, V. I., Bokii, N. G., Prikhot'ko, A. F., and Struchkov, Yu. T., *Zh. Strukt. Khim.* **16**, 252 (1975).
346. Barrans, Y., Pereyre, M., and Rahm, A., *J. Organomet. Chem.* **125**, 173 (1977).
346a. Harrison, P. G., and Molloy, K., *J. Organomet. Chem.* **152**, 53 (1978).
347. Gynane, M. J. S., Lappert, M. F., Miles, S. J., Carty, A. J., and Taylor, N. J., *J. Chem. Soc., Dalton Trans.* p. 2009 (1977).
348. Clarke, P. L., Cradwick, M. E., and Wardell, J. L., *J. Organomet. Chem.* **63**, 279 (1973).
349. Bokii, N. G., Struchkov, Yu. T., Kravtsov, D. N., and Rokhlina, E. M., *J. Struct. Chem. (USSR)* **15**, 424 (1974).
350. Menzebach, B., and Bleckmann, P., *J. Organomet. Chem.* **91**, 291 (1975).
351. Jacobsen, H.-J., and Krebs, B., *J. Organomet. Chem.* **136**, 333 (1977).
352. Kobelt, D., Paulus, E. F., and Scherer, H., *Acta Crystallogr., Sect. B* **28**, 2323 (1972).
353. Petrosyan, V. S., and Reutov, O. A., *Pure Appl. Chem.* **37**, 147 (1974).
354. Buckle, J., Harrison, P. G., King, T. J., and Richards, J. A., *J. Chem. Soc., Dalton Trans.* p. 1552 (1975).
355. Chih, H., and Penfold, B. R., *J. Cryst. Mol. Struct.* **3**, 285 (1973).
356. Boer, F. P., Doorakian, G. A., Freedman, H. H., and McKinley, S. V., *J. Am. Chem. Soc.* **92**, 1225 (1970).
347. Van Koten, G., Noltes, J. G., and Spek, A. L., *J. Organomet. Chem.* **118**, 183 (1976).
358. Bancroft, G. M., Davies, B. W., Payne, N. C., and Sham, T. K., *J. Chem. Soc., Dalton Trans.* p. 973 (1975).
359. King, T. J., and Harrison, P. G., *J. Chem. Soc., Dalton Trans.* p. 2298 (1974).
360. Buttenshaw, A. J., Duchêne, M., and Webster, M., *J. Chem. Soc., Dalton Trans.* p. 2230 (1975).

361. Cunningham, D., Douek, I., Frazer, M. J., McPartlin, M., and Matthews, J. D., *J. Organomet. Chem.* **90,** C23 (1975).
362. Graziani, R., Bombieri, G., Forsellini, E., Furlan, P., Peruzzo, V., and Tagliavini, G., *J. Organomet. Chem.* **125,** 43 (1977).
363. Webster, M., Mudd, K. R., and Taylor, D. J., *Inorg. Chim. Acta* **20,** 231 (1976).
364. Fitzsimmonds, B. W., Othen, D. G., Shearer, H. M. M., Wade, K., and Whitehead, G., *J. Chem. Soc., Chem. Commun.* p. 215 (1977).
365. Holloway, J. H., McQuillan, G. P., and Ross, D. S., *J. Chem. Soc. A* p. 2505 (1969).
366. Schlemper, E. O., *Inorg. Chem.* **6,** 2012 (1967).
367. Harrison, P. G., King, T. J., and Phillips, R. C., *J. Chem. Soc., Dalton Trans.* p. 2317 (1976).
368. Lindley, P. F., and Carr, P., *J. Cryst. Mol. Struct.* **4,** 173 (1974).
369. Miller, G. A., and Schlemper, E. O., *Inorg. Chem.* **12,** 677 (1973).
370. Harrison, P. G., King, T. J., and Richards, J. A., *J. Chem. Soc., Dalton Trans.* p. 1723 (1974).
371. Coghi, L., Pelizzi, C., and Pelizzi, G., *Gazz. Chim. Ital.* **104,** 873 (1974).
372. Schlemper, E. O., and Hamilton, W. C., *Inorg. Chem.* **5,** 995 (1966).
373. Allen, F. H., Lerbscher, J. A., and Trotter, J., *J. Chem. Soc. A* p. 2507 (1971).
374. Alcock, N. W., and Sawyer, J. F., *J. Chem. Soc., Dalton Trans.* p. 1090 (1977).
375. Davies, A. G., Milledge, H. J., Puxley, D. C., and Smith, P. J., *J. Chem. Soc. A* p. 2862 (1970).
376. Chow, Y. M., *Inorg. Chem.* **9,** 794 (1970).
377. Foder, R. A., and Sheldrick, G. M., *J. Organomet. Chem.* **22,** 611 (1970).
378. Bokii, N. G., Struchkov, Yu. T., and Prokof'ev, A. K., *J. Struct. Chem.* **13,** 619 (1972).
379. Harrison, P. G., King, T. J., and Richards, J. A., *J. Chem. Soc., Dalton Trans.* p. 826 (1975).
380. Kimura, T., Yasuoka, N., Kasai, N., and Kakudo, M., *Bull. Chem. Soc. Jpn.* **45** 1649 (1972).
381. Calligaris, M., Nardin, G., and Randaccio, L., *J. Chem. Soc., Dalton Trans.* p. 2003 (1972).
382. Harrison, P. G., and Mangia, A., *J. Organomet. Chem.* **120,** 211 (1976).
383. Lecomte, C., Protas, J., and Devaud, M., *Acta Crystallogr., Sect. B* **32,** 923 (1976).
384. Brownlee, G. S., Walker, A., Nyburg, S. C., and Szymański, J. T., *J. Chem. Soc., Chem. Commun.* p. 1073 (1971).
385. Schlemper, E. O., personal communication, quoted in: Naik, D. V., and Curran, C., *Inorg. Chem.* **10,** 1017 (1971).
386. Naik, D. V., and Scheidt, W., *Inorg. Chem.* **12,** 272 (1973).
387. Coghi, L., Pelizzi, C., and Pelizzi, G., *J. Organomet. Chem.* **114,** 53 (1976).
388. Mangia, A., Pelizzi, C., and Pelizzi, G., *J. Chem. Soc., Dalton Trans.* p. 2557 (1973).
389. Mackay, C. A., Ph.D. Thesis, University of London, 1973.
390. Preut, H., Huber, F., Haupt, H.-J., Cafalù, R., and Barbieri, R., *Z. Anorg. Allg. Chem.* **410,** 88 (1974).
391. Evans, D. L., and Penfold, B. R., *J. Cryst. Mol. Struct.* **5,** 93 (1975).
392. Domingos, A. M., and Sheldrick, G. M., *J. Chem. Soc., Dalton Trans.* p. 470 (1974).
393. Ashmore, J. P., Chivers, T., Kerr, K. A., and van Roode, J. H. G., *Inorg. Chem.* **16,** 191 (1977).
394. Domingos, A. M., and Sheldrick, G. M., *Acta Crystallogr., Sect. B* **30,** 519 (1974).
395. Ewings, P. F. R., Harrison, P. G., King, T. J., Phillips, R. C., and Richards, J. A., *J. Chem. Soc., Dalton Trans.* p. 1950 (1975).

396. Domingos, A. M., and Sheldrick, G. M., *J. Organomet. Chem.* **69**, 207 (1974).
397. Hall, J. B., and Britton, D., *Acta Crystallogr., Sect. B* **28**, 2133 (1972).
398. Okawara, R., and Wada, M., *Adv. Organomet. Chem.* **5**, 150 (1967).
399. Ho, B. Y. K., Zubieta, J. A., and Zuckerman, J. J., *J. Chem. Soc., Chem. Commun.* p. 88 (1975).
400. Harrison, P. G., ref. *31*, p. 258.
401. Clark, H. C., personal communication, quoted in: Potts, D., Sharma, H. D., Carty, A. J., and Walker, A., *Inorg. Chem.* **13**, 1205 (1974).
402. Hengel, R., Kunze, U., and Straehle, J., *Z. Anorg. Allg. Chem.* **423**, 35 (1976).
403. Sheldrick, G. M., and Taylor, R., *Acta Crystallogr., Sect. B* **33**, 135 (1977).
404. Ginderow, G., and Huber, M. M., *Acta Crystallogr., Sect. B* **29**, 560 (1973).
405. Harrison, P. G., Phillips, R. C., and Richards, J. A., *J. Organomet. Chem.* **114**, 47 (1976).
406. Domingos, A. M., and Sheldrick, G. M., *J. Chem. Soc., Dalton Trans.* p. 477 (1974).
407. Pommier, J.-C., Mendes, F., and Valade, J., *J. Organomet. Chem.* **55**, C19 (1973).
408. Kunchur, N. R., and Borhani, S., *J. Appl. Crystallogr.* **9**, 508 (1976).
409. Williams, D., *Aust. Def. Stand. Lab. Ann. Rep.* p. 37 (1967/1968).
410. Huber, F., Haupt, H.-J., Preut, H., Barbieri, R., and Lo Giudice, M. T., *Z. Anorg. Allg. Chem.* **432**, 51 (1977).
411. Preut, H., Haupt, H.-J., Huber, F., Cefalù, R., and Barbieri, R., *Z. Anorg. Allg. Chem.* **407**, 257 (1974).
412. Preut, H., Huber, F., Barbieri, R., and Bertazzi, N., *Z. Anorg. Allg. Chem.* **423**, 75 (1976).
413. Glidewell, C., and Liles, D. C., *Acta Crystallogr., Sect. B* **34**, 129 (1978).
414. Bokii, N. G., Struchkov, Yu. T., Kravtsov, D. N., and Rokhlina, E. M., *J. Struct. Chem. (USSR)* **14**, 458 (1973).
415. Domingos, A. M., and Sheldrick, G. M., *J. Organomet. Chem.* **67**, 257 (1974).
416. Tarkhlova, T. N., Chuprunov, E. V., Simonov, M. A., and Belov, N. V., *Kristallografiya* **22**, 1004 (1977).
417. Bokii, N. G., Struchkov, Yu. T., Korolikov, V. V., and Tolstaya, T. P., *Koord. Khim.* **1**, 1144 (1975).
418. Uber, W., Stegmann, H. B., Scheffler, K., and Straehle, J., *Z. Naturforsch., Teil B* **32**, 355 (1977).
419. Kulishov, V. I., Rodé, G. G., Bokii, N. G., Prikhot'ko, A. F., and Struchkov, Yu. T., *J. Struct. Chem. (USSR)* **16**, 227 (1975).
420. Weidenborner, J. E., Larrabee, R. B., and Bednowitz, A. L., *J. Am. Chem. Soc.* **94**, 4140 (1972).
421. Pelizzi, C., Pelizzi, G., and Tarasconi, P., *J. Chem. Soc., Dalton Trans.* 1935 (1977).
422. Preut, H., Haupt, H.-J., and Huber, F., *Z. Anorg. Allg. Chem.* **396**, 81 (1973).
423. Nardelli, M., Pelizzi, C., Pelizzi, G., and Tarasconi, P., *Z. Anorg. Allg. Chem.* **431**, 250 (1977).
424. Howard, J. A., Kellett, S. C., and Woodward, P., *J. Chem. Soc., Dalton Trans.* p. 2332 (1975).
425. Gilmore, C. J., and Woodward, P., *J. Chem. Soc., Dalton Trans.* 1387 (1972).
426. Sweet, R. M., Fritchie, Jr., C. J., and Schunn, R. A., *Inorg. Chem.* **6**, 749 (1976).
427. Harrison, P. G., King, T. J., and Richards, J. A., *J. Chem. Soc., Dalton Trans.* p. 2097 (1975).
428. Shklover, V. E., Skripkin, V. V., Gusev, A. I., and Struchkov, Yu. T., *Zh. Strukt. Khim.* **13**, 744 (1972).

429. Pomeroy, R. K., Vancea, L., Calhoun, H. P., and Graham, W. A. G., *Inorg. Chem.* **16,** 1508 (1977).
430. Weaver, J., and Woodward, P., *J. Chem. Soc., Dalton Trans.* p. 1060 (1973).
431. Boer, F. P., Tsai, J. H., and Flynn, Jr., J. J., *J. Am. Chem. Soc.* **92,** 6092 (1970).
432. Folting, K., personal communication, quoted in: Garrou, P., and Hartwell, G. E., *J. Chem. Soc., Chem. Commun.* p. 881 (1972).
433. Collman, J. P., Murphy, D. W., Fleischer, E. B., and Swift, D., *Inorg. Chem.* **13,** 1 (1974).
434. Sasse, H. E., and Ziegler, M. L., *Z. Anorg. Allg. Chem.* **402,** 129 (1973).
435. McPhail, A. T., personal communication, quoted in: Garner, C. D., and Hughes, B., *J. Chem. Soc., Dalton Trans.* p. 1306 (1974).
436. Brice, M. D., and Cotton, F. A., *J. Am. Chem. Soc.* **95,** 4529 (1973).
437. Harrison, P. G., King, T. J., Richards, J. A., and Phillips, R. C., *J. Organomet. Chem.* **116,** 307 (1976).
438. Schumann, H., Schumann-Ruidisch, I., and Schmidt, M., *in* "Organotin Compounds," (A. K. Sawyer, ed.), Vol. 2, p. 318. Dekker, New York, 1971.
439. Smart, L. E., and Webster, M., *J. Chem. Soc., Dalton Trans.* p. 1924 (1976).
440. Harrison, P. G., Molloy, K., Phillips, R. C., Smith, P. J., and Crowe, A. J., *J. Organomet. Chem.* **160,** 421 (1978).
440a. Aslanov, L. A., Attiya, V. M., Ionov, V. M., Permin, A. B., and Petrosyan, V. S., *Zh. Strukt. Khim.* **18,** 1113 (1977).
440b. Harrison, P. G., and Molloy, K., *J. Organomet. Chem.* **152,** 63 (1978).
441. Aslanov, L. A., Ionov, V. M., Attiya, V. M., Permin, A. B., and Petrosyan, V. S., *J. Organomet. Chem.* **144,** 39 (1978).
442. Randaccio, L., *J. Organomet. Chem.* **55,** C58 (1973).
443. Isaacs, N. W., and Kennard, C. H. L., *J. Chem. Soc. A* p. 1257 (1970).
444. Calligaris, M., Randaccio, L., Barbieri, R., and Pellerito, L., *J. Organomet. Chem.* **76,** C56 (1974).
445. Garner, C. D., Hughes, B., and King, T. J., *J. Chem. Soc., Dalton Trans.* 562 (1975).
446. Nardelli, M., Pelizzi, C., and Pelizzi, G., *J. Organomet. Chem.* **112,** 263 (1976).
447. Nardelli, M., Pelizzi, C., and Pelizzi, G., *J. Organomet. Chem.* **125,** 161 (1977).
448. Pelizzi, C., Pelizzi, G., and Tarasconi, P., *J. Organomet. Chem.* **124,** 151 (1977).
449. Pelizzi, G., *Inorg. Chim. Acta* **24,** L31 (1977).
450. Kolakowski, A., and Kolakowski, B., *J. Appl. Crystallogr.* **10,** 494 (1977).
451. Midollini, S., Orlandini, A., and Sacconi, L., *Cryst. Struct. Commun.* **6,** 733 (1977).
452. Bancroft, G. M., "Mössbauer Spectroscopy," McGraw-Hill, New York, 1973.
453. Herber, R. H., and Leahy, M. F., ref. *31,* p. 155.
454. Cotton, J. D., Davidson, P. J., Lappert, M. F., Donaldson, J. D., and Silver, J., *J. Chem. Soc., Dalton Trans.* p. 2286 (1976).
455. Stöckler, H. A., and Sano, H., *Trans. Faraday Soc.* **64,** 577 (1968).
456. Ali, K. M., Cunningham, D., Donaldson, J. D., Frazer, M. J., and Senior, B. J., *J. Chem. Soc. A* p. 2836 (1969).
457. de Vries, J. L. K. F., and Herber, R. H., *Inorg. Chem.* **11,** 2458 (1972).
458. May, J. C., Petridis, D., and Curran, C., *Inorg. Chim. Acta* **5,** 511 (1971).
459. May, J. C., and Curran, C., *J. Organomet. Chem.* **39,** 289 (1972).
460. Harrison, P. G., Phillips, R. C., and Thornton, E. W., *J. Chem. Soc., Chem. Commun.* p. 603 (1977).
461. Nardelli, M., Pelizzi, C., and Pelizzi, G., *J. Chem. Soc., Dalton Trans.* p. 131 (1978).
462. D'yachenko, O. A., Zolotoi, A. b., Atovmyan, L. O., Mirskov, R. G., and Voronkov, M. G., *Dokl. Akad. Nauk SSSR* **237,** 863 (1977).

463. van Koten, G., Jastrzebski, J. T. B. H., Noltes, J. G., Spek, A. L., and Schoone, J. C., *J. Organomet. Chem.* **148**, 233 (1978).
464. van Koten, G., Jastrzebski, J. T. B. H., Noltes, J. G., Pontenagel, W. M. G. F., Kroon, J., and Spek, A. L., *J. Am. Chem. Soc.* **100**, 5021 (1978).
465. Liebich, W., and Tomassini, M., *Acta Crystallogr., Sect. B* **34**, 944 (1978).
466. Cremer, J. E., *Biochem. J.* **68**, 685 (1958).
467. Kimmel, E. C., Fish, R. H., and Casida, J. E., *J. Agric. Food Chem.* **25**, 1 (1977).
468. Meynier, D., Doctoral Dissertation, University of Toulouse, France (1955).
469. Smith, P. J., and Smith, L., *Chem. Br.* **11**, 208 (1975).
470. Smith, P. J., "Toxicological Data on Organotin Compounds," Int. Tin Res. Inst. Publ. No. 538 (1978).
471. Barnes, J. M., and Magos, L., *Organomet. Chem. Rev.* **3**, 137 (1968).
472. Ascher, K. R. S., and Nissim, S., *World Rev. Pest Control* **3**, 188 (1964).
473. Sijpesteijn, A. K., Luijten, J. G. A., and van der Kerk, G. J. M., in "Fungicides, An Advanced Treatise," (D. C. Torgeson, ed.), Vol. II, p. 331. Academic Press, New York, 1969.
474. van der Kerk, G. J. M., and Luijten, J. G. A., *J. Appl. Chem.* **4**, 314 (1954).
475. Evans, C. J., *Tin Its Uses* **86**, 7 (1970).
476. Evans, C. J., *Tin Its Uses* **110**, 6 (1976).
477. Aldridge, W. N., ref. *31*, p. 186.
478. Hall, W. T., and Zuckerman, J. J., *Inorg. Chem.* **16**, 1239 (1977).
479. Farrow, B. G., and Dawson, A. P., *Eur. J. Biochem.* **86**, 85 (1978).
480. Cain, K., and Griffiths, D. E., *Biochem. J.* **162**, 575 (1977).
481. Frankel, M., Gertner, D., Wagner, D., and Zilkha, A., *J. Org. Chem.* **30**, 1596 (1965).
482. Ho, B. Y. K., and Zuckerman, J. J., *Inorg. Chem.* **12**, 1552 (1973).
483. Wirth, H. O., Lorenz, H. J., and Friedrich, H.-H., U.S. Patent 3,933,877 (1976).
484. Luijten, J. G. A., Janssen, M. J., and van der Kerk, G. J. M., *Recl. Trav. Chim. Pays-Bas* **81**, 202 (1962).
485. Rose, M. S., in "Pesticide Terminal Residues," (A. S. Tahori, ed.). Butterworth, London, 1971.
486. Elliott, B. M., and Aldridge, W. N., *Biochem. J.* **163**, 583 (1977).
487. Lock, E. A., and Aldridge, W. N., *J. Neurochem.* **25**, 871 (1975).
488. Rose, M. S., and Lock, E. A., *Biochem. J.* **120**, 151 (1970).
489. Cardarelli, N. F., "Controlled Release Molluscicides," Environ. Manage. Lab. Mon., p. 34. Univ. Akron, Ohio, 1977.
490. Cain, K., Partis, M. D., and Griffiths, D. E., *Biochem. J.* **166**, 593 (1977).
490a. Tan, L. P., Ng, M. L., and Kumar Das, V. G., *J. Neurochem.* (1978). **31**, 1035 (1978).
491. Elliott, B. M., Aldridge, W. N., and Bridges, J. W., *Biochem. J.,* **177**, 461 (1949).
492. van Koten, G., and Noltes, J. G., ref. *31*, p. 275.
493. Tzschach, A., Reiss, E., Held, P., and Bollmann, W., E. Ger. Patent 63,490 (1968).
494. Aldridge, W. N., Casida, J. E., Fish, R. H., Kimmel, E. C., and Street, B. W., *Biochem. Pharmacol.* **26**, 1997 (1977).
495. Ascher, K. R. S., and Nemny, N. E., *Experientia* **32**, 902 (1976).
496. Abbas, S. Z., and Poller, R. C., *Polymer* **15**, 543 (1974).
497. Omae, I., *Rev. Silicon, Germanium, Tin, Lead Compd.* **1**, 59 (1974).
498. Poller, R. C., ref. *31*, p. 177.
499. Stoner, H. B., Barnes, J. M., and Duff, J. I., *Br. J. Pharmacol.* **10** 16 (1955).
500. Casida, J. E., Kimmel, E. C., Holm, B., and Widmark, G., *Acta Chem. Scand.* **25**, 1497 (1971).

501. Bridges, J. W., Davies, D. S., and Williams, R. T., *Biochem. J.* **105,** 1261 (1967).
502. Blair, E. H., *Environ. Qual. Saf., Suppl.* **3,** 406 (1975).
503. Fish, R. H., ref. *31,* p. 197.
504. Freitag, K. D., and Bock, R., *Pestic. Sci.* **5,** 731 (1974).
505. Barnes, R. D., Bull, A. T., and Poller, R. C., *Pestic. Sci.* **4,** 305 (1973).
506. Chapman, A. H., and Price, J. W., *Int. Pest. Control* **14,** 11 (1972).
507. Cenci, P., and Cremonini, B., *Ind. Sacc. Ital.* **62,** 313 (1969).
508. Getzendaner, M. E., and Corbin, H. B., *J. Agric. Food Chem.* **20,** 881 (1972).
509. Waggon, H., and Jehle, D., *Nahrung* **19,** 271 (1975).
510. Chapman, A. H., personal communication.
511. Klotzer, D., and Thust, U., *Chem. Tech. (Leipzig)* **28,** 614 (1976).
512. Smith, P. J., Crowe, A. J., Allen, D. W., Brooks, J., and Formstone, R., *Chem. Ind. (London)* 874 (1977).
513. Mazaev, V. T., Golovanov, O. V., Igumnov, A. S., and Tsay, V. H., *Gig. Sanit.* p. 17 (1976).
514. Woggon, H., Uhde, W.-J., and Säuberlich, H., *Ernaehrungforschung* **16,** 645 (1971).
515. Akagi, H., and Sakagami, Y., *Koshu Eiseiin Kenkyu Hokoku* **20,** No. 1 (1971).
516. Sheldon, A. W., *J. Paint Technol.* **47,** 54 (1975).
517. Ridley, W. P., Dizikes, L. J., and Wood, J. M., *Science* **197,** 329 (1977).
518. Dizikes, L. I., Ridley, W. P., and Wood, J. M., *J. Am. Chem. Soc.* **100,** 1010 (1978).
519. Wood, J. M., Cheh, A., Dizikes, L. J., Rakow, S., and Lakowicz, J. R., *Fed. Proc.* **37,** 16 (1978).
520. Huey, C., Brinckman, F. E., Grim, S., and Iverson, W. P., *Proc. Int. Conf. Transp. Persist. Chem. Aquat. Ecos., 1974,* pp. 73–78.
521. Ridley, W. P., *J. Freshwater* **1,** 7 (1977).
522. Frankland, E., *Justus Liebigs Ann. Chem.* **71,** 171, 212 (1849).
523. Kumar Das, V. G., and Cheong, C. K., *Planter (Kuala Lumpur)* **51,** 355 (1975).
524. Gitlitz, M. H., ref. *31,* p. 167.
525. Hunter, R. C., *Environ. Health Perspect.* **14,** 47 (1976).
526. Huber, G., *Tin Its Uses* **113,** 7 (1977).
527. Nösler, H. G., *Gesund. Desinfekt.* **62,** 10, 65, 175 (1970).
528. Verdicchio, R. J., *Proc. Chem. Spec. Manufact. Assoc. Ann. Meet.* **56,** 114 (1969).
529. Hueck, H. J., and Luijten, J. G. A., *J. Soc. Dyers Colour.* **74,** 476 (1958).
530. Richardson, B. A., *Proc. Br. Wood Preserv. Assoc. Annu. Conven., Cambridge* p. 37 (1970).
531. Evans, C. J., *Tin Its Uses* **115,** 11 (1978).
532. Richardson, B. A., *Proc. Int. Biodet. Symp., 1st 1968* p. 498.
533. M and T Chemicals, Inc., Ger. Offen. 2,617,821 (1976).
534. Richardson, B. A., and Cox, T. R. G., *Tin Its Uses* **102,** 6 (1974).
535. Richardson, B. A., *Stone Ind.* **8,** 2 (1973).
536. Crowe, A. J., and Smith, P. J., *Inorg. Chim. Acta* **19,** L7 (1976).
537. Crowe, A. J., Hill, R., and Smith, P. J., *Proc. Int. Control Rel. Pestic. Symp., 5th,* 1978.
538. Shioyama, H., Kuriyama, Y., and Suzuki, R., Jpn. Patent 18,489 (1976).
539. Blunden, S., Chapman, A. H., Crowe, A. J., and Smith, P. J., *Int. Pest Control* **20,** 5 (July/Aug., 1978).
540. Bennett, R. F., and Zedler, R. J., *J. Oil Colour Chem. Assoc.* **49,** 928 (1966).
541. Evans, C. J., and Smith, P. J., *J. Oil Colour Chem. Assoc.* **58,** 160 (1975).
542. Chandler, R. H., and Chandler, J., *Bibliogr. Paint Technol.* **29,** Chandler, R. H., Ltd., Braintree, Essex, England, 1977.

543. Monaghan, C. P., Hoffman, J. F., O'Brien, E. J., Frenzel, L. M., and Good, M. L., *Proc. Int. Control Rel. Pestic. Symp., 4th, 1977.*
544. Perkin, A., and Poller, R. C., *Int. Pest Control* **19,** 1 (July/Aug., 1977).
545. Frimann-Dahl, C., Technical Data Sheet: "Combiprimer Systemet." Oslo, Norway, 1977.
546. King, S., *Tin Its Uses* **121,** 7 (1979).
547. Phillip, A. T., Bocksteiner, G., Pettis, R. W., and Glew, G. W., Austral. Patent 56,893 (1974).
548. Montemarano, J., and Dyckman, E. J., *J. Paint Technol.* **47,** 59 (1975).
549. Montemarano, J., Cohen, S. A., and Fischer, E. C., *Proc. Int. Control Rel. Pestic. Symp., 2nd, 1975.*
550. Cardarelli, N. F., "Controlled Release Pesticide Formulations." Chemical Rubber Co., Cleveland, Ohio, 1976.
551. Bell, D. R., *Br. J. Hosp. Med.* p. 29 (Jan., 1974).
552. Chliamovitch, Y.-P., and Kuhn, C., *J. Fish Biol.* **10,** 575 (1977); Chliamovitch, Y.-P., Ph.D. Thesis, Univ. Geneva, 1978.
553. Klimsch, P., *Plasty Kauc.* **24,** 380 (1977).
554. Weisfeld, L. B., and Witman, R. C., U.S. Patent 3,810,868 (1974).
555. Lanigan, D., and Weinberg, E. L., *ref. 31,* p. 134.
556. Luijten, J. G. A., and Pezarro, S., *Br. Plast.* **30,** 183 (1957).
557. Mazur, H., *Rocz. Pzh.* **22,** 39 (1971).
558. Frey, H. H., and Dörfelt, C., Ger. Patent 1,160,177 (1958).
559. Smith, H. V., Technical Data Sheet: "New Lauryltin Compounds." Petit-Lancy, Switzerland, June 23rd, 1976.
560. Jones, K., Biddle, K. D., Das, A. K., and Emblem, H. G., *Int. Symp. Organomet. Coord. Chem. Germanium, Tin, Lead, 2nd 1977;* Das, A. K., Jones, K., and Emblem, H. G., Br. Patent 1,494,209 (1977); Jones, K., Biddle, K. D., Das, A. K., and Emblem, H. G., *Tin Its Uses* **119,** 10 (1979).
561. M and T Chemicals Inc., Technical Data Sheet: "FASCAT 4101 Monobutyltin Esterification Catalyst." Rahway, New Jersey, 1977.
562. Suzukawa, T., *Seramikkusu* **4,** 852 (1969).
563. Fuller, M. J., *Tin Its Uses* **103,** 3 (1975).
564. Kane, J., Schweizer, H. P., and Kern, W., *J. Electrochem. Soc.* **122,** 1144 (1975).
565. Baliga, B. J., and Ghandhi, S. K., *J. Electrochem. Soc.* **123,** 941 (1976).
566. M and T Chemicals, Inc., U.S. Patent 4,130,673 (1978).

TRANSITION METAL VAPOR CRYOCHEMISTRY

WILLIAM J. POWER* and GEOFFREY A. OZIN

Lash Miller Chemical Laboratories and Erindale College, University of Toronto, Toronto, Ontario, Canada

I. Introduction	80
II. Atomic, Diatomic, and Cluster Species	81
A. Sc, Ti, and V Studies	83
B. Nb and Mo Studies	85
C. Co, Rh, and Ir Studies	86
D. Ni, Pd, and Pt Studies	89
E. Cu, Ag, and Au Studies	92
F. Bimetallic Clusters	96
III. New Cluster Techniques	101
A. Cryophotoclustering	101
B. Selective Naked-Cluster Cryophotochemistry	103
C. Relative Extinction-Coefficient Measurements for Naked Silver Clusters by Photoaggregation Techniques	106
D. Photonucleation Kinetic Studies	107
E. Photomanipulation of Cluster Distributions	107
F. Photoselective Bimetallic Aggregation	108
IV. Cluster Complexes	114
A. CO Species	115
B. O_2 Species	118
C. C_2H_4 Species	120
V. Classical Inorganic Ligands	130
A. CO Complexes	130
B. O_2 Complexes and Oxides	138
C. N_2 Complexes and Nitrides	140
D. Miscellaneous Complexes	143
VI. Organometallic Complexes	145
A. Arene Complexes	145
B. Alkene and Alkyne Complexes	149
C. Oxidative Addition Reactions	158
D. Organic Reactions	160
E. Miscellaneous Organometallic Species	163
VII. Conclusion	166
Addendum	166
Addendum References	168
References	169

* Present address: Imperial Oil Limited, Sarnia, Ontario, Canada.

I. Introduction

When we were invited to write a review on an "in vogue" research topic, especially one that had previously been surveyed, there was an initial tendency to try to justify the effort. The burgeoning field of metal-vapor cryochemistry presents little difficulty in this regard, as the past two years have again witnessed continued growth of interest, impressive accomplishments, and the development of a large body of new literature. We therefore consider that a general, state-of-the-art overview of the topic examining selected, research highlights since 1976, rather than presenting comprehensive tabulations of papers, would be both timely and appropriate.

Perusal of the recent literature revealed that Timms and Turney (*179*) had initially evaluated the organometallic, synthetic versatility of the metal-atom technique both for main-group- and transition-metal vapors, and that Burdett (*21*) had comprehensively covered the metal-atom and photochemical aspects of matrix-entrapped, metal carbonyl and related complexes. A conference report on the matrix photochemistry of metal carbonyls has also appeared (*191*), as well as a survey of the spectroscopic aspects of metal-atom chemistry (*99*). A significant indicator of the impressive recognition that metal-atom and matrix chemistry have earned may be seen from the list of recent recipients of Coblentz, Corday–Morgan, and Meldola awards (*3*). Ozin (*100*) has presented a view of the "metal cluster–metal surface analogy" through the eye of matrix chemistry. Klabunde (*64*) has written on the organic chemistry of metal vapors, with particular emphasis on active metal slurries. The synthetic and catalytic significance of Klabunde's (*64*) approach to finely dispersed, metal clusters should be compared with the Wada (*196*) type of gas-evaporation technique and the Rieke (*158*) type of metal halide–alkali metal, solution-reduction method for generating ultrafine, metal particles. Complete works have now been devoted to the subject of matrix cryochemistry, as seen in a monograph by Craddock and Hinchcliffe (*23*) and the more comprehensive coverage of Moskovits and Ozin (*91*).

In its more recent phase of development, the multidisciplinary nature of metal-vapor cryochemistry is becoming evident with, for example, chemical physicists attempting to explain subtle, spectroscopic phenomena associated with matrix-entrapped, metal atomic species (*75–77*). A clear display of renewed physics interest in the field may be seen from a glance at the proceedings of the International Conference of Matrix Isolation Spectroscopy (*Ber. Bunsenges. Phys. Chem.*, January, 1978). In addition, matrix reactions are providing unique, synthetic pathways to "experimental, chemisorption models" for theoretical chemists to calculate on, with, for example, $Ni(C_2H_4)/Ni_2(C_2H_4)$ (*101*) providing the basis for a GVB–CI–MO analysis (*101*), coordinatively unsaturated, metal carbonyl fragments being models for Hoffman and Elian's (*53*) stereochemical predictions, and naked metal-cluster species providing the stimulus for EHMO (*2*) and SCF–Xα–SW (*102, 113*) computations for the determination of electronic, spectral, and bonding properties of small, ligand-free, metal clusters, as well as the evolution of bulk-metal properties.

Another interdisciplinary thrust, particularly from the groups of Ozin (*100, 103*) and Moskovits (*91, 95, 96a*), is the correlation of the bonding of "supercoordinatively" unsaturated, M_nL matrix-stabilized, cluster systems with that of L chemisorbed onto the corresponding bulk metal, M. It is becoming fairly evident that complexes synthesized under matrix conditions, particularly with CO, N_2, O_2, olefin, and acetylene ligands, may be used [with caution (*96b*)] as chemisorption models for evaluating localized-bonding aspects of these same ligands adsorbed on metal surfaces. Furthermore, the complexes may often be considered to be model representations for reactive intermediates involved in heterogeneously catalyzed reactions (*105, 132, 136*). Bimetal-vapor cocondensations and bimetallic, photoaggregation, matrix experiments (*57, 78, 89, 113–116, 203*) have pointed the way to an extremely promising method for securing spectroscopic and chemical data on mixed metal-cluster systems. Such data are likely to throw considerable light on a number of unanswered questions in the field of alloy and bimetallic-cluster catalysis (*167*).

From this brief exposure to the pervasive character of metal-vapor cryochemistry and matrix-isolation spectroscopy, it is clear that a field of multidisciplinary concern is surveyed. However, in this review, we shall focus on those accomplishments of greatest interest to the inorganic community in general, while pointing out, where applicable, the relevance of the results to other branches of chemistry and physics.

II. Atomic, Diatomic, and Cluster Species

It is now fairly well established that atomic and few-atom cluster arrays can be generated and trapped in weakly interacting matrices (*91*), and subsequently scrutinized by various forms of spectroscopy. Up to this time, IR–Raman–UV–visible absorption and emission–esr–MCD–EXAFS–Mössbauer methods have been successfully applied to matrix-cluster samples. It is self-evident that an understanding of the methods of generating and identifying these species is a prerequisite for their subsequent use in the synthesis of mononuclear, binuclear, or higher, cluster complexes, and so we shall briefly dwell on this point.

The problem of assigning observed atomic, and cluster-matrix, spectral features to individual species of known nuclearity was treated by Moskovits and Hulse (92), and the results were first applied to the Cr_2 (106) diatomic. In the majority of cluster experiments referred to in this Chapter, the growth–decay behavior of vibrational, or electronic, spectral features as a function of matrix–metal concentration must be analyzed, the objective being the determination of cluster nuclearity through an analysis of the deposition process. Two models were investigated, statistical and kinetic. The statistical model (92) attempted to simulate the occurrence of neighboring metal-atoms in a matrix by means of a Monte Carlo calculation, by way of the random placement of atomic reagent within the lattice confines of a face-centered-cubic network of inert-gas atoms. Counting procedures established that the concentration ratio, $[M_n]/[M]$, is roughly proportional to the metal concentration raised to the power $(n - 1)$, up to ~10% of total metal. The steady-state, kinetic model (92) involved metal-atom encounters in the reaction zone at the matrix surface during matrix deposition. This fluid surface-region, being quite mobile, facilitates metal-atom diffusion and aggregation, which can be expected to ensue up to the point of matrix solidification. After this point, bulk diffusion of atoms and clusters is assumed to contribute negligibly to further cluster growth. Simulation of the large number of metal-atom encounters in the semi-liquid region was achieved by assembling a set of simultaneous, linear, differential equations to describe the nucleation sequence.

$$M \xrightarrow[M]{k_1} M_2 \xrightarrow[M]{k_2} M_3 \xrightarrow[M]{k_3} \cdots \cdots \qquad (1)$$

Solutions were obtained, either analytically or numerically, on a computer. The quenched-reaction, kinetic model considered that the nucleation sequence of reactions evolves to some time τ_q (the quenching time) and then promptly halts. Both kinetic models yield a result having the same general form as the statistical model, namely,

$$[M_n]/[M] \; \widetilde{\propto} \; [M_0]^{(n-1)}, \qquad (2)$$

although Moskovits and Hulse's study (92) indicated that the quenched-reaction model fits the available experimental data slightly better than the steady-state model. Thus, the slope of a log $([M_n]/[M])$ versus log $[M_0]$ plot (where $[M_0]$ is the total metal concentration) provides a convenient and direct analytical route to metal-cluster size. By similarly analyzing the nucleation sequence in the presence of various concentrations of a reactive ligand L, the analogous relationship, namely,

$$[M_nL_m]/[ML_p] \tilde{\propto} [M_0]^{(n-1)}, \qquad (3)$$

is arrived at; that is, the absorbance of a cluster complex containing n metal atoms, when normalized with respect to that of a mononuclear complex, is approximately proportional to the metal concentration raised to the power $(n-1)$. Consequently, log–log plots for suspected, cluster species M_nL_m relative to known mononuclears ML_p as a function of the total, metal concentration provide access to cluster-size evaluation. The ultility of these expressions, both for naked metal clusters and for cluster complexes, will become evident throughout this section.

A. Sc, Ti, AND V STUDIES

Ozin *et al.* (*107, 108*) performed matrix, optical experiments that resulted in the identification of the dimers of these first-row, transition metals. For Sc and Ti ($4s^23d^1$ and $4s^23d^2$, respectively), a facile dimerization process was observed in argon. It was found that, for Sc, the atomic absorptions were blue-shifted 500–1000 cm^{-1} with respect to gas-phase data, whereas the extinction coefficients for both Sc and Sc_2 were of the same order of magnitude, a feature also deduced for Ti and Ti_2. The optical transitions and tentative assignments (based on EHMO calculations) are summarized in Table I.

A somewhat more detailed study of vanadium atoms and dimers has also appeared (*108*). Figure 1 shows the UV–visible spectra of V and V_2 as a function of vanadium concentration. Figure 2 shows a typical, metal-concentration plot illustrating the aforementioned kinetic anal-

TABLE I

OBSERVED AND CALCULATED OPTICAL SPECTRA FOR DISCANDIUM AND DITITANIUM (*107*)

	Optical transition	Obs. (cm^{-1})	Calc. (cm^{-1})
Sc_2	$1\sigma_g \rightarrow 1\sigma_u$	15 100	15 492
	$1\sigma_g \rightarrow 2\sigma_u$	21 050	19 880
	$1\sigma_g \rightarrow 2\pi_u$	29 850	29 558
Ti_2	$1\sigma_g \rightarrow 1\sigma_u$	16 020	15 355
	$1\sigma_g \rightarrow 2\sigma_u$	18 310	16 654
	$1\sigma_g \rightarrow 2\pi_u$	23 250	23 307

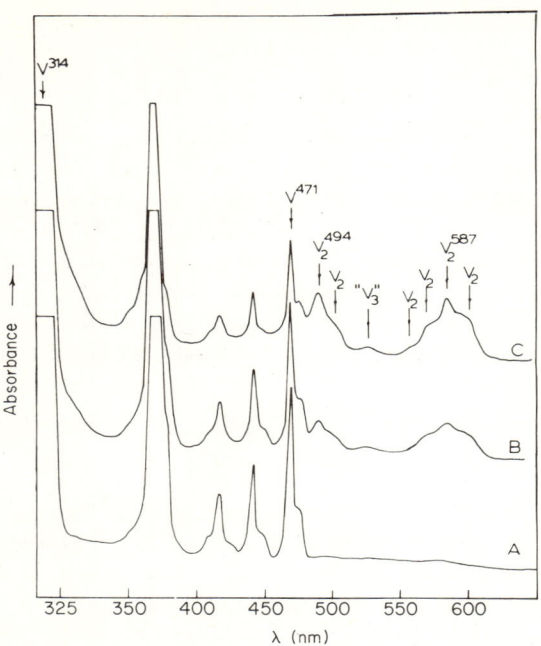

FIG. 1. The UV-visible spectrum of V in Ar matrices at 6–10 K; A, at low V concentrations, showing isolated V atoms, and B and C, at progressively higher V concentrations, showing both V atoms and the growth of V_2 molecules (108).

ysis, and identifying at least two absorptions as being associated with V_2. The observed, metal-concentration behavior implies that the proposal of dimer formation on the matrix surface, rather than in the gas phase, is a reasonable assumption; in addition, this study was instructive in demonstrating the efficiency of matrix gases of higher molecular weight in quenching this process, that is, fewer dimers formed in Xe and Kr than in Ar under comparable, deposition conditions. The extinction coefficient ratio $\epsilon_V/\epsilon_{V_2}$ was found to be roughly 3.0, in agreement with the results of other studies (91).

Vanadium atom depositions were further studied in alkane matrices (109) in an effort to observe the influence of other low-temperature, matrix environments on the optical spectra and clustering properties of metal atoms. Thus, vanadium atoms were deposited with a series of normal, branched, and cyclic alkanes over a wide range of temperature. The atomic spectra were somewhat broadened compared to those in argon, but the matrix-induced, frequency shifts from gas-phase values were smaller. As shown in Fig. 3, these shifts decrease with in-

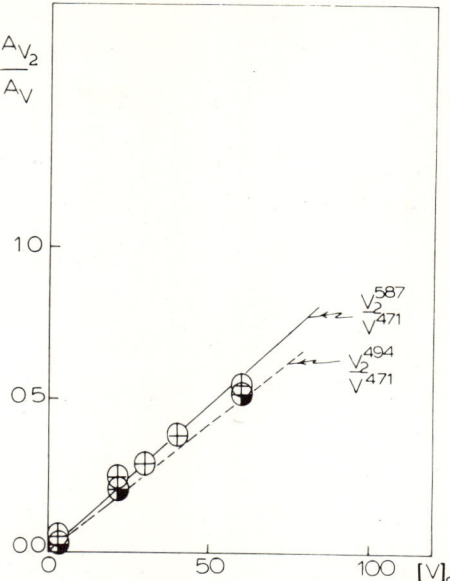

Fig. 2. A plot of the absorbances of typical lines attributed to V_2, relative to that of a V resonance absorption as a function of the V metal deposition rate at constant Ar deposition rate (108).

creasing chain-length for n-alkanes. In general, under similar experimental conditions, alkane matrices are more efficient than argon for the isolation of atoms over diatomic, and higher, nuclearity clusters, and the isolation efficiency increases with increasing chain-length of the n-alkane matrix-material up to decane. These data may prove useful in tailoring the production of mononuclear complexes for those metals for which dimerization is particularly facile, e.g., Co and Rh. In addition, the higher-temperature, diffusion properties observed for vanadium atoms in these supports may permit the design of specific, metal-atom syntheses which, although kinetically impeded in noble-gas matrices (4.2–40 K), may well proceed in higher-temperature, alkane supports (40–300 K). The potential application of such ideas is unbounded, limited only by the imagination of the metal-atom chemist.

B. Nb and Mo Studies

By use of quantitative, metal-atom, matrix-cocondensation techniques and the kinetic analysis previously discussed, the dimeric spe-

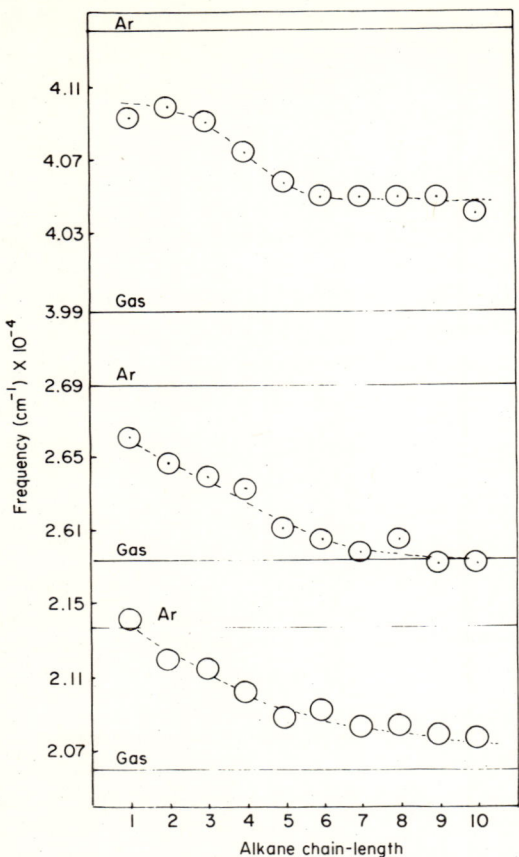

Fig. 3. Frequencies of three representative vanadium atom absorptions in different open-chain C_nH_{2n+2} matrices (where $n = 1-10$), compared to gas-phase and argon-isolated atomic vanadium (109).

cies Nb_2 and Mo_2 were generated (110) and their optical spectra (200–700 nm) established. The spectral assignments of Nb_2 and Mo_2 are summarized in Table II, where they are also compared with their first-row congeners V_2 and Cr_2. Interestingly, EHMO and $X\alpha$ calculations suggest that a closed-shell configuration for Nb_2 and Mo_2 (at the bond distances listed in Table II) would lead to formally high-order, "pentuple" and "sextuple" bonds, respectively.

C. Co, Rh, AND Ir Studies

The metal concentration, matrix, and temperature effects that favor clustering of the cobalt group of metal atoms have been assessed by

Ozin, Hanlan, and Power, using optical spectroscopy (49, 121). In view of the marked temperature-effect observed for the cobalt system, we shall focus on this cluster system here. Evidence for cobalt-atom aggregation at the few-atom extreme first came from a comparison of the optical data for Co:Ar $\simeq$ 1:10^4 mixtures recorded at 4.2 and 12 K (see Fig. 4). A differential of roughly 8 K in this cryogenic-temperature regime was sufficient to cause the dramatic appearance of an entirely new set of optical absorptions in the regions 320–340 and 270–280 nm (see Fig. 4). Matrix variation, from Ar, to Kr, to Xe, helped clarify atom–cluster, band-overlap problems (see Fig. 5).

The effect of deposition temperature on the clustering ability of Co atoms to form small, cobalt clusters was most revealing in terms of optical assignments, as well as activation-energy considerations. A series of runs with Co:Ar $\simeq$ 1:10^4 deposited at 4.2, 12, 20, 25, 30, and 35 K (see Fig. 4) nicely demonstrated the gradual progression from isolated Co atoms to mixtures of Co/Co$_2$ to mixtures of Co/Co$_2$/Co$_3$. The most

TABLE II

Observed and Calculated Optical Spectra for
V$_2$, Cr$_2$, Nb$_2$, and Mo$_2$ (110)

	r_e (pm)	D_e (eV)	Assignment	Spectra (nm) Obs.	Spectra (nm) Calc.[a]
V$_2$,	190	0.99	$1\sigma_g \to 2\sigma_u$	494	521
			$1\sigma_g \to 1\sigma_u$	558	558
			$1\sigma_g \to 2\pi_u$	[b]	350
			$2\sigma_g \to 2\pi_u$	588	607
			$1\pi_u \to 3\sigma_g$	[b]	350
Cr$_2$,	180	2.65[c]	$1\sigma_g \to 2\sigma_u$	456	494
			$1\sigma_g \to 2\pi_u$	340	309
			$2\sigma_g \to 2\pi_u$	[b]	389
			$1\pi_u \to 3\sigma_g$	260	261
Nb$_2$,	220	1.52	$1\sigma_g \to 2\sigma_u$	660	612
			$1\sigma_g \to 2\pi_u$	420	512
			$1\pi_u \to 3\sigma_g$	280	269
Mo$_2$,	210	3.45[c]	$1\sigma_g \to 2\sigma_u$	512	500
			$1\sigma_g \to 2\pi_u$	308	362
			$2\sigma_g \to 2\pi_u$	[d]	458
			$1\pi_u \to 3\sigma_g$	232	186

[a] Only electric-dipole- and spin-allowed electronic transitions in the proximity of the observed transitions are tabulated. [b] Anticipated overlap with atomic absorptions.
[c] Note that, because of the different approximations inherent in EH and Xα MO calculations, these assignments do not correlate exactly with those presented in Table VI.
[d] The predicted $2\sigma_g \to 2\pi_u$ transition at 458 nm and $1\delta_g \to 2\pi_u$ at 490 nm are either too weak to be observed, or accidentally coincident with the Mo$_2$ absorption at 512 nm.

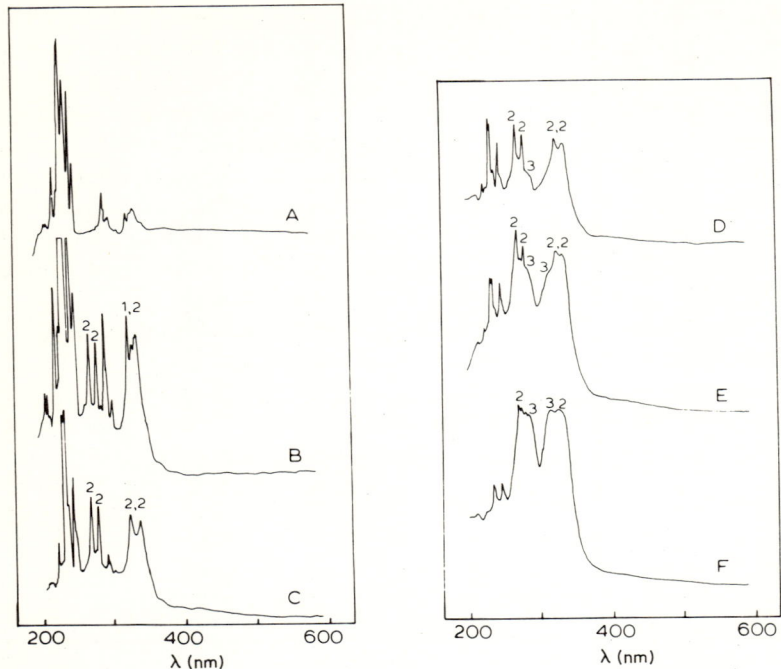

FIG. 4. The ultraviolet–visible absorption spectra of Co:Ar ≃ 1:10^4 mixtures deposited at (A) 4.2 K, (B) 10–12 K, (C) 20 K, (D) 25 K, (E) 30 K, and (F) 35 K, showing the progression from isolated Co atoms to Co/Co$_2$ and Co/Co$_2$/Co$_3$ mixtures (49, 154).

dramatic realization of the presence of the third Co species is seen in the 35 K deposition (see Fig. 4), where the strongest, Co atomic-resonance lines have intensities close to zero, while almost equal absorbances of Co$_2$ and Co$_3$ remain. A large number of experiments at various temperatures and cobalt-concentration matrices were performed, and these generally supported the contention of *three* distinguishable, small, cobalt-cluster species ascribed to Co/Co$_2$/Co$_3$. Preliminary, spin-unrestricted, SCF–Xα–SW calculations (148) of the optical spectrum for Co$_2$ were found to agree quite well with the observed spectra, as may be seen by reference to Table III.

Iridium, the heaviest element of the cobalt group, was found to display the least tendency towards cluster formation at 10–12 K (49), which, as already mentioned, was quite facile for Co and Rh (49). Considering the plethora of sharp, well-defined, atomic-resonance lines observed for Ir (see Fig. 6) compared to those of Co and Rh, the remarkably impressive correlation with the representation of the gas-

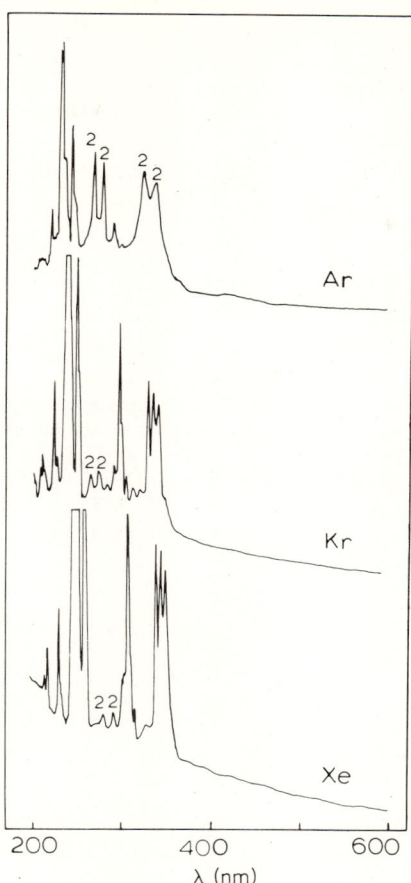

FIG. 5. Correlation of the UV–visible spectra of Co atoms and Co_2 molecules isolated in Ar, Kr, and Xe matrices under identical conditions of temperature and gas and metal deposition rates (49, 154).

phase, atomic-emission spectrum reported by Gruen et al. (48) and Moore (90) is interesting. Attempts to generate diiridium or higher iridium clusters in Ar, Kr, and Xe have thus far proved unsuccessful (49).

D. Ni, Pd, AND Pt STUDIES

The optical spectra of the nickel-triad metals have been reinvestigated (111) in Ar, Kr, and Xe matrices, and, although the data for Ni and Pt atoms correlated well both with previous studies and with the

TABLE III

OBSERVED AND CALCULATED (SCF-Xα-SW MO) OPTICAL
SPECTRUM OF DICOBALT,[a] Co_2 (148)

	Obs. (nm)			Calc. (nm)[b]	Assignment[c,d]
Ar	Kr	Xe	CH_4		
340[e]	321	—	345	353 } 355 }	$2\sigma_g^+ \to 2\pi_u$ $1\sigma_g^+ \to 2\pi_u$
320	312	328	325	324	$1\sigma_u^+ \to 3\sigma_g^+$
280[e]	284	290	283	305	$1\sigma_g^+ \to 2\sigma_u^+$
270	274	279	274	247 } 252 }	$1\sigma_g^+ \to 2\pi_u$ $1\pi_u \to 3\sigma_g^+$

[a] Cobalt–cobalt bond-length, 232 pm. [b] $2\sigma_g^+ \to 2\sigma_u^+$, calculated at 484 nm, which is in the region of extremely weak absorptions, possibly associated with Co_2. [c] Ground-state configuration $(1\sigma_g)^2(1\pi_u)^4(1\delta_g)^4(2\sigma_g)^2(1\sigma_u)^4(1\pi_g)^2$. [d] Preliminary values obtained from ground-state transition-energies; the detailed, spin-unrestricted, transition-state calculation will be reported later (148). [e] Narrow-band, continuous photoexcitation in these bands causes photobleaching of all of the dicobalt absorptions.

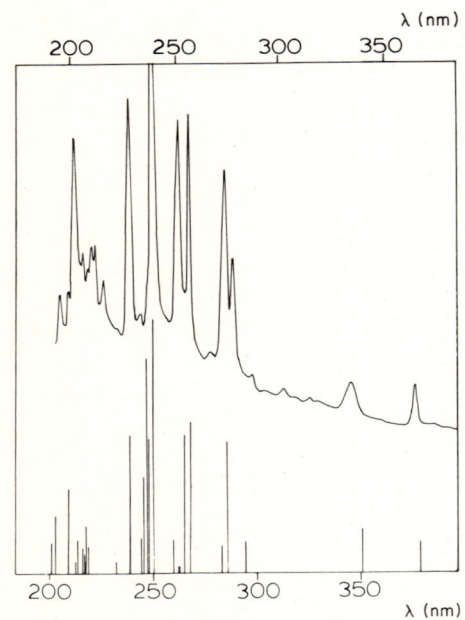

FIG. 6. Optical spectrum of Ir atoms isolated in solid Ar at 10–12 K, compared to the gas-phase atomic transitions of Ir. The stick heights correspond to reported oscillator strengths of gaseous Ir atoms (49).

gas-phase, atomic spectra, the results for Pd were inconsistent with earlier reports (82). It was found that the absorbances originally assigned to Pd atoms at wavelengths greater than 250 nm were actually associated with dinitrogen complexes of Pd (an atmospheric-impurity problem), and the authentic, optical spectrum of Pd atoms was found at higher energy ($\lambda < 250$ nm). Interestingly, shifts in the optical spectra of these metals as a function of inert-gas matrix suggested that Xe is not quite so "inert" as Ar or Kr. The spectral changes on passing from Ar, to Kr, to Xe, are much more drastic for Pd and Pt than for Ni, and it is possible that these metals, being somewhat more polarizable than Ni ("soft," in the Pearson sense), may be considered to be forming a "weak complex" with Xe. Such a weak, bonding interaction had been postulated by Turner and Perutz (192), who, on photodissociating $M(CO)_6$ to $M(CO)_5$, found (by UV–visible spectroscopy) that the matrix interacted with the metal atom at the vacant coordination-site. Thus, it is at times necessary to exercise caution in regard to the meaning of the word "inert."

The UV–visible spectra of Ni_2 and Ni_3 have also been identified in argon matrices (93); Ni_2 absorbed at 377, 529, and 410 nm, with vibronic structure on the first two bands, and with spacing of ~ 330 cm^{-1}, and Ni_3 absorbed at 420 and 480 nm, the latter band showing vibrational spacing of ~ 200 cm^{-1}. Higher-nuclearity clusters were observed, but not characterized. After prolonged warm-up of these matrices, nickel colloid was formed (93).

The metal-atom technique and its application in the formation of cluster species has been extended to the clustering of nickel atoms in organic media (66), in that codeposition of metal vapors with certain organic solvents allows the formation of very reactive, high-surface-area, metal slurries (67). The mechanism suggested for the formation of these nickel clusters, summarized in Scheme 1, is made up of a number of discrete stages, beginning with weak complex-formation. The nature of the clusters formed depends on the solvent used and their surface areas range from 45–400 m^2/g (compared to 80–100 m^2/g for Raney Ni) (66). A more detailed study of the reaction of nickel atoms with pentane (68) showed that, during warm-up, the nickel clusters react with the organic compound to yield a "pseudo-nickel organometallic" that has Ni:C:H ratios of 2–5:1:2, is thermally stable to $>200°C$, and is an extremely active, hydrogenation catalyst. It was suggested (68) that the reaction occurs at $<-100°C$, probably by oxidative addition to C–H and C–C bonds, to afford the "pseudo complexes." Catalytic studies (69) showed that a Ni–toluene product is also a hydrogenation catalyst, and that Ni–THF is fairly unreactive,

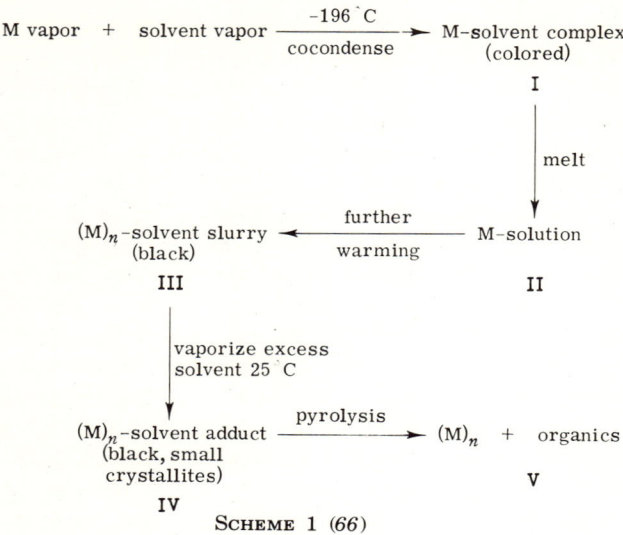

SCHEME 1 (66)

either as a hydrogenation or an olefin-isomerization catalyst. It is informative to contrast Klabunde and Davis's studies (69) with those of Rieke et al. (159), who produced similar reactive slurries by the reduction of a metal salt in an ethereal or hydrocarbon solvent with an alkali metal. These nickel-triad slurries were found to insert readily into R–X bonds, a reaction known in metal-atom chemistry for many years (168).

E. Cu, Ag, and Au Studies

Copper clusters containing two to four atoms have been formed (94) in argon and methane, whereas large, colloidal-copper particles resulted in dodecane matrices (94). The authors suggested that the "birth" of the band structure of copper is clearly visible on passing from the dimer to the tetramer, with Cu_4 already possessing many of the features of the bulk metal (94).

Considerable effort has been expended on Ag atoms and small, silver clusters. Bates and Gruen (10) studied the spectra of sputtered silver atoms (a metal target was bombarded with a beam of 2-keV, argon ions produced with a sputter ion-gun) isolated in D_2, Ne, and N_2. They found that an inverse relationship between Z_{eff} of the metal atom and the polarizability of rare-gas matrices (as determined from examination of spin–orbit coupling and crystal-field parameters) could be extended to include silver atoms in D_2 and N_2 matrices, thereby allowing a detailed evaluation of atom–matrix interactions. The correlation breaks down for Ne, presumably because the size of a substitutional site and the polarizability of Ne are such as to cause multiple-site occupation. They also reported (10) data for Ag_2, which has now been studied in detail in the gas phase (18) [although some of the Ag_2 assignments were confused with unsuspected Ag_3, also present in the matrix (112)].

Ozin and Huber (112) synthesized and characterized very small silver particles, Ag_n ($n = 2–5$) by conventional deposition methods, as well as by a novel technique that they have termed "cryophotoaggregation." This study will be discussed in detail in Section III. Of interest here is a study of silver atoms and small, silver clusters entrapped in ice and high-molecular-weight paraffin ($n\text{-}C_{22}H_{46}$, $n\text{-}C_{32}H_{66}$) matrices (146) (see Figs. 7 and 8, and Tables IV and V). Besides the intriguing, multiple-site (solvation) occupancy of atomic silver in ice matrices, and their thermal and photochemical interconvertibility, their extremely

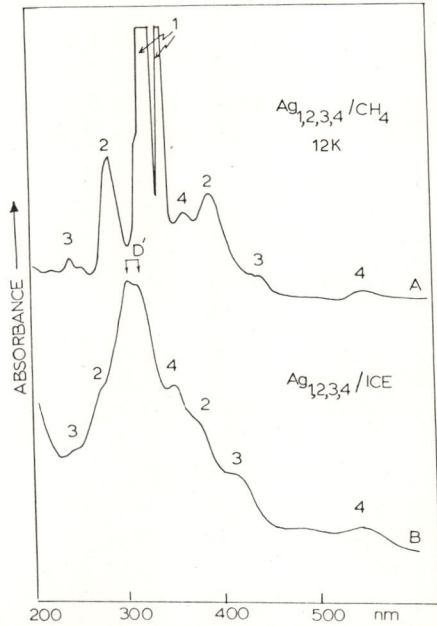

FIG. 7. The optical spectra obtained on depositing (A) silver vapor with methane at 10–12 K at $Ag:CH_4 \simeq 1:10^2$, and (B) silver vapor with water vapor at 10–12 K at $Ag:H_2O \simeq 1:10^2$, brief warming to 77 K, and recooling to 10–12 K for spectral recording (146).

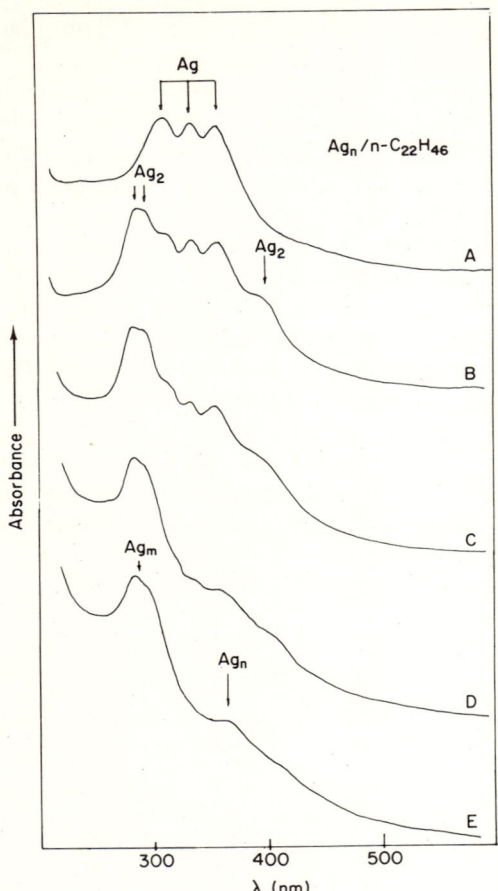

FIG. 8. The optical spectra obtained on depositing silver vapor with n-$C_{22}H_{46}$ vapor at 10–12 K, and (A) Ag: n-$C_{22}H_{46} \simeq 1:10^4$, (B) Ag: n-$C_{22}H_{46} \simeq 1:10^3$, and (C)–(E) showing the progress of annealing (B) at 30, 40, and 80 K, respectively (146).

high, bulk diffusion–aggregation behavior (0.5 T_{mp}) is a most noteworthy feature in terms of potential, supported-cluster, catalyst fabrication (146). In contrast, the glassy, open structures of the silver atom–wax-quenched, condensed films, having considerable chain-mobility built in, even at cryogenic temperatures, resulted in a matrix having poor trapping ability for atoms on deposition, and surprisingly low, bulk diffusion–aggregation temperatures (0.1 T_{mp}) during matrix annealing (146). Aside from inhomogeneous band-broadening effects of the silver, atomic and cluster, optical absorptions in ice and wax matrices, it is

TABLE IV

The Optical Spectra of $Ag_{1,2,3,4}$ in Ice and Methane Matrices at 10–12K (146)

Ice matrices (nm)	Methane matrices (nm)	Ag_n cluster assignments in ice
542	528	Ag_4
412	428–418	Ag_3
375–385	380	Ag_2 + Ag (site A')
353	360	Ag_4 + Ag (site B')
333	333	Ag (site C')
316	320	Ag (site D') + Ag (site C')
301	310	Ag (site D')
270	278	Ag_2
245	252–238	Ag_3

significant to note the relatively minor frequency-perturbations on passing from an environment that could be regarded as a nonpolar paraffin to a highly dipolar, ice lattice (146). While discussing the synthesis of small silver particles, brief mention should be made of the isolation of Ag_6 molecules (165), formed within cubes of eight Ag^+ ions in fully Ag^+-exchanged zeolite A that had been dehydrated under vacuum at elevated temperatures. A crystal structure is shown in Fig. 9.

Techniques other than UV–visible spectroscopy have been used in matrix-isolation studies of Ag; see, for example, some early ESR studies by Kasai and McLeod (56). The fluorescence spectra of Ag atoms isolated in noble-gas matrices have been recorded (76, 147), and found to show large Stokes shifts when optically excited via a $^2S_{1/2} \rightarrow {}^2P_{1/2,3/2}$ atomic transition which is threefold split in the matrix by spin–orbit and vibronic interactions. The large Stokes shifts may be explained in terms of an excited state silver atom–matrix cage complex; in this

TABLE V

Optical Spectra of Ag Atoms and Ag_2 Dimers in Methane and High-Molecular-Weight Paraffin Wax (n-$C_{22}H_{46}$, n-$C_{32}H_{66}$) Matrices (146) at 10–12 K

CH_4	n-$C_{22}H_{46}$	n-$C_{32}H_{66}$	Assignment
380	390	390	Ag_2
333	357	360	Ag
320	334	338	Ag
310	312	315	Ag
278	298	300	Ag_2
	284	284	Ag_2

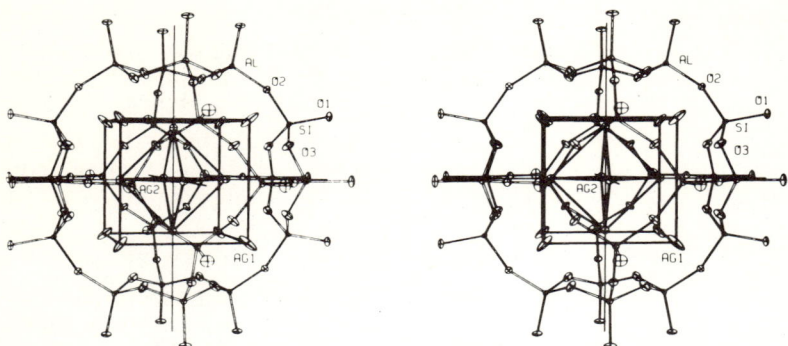

Fig. 9. A stereoview of a sodalite unit containing an octahedral Ag_6 molecule within a cube of eight Ag^+ ions is shown, using ellipsoids of 20% probability. The ions at Ag(3) occupy 8-ring sites and are not shown (*165*).

way, the ensuing photoaggregation observed by Ozin and Huber (*112*) may be rationalized (see later). Raman data have also been gathered for Ag_2 and Ag_3 (*163*) in solid Kr, a number of bands being observed below 220 cm^{-1}. The two strongest, at 194 and 120 cm^{-1}, were assigned to Ag_2 and Ag_3, respectively, and the authors speculated that the observation of only one band for Ag_3 suggests that the trimer is linear, in agreement with previous MO calculations (*5*). However, laser-Raman investigations of this system unveiled unsuspected, laser photoaggregation and resonance Raman-fluorescence complications that have necessitated a re-evaluation of the silver-cluster, vibrational assignments (*164*).

Theoretical analyses (*75–77*) of the matrix-induced changes in the optical spectra of isolated, noble-metal atoms have also been made. The spectra were studied in Ar, Kr, and Xe, and showed a pronounced, reversible-energy shift of the peaks with temperature. The authors discussed the matrix influence in terms of level shift-differences, as well as spin–orbit coupling and crystal-field effects. They concluded that an increase in the matrix temperature enhances the electronic perturbation of the entrapped atom, in contrast to earlier prejudices that the temperature dilation of the surrounding cage moves the properties of the atomic guest towards those of the free atom.

F. Bimetallic Clusters

In 1977, a number of groups independently demonstrated that matrix-isolation methods could be used for generating, isolating, and spectroscopically probing small, bimetallic clusters. For example, Zmbova et al. (203) formed AuLi via the simultaneous vaporization of Au and Li from separate Knudsen cells. They monitored the product by IR spectroscopy in various matrix supports (see Fig. 10). The vibrational frequencies for AuLi were found to be strongly dependent on the kind of matrix gas used and are split into two components for both istopomers, Au^7Li (A) and Au^6Li (B), in all cases (multiple trapping site effect). Using an empirical relationship, they estimated the gas phase frequencies ν_{Au-^6Li} at 746 ± 10 cm^{-1}, and ν_{Au-^7Li} at 705 ± 10 cm^{-1}, with k_{AuLi} = 1.90 ± 0.03 mdyn/100 pm.

Simultaneous with this work, Ozin and co-workers were independently investigating other bimetallic combinations. When Cr and Mo were cocondensed (133) together in Ar, using the apparatus shown in Fig. 11, a controlled pathway to CrMo was found. This molecule had previously been observed in the gas phase (30) from flash photolysis of a mixture of $Cr(CO)_6$ and $Mo(CO)_6$ vapors. The molecule was identified (UV–visible spectroscopy) by a series of Cr/Mo/Ar concen-

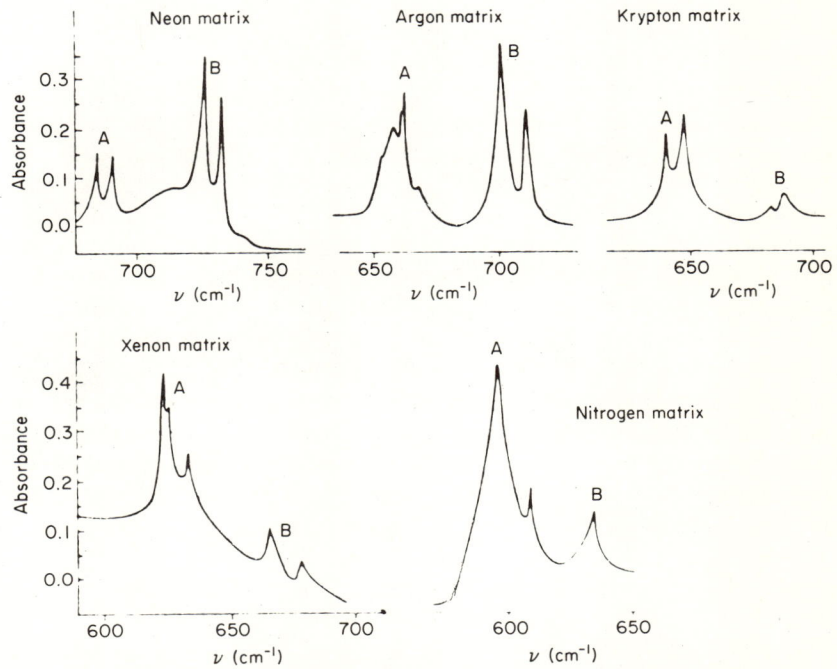

FIG. 10. IR absorption spectra of the species AuLi in various matrices (203).

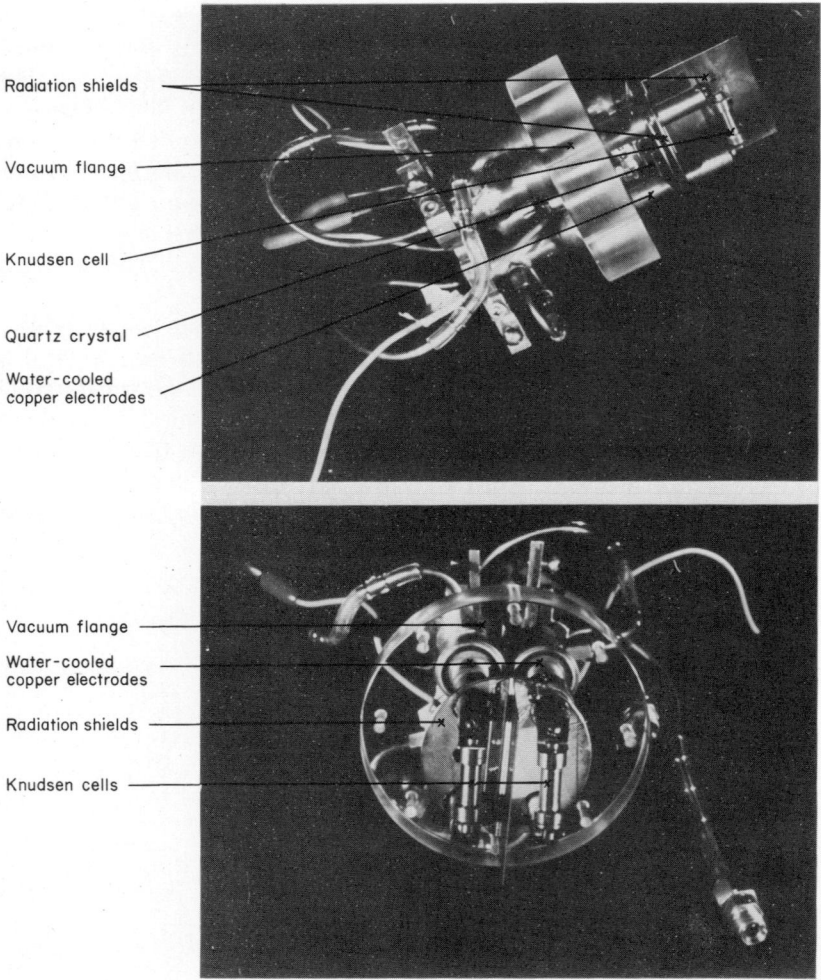

Fig. 11. Photograph of the four-electrode, vacuum flange and dual, quartz crystal, microbalance assembly, (A) side view, and (B) front view, used for mixed Cr atom, Mo atom matrix depositions with simultaneous monitoring of the individual metal flows. (The resolution of the microbalance is $\sim 10^{-8}$ g) (113).

tration experiments; the spectra of Cr_2, Mo_2, and CrMo are summarized in Table VI, along with assignments based on SCF–Xα–SW calculations (113). Interestingly, there is a close correspondence between the energies of the observed transitions of CrMo and those calculated under the assumption that

$$\nu_{CrMo} = 0.5(\nu_{Cr_2} + \nu_{Mo_2}). \quad (4)$$

Such an averaging effect of the transition energies of CrMo relative to Cr_2 and Mo_2 is intuitively understandable, as the electronic ground-states of Cr and Mo atoms are both $ns^1(n-1)d^5$ and those of Cr_2 and Mo_2 are both considered to be $1\sigma_g{}^2 \, 1\pi_u{}^4 \, 2\sigma_g{}^2 \, 1\delta_g{}^4$. Furthermore, the Cr 4s, 3d and the Mo 5s, 4d atomic orbitals, considered to be the main contributors to the metal–metal bonding in $Cr_2/CrMo/Mo_2$ are known to have similar energies. Further discussion of these bimetallics formed by cryophotoclustering methods will be found in Section III.

Montano (89) developed matrix Mössbauer spectroscopy to the point of being able to identify FeMn, FeCo, FeNi, and FeCu bimetallics. In these combinations, an increase in the electron density at the ^{57}Fe nu-

TABLE VI

Electronic Transitions of Cr_2, CrMo, and Mo_2, as Calculated by the SCF–Xα–SW Method[a,b] (113)

Transition	Molecule	Calc.[c]	Exptl.
$1\delta_g \to 1\delta_u$	Cr_2	12.7	
	Mo_2	14.4	
$2\sigma_g \to 1\sigma_u$	Cr_2	24.9	21.7
$2\sigma \to 3\sigma$	CrMo	22.0	20.6
$2\sigma_g \to 1\sigma_u$	Mo_2	21.8	19.4
$2\sigma_g \to 2\sigma_u$	Cr_2	28.8	29.4
$2\sigma \to 4\sigma$	CrMo	28.0	31.1?
$2\sigma_g \to 2\sigma_u$	Mo_2	30.1	32.5
$1\pi_u \to 1\pi_g$	Cr_2	32.0	[d]
$1\pi \to 2\pi$	CrMo	31.9	[d]
$1\pi_u \to 1\pi_g$	Mo_2	40.1	[d]
$1\sigma_g \to 1\sigma_u$	Cr_2	36.5	38.0
$1\sigma \to 3\sigma$	CrMo	38.1	40.3
$1\sigma_g \to 1\sigma_u$	Mo_2	45.4	43.1
$1\sigma_g \to 2\pi_u$	Cr_2	40.2	[e]
	Mo_2	47.1	[e]

[a] Band positions are given in $cm^{-1} \times 10^3$. The calculated values were obtained by using the relation 1 eV = 8.06548×10^3 cm^{-1}. [b] See footnote c in Table II. [c] Explicit predictions of the singlet (spin-allowed) transitions, except for the $1\delta_g \to 1\delta_u$ and $1\pi_u \to 1\pi_g/1\pi \to 2\pi$ excitations, where the predictions are of the average transition-energy to the four singlet-states arising from the excited, orbital configuration. [d] Either too weak to be observed or obscured by free atom absorptions. [e] Could possibly be associated with the weak bands observed at 268 nm for Cr_2 and 226 nm for Mo_2, respectively. If this is, in fact, true, the weak band at ~255 nm for CrMo could be the corresponding $1\sigma \to 3\pi$ transition.

cleus was observed on going from Cu to Ni, with the FeNi molecule having the weakest bond. In FeNi, the quadrupole splitting of the Mössbauer spectrum is half the value of Fe_2, indicating a different, orbital ground-state for FeNi. By contrast, FeCo is more strongly bound, this being reflected in a more positive, isomer shift. A typical spectrum is shown in Fig. 12.

Kasai and McLeod (57, 58) also studied a series of bimetallic diatomics, AgM (M = Mg, Ca, Sr, Be, Zn, Cd, or Hg), by ESR spectroscopy. For all of these species, the hyperfine coupling to the Ag nucleus was found to be isotropic. It was shown that the unpaired electron resides in an orbital resulting essentially from an anti-bonding combination of the valence s orbitals of the Ag and M atoms. A typical spectrum is shown in Fig. 13.

Some macroscale, bimetal-vapor, cryochemical reactions involving group IA/IB–ammonia cocondensations have pointed the way to solvated, transition-metal anions of the type M^-. Thus, in sharp contrast to the Li/Au/Ar, 10–12 K, matrix reactions of Zmbova et al. (203) that led to the IR detection of molecular ^{6,7}LiAu, Lagowski and Peer (79b) discovered that simultaneous $M/Au/NH_3$ cocondensations at 77 K (M = Li, K, Rb, or Cs) resulted in solutions at −65°C that display gradual loss of the 1850-nm, solvated-electron absorption, with concomitant growth of an intense, UV absorption in the range of 277 to 289 nm, having ϵ in the range of 1.8 to 7×10^4 L mol^{-1} cm^{-1}. The

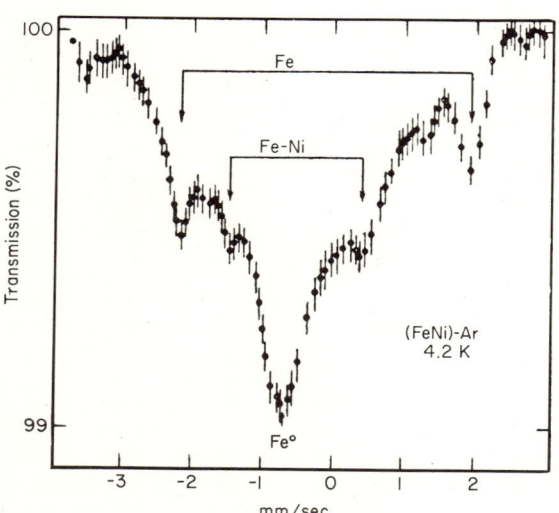

FIG. 12. Mössbauer spectrum of an argon matrix containing iron and nickel (89).

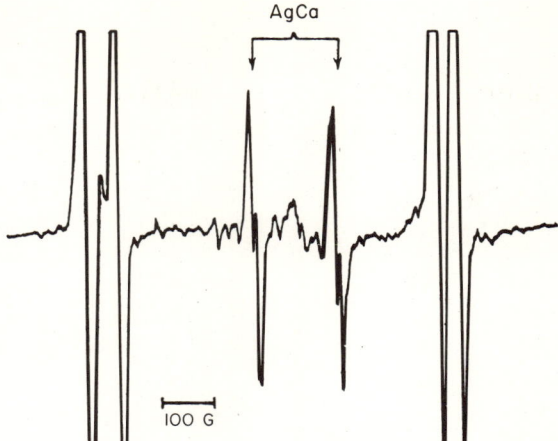

FIG. 13. ESR spectrum of AgCa generated in an argon matrix (58).

chemical behavior of these gold solutions, and the large extinction- and temperature-coefficients of the gold-related band, indicated the presence of a solvated, gold anion Au_y^{x-}. In view of the results of a Zintl type of potentiometric titration that favored an Au_x^{x-} stoichiometry, together with small, alkali-cation effects, and UV-absorption extinction coefficients approximately the same as that of the solvated electron, Lagowski and Peer (79b) promoted the idea of the existence of a $5d^{10}6s^2$ Au^- solvated, gold anion.

III. New Cluster Techniques

Cryophotochemical techniques have been developed that (i) allow a controlled synthetic approach to mini-metal clusters (112), (ii) have the potential for "tailor-making" small, bimetallic clusters (mini-alloy surfaces) (114, 116), (iii) permit the determination of relative extinction-coefficients for naked-metal clusters (149), and (iv) allow naked-cluster, cryophotochemical experiments to be conducted in the range of just a few atoms or so (112, 150, 151).

A. Cryophotoclustering

The cryophotoaggregation phenomenon was first observed for Ag atoms (112) entrapped in Ar at 10–12 K (see Fig. 14). The trick essentially involves narrow-band, continuous irradiation into the

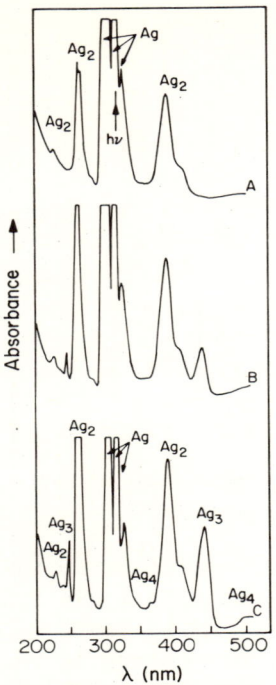

FIG. 14. UV–visible spectra of Ag/Ag$_2$/Ar generated from Ag:Ar ≃ 1:10^3 depositions: (A) at 10–12 K; (B) after 15-min, 10–12 K photolysis at 315 nm; (C) after an additional 45-min, 10–12 K photolysis at 315 nm, showing the decay of atomic Ag and the concomitant growth of Ag$_2$ and Ag$_3$ (*112*).

atomic-resonance absorptions of matrix-entrapped, metal atoms. In the silver system, photoclustering was initiated by excitation of any one of the components of the $^2S_{1/2} \rightarrow {}^2P_{1/2,3/2}$ atomic resonance. It may be envisaged that, following atomic excitation, some nonradiative, electronic-to-lattice, phonon energy transfer, local warming and softening of the surrounding matrix-cage, atom photomobilization (bulk diffusion), and subsequent photoaggregation to diatomic and higher metal clusters may occur. Cluster sizes were found to be dependent on the metal concentration, the matrix material, the matrix temperature, the excitation wavelength, and the light-intensity. Controlled photoaggregation up to Ag$_5$ has been observed in this way (*112*), and the process appears to be amenable to analysis by conventional, solid-state, diffusion theory (*149*). Cluster size-determination appears to be tractable by these methods (see later) (*112, 149*).

B. Selective Naked-Cluster Cryophotochemistry

The copper system appears to behave similarly to the silver system, and it may be used here in order to illustrate the idea of "selective, naked-cluster cryophotochemistry" (150, 151). A typical series of optical-spectral traces that illustrate these effects for Cu atoms is given in Fig. 15, which shows the absorptions of isolated Cu atoms in the presence of small proportions of Cu_2, and traces of Cu_3 molecules. Under these concentration conditions, the outcome of 300-nm, narrow-band photoexcitation of atomic Cu is photoaggregation up to the Cu_3 stage. The growth–decay behavior of the various cluster-absorptions allows unequivocal pinpointing of UV–visible, electronic transitions associated with Cu_2 and Cu_3 (150). With the distribution of $Cu_{1,2,3}$ shown in Fig. 15, 370-nm, narrow-band excitation of Cu_2 can be considered. Immediately apparent from these optical spectra is the growth ($\sim$10%) of the Cu atomic-resonance lines. Noticeable also is the concomitant

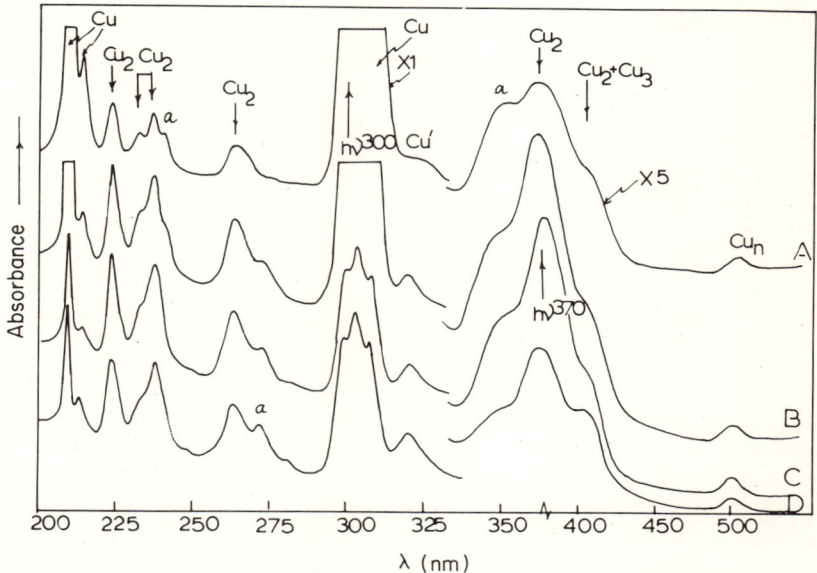

Fig. 15. The optical spectrum of Cu:Ar $\simeq$ 1:10^4 at 10–12 K, (A) showing isolated Cu atoms and Cu_2 molecules; (B), (C) photoaggregation as the result of two 30-min irradiations in the resonance lines of Cu atoms at 302 nm, (D) photodissociation of Cu_2 resulting from a 30-min irradiation at the 370-nm band of Cu_2. The features marked "a" are thought to arise from secondary trapping sites of Cu_2. Note the scale change between 325 and 400 nm (150).

growth of Cu_3, and yet *decay* of Cu_2, under these matrix-concentration conditions (*150*). Similar effects have been observed for Ag_2 (*150*). The Cu_2 and Ag_2 visible cryophotochemistry observed may be rationalized in terms of a highly selective, matrix-induced, photodissociation step for which a mechanism involving formation of an excited Cu (Ag) atom–matrix cage complex has been proposed. SCF–Xα–SW MO calculations for Cu_2 (and Ag_2) (*150*) confirmed that the wavelengths used for the photoexcitations are essentially those of the electronic transitions from the main bonding to the main antibonding orbital (i.e., s$\sigma_g \rightarrow$ sσ_u), thus providing a description of the excited states that is useful in accounting for the observed photochemistry (*112, 150*) (see Table VII).

Selective, trisilver, cryophotochemical transformations have also been observed that involve HOMO–LUMO visible excitation (420–440 nm, depending on the matrix support) (*151*). A typical series of optical traces that depict the outcome of Ag_3/Kr, 423-nm excitation at 10–12 K is illustrated in Fig. 16. These data show that Ag_3 absorptions at 423/247 nm may be selectively photoannihilated simultaneously with the

TABLE VII

CALCULATED AND EXPERIMENTAL ELECTRONIC SPECTRA OF Cu_2 AND Ag_2 (cm^{-1} × 10^3) (*150*)

		Experimental			
Transition	Calculated[a]	Gas[b]	Ar	Kr	Xe
		Cu_2			
$1\pi_g \rightarrow 2\sigma_u$	24.1 (0.02)	20.4	25.0	25.0	25.0
$2\sigma_g \rightarrow 2\sigma_u$	26.5 (0.32)	21.7	27.0	27.0	27.8
$1\pi_g \rightarrow 2\pi_u$	35.2 (0.15)		38.2	37.0	35.1
$2\sigma_g \rightarrow 2\pi_u$	37.2 (1.02)		41.7/42.4	41.5/42.0	40.5/41.0
$1\delta_g \rightarrow 2\pi_u$	39.5 (0.13)		43.1	42.7	41.7
$1\sigma_g \rightarrow 2\sigma_u$	39.8 (0.36)		44.8	44.0	43.1
$1\sigma_u \rightarrow 3\sigma_g$	43.7 (0.03)				
		Ag_2			
$2\sigma_g \rightarrow 2\sigma_u$	22.0 (0.64); 25.7 (0.63)	23.0	24.3/25.8	25.6	25.6
$2\sigma_g \rightarrow 2\pi_u$	33.2 (1.33); 33.6 (1.37)	37.6	37.8/38.3	35.5/37.0	34.5/35.3
$1\pi_g \rightarrow 2\sigma_u$	43.5 (0.03); 45.6 (0.04)	40.2	44.1	45.0	46.1

[a] All spin- and dipole-allowed transitions below 48 and 51 cm^{-1} × 10^3 for Cu_2 and Ag_2, respectively. For Ag_2, the first value is for 284 pm and the second for 247 pm. Oscillator strengths are given in parentheses. [b] 0 → 0 transitions, from "Spectroscopic Data Relating to Diatomic Molecules," Pergamon Press, New York, 1970, for Cu_2, and from *J. Mol. Spectrosc.* **69**, 25 (1978) for Ag_2. The weak B ← X and D ← X bands of Ag_2 at 35.8 and 39.0 cm^{-1} × 10^3, believed to be due to forbidden transitions, are omitted.

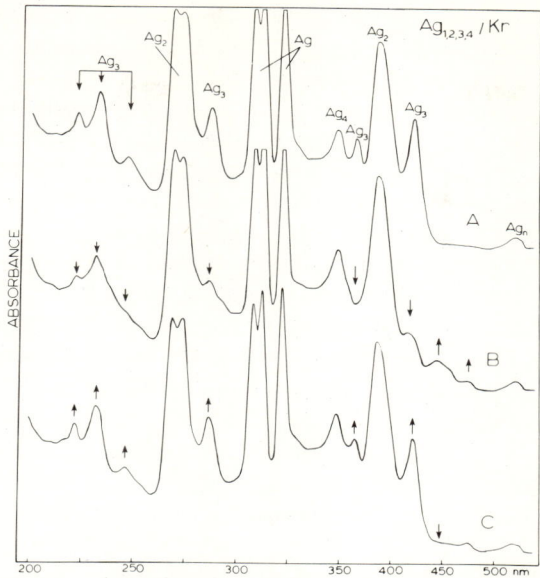

FIG. 16. The UV–visible spectra of $Ag_{1,2,3,4}$/Kr mixtures (Ag/Kr ≃ $1/10^2$) at 10–12K: (A) After a 30-min irradiation centered at the atomic resonance absorption lines. (B) The outcome of a 10-min, 423-nm Ag_3 irradiation, showing major decay of the bands associated with Ag_3 (indicated by arrows) and the appearance of two new bands near 450 nm. (C) The result of a 5-min, 25K bulk thermal annealing period, showing regeneration of the original Ag_3 spectrum and loss of the new band near 445 nm (*151*).

growth of a new band-system around 450 nm. A number of aspects of this selective, photobleaching phenomenon are significant. Firstly, certain absorptions of the original, Ag_3, optical spectrum in Kr (233 and 223 nm) displayed only slight changes on 423-nm, Ag_3 photoexcitation. Secondly, the absorptions associated with other silver species, $Ag_{1,2,4}$, co-trapped with Ag_3 remained essentially unchanged with respect to Ag_3 photoexcitation. Thirdly, higher silver-cluster absorptions belonging to $Ag_{5,6}$ remained spectroscopically invisible. Fourthly, in the densely packed, cluster-overlap region (330–370 nm), where at least one Ag_3 absorption was suspected, concomitant decay of the 368-nm absorption to disappearance, along with the photobleaching of the 423/247 nm Ag_3 absorptions just listed, were observed. Finally, it was discovered that this highly discriminative photobleaching of *part* of the Ag_3 optical spectrum could be approximately reversed by 25K thermal annealing, while leaving the other Ag_n spectral lines largely unaffected (see Fig. 16).

In brief, one possible rationale for the 423-nm, Ag_3/Kr photochemistry may be formulated in terms of a photoisomerization process whereby a specific, Ag_3, geometrical isomer (for example, linear) is selectively excited to an electronic state that allows conversion into, and trapping of, a metastable, geometric isomer (for example, nonlinear) of Ag_3, all localized within the matrix cage. It would appear at this stage that an alternative explanation, in terms of the formation of $Ag_2 + Ag$ (just outside the matrix cage) and re-formation of Ag_3 on annealing, is unlikely, in view of the prior knowledge that the $\epsilon_2:\epsilon_1$ and $\epsilon_3:\epsilon_1$ extinction-coefficient ratios are very close to unity in Kr matrices (*149*). Therefore, a primary, $Ag_3 \rightarrow Ag_2 + Ag$, photodissociative step should have been detected through substantial variations in the intensity of the Ag and Ag_2 bands, but this was not found in practice (see ref. *151* for further details).

C. Relative Extinction-Coefficient Measurements for Naked Silver Clusters by Photoaggregation Techniques

Low-temperature, photoaggregation techniques employing ultraviolet–visible absorption spectroscopy have also been used to evaluate extinction coefficients relative to silver atoms for diatomic and triatomic silver in Ar and Kr matrices at 10–12 K (*149*). Such data are of fundamental importance in quantitative studies of the chemistry and photochemistry of metal-atom clusters and in the analysis of metal-atom recombination-kinetics. In essence, simple, mass-balance considerations in a photoaggregation experiment lead to the following expression, which relates the decrease in an atomic absorption to increases in diatomic and triatomic absorptions in terms of the appropriate extinction coefficients.

$$(A_1' - A_1'') = 2\, \epsilon_1/\epsilon_2(A_2'' - A_2') + 3\, \epsilon_1/\epsilon_3(A_3'' - A_3') \tag{5}$$

It is prearranged in this analysis that M_4 and higher clusters are not produced in significant proportions. The symbol A_n' represents the absorbance due to M_n at time t', A_n'' is the absorbance due to M_n at time t'', and ϵ_n represents the molar extinction coefficient for M_n. Dilute conditions and short irradiation times yield ϵ_1/ϵ_2 directly, whereas longer irradiation times allow solving for ϵ_1/ϵ_3. However, the advantage of the photoaggregation method is that only one deposition is needed for each ϵ_1/ϵ_2 or ϵ_1/ϵ_3 determination, so that the method is much more convenient and considerably more accurate, as mass balance of the total metal is maintained after each irradiation, thus eliminating the need for multiple, quantitative depositions.

The relative extinction-coefficients for $Ag_{1,2,3}$ determined by photoaggregation procedures were found not to be strongly matrix-dependent (see Table VIII). Moreover, the results for Ag_2 were in good agreement with those obtained by quantitative, metal-atom deposition-techniques.

D. Photonucleation Kinetic Studies

Kinetic studies of the cryophotoclustering process are now in progress. Preliminary results indicate that, under certain conditions, the rates of formation of diatomic and triatomic silver may usefully be approximated by simple, second-order kinetics (149). A simple analysis predicts that the slope of a $\log[Ag_n]/[Ag]$ versus $\log(t)$ plot, where Ag_n and Ag represent absorbances, and t represents the irradiation time, should have a value close to 1.0 for $n = 2$, and 2.0 for $n = 3$ (149). A typical plot is shown in Fig. 17. The observed slopes, 0.9/1.0 and 2.1/2.2, support the Ag_2 and Ag_3 assignments for the run indicated in Fig. 18, and correlate exactly with earlier assignments based on Ag-atom concentration experiments.

E. Photomanipulation of Cluster Distributions

It has also been demonstrated that cryophotoaggregation experiments involving matrix-entrapped silver atoms can possibly be tailored to the point of generating almost pure disilver clusters, as well as cluster distributions that are inaccessible by conventional, deposition and bulk-annealing procedures (152). By carefully selecting the silver concentration and matrix support, it may be arranged that substantial conversion of Ag atoms into Ag_2 occurs, with only slight conversion losses to Ag_3 or higher silver clusters. Figure 19 demonstrates this remarkable, atom–diatom photoredistribution-reaction, where a Kr matrix

TABLE VIII

Relative Extinction Coefficients[a,b] for Ag_2 and Ag_3

	Peak height	Peak area
$\epsilon_1^{315}/\epsilon_2^{260}$ (Ar)	0.8 ±0.2	0.40 ±0.05
$\epsilon_1^{315}/\epsilon_3^{245}$ (Ar)	1.2 ±0.5	0.60 ±0.30
$\epsilon_1^{323}/\epsilon_2^{270}$ (Kr)		0.43 ±0.05

[a] The corresponding wavelengths (in nm) in Ar and Kr matrices are indicated as superscripts. The uncertainty limits represent estimated upper and lower bounds.
[b] Determined by photoaggregation procedures (149).

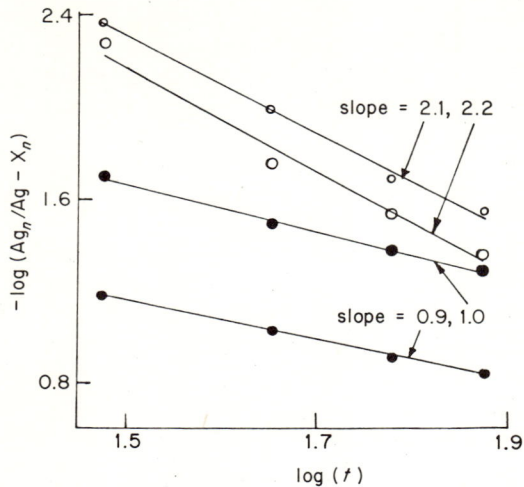

FIG. 17. Kinetic plots showing a linear dependence on irradiation time (305 nm) of the absorbance ratios $Ag_2^{263nm}:Ag^{300nm}$ and $Ag_2^{390nm}:Ag^{300nm}$ (○) and a linear dependence on the square of the irradiation time of the absorbance ratios $Ag_3^{245nm}:Ag^{300nm}$ and $Ag_3^{440nm}:Ag^{300nm}$ (●), as predicted from the simple kinetic analysis. The quantities X_n were chosen in order to shift the $Ag_2:Ag$ vs. t and $Ag_3:Ag$ vs. t^2 plots through the origin (149).

containing dominant amounts of Ag_2 has been produced. The potential of the method for generating very narrow distributions of Ag_2/Ag_3 in the absence of Ag_1 has been realized in cyclooctane matrices (see Fig. 20), and of $Ag_2/Ag_3/Ag_4$, with very little Ag_1, in methane matrices (see Fig. 21). The ramifications of these kinds of experiments in terms of the spectroscopy of individual clusters and the possible fabrication of highly potent and selective diatom and diatom–triatom cluster catalysts are clearly considerable.

F. PHOTOSELECTIVE BIMETALLIC AGGREGATION

We have already shown how simultaneous codeposition of two metals in inert-gas matrices can lead to the formation of mixed-metal dimers. As in the case of silver, it was found that irradiation into the atomic absorptions of Cr or Mo results in formation of their respective dimers and trimers (114). In addition to this, however, irradiation into the atomic resonances of the two metals in the presence of each other results (114) in formation of the mixed-metal species CrMo, Cr_2Mo, and $CrMo_2$. It would seem that selective irradiation into the 300–400-nm bands of atomic Cr or Mo excites the $3d^54p^1, 3d^44s^14p^1$, or $4d^55p^1$,

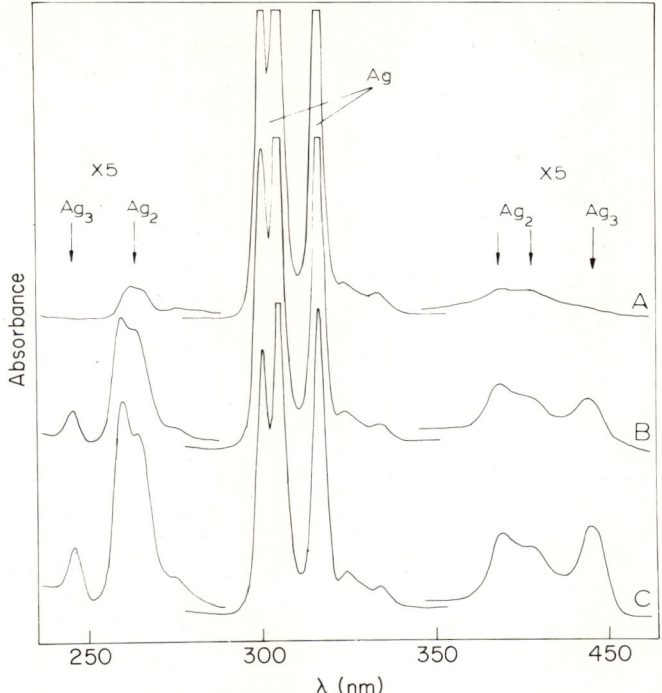

FIG. 18. UV–visible spectra of $Ag_{1,2,3}$/Ar mixtures (Ag:Ar $\simeq$ 1:10³) at 10–12K. Note the growth of Ag_2 and Ag_3 clusters and loss of Ag atoms as a result of 305 nm, Ag atom excitation. Spectra A, B, and C represent irradiation times of 0, 1, and 4 min, respectively (*149*).

$4d^45s^15p^1$ states, respectively, which subsequently decay and transfer energy to the surrounding, matrix cage. It seems likely, therefore, that part, or all, of the electronic energy of these states is channeled into lattice vibrational-energy and translational energy of the caged atoms, the results being photoinduced, bulk diffusion and aggregation, to afford the mixed-cluster species observed. This concept points to the likelihood of localized excitation, matrix cage-softening, and short-range, bulk diffusion, rather than of extensive matrix-softening. Further evidence in support of this possibility stems from consecutive (instead of simultaneous) deposition of Mo and Cr, followed by selective, atomic excitation. Under these conditions, formation of mixed-metal species was not observed.

The selectivity of the photochemical events was demonstrated in a more detailed study of the Cr/Mo system (*115*). Starting with matrices

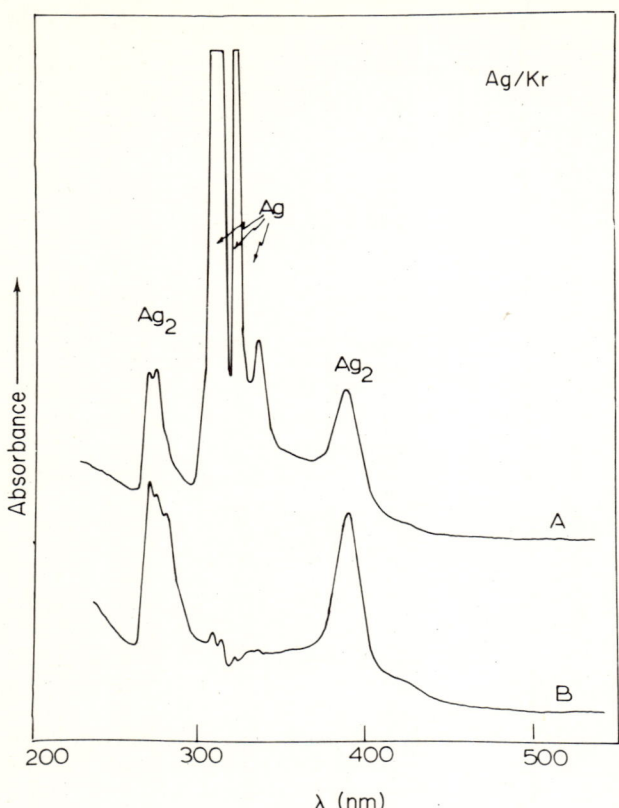

FIG. 19. The optical spectrum of the products of a Ag:Kr ≃ 1:10⁴ cocondensation reaction, (A) after deposition at 10–12 K, and (B) after 60-min, narrow-band (8 nm), 325-nm continuous irradiation from an Oriel 500-W xenon lamp–Schoeffel monochromator assembly (152).

(Kr or Ar at 12 K) that clearly displayed the characteristic absorptions of Cr, Mo, Cr_2, Mo_2, CrMo, Cr_3, Cr_2Mo, $CrMo_2$, and Mo_3, the concept of photoselectivity could be explored as follows. Upon irradiation at 295 nm (Mo atoms), major decay of Mo absorptions was observed, and selective growth of CrMo, $CrMo_2$, Mo_2, and Mo_3 occurred. Only a slight decay of Cr was noticed; this was probably the result of formation of CrMo, while Cr_2 and Cr_3 remained approximately unchanged. This and similar results obtained upon 335-nm (Cr atom) irradiation (where the behavior of the metals is interchanged) argue strongly in favor of selective photoexcitation and photomobilization of the irradiated species.

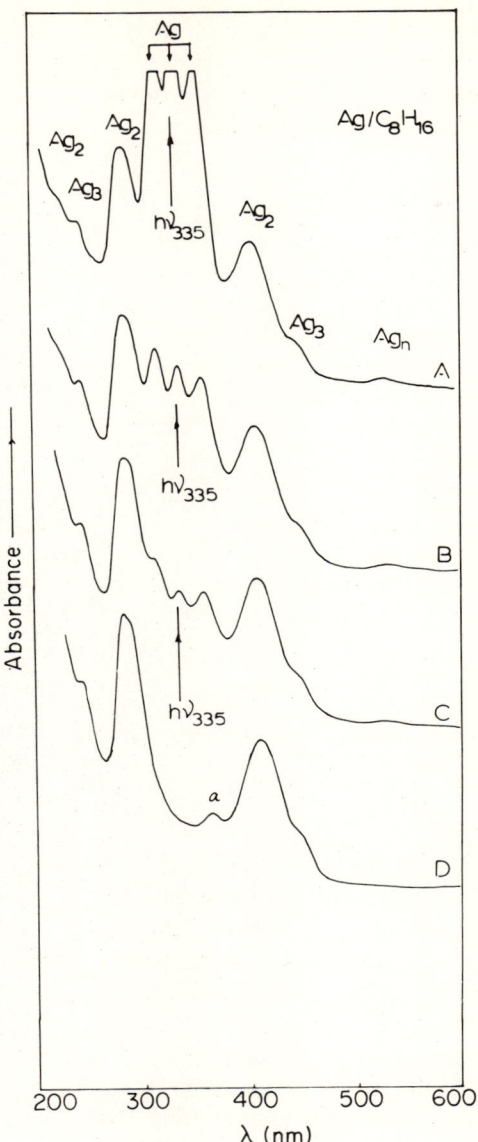

FIG. 20. The optical spectrum of the products of a Ag:C_8H_{16} (cyclooctane) $\simeq 1:10^3$ cocondensation reaction, (A) after deposition at 10–12 K, and (B)–(D) after 15-, 30-, and 60-min, narrow-band (8 nm), 335-nm, continuous excitation (152).

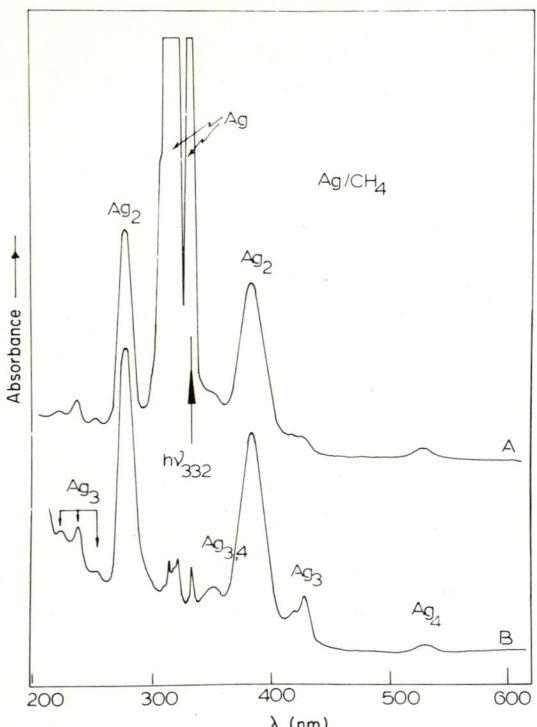

FIG. 21. The optical spectrum of the products of a Ag:CH$_4$ ≃ 1:10^3 cocondensation reaction, (A) after deposition at 10–12 K, and (B) after 30-min, narrow-band (8 nm), 332 nm, continuous irradiation (152).

This behavior is illustrated in Fig. 22, and summarized schematically in Fig 23. It is also noteworthy that nonselective photonucleation may also be arranged by irradiation into Cr/Mo overlap regions (350 nm), upon which, growth of all of the cluster species expected was observed, concomitant with the decay of both the Cr and the Mo resonances (115).

The photoaggregation technique has been extended to the Ag/Cr system (116), where the naked, bimetallic species AgCr and Ag$_2$Cr can be selectively photogenerated, and identified, in the presence of the unimetallic, parent clusters, with photoselectivity reminiscent of that of the Cr/Mo system (115). A typical trace is shown in Fig. 24.

From these early successes, it is evident that this new photochemical technique is likely to find a wide range of applications in the continuing quest to thoroughly understand the electronic, geometric, chemi-

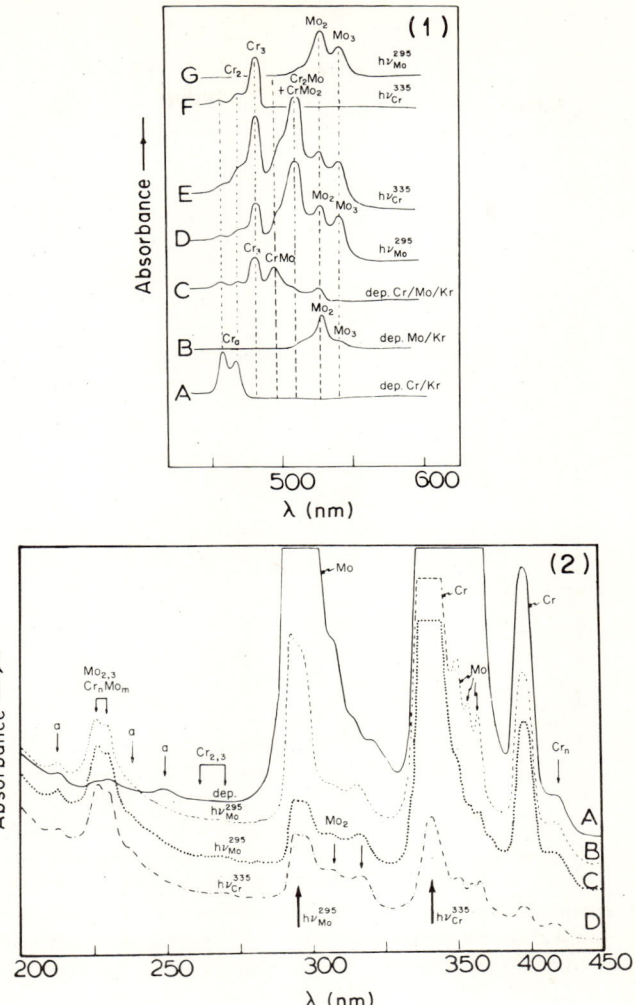

FIG. 22. (1) The low-energy optical spectra of (A) $Cr:Kr \simeq 1:10^3$, (B) $Mo:Kr \simeq 1:10^3$, and (C) $Cr:Mo:Kr \simeq 1:1:10^3$ mixtures, all deposited at 10–12 K, showing the characteristic absorptions of Cr_2, Cr_3, $CrMo$, Cr_2Mo, $CrMo_2$, Mo_2, and Mo_3; (D) the result of 10-min, $h\nu_{Mo}^{295}$ photolysis of sample (C); (E) the result of 5-min, $h\nu_{Cr}^{335}$ photolysis of sample (D); (F) the result of 3-min, $h\nu_{Cr}^{335}$ photolysis of sample (A); (G) the result of 30-min, $h\nu_{Mo}^{295}$ photolysis of sample (B). (2) (A) The high-energy, optical spectra of $Cr:Mo:Kr = 1:1:10^3$ mixtures deposited at 10–12 K, showing the known absorptions of Cr, Mo, Cr_2, Mo_2, and Mo_3; (B) and (C), the result of 50- and 130-min $h\nu_{Mo}^{295}$ photolysis of sample (A); and (D) the result of a further, 22-min, $h\nu_{Cr}^{335}$ photolysis of sample (C). Note that the bands labeled "a" could be associated with mixed clusters Cr_nMo_m (115).

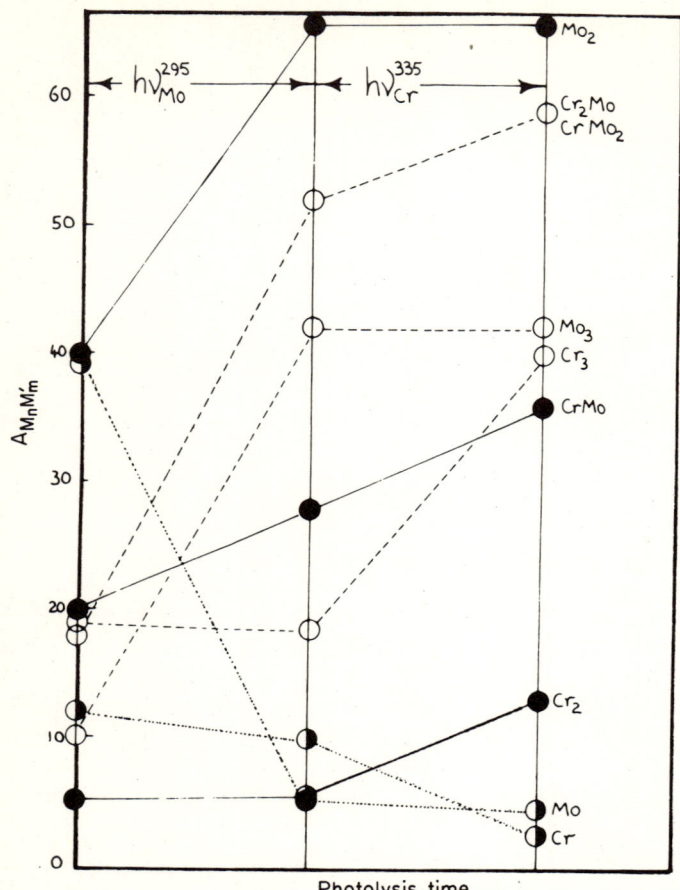

FIG. 23. A schematic representation of the photoselectivity experiments involving (Cr:Mo:Kr ≃ 1:1:10³) mixtures deposited at 10–12 K and then sequentially subjected to $h\nu_{Mo}^{295}$ and $h\nu_{Cr}^{335}$ photolyses. Note that the Cr and Mo atom absorbances are recorded on a scale that is 1/10th that used for the Cr_nMo_m clusters (115).

sorptive, and catalytic properties of unimetallic and bimetallic clusters as a function of size and composition.

IV. Cluster Complexes

Strictly speaking, a cluster complex, as generally considered in organometallic chemistry, consists of a framework of more than two transition-metal atoms. However, in this Section, we shall ignore tra-

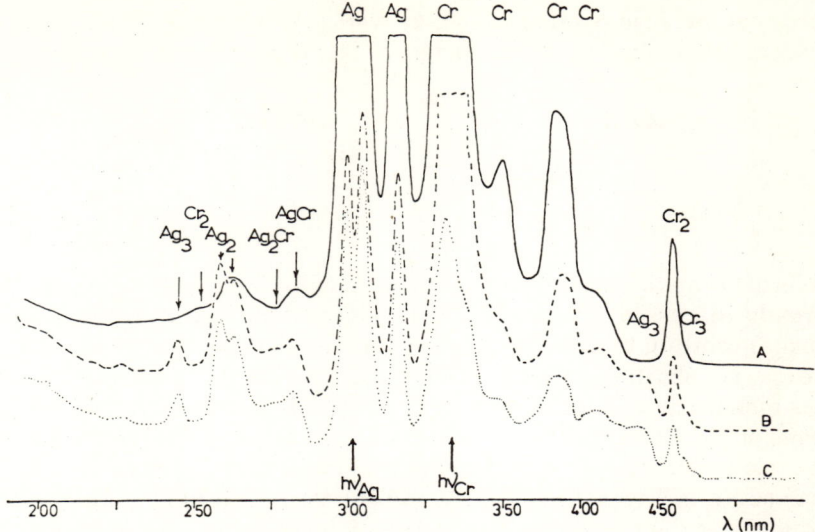

FIG. 24. The UV–visible spectra of a Cr:Ag:Ar ≃ 1:1:10³ mixture (A) deposited at 10–12 K, showing Cr, Ag, Cr_2, Ag_2 and AgCr; (B) after Ag atom 305-nm photoexcitation, showing the growth of Ag_2, Ag_3, AgCr, and Ag_2Cr; and (C) after Cr atom 350-nm photoexcitation and 30-min relaxation time, showing the growth of Cr_3 and AgCr (116).

dition, and arbitrarily include dimer species, as they constitute a logical extension of the discussion presented in Sections II and III.

One of the major attractions in the metal-atom synthesis of dimer and cluster species is the ability to isolate highly unsaturated species, M_nL_m, that may then be considered to be models for chemisorption of the ligand, L, on either a bare, or a supported, metal surface (100). It is quite informative to compare the spectral properties of these finite cluster-complexes to those of the corresponding, adsorbed surface-layers (100), in an effort to test localized-bonding aspects of chemisorption, and for deciphering UPS data and vibrational-energy-loss data for the chemisorbed state. At times, the similarities are quite striking.

A. CO Species

Nickel cluster carbonyls containing up to three nickel atoms were formed by cocondensing monatomic nickel vapor with CO in Ar at cryogenic temperatures (95) as an extension of DeKock's (26) earlier work on mononuclear nickel carbonyls. By using considerably higher metal:matrix ratios, "low-coverage," CO chemisorption models were formed; these clusters included monocarbonyls having two-center, and three-center, bridge-bonded CO, as well as linear CO. The structures illustrated, including three forms of $Ni_3(CO)$, were assigned on the

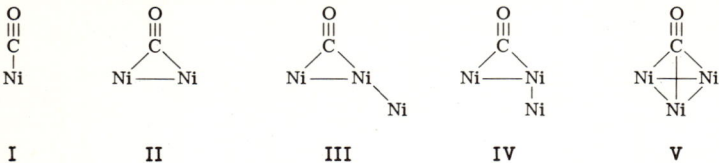

basis of (a) metal-concentration studies, using the kinetic analyses (92) already discussed, (b) mixed $^{12}CO/^{13}CO$ isotope experiments, and (c) normal-coordinate analyses of the vibrational data. The complexity of the spectra obtained is illustrated in Fig. 25, and the spectra of the various monocarbonyls were assigned as in Table IX. Observation of three forms of $Ni_3(CO)$ implies the presence of three forms of Ni_3 in the matrix. Considering MO calculations (2) that suggested that the stable form of Ni_3 is linear, the other trinuclear forms might be a kinetic consequence of the low temperature at which the complexes were synthesized.

The carbonyl clusters provide an interesting set of models for the chemisorption of CO on nickel. It is very interesting that, for the $Ni_n(CO)$ assignments, a plot of ν_{CO} versus $1/n$ for the three-center-bonded CO moieties extrapolates to 1950 cm^{-1} for $n = \infty$ (the "chemi-

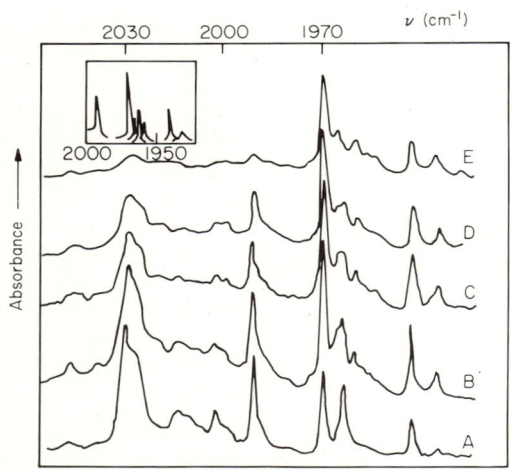

FIG. 25. Portions of the IR spectra of the products of the cocondensation of Ni atoms with CO/Ar (1:250). A through E refer to increasing, total-metal concentration. The inset is a curve-resolved version of spectrum B (95).

TABLE IX

νCO Stretching Frequencies for Various Ni_nCO Species (95)

Complex	$\nu CO(cm^{-1})$	Notation
NiCO	1996	I
Ni_2CO	1973	II
Ni_3CO	1969	III
Ni_3CO	1963	IV
Ni_3CO	1938	V

sorbed" limit), which corresponds to the center of the spectrum of CO on Ni (95). A different extrapolation, from NiCO, to Ni_3CO, to Ni_nCO, yields a value (1905 cm^{-1}) close to that observed for low-coverage, chemisorbed CO on Ni. Such convergence behavior suggests that CO interacting with bulk nickel bonds in a manner similar to that found in stable, metal carbonyls and, moreover, supports the proposal that both terminal- and multi-center-bonded CO occur on nickel surfaces.

Moskovits and Hulse (96) also investigated the interaction of CO with small, copper clusters of known size. Species absorbing at 2128, 2117, 2103, and 2011 cm^{-1} were found, from metal-concentration studies and mixed-isotope experiments, to be associated with finite complexes of stoichiometry Cu_2CO, Cu_3CO, Cu_4CO, and CuCO, the last having been identified previously (117). Annealing the matrix containing these species to 32 K results in the formation of a larger, copper carbonyl cluster species (ν_{CO} = 2090 cm^{-1}) of indeterminate stoichiometry, the IR spectrum of which closely resembles that of CO chemisorbed on bulk copper (155). A ν_{CO} trend similar to that of Ni_nCO (95) was found for Cu_nCO on extrapolating to $n = \infty$, implying that the bonding of CO to Cu [and other transition metals (95)] is localized. The fact that the CO stretching-frequency of Cu_4CO lies close to that of CO chemisorbed on Cu (155) suggested that four copper atoms provide a good model for the bulk-copper surface.

In this context, it is noteworthy that the IR spectrum of bridge-bonded $Rh_2(CO)_8$ [generated from $Rh(CO)_4$ matrix-dimerization reactions and Rh/CO matrix-concentration studies (153)] shows bands in the terminal region at 2060, 2043, and 2038 cm^{-1}, and in the bridge region at 1852 and 1830 cm^{-1}, that are *very close to the proposed, terminal-bridged, chemisorbed, CO species on clean Rh films and supported Rh clusters (4, 34, 50, 162)* (see Table X). Moreover, the matrix, ν_{CO} frequency observed for RhCO (2013 cm^{-1}) (154) falls between the ν_{CO} values for the proposed terminal and bridge sites of chemisorbed CO on

TABLE X

CO Chemisorption Data for Supported Rhodium Samples (154)

Rh/SiO$_2$ (Guerra[a])	Rh/Al$_2$O$_3$			Assignment	Rh$_2$(CO)$_8$ (Ozin[e])	Assignment
	(Garland[b])	(Rothschild[c])	(Arai[d])			
2080	2095	2100	2108	Rh(CO$_{chemis}$)$_2$	2060	νCO_t
2020–1990[f]	2027	2030	2065		2043	νCO_t
2065–2040	2062–2045	2000	2040	Rh(CO$_{chemis}$)$_2$	2038	νCO_t
1900–1890	1925	1900–1850	1860	Rh$_2$(CO$_{chemis}$)	1852	νCO_b
					1830	νCO_b

[a] Ref. (48b). [b] Ref. (34). [c] Ref. (162). [d] Ref. (4). [e] Ref. (153). [f] May be partially a SiO$_2$–CO species.

both supported and unsupported Rh. Both of these sets of data point to a localized description for CO chemisorbed on Rh, and provide additional evidence for the surface vibrational-assignments.

A kinetic study (118) of the controlled formation of a diatomic species, according to the equation

$$2\ Ag(CO)_3 \rightarrow Ag_2(CO)_6 \quad (6)$$

has been conducted at temperatures ranging from 30 to 37 K. The kinetics of the reaction appear to be diffusion-controlled in solid CO matrices, with the diffusion coefficient of Ag(CO)$_3$ found to have a value of 7×10^{-16} cm^2/sec at 35 K; the activation energy of the diffusion process was calculated to be 1.9 kcal/mol. The dinuclear carbonyl was found to be unstable, even at these cryogenic temperatures, decomposing, presumably, to silver dimers, or higher clusters, or both. The study is interesting, in that it presented a mechanism (bimolecular, diffusion-controlled reaction) for the formation of a binuclear cluster-species in an annealed matrix, and, moreover, illustrated the feasibility of matrix kinetics studies involving highly reactive intermediates.

B. O$_2$ Species

The interaction of small, well defined, rhodium clusters, Rh$_2$ and Rh$_3$, with O$_2$ has been investigated (120) by matrix infrared, and UV-visible, spectroscopy, coupled with metal/O$_2$ concentration studies, warm-up experiments, and isotopic oxygen studies. A number of binuclear O$_2$ complexes were identified, with stoichiometries Rh$_2$(O$_2$)$_n$, $n = 1–4$. In addition, a trinuclear species Rh$_3$(O$_2$)$_m$, $m = 2$ or 6, was identified. The infrared data for these complexes, as well as for the mononuclear complexes Rh(O$_2$)$_x$, $x = 1–2$ (119), are summarized in Table XI. Metal-concentration plots that led to the determination of

TABLE XI

Observed Infrared Spectra for $Rh_x(O_2)_y$ Species Formed in O_2 and O_2/Ar Matrices (120)

O_2 matrices	Obs. O_2/Ar matrices	Designation	Assignment [see ref. (120)]
1266	1278 \} 1268	D_1'	$Rh_2(O_2)$
1130	1126	T	$Rh_3(O_2)_{2 \text{ or } 6}$
1120sh	1115		$Rh_3(O_2)_n$?
	1081	D_3	$Rh_2(O_2)_3$
1075	1076	D_4	$Rh_2(O_2)_4$
1038	1048	M_2	$Rh(O_2)_2$
	922	D_1	$Rh_2(O_2)$
	908	M_1	$Rh(O_2)$
	902	D_2	$Rh_2(O_2)_2$
	890	D_3	$Rh_2(O_2)_3$

the metal nuclearity in some of these complexes are illustrated in Fig. 26.

The compound $Rh_2(O_2)_4$ may be regarded as the metal–metal-bonded dimer of the parent monomer $Rh(O_2)_2$, presumably formed by a mechanism similar to that described for $Ag_2(CO)_6$ (118). The lower-stoichiom-

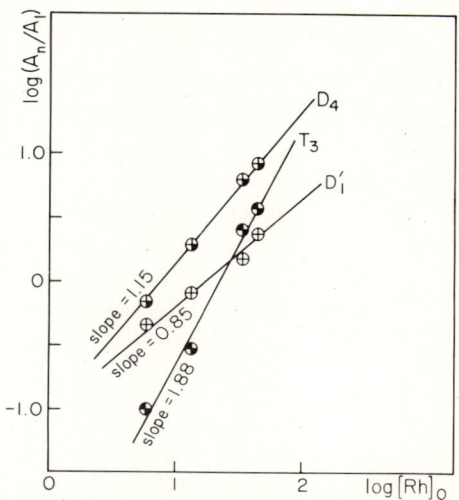

FIG. 26. A log–log plot of the ratio of the absorbances of lines attributed to $Rh_x(O_2)_y$ species to that of a $Rh(O_2)_2$ reference absorption as a function of the total rhodium concentration $[Rh]_0$ at constant dioxygen deposition-rate (120).

etry binuclears may be visualized as deriving from $Rh_2(O_2)_4$ by successive O_2 stripping. In addition, it would appear that $Rh_3(O_2)_m$ is also derived from $Rh(O_2)_2$, and, hence, it is probably best formulated as a triangular cluster, $Rh_3(O_2)_6$, but this has not been confirmed. All of these dioxygen complexes show ν_{OO} (see Table XI) in the range normally expected for coordinated $O_2^{\delta-}$, where $1 \leq \delta \leq 2$. There is, however, another binuclear complex, $Rh_2(O_2)$, having $\nu_{O-O} = 1275$–1265 cm^{-1}, that evidently has much less charge transferred from the dirhodium site to the dioxygen, and, hence, the O_2 is more weakly coordinated. Interestingly, no evidence for bridging dioxygen was found, although the latter complex, with its weak $Rh_2 \cdots O_2$ interaction, has been proposed as a model for "physisorbed" O_2 on bulk Rh in a manner such as that depicted.

$$\begin{array}{c} O{\overset{\textstyle O}{\diagup}} \\ \vdots \\ Rh \text{———} Rh \end{array} \qquad (7)$$

Although there is a severe paucity of vibrational data for the molecular form of O_2 chemisorbed on rhodium surfaces, it is possible to visualize the dinuclear and trinuclear complexes as models for the associative chemisorption of O_2 on rhodium. The ν_{O-O} values of the complexes $Rh_2(O_2)_{1,2}$ show little

$$\begin{array}{cc} O{\overset{\textstyle O}{\diagup}} & O{\overset{\textstyle O}{\diagup}} \qquad O{\overset{\textstyle O}{\diagdown}} \\ Rh\text{—}Rh & Rh\text{—}Rh \\ (922 \text{ cm}^{-1}) & (902 \text{ cm}^{-1}) \end{array} \qquad (8)$$

perturbation from $Rh(O_2)$ (908 cm^{-1}), and, hence, it may be argued that they might be useful "low- and high-coverage" models, respectively. The small frequency-shift on passing from one to two coordinated dioxygens, and the similarity to the mononuclear, imply minimal, nearest-neighbor, dioxygen-coupling effects and minimal perturbation effects of adding a second Rh atom. Such minor, intraligand shifts with metal stoichiometry are also noted with such ligands as C_2H_4 (101) and, as already mentioned, CO (95, 96). Insensitivity of this type argues, for these simple molecules, in favor of localized bonding to metal surfaces.

C. C_2H_4 Species

A wide range of ethylene complexes, both mononuclear and higher cluster in nature, have been synthesized, and studied, by the metal atom-matrix technique. In this Section, we shall focus on the reactions of Co (*121*), Ni (*101, 123*), and Cu (*122, 124*) with ethylene and, although this Section deals mainly with dimer and cluster species, we shall also include, at this point, a discussion of the mononuclear complexes of these metals. To a large extent this is necessary, as, in order to characterize the clusters, it is first essential to identify the mononuclears. In addition, the spectral trends on progressing from one, to two, to more, metal atoms are quite informative.

The cryochemical reaction of copper atoms with C_2H_4 and C_2H_4/Ar mixtures at 10–12 K, using copper-concentration conditions (*94*) that favor mononuclear reaction-products, gives rise (*124*) to three highly colored, binary, copper–ethylene complexes, $Cu(C_2H_4)_n$, $n = 1-3$. These stoichiometries were confirmed by vibrational and electronic spectroscopy, taken in conjunction with metal and ligand concentration experiments and $^{12}C_2H_4/^{13}C_2H_4$ mixed-isotope, concentration studies. These complexes are less stable than the analogous nickel species (*123*), with $Cu(C_2H_4)_3$ being converted into $Cu_2(C_2H_4)_m$ (m not defined) at low temperatures (*122*), whereas $Ni(C_2H_4)_3$ is stable to 0°C (*199*). This instability is considered to be, in part, a manifestation of their paramagnetic character and tendency toward dimerization, as well as their inherent, thermal lability. These mononuclear species all display intense visible and ultraviolet absorptions that monotonically red- and blue-shift, respectively, with increasing olefin stoichiometry. The visible transitions have been assigned, by $X\alpha$ calculations (*148*), as metal-to-ligand charge-transfer involving a mainly localized, metal s-electron, whereas the ultraviolet transitions may be considered to be metal-to-ligand charge-transfer involving mainly localized metal d-electrons. The optical results are summarized in Table XII.

Quantitative Cu/C_2H_4 concentration experiments and controlled annealing of matrices containing $Cu(C_2H_4)_{2 or 3}$ in the 30–45K range show convincing evidence for a transformation to a dinuclear species, $Cu_2(C_2H_4)_m$ where m was not defined. Such a transformation is well il-

TABLE XII

Ultraviolet Spectral Data[a] for $M(C_2H_4)_m$ and $M_2(C_2H_4)_m$ (where M = Co, Ni, or Cu, and $m = 1$ or 2) (*121*)

Complex	Co	Ni	Cu
$M(C_2H_4)$	375	320	382
$M(C_2H_4)_2$	~280[b]	280	276
$M_2(C_2H_4)$	240	240	240
$M_2(C_2H_4)_2$	225		

[a] Units in nm. [b] Appears as a broad shoulder on the 240-nm absorption of $Co_2(C_2H_4)$.

lustrated by the UV–visible spectra shown in Fig. 27. Although the shifts in the optical spectra between mononuclears and dinuclears are large, the IR spectra do not show such a large effect of metal nuclearity, with $\nu_{C=C}$ shifts of only a few cm^{-1}. This behavior is reminiscent of the effects of Rh cluster size in rhodium–dioxygen complexes (119, 120), where the ν_{O-O} shifts vary little on moving from one to three Rh atoms.

These binuclear copper–ethylene complexes do not show very great thermal stability as discrete species. Rather, on warming of matrices containing these complexes to >50 K, ethylene dissociation and controlled copper-clustering occur, and, at <104 K, very small Cu_n species ($n < \sim 10$) may be observed in pure C_2H_4. The data did not, however, permit an unambiguous, size-frequency determination of the distribution of the Cu_n species; a mixture of species is evidently present. In dilute C_2H_4/Ar mixtures, such clustering is even more pronounced. Optical monitoring of these decomposition reactions is shown in Fig. 28.

Nickel atoms have also been allowed to react with C_2H_4 under cryogenic conditions (101, 123). Depending on the metal-concentration conditions and the deposition temperature, either mononuclear species, $Ni(C_2H_4)_n$, $n = 1-3$ (123), or multinuclear species, $Ni_2(C_2H_4)_m$, $m = 1-2$, and $Ni_3(C_2H_4)_l$, may be isolated. Unlike the copper complexes, these species are all colorless; the mononuclear ethylene complexes each dis-

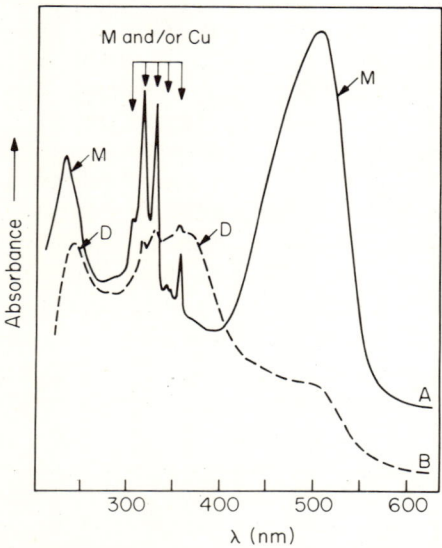

FIG. 27. The UV–visible spectrum of the products of the Cu:$C_2H_4 \simeq 1:10^4$ cocondensation reaction (A) at 10–12K, showing $Cu(C_2H_4)_3$ (M), and (B) after warm-up to 50 K, showing the growth of the 365/240 nm absorptions of $Cu_2(C_2H_4)_m$ (D) (122).

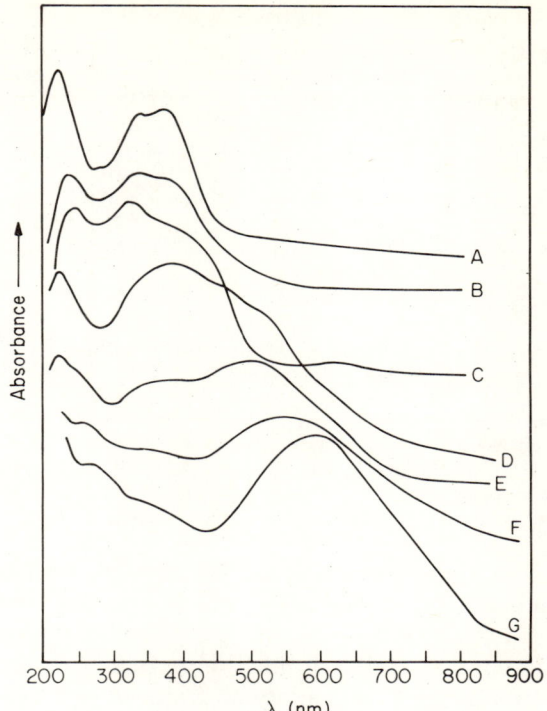

FIG. 28. The optical spectra of the products of the decomposition of $Cu_2(C_2H_4)_m$ (A)–(C) warmed from 50 to 100 K from pure C_2H_4 matrices; (D)–(G) warmed from 45 to 70 K from concentrated C_2H_4 : Ar $\simeq$ 1 : 10 matrices, showing the temperature–time evolution of "growing" copper clusters in the size regime less than 1.0 nm (122).

play one UV absorption that blue-shifts with increasing ethylene stoichiometry. The optical spectra, as with copper, show drastic changes on going from one nickel atom to two in an ethylene complex, but, once again, the IR spectra show only minor shifts in the $\nu_{C=C}$ and δ_{CH_2} modes (see Table XIII). [It should be mentioned that all of the ethylene complexes discussed in this Section yielded vibrational spectra consistent with their formulation as π-complexes (83). No evidence was found for di-σ-type species.] The spectroscopic data are summarized in Tables XII and XIII.

GVB–CI MO calculations have been performed on $Ni(C_2H_4)$ and $Ni_2(C_2H_4)$ (39, 101) in an effort to determine the electronic effects responsible for the shifts in the UV spectra and the nature of the bonding interactions between Ni and C_2H_4. In essence, the two major points of interest in this Ni/C_2H_4 study were (1) the minimal perturbation of

TABLE XIII

Comparison of HRELS[a] for C_2H_4 on Ni(111) with $Ni_n(C_2H_4)$ Chemisorption Models (121)

HRELS C_2H_4/Ni(111) (Ibach, Demuth, Lehwald[b])	$Ni(C_2H_4)$ Ar/10–12K (Ozin, Power, Huber[c])	$Ni_2(C_2H_4)$ Ar/10–12K (Ozin, Power, Upton, Goddard[d])	Assignment
2940	2963	2880, 2908	νCH_2
2690[e]			νCH_2[e]
1500[f]	1499	1488	$\nu C=C$
1430	[g]	[g]	δCH_2
1090	1160	1208, 1180	$\rho_w CH_2$
880	902	910	$\rho_r CH_2$
440	376	416 or 446[h]	νNiC

[a] HRELS = high-resolution, electron-energy-loss spectroscopy. [b] *Surf. Sci.* (in press). [c] Ref. (*123*). [d] Ref. (*101*). [e] Softened νCH_2 surface-mode. [f] Weak band observed around 1500 cm^{-1}; could be a surface-dipole-forbidden, $\nu_{C=C}$ mode. [g] Hidden under intense δCH_2 mode of free C_2H_4 in the matrix. [h] One of these bands belongs to $Ni_2(C_2H_4)_2$.

the coordinated, ethylene intraligand, vibrational spectrum on placing the second nickel atom on $Ni(C_2H_4)$, and (2) the observation of a UV transition for $Ni(C_2H_4)$ at 320 nm that blue-shifts to 243 nm on passing to $Ni_2(C_2H_4)$. In chemisorption terms, the perturbation of the electronic structure of a π-bonded, C_2H_4 moiety on a single, nickel-atom site by a neighboring Ni atom is observed.

To be brief, the calculations showed that, for both complexes, the ethylene geometry was only weakly perturbed, with R(CC) = 132 pm and R(NiC) = 207 pm. The C–H bonds were found to have a bend-back angle of only 2°. Upon coordination of the second nickel atom, the Ni–C_2H_4 binding-energy increased from 14.2 kcal for $Ni(C_2H_4)$ to 27.2 kcal for $Ni_2(C_2H_4)$. The GVB–CI transition-state calculations permitted making of tentative assignments for the optical spectra, and it was suggested that these excitations might best be described as MLCT [actually, Ni (3d → 4p), but with some mixing of $4p_y$ with π^* of C_2H_4]. The blue shift on going from $Ni(C_2H_4)$ to $Ni_2(C_2H_4)$ results from the greater stabilization of the 3d levels in the binuclear. The minimal, geometrical perturbation of the coordinated ethylene as a function of nickel cluster-size is, as mentioned, reflected in the insensitivity of the intra ligand, vibrational spectra.

Cobalt atom reactions with ethylene were also studied (*121*). By using techniques similar to those described for Cu (*122*) and Ni (*101*), it has proved possible to synthesize a novel series of mononuclear and binuclear cobalt–ethylene complexes, $Co(C_2H_4)_l$, $l = 1$, or 2, and

$Co_2(C_2H_4)_m$, $m = 1$, or 2, as well as a suspected tetranuclear species $Co_4(C_2H_4)_n$. The routes to these complexes are summarized in Schemes 2 and 3, which also include the optical and vibrational data pertinent to the complexes. The formation of the dinuclear species was facilitated by the ready formation of Co_2 in Ar matrices, as shown by Hanlan (49). With respect to the mononuclears, the lack of evidence for a species of stoichiometry $Co(C_2H_4)_3$ reported by Timms and Piper (186) is interesting, as is, in this system, the stability of $Co_2(C_2H_4)_2$ at 80 K in the absence of any matrix (i.e., C_2H_4 pumped off under dynamic vacuum). Warming this complex to 100K resulted in its dimerization to $Co_4(C_2H_4)_n$, which further clustered at 150 K; the final product decomposed [presumably, to Co_n, similar to the Cu system (122)] at 190 K [coincidentally, the decomposition temperature reported for $Co(C_2H_4)_3$ (186)]. The conversion from $Co_2(C_2H_4)_2$ to $Co_4(C_2H_4)_n$ is shown in Fig. 29.

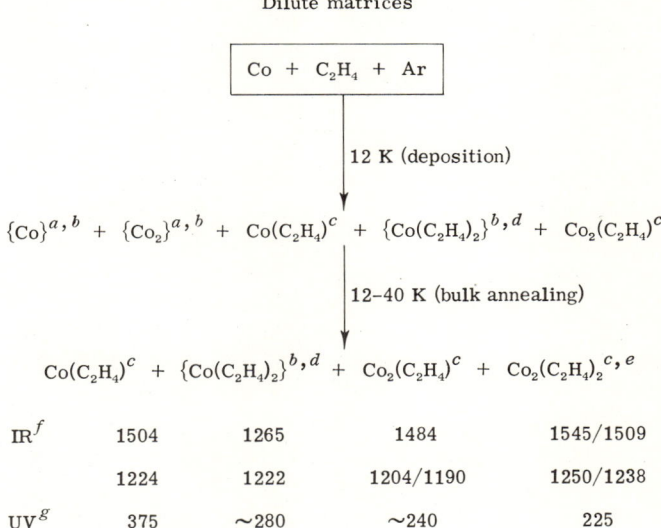

[a] Seen in UV-visible. [b] The braces indicate minor products under these reaction conditions. [c] $^{12}C_2H_4/^{13}C_2H_4/(Ar)$ isotopic confirmation. [d] Seen in 1/20–1/50 range; highly dependent on deposition conditions. No isotopic confirmation (see text). [e] Predominates at high Co concentrations and after 35 K, bulk annealing experiments. [f] Units in cm^{-1}. [g] Units in nm.

SCHEME 2. Dilute matrices (121).

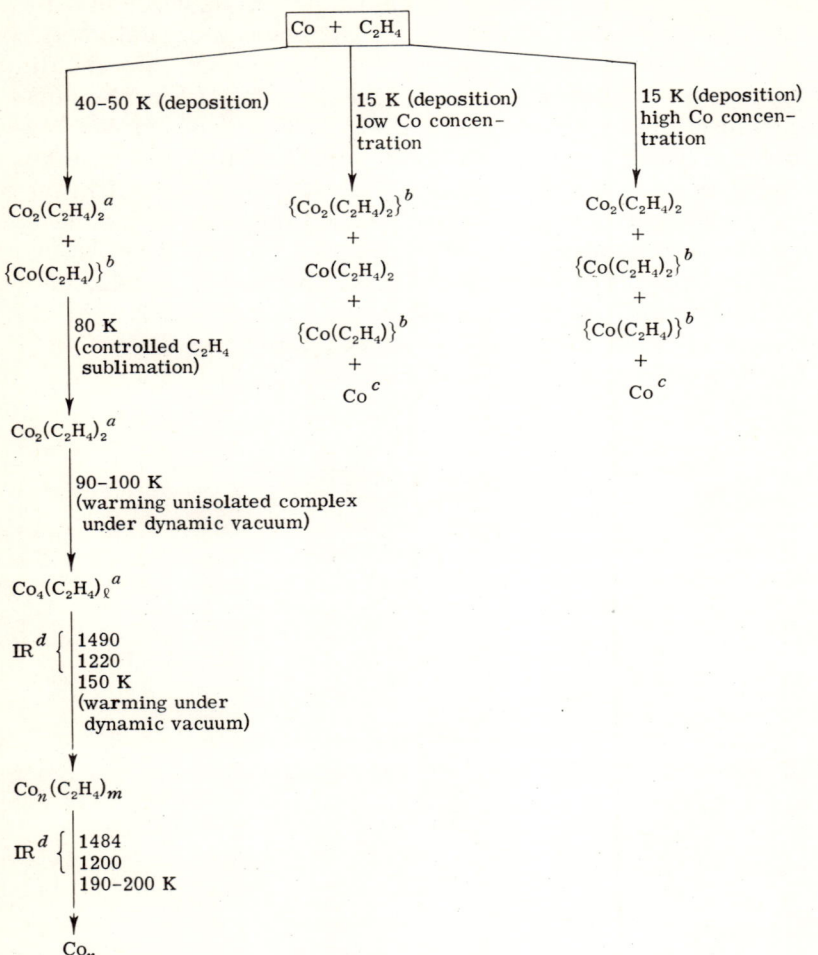

[a] $^{12}C_2H_4/^{13}C_2H_4$ isotopic confirmation. [b] Braces indicate minor products under these reaction conditions. [c] Seen in UV-visible spectrum. [d] Units in cm^{-1}

SCHEME 3. Pure ethylene matrices (121).

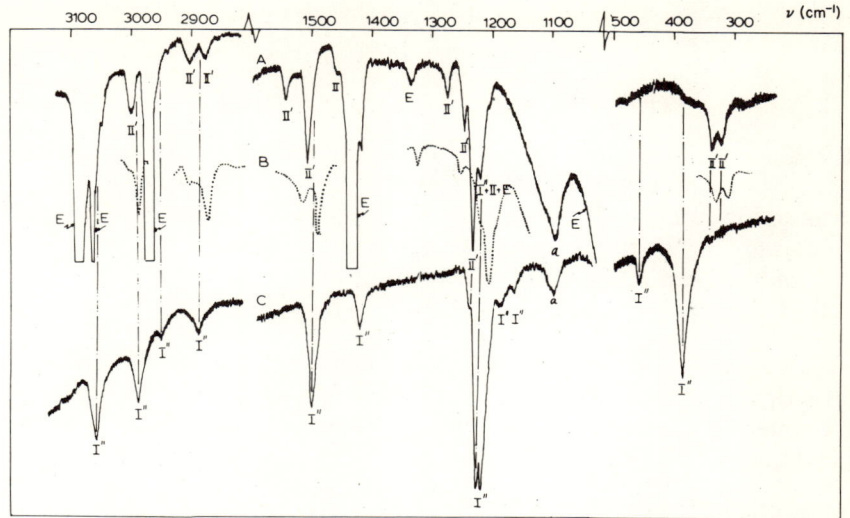

FIG. 29. The matrix IR spectra of Co atoms (A) deposited with $^{12}C_2H_4$ at 40–50 K, (B) deposited with $^{13}C_2H_4$ (91%) at 40–50 K, and (C) the spectrum obtained after controlled, $^{12}C_2H_4$ sublimation at 75–80 K from the matrix in (A) and further warming to 90–100 K, showing the transformation $Co_2(C_2H_4)_2$ to $Co_4(C_2H_4)_n$, where I = $Co(C_2H_4)$; II = $Co(C_2H_4)_2$; I' = $Co_2(C_2H_4)$; II' = $Co_2(C_2H_4)_2$; II" = $Co_4(C_2H_4)_n$; E = unreacted C_2H_4 (121).

At this stage, it need hardly be said that the optical spectra for these complexes show a more significant dependence on olefin stoichiometry and metal nuclearity than the vibrational spectra. The exception to this statement is the noticeably large sensitivity of the ν_{M-C} modes both to metal-cluster size and the nature of the metal. For example, on passing from $Ni(C_2H_4)$ to $Ni_2(C_2H_4)_{1,2}$, the observed ν_{Ni-C} stretching modes shift from 376 to 416 and 446 cm^{-1}, respectively; in the case of $Co_2(C_2H_4)_2$ and $Co_4(C_2H_4)_n$, the observed ν_{Co-C} modes shift from 332, 314, to 384, 460 cm^{-1}. The higher frequencies for the nickel complexes are in line with the correspondingly greater thermal stabilities with respect to the cobalt complexes.

At this point, it seems appropriate to consider the Co–, Ni–, and Cu–ethylene system as a whole, both to rationalize any spectral trends as a function of metal, and to evaluate the use of these complexes as localized-bonding models for chemisorption of C_2H_4.

In spite of the general insensitivity of the intraligand, vibrational modes to cluster size, there are a few specifics worth mentioning, particularly with respect to $\nu_{C=C}$ (ignoring effects of coupling with δ_{CH_2}

modes). The first point relates to the observed $\nu_{C=C}$ orderings

$$\begin{array}{l}\text{Co}(C_2H_4) > \text{Co}(C_2H_4)_2 \\ \text{Ni}(C_2H_4)_3 > \text{Ni}(C_2H_4) > \text{Ni}(C_2H_4)_2 \\ \text{Cu}(C_2H_4)_3 > \text{Cu}(C_2H_4)_2 > \text{Cu}(C_2H_4)\end{array} \quad (9)$$

The amonotonic order for Co and Ni, compared to the monotonic order for Cu, may best be rationalized in terms of a finite, positive $k_{C=C,C=C}$ interaction force constant for Co and Ni, rather than as the outcome of amonotonicity in the principal $k_{C=C}$ force-constants.

A second significant point concerning the $\nu_{C=C}$ is the order

$$\text{Co}(C_2H_4) > \text{Ni}(C_2H_4) > \text{Cu}(C_2H_4), \quad (10)$$

which is consistent with the GVB–CI idea that the main bonding interaction for these complexes is $\pi(C_2H_4) \rightarrow \sigma(sp)M$ bonding.

Some other interesting parallels in the $\nu_{C=C}$ frequencies are the following.

$$\begin{array}{ll} \text{Co}\text{—}\| \quad 1504 \text{ cm}^{-1} & \text{Ni}\text{—}\| \quad 1499 \text{ cm}^{-1} \\ \text{Co}\text{—}\text{Co}\text{—}\| \quad 1484 \text{ cm}^{-1} & \text{Ni}\text{—}\text{Ni}\text{—}\| \quad 1488 \text{ cm}^{-1} \\ \|\text{—}\text{Co}\text{—}\text{Co}\text{—}\| \quad 1508 \text{ cm}^{-1} & \|\text{—}\text{Ni}\text{—}\text{Ni}\text{—}\| \quad 1504 \text{ cm}^{-1} \end{array} \quad (11)$$

The red shift from $M(C_2H_4)$ to $M_2(C_2H_4)$ is probably the result of lessened π-charge density on the ligand in the dimer; with charge density polarized (GVB–CI) away from the ethylene, it can more closely approach the metal atom, with subsequent, greater π-donation and, hence, a lower $\nu_{C=C}$. The subsequent blue-shift on adding the second C_2H_4 ligand most probably arises from increased, charge-repulsion effects between the π-electron densities of the ethylenes and the mainly s-s charge-density localized in the metal–metal bond. This would result in weaker π-bonding, which is manifested in the increased $\nu_{C=C}$.

The order of the UV transitions is as follows:

$$\text{Co}(C_2H_4)(375 \text{ nm}) < \text{Ni}(C_2H_4)(320 \text{ nm}) > \text{Cu}(C_2H_4)(382 \text{ nm}). \quad (12)$$

Assuming that the $\text{Ni}(C_2H_4)$ transition is MLCT (d $\rightarrow$ p), a metal-localized, d $\rightarrow$ p energy ordering in $M(C_2H_4)$ of the form Co < Ni > Cu must be accounted for. In terms of effective nuclear charge–orbital penetration arguments, the d $\rightarrow$ p gap would initially be expected to take on the order Co < Ni < Cu. This is evidently an oversimplification, and account must be taken of the effect of a monotonically increasing, $\pi(C_2H_4) \rightarrow \sigma(sp)M$, charge donation from C_2H_4 to the metal; this will clearly have an effect opposite to Z^*/penetration effects alone, and could well account for the "anomalously" low energy of the d $\rightarrow$ p excitation observed for $\text{Cu}(C_2H_4)$, and the resulting, amonotonic, d $\rightarrow$ p, metal-localized ordering, Co < Ni > Cu.

The utility of any or all of these complexes as models for the chemisorption of ethylene on the pertinent metals is best illustrated by reference to Table XIII, which compares the vibrational data of C_2H_4 on the Ni(111) surface (27) to that of $Ni_2(C_2H_4)$ (101) and $Ni(C_2H_4)$. The similarity of the data presented indeed suggests that these matrix-isolated species may be regarded as such models. The spectrum of chemisorbed C_2H_4 consists of a broad band at 2940 cm^{-1} that would correspond to bands observed in the 3050–2875-cm^{-1} region of the finite-cluster species. The crucial, $\nu_{C=C}$ stretching modes of π-bonded ethylene seem relatively insensitive to cluster size or ethylene stoichiometry; this suggests that the δ_{CH_2} modes around 1200 cm^{-1} and the ν_{M-C} modes around 450–300 cm^{-1} are likely to be most informative in terms of localized-bonding discussions. In particular, it would appear that increasing the cluster size from $1 \rightarrow 2 \rightarrow 4$ causes a monotonic blue-shift of ν_{M-C} towards the value observed for the corresponding, ethylene-chemisorption mode, being already reasonably close for 2 to 4 metal-atom clusters. Broadly speaking, this would imply that it is possible to represent π-chemisorbed ethylene by a finite, cluster–ethylene complex of this nuclearity.

At this time, there is no obvious way of ascertaining whether or not matrix-isolated complexes are the best models for the interaction of small molecules with metal surfaces, clean or supported. Throughout the previous section, we gave an indication of the similarities of the spectra obtained in the two distinct types of experiments, and, although the correspondences are striking, there is no readily accessible, one-to-one correlation. It is, however, evident that the general idea is sound, and, if nothing else, the synthesis of such model complexes should serve to alert the surface and heterogeneous catalyst chemist to the availability of such organometallic data. Nevertheless, it must be borne in mind that the vibrational data for such "chemisorption models" as $M(C_2H_4)_n$ and $M_2(C_2H_4)_m$ can only be used as a guide to an understanding of the geometric and electronic structures of the π-chemisorbed form of ethylene on metal surfaces. The same idea applies to the other molecules under investigation. The clear distinction, for instance, between π-coordination and metallocyclopropane

$$\begin{array}{ccc} \overline{\underset{\text{M M M}}{\overline{\vdots}}} & \text{or} & \overline{\underset{\text{M M M}}{\triangledown}} \\ \pi \text{ model} & & \text{Metallo-} \\ & & \text{cyclopropane} \\ & & \text{model} \end{array} \qquad (13)$$

bonding found for $Ni(C_2H_4)$ may not persist for the metal surface, owing to the effect of long-range interactions with other metal atoms on and near the surface. Indeed, the type of bonding could depend on the surface site and may be radically modified for sites at edges or corners, rather than on plateaus (*174, 175*).

An indication of growing interdisciplinary interest in the field is illustrated in a review on new perspectives in surface chemistry and catalysis by Roberts (*160*), who discussed the interaction of N_2 with iron surfaces. In so doing, he referred to the $Fe_n(N_2)_m$ matrix Mössbauer work of Barrett and Montano (*7*), which showed that molecular nitrogen only bonds to iron when the latter is present as a dimer. As the chemisorption studies (*161*) indicated that N_2 is absorbed on single-atom sites, Roberts suggested (*160*), of the matrix data (*7*), "if this is correct, then our assignment of the N(1s) peak at 405 eV to end-on chemisorbed N_2 will require further investigation." Other reviews that consider matrix-isolation techniques for chemisorption simulation are collected in footnote *a*.

V. Classical Inorganic Ligands

Both the metal-atom and matrix techniques have many applications (*91*) in the study of complexes having such classical, inorganic ligands as CO, N_2, O_2, or phosphines. From a metal-atom point of view, novel complexes have been synthesized that have not been readily accessible via normal, chemical-synthesis techniques. The entrapment of such species permits both a rationalization of their spectroscopic and chemical properties and an evaluation of their stability.

A. CO Complexes

Two separate publications (*125, 126*) described the synthesis of a number of carbonyl complexes of vanadium. The mononuclear species $V(CO)_n$, $n = 1-6$, have all been identified by using CO matrix-dilution experiments and mixed $^{12}CO-^{13}CO$ isotope experiments while main-

---

[a] 1. The Characterization and Properties of Small Metal Particles. Y. Takasu and A. M. Bradshaw, *Surf. Defect. Prop. Solids* p. 401 (1978). 2. Cluster Model Theory. R. P. Messmer, *in* "The Nature of the Chemisorption Bond" G. Ertl and T. Rhodin, eds. North-Holland Publ., Amsterdam, 1978. 3. Clusters and Surfaces. E. L. Muetterties, T. N. Rhodin, E. Band, C. F. Brucker, and W. R. Pretzer, Cornell National Science Center, Ithaca, New York, 1978. 4. Determination of the Properties of Single Atom and Multiple Atom Clusters. J. F. Hamilton, *in* "Chemical Experimentation Under Extreme Conditions" (B. W. Rossiter, ed.) (Series, "Physical Methods of Organic Chemistry"), Wiley (Interscience), New York (1978).

taining mononuclear metal-concentration conditions. The geometries of the carbonyls were assigned (126) as follows: $V(CO)_5$, D_{3h}; $V(CO)_4$, D_{4h} or T_d; $V(CO)_3$, D_{3h}; and $V(CO)_2$, three isomers, C_{2v}, C_{2h}, and $D_{\infty h}$. These geometries have been compared with theoretical predictions (21a, 54). An interesting point concerning V(CO) was the anomalously high, ν_{CO} stretching-frequency (1904, 1890, and 1868 cm^{-1} in Ar, Kr, and Xe, respectively) when compared to Fe(CO), Co(CO), Ni(CO), and Cu(CO) (1898, 1954, 1996, and 2010 cm^{-1}, respectively) (127). Molecular-orbital calculations suggested that the high ν_{CO} can be taken as strong evidence that the predicted Jahn–Teller instability of the molecule resulted in the formation of a nonlinear monocarbonyl.

The "saturated" species $V(CO)_6$ and $V_2(CO)_{12}$ were directly synthesized (28, 125) in CO, and CO-doped rare-gas, matrices by standard, metal-atom techniques. $V(CO)_6$ was investigated by IR and UV–visible spectroscopy, and $V_2(CO)_{12}$ by IR spectroscopy. The distortion indicated for $V(CO)_6$ in pure CO was found to persist in Ne, Ar, Kr, and Xe matrices, but with a magnitude of the same order as, or less than, a matrix-site splitting, as seen by comparison with regular O_h $M(CO)_6$ complexes in solid, noble-gas matrices. The polarizability–frequency plots suggested that the larger splitting in CO results from a matrix split superimposed on a genuine, molecular distortion. As shown in Fig. 30, two binuclear complexes could be generated in pure CO matrices, one of which (D_1) is more thermally stable than the other (D_2). By quantitative, V-concentration studies, the nuclearity was confirmed. From warm-up studies and isotope experiments, and by comparison with the isoelectronic, complex anion $[V_2(CO)_8(CN)_4]^{4-}$ (156), the more stable dimer D_1 was formulated as $(OC)_5V(\mu CO)_2V(CO)_5$, containing two equivalent, vanadium atoms and two bridging, CO groups.

The matrix UV–visible data for $M(CO)_6$ (M = Ti, V, or Cr) and $M(N_2)_6$ (M = Ti, V, or Cr) (128) have been obtained, and the ligand field and charge-transfer spectra analyzed and assigned. These data are summarized in Table XIV. Charge transfer from the ligand, N_2 or CO, was observed, with the former at lower energy. The interpretation of the data implied that N_2 is a weak ligand compared to CO, with both σ-donation and π-acceptance being less. A molecular-orbital analysis indicated that the lower the first-order energy of the acceptor orbital and the better its overlap, the greater the degree of back-donation in the ground state and the higher the energy of the charge-transfer, excited state. In this way, the lower-energy charge-transfer to N_2 compared to CO may be rationalized. In addition, $10D_q$ for N_2 is lower than for CO and, in these complexes, is relatively insensitive to the metal.

The magnetic circular dichroism (MCD) spectrum of matrix-isolated

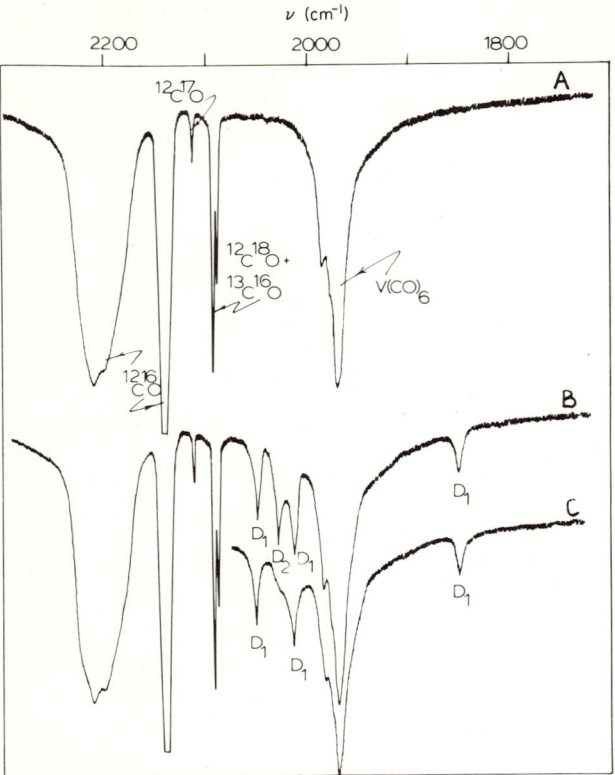

FIG. 30. IR spectra of the cocondensation reaction of V atoms with CO at 6–10 K: (A) at low V concentrations; (B) at high V concentrations; (C) at the same concentrations as (B), but after warming to 35 K (*125*).

$V(CO)_6$ generated by cocondensing presynthesized $V(CO)_6$ with N_2 at 10 K has been observed (*44, 45*). As suggested in a metal-atom study (*125*), the results indicated that a static, Jahn–Teller distortion is present. Matrix MCD also proved useful in confirming the predicted paramagnetism of $Fe(CO)_4$ (*45*) (produced by photolysis of $Fe(CO)_5$). In addition, matrix MCD was used to detect such paramagnetic species as MnO_2Cl_2 in the presence of MnO_3Cl (*45*).

Nonmetal-atom, matrix-isolation spectroscopy has proved useful in structure and isomer determination of stable, metal carbonyls. $Fe(CO)_4(NO)$ was investigated (*157*) in low-temperature matrices with ^{13}CO enrichment, and it was demonstrated that the IR spectrum is consistent with C_{2v} symmetry (trigonal bipyramid with an equatorial NO), in agreement with X-ray studies (*55*). The work resolves the dis-

TABLE XIV

LIGAND FIELD AND CHARGE-TRANSFER TRANSITION-ENERGIES FOR $M(N_2)_6$ AND $M(CO)_6$ (WHERE M = Ti, V, Cr) (128)

	M = Ti	M = V	M = Cr	
		$M(N_2)_6$		
d-d	20 833	19 455	20 576	
	24 096	23 585	23 474	
Charge transfer	26 738	26 740 ⎫	27 397	$t_{2g} \to t_{1u}$ CT1(N_2)
		29 155 ⎭		
	30 581	31 646 ⎫	32 680	$t_{2g} \to t_{2u}$ CT2(N_2)
		33 557 ⎭	34 602	
		35 461	40 000	$t_{2g} \to t_{2g}^*$ (?)
		$M(CO)_6$		
d-d	27 174	25 840	31 350[a]	
	29 762	30 478		
Charge transfer		32 154 ⎫		
	33 898	33 784 ⎬	35 780[a]	$t_{2g} \to t_{1u}$ CT1(CO)
		37 453 ⎭		
	38 023	40 650 ⎫	44 480[a]	$t_{2g} \to t_{2u}$ CT2(CO)
		44 248 ⎭		
			51 280[a]	$t_{2g} \to t_{2g}^*$

[a] H.B. Gray and N.A. Beach, *J. Am. Chem. Soc.* **85,** 2922 (1963).

parity between previous solution IR studies, where C_{3v} symmetry and an axial NO were suggested (55, 177), and the X-ray studies (55).

A partially oriented (using polarized, visible photolysis) sample of $Fe(CO)_5$ in solid CO at 20 K has been prepared (193). It was subsequently found that *no change* in the polarization properties of the system occurred during several hours of spectroscopic observation. It was concluded that the fluxionality of $Fe(CO)_5$ had been quenched under these conditions, as, were this not the case, maintenance of polarization for more than a fraction of a second would be impossible.

$Co_2(CO)_8$ has also been studied in low-temperature matrices (19, 20), the photochemical behavior of which led to the identification of three isomeric forms of the dimer complex (19). Two of these are the accepted forms, **1** and **2,** whereas the third has no bridging, CO ligands. The structure most

(14)

1　　　　2　　　　3

compatible with the IR spectrum of the third isomer (3) was D_{2d}, in which the Co–Co bond axis lies in the plane of the trigonal, bipyramidal arrangement at each metal. Moreover, the isomers were found to be readily interconvertible: the nonbridged, D_{3d} isomer is transformed into the bridged form in hexane matrices, with $\Delta G^{\ddagger} = 6.4 \pm 0.4$ kcal/mol at 84 K. A separate study (17), using high temperatures and pressures, also showed the existence of a third isomer of $Co_2(CO)_8$.

On photolyzing $Co_2(CO)_8$ in the matrix (20), a number of photoproducts could be observed. The results of these experiments are summarized in Scheme 4, which illustrates the various species formed. Of particular interest is the formation of $Co_2(CO)_7$ on irradiation of $Co_2(CO)_8$ in CO (254 nm), as this species had not been characterized in the metal-atom study of Hanlan et al. (129). Passage of $Co_2(CO)_8$ over an active, cobalt-metal surface before matrix isolation causes complete decomposition. On using a less active catalyst, the IR spectrum of $Co(CO)_4$ could be observed. An absorption due to a second decomposition product, possibly $Co_2(CO)_6$, was also noted.

Using Ag atom cocondensations, as well as other standard, matrix-characterization techniques, the silver carbonyls $Ag(CO)_n$, $n = 1-3$, and $Ag_2(CO)_6$ were synthesized (130). An illustration of the $^{12}CO-^{13}CO$ stoichiometry confirmations for the mononuclears is shown in Fig. 31. When $Ag(CO)_3$ is synthesized in pure CO, the vibrational data are consistent with a slightly distorted, trigonal planar structure. However, this is a matrix effect, because, in Ar, Kr, or Xe, a D_{3h} structure is observed. The esr spectrum of $Ag(CO)_3$ in CO/Ar matrices supports this description ($g_{\parallel} = 2.012$ and $g_{\perp} = 1.995$), whereas the corresponding, optical spectra are consistent with that expected for a D_{3h} species having a $^2A_2''$ ground-state. Unlike the analogous, Cu complex, $Ag_2(CO)_6$ could not be synthesized from the matrix reaction of Ag

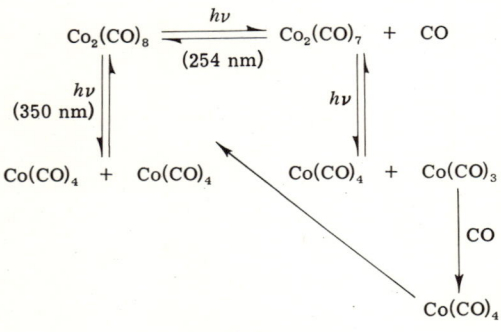

SCHEME 4

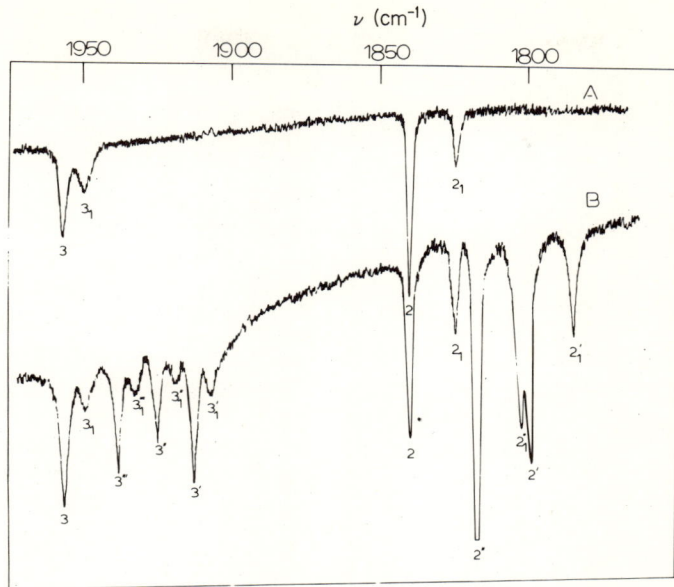

FIG. 31. The matrix IR spectrum of the products of the cocondensation reaction of Ag atoms with (A) $^{12}C^{16}O:Ar \simeq 1:100$, and (B) $^{12}C^{16}O:^{12}C^{18}O:Ar \simeq 1:1:250$ matrices at 10–12K, showing Ag(CO)$_3$ and Ag(CO)$_2$ (where 2 and 3 = site 1; 2_1 and 3_1 = site 2) (130).

and CO at high silver concentrations. However, as already discussed, Ag(CO)$_3$ dimerizes in CO matrices at 30–35 K, to yield Ag$_2$(CO)$_6$ (118).

Binary gold carbonyls Au(CO)$_{1,2}$ have also been synthesized (131), and characterized by using ^{12}CO–^{13}CO isotopic substitution. The IR data for Au(CO)$_2$ in rare gas matrices favor a linear D$_{\infty h}$ structure. Detailed investigation of the complexes revealed a variety of interesting site-effects and matrix-induced frequency-shifts. However, unusual, vibrational-isotope patterns were observed for the product formed in $^{12}C^{16}O/^{13}C^{16}O$, $^{12}C^{16}O/^{12}C^{18}O$, and $^{12}C^{16}O/^{13}C^{18}O$ mixtures, as shown in Fig. 32. Ten distinct, mixed isotopic molecules were observed, containing nonequivalent carbonyl ligands. These data could be interpreted in terms of an isocarbonyl(carbonyl)gold complex, (OC)Au(OC), a linkage isomer of bis(carbonyl)gold. A further, spectroscopic difference is illustrated in the optical spectra (see Fig. 33), which show a large shift in one particular UV absorption and a small shift in the intense, visible absorption. It would seem that the existence of the isocarbonyl may be a consequence of the head-to-tail, orientational requirements of the CO molecules in the fcc lattice of crystalline CO, rather than of an inher-

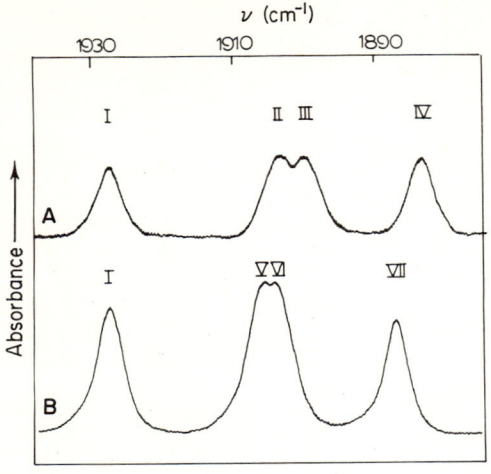
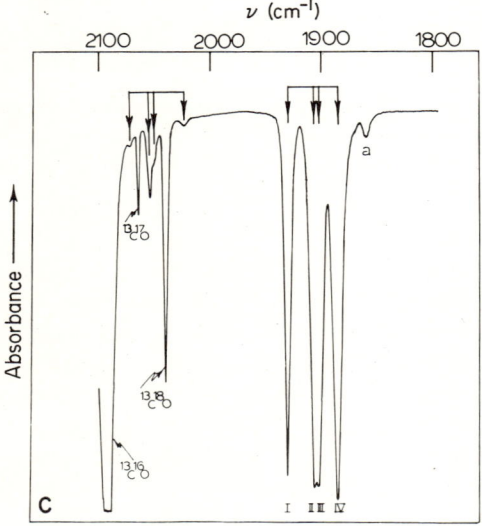

Fig. 32. Matrix IR spectrum of Au atoms deposited with (A) a $^{12}C^{16}O:^{13}C^{16}O \simeq 1:1$ mixture, and (B) a $^{12}C^{16}O:^{12}C^{18}O \simeq 1:1$ mixture at 6–10 K, showing the presence of the seven, distinct, mixed isotopic molecules $Au(^{12}C^{16}O)(^{16}O^{12}C)$ (I), $Au(^{12}C^{16}O)(^{16}O^{13}C)$ (II), $Au(^{13}C^{16}O)(^{16}O^{12}C)$ (III), $Au(^{13}C^{16}O)(^{16}O^{13}C)$ (IV), $Au(^{12}C^{16}O)(^{18}O^{12}C)$ (V), $Au(^{12}C^{18}O)(^{16}O^{12}C)$ (VI), and $Au(^{12}C^{18}O)(^{18}O^{12}C)$ (VII). Curve C is the same as curve A, but with a complete scan after a long deposition-time, showing the quartet isotopic pattern associated with the weak, high-frequency absorptions at 2072, 2054, 2050, and 2024 cm^{-1} (note that, under these conditions, weak absorptions due to traces of free $^{13}C^{17}O$ and $^{13}C^{18}O$ in the commercial $^{12}C^{16}O/^{13}C^{16}O$ mixture were observed in the 2100-cm^{-1} region) (*131*).

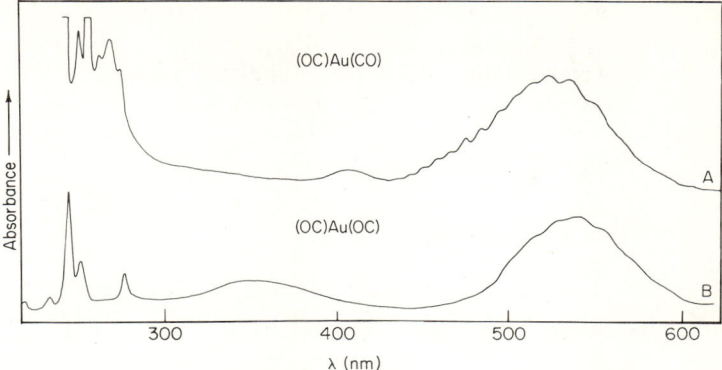

FIG. 33. UV–visible spectrum of Au atoms deposited with (A) a CO : Ar ≃ 1 : 5 mixture at 6–10 K, and (B) pure CO at 6–10 K (131).

ent preference for this mode of bonding. [It should be mentioned, however, that the evidence is not incontrovertible and, in fact, Burdett (21) concluded from the same data that the spectra simply represent a matrix-perturbed, bis(carbonyl) species.]

Further reactions of gold atoms (132) with mixed O_2/CO matrices led to some interesting results. By cocondensing Au atoms with equimolar mixtures of O_2/CO at 10–12K, a single compound was isolated which, by means of $^{13}C^{16}O/^{16}O_2$, $^{12}C^{16}O/^{18}O_2$, and $^{12}C^{16}O/^{13}C^{18}O/^{16}O_2/^{16}O^{18}O/^{18}O_2$ isotope studies, was best formulated as

$$O\equiv C-Au\underset{\underset{O}{\overset{\|}{C}}}{\overset{O}{\diagup}}\diagdown O \tag{15}$$

a peroxyformate complex of Au(II). At 30–40 K, this complex decomposes to CO_2, a reaction that occurs, as shown by IR spectroscopy, by cleavage of the peroxy group as the primary step in a two-step fragmentation process. The existence of a second species as an intermediate suggests the following decomposition-pathway.

$$O\equiv C-Au\underset{O}{\overset{O-\overset{\overset{O}{\|}}{C}}{\diagup}}\diagdown O \xrightarrow{30\ K} O\equiv C-Au=O \xrightarrow{40\ K} Au_n \tag{16}$$
$$+\qquad\qquad +$$
$$CO_2\qquad\qquad CO_2$$

This reaction system illustrates that metal-atom reactions can proceed in unexpected directions, in this case, the oxidation of CO to CO_2.

As a byproduct of the $Au/CO/O_2$ study, the interaction of Au atoms with CO_2 was studied (132). Two compounds, one formed on deposition, and the other after warm-up to 40–45 K, were observed. The corresponding IR spectra only show weakly perturbed CO_2 lines, and, hence, the low-temperature species is probably $Au(CO_2)$, with a weak gold–CO_2 interaction. The warm-up product was not characterized. In a similar way, Ag atoms reacted with CO_2 to form a single compound best formulated as $Ag(CO_2)$ (133), a weak π-complex.

B. O_2 Complexes and Oxides

Manganese atoms were cocondensed with O_2, N_2O, or O_3, to afford MnO, MnO_2, MnO_3, and MnO_4 (197). The actual reactions involved were

$$Mn + O_2/Ar \xrightarrow[\text{vuv}]{h\nu} MnO + MnO_2 \tag{17}$$

$$Mn + N_2O/Ar \xrightarrow[\text{Hg}]{h\nu} MnO \tag{18}$$

$$Mn + O_3/Ar \rightarrow MnO + MnO_2 \tag{19}$$

$$Mn + O_3/Ne \rightarrow MnO_3 \tag{20}$$

$$Mn + O_3/Ne \text{ or } Ar \rightarrow MnO_4 \tag{21}$$

The compounds formed were studied by esr spectroscopy, with the magnetic parameters being used to determine the geometries. MnO_2 is linear, whereas MnO_3 is trigonal planar (D_{3h}), and MnO_4 is distorted tetrahedral with C_{3v} symmetry.

In a preliminary report (1), Fe atoms were reacted with O_2, leading to formation of $Fe(O_2)$, a cyclic isosceles (C_{2v}) species, as suggested by mixed isotope experiments. Reaction of Fe atoms with N_2O resulted in formation of FeO. A feature at 887 cm^{-1}, assigned to a Fe/N_2 complex, is probably erroneous, and may be an iron nitride species. In the same triad, the MCD spectrum of matrix-entrapped OsO_4 was studied (46). The spectrum was found to be similar to that of MnO_4^- in a solid lattice, and was assigned accordingly.

The Rh/O_2 system (119, 120) has already been discussed with particular reference to the multinuclear species. To summarize, at low Rh concentrations, the species $Rh(O_2)_{1\text{and}2}$ can be isolated. The optical spectra of $Rh(O_2)_2$ and $Pd(O_2)_2$ were compared (see Fig. 34), and the spectral shifts interpreted in light of qualitative, MO considerations. The absence of any visible absorption argues against an $M^{2+}(O_2^-)_2$ formulation, and the UV-spectral shifts are consistent with a LMCT assignment. Thus, although the π and π^* O_2 levels remain at about the same energy in the two complexes, the metal d-orbitals stabilize on going from Rh to Pd. Hence a transition from a low-lying, O_2 π orbital

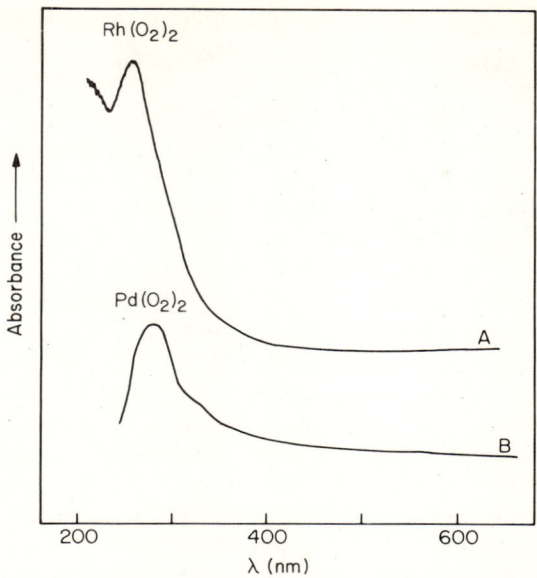

FIG. 34. UV–visible spectrum of the products formed when (A) Rh atoms, and (B) Pd atoms are cocondensed with $^{16}O_2$ at 10–12 K and M : $^{16}O_2 \simeq$ 1 : 10^5 (119).

to a vacancy in the upper M—O_2 levels would red-shift in the observed order (119, 120).

A number of investigations of the copper-group oxides and dioxygen complexes have been reported. The electronic spectra of CuO, AgO, and AuO were recorded in rare-gas matrices (9), and it was found that the three oxides could be formed effectively by cocondensation of the metal atoms with a dilute, oxygen matrix, followed by near-ultraviolet excitation. The effective wavelengths for CuO or AgO formation were $\lambda \geq 300$ nm and for AuO was $\lambda \geq 200$ nm. In addition, the laser fluorescence spectrum of CuO in solid Ar has been recorded (97).

The metal-atom reactions of Cu (98), Ag (134), and Au (135) with O_2 provided interesting results, especially when these were compared with the results from the nickel triad (137). As shown by standard matrix-techniques, Ag forms two O_2 complexes that are best formulated as $Ag^+O_2^-$ and $Ag^+O_4^-$, based on the absence of visible absorptions and the similarity of the IR spectra to those of $Cs^+O_2^-$ and $Cs^+O_4^-$ (3a,b). The UV absorptions for Ag(O_2) and Ag(O_4), at 275 and 290 nm, respectively, could be associated with the O_2^- and O_4^- anions. The shifts in the IR spectra on going from Ag(O_2) to Ag(O_4) also argue against an (O_2)Ag(O_2) formulation for the latter complex, being in the opposite sense to those observed for Pd(O_2)$_{1\ \text{and}\ 2}$ (137). In contrast, whereas

$Cu^+O_2^-$ may be formulated in the same way, $Cu(O_2)_2$ is best regarded as a bis(dioxygen) species [actually, $Cu^{2+}(O_2^-)_2$ (98)], rather than as a complex of O_4^-. The geometry of $Ag^+O_4^-$ could not determined from the spectroscopic data; however, the results of MO calculations on $Na^+O_4^-$ (3a, 3b) favored a puckered, five-membered ring. These differences between Cu, Ag, and Au may possibly be explained by noting that the unfavorably high, second ionization-potential for Ag might preclude the formation of an Ag(II) dioxygen complex. This argument is somewhat strengthened by the fact that gold forms only $Au(O_2)$ (135). The green color of $Au(O_2)$ suggests a zero-valent, gold formulation, where the high, first-ionization potential of Au (9.22 eV) precludes oxidation to Au(I) under cryogenic reaction-conditions ($IP_{Cu} = 7.72$, $IP_{Ag} = 7.57$ eV).

On cocondensing Ag atoms with mixed CO/O_2 matrices (136), a new species $(OC)Ag(O_2)$ was observed, most probably containing O_2 bound in a side-on fashion ($^{16}O_2/^{18}O_2$ isotope data). The appearance of the carbonyl stretching-mode (2165 cm^{-1}) in a region above that of free CO (2138 cm^{-1}), and the O_2 stretching mode at about the frequency of the superoxide anion (1110 cm^{-1}), when taken in conjunction with an observed UV absorption (285 nm) close to that of free O_2^- and the absence of any visible absorption, provided convincing evidence in favor of the tight, ion-pair formulation $(OC)Ag^+(O_2^-)$, in which the silver atom may be considered to act as a strong σ-acceptor, and yet a weak π-donor with respect to CO. The $(OC)Ag^+(O_2^-)$ complex, unlike the peroxyformate product of the $Au/CO/O_2$ system (132), does not act as a precursor to CO_2.

Work has also been conducted that involved the investigation, via infrared spectroscopy, of matrix-isolated, plutonium oxides (40), with the appropriate precautions being taken because of the toxicity of plutonium and its compounds. A sputtering technique was used to vaporize the metal. The IR spectra of PuO and PuO_2 in both Ar and Kr matrices were identified, with the observed frequencies for the latter (794.25 and 786.80 cm^{-1}, respectively) assigned to the ν_3 stretching-mode of $Pu^{16}O_2$. Normal-coordinate analysis of the PuO_2 isotopomers, $Pu^{16}O_2$, $Pu^{18}O_2$, and $Pu^{16}O^{18}O$ in Ar showed that the molecule is linear. The PuO molecule was observed in multiple sites in Ar matrices, but not in Kr, with $Pu^{16}O$ at 822.28 cm^{-1} in the most stable, Ar site, and at 817.27 cm^{-1} in Kr. No evidence for PuO_3 was observed.

C. N_2 Complexes and Nitrides

Titanium atoms have been cocondensed with CO and N_2 matrices (138) and the products identified as $Ti(CO)_6$ and $Ti(N_2)_6$. The IR data for the two species indicated that they are structurally related, and are subject to a Jahn–Teller distortion. The 40-cm^{-1} split observed for the T_{1u} ν_{CO} stretching-mode in solid CO is retained in noble-gas matrices, a fact that, coupled with the linear, Buckingham plots (see Fig. 35) for both components, supports the designation of an inherent, molecular distortion for Ti(CO)$_6$ [and Ti(N$_2$)$_6$], rather than a matrix effect. However, the magnitude of the distortion is probably quite small, as the electronic spectra of both species display the gross features expected for low-spin, d^4 octahedral complexes (128).

Under similar reaction-conditions, the vanadium species V(N$_2$)$_6$ (139) has been isolated. In addition, a species V$_2$(N$_2$)$_n$ (n probably = 12) was observed (139). The metal nuclearity was established by the standard, metal-concentration techniques. A comparison of the optical spectra of V(N$_2$)$_6$ and V(CO)$_6$ (128) suggested that these molecules have very similar, electronic properties, and the data clearly established that N$_2$ is a strong, field ligand in its bonding properties. Interestingly, atomic V could be isolated in N$_2$ matrices from 8–12K co-

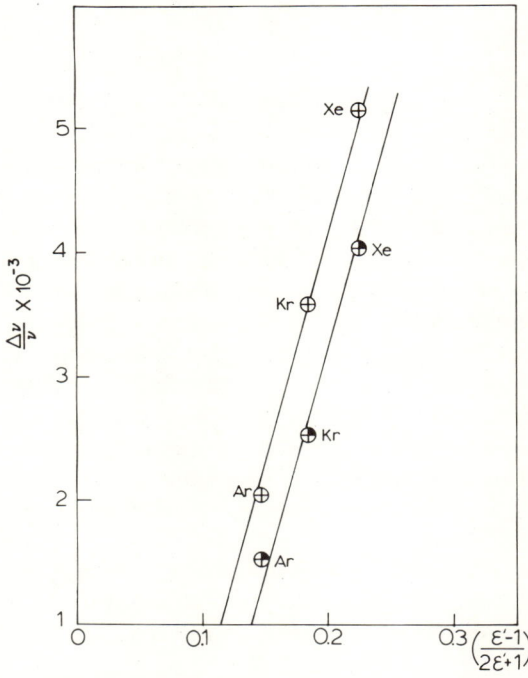

FIG. 35. Buckingham plot for both components of the doublet splitting of Ti(CO)$_6$ in Ar, Kr, and Xe matrices (138).

depositions, and yet $V(N_2)_6$ is preferentially formed at 20–25 K; this demonstrated the subtle temperature-dependence of product yield when direct syntheses are performed with metal atoms.

Dinitrogen has also been reacted with atomic Nb (produced from a sputtering source) (47), a number of $\nu_{N\equiv N}$ stretching-modes being observed. The spectra were too complicated to permit full, stoichiometric and structural identification, but $Nb(N_2)$ and a D_{4h} square-planar $Nb(N_2)_4$ were tentatively assigned. The authors deduced a number of conclusions. (1) Geometrical isomers exist for the higher stoichiometries, e.g., D_{3h} and C_{2v} for $Nb(N_2)_3$. (2) Both "sideways" and "end-on" bonding of N_2 to Nb result in more than one absorption peak, even for the lower stoichiometry species, such as $Nb(N_2)$. (3) There is an unusually strong interaction between the complexes and the matrix environment, resulting in multiple sites and, hence, multiple absorptions. (4) Some of the complexes are multinuclear. (5) Some of the complexes contain a ligand other than N_2 (perhaps a CO impurity). (6) The warm-up data can be explained by such reactions as

$$Nb_y(N_2)_x + N_2 \rightarrow Nb_y(N_2)_{x+1}, \tag{22}$$

rather than

$$Nb_y(N_2)_x + Nb \rightarrow Nb_{y+1}(N_2)_x. \tag{23}$$

Clearly, more work needs to be done with this system.

The products of the cocondensation of Cr atoms with N_2 at 10–12 K have been investigated (29) using IR and UV–visible spectroscopy, and the products of similar N_2/Ar depositions were monitored by IR spectroscopy. From ligand-concentration experiments and annealing studies, six mononuclear species, $Cr(N_2)_n$ ($n = 1-6$) were identified. It was suggested that $Cr(N_2)$ and $Cr(N_2)_2$ are linear, with $C_{\infty v}$ and $D_{\infty h}$ symmetries, respectively. $Cr(N_2)_3$ is probably C_{2v}, whereas $Cr(N_2)_4$ is either T_d or D_{4h}. $Cr(N_2)_5$ is a tetragonal pyramid, C_{4v}, whereas $Cr(N_2)_6$ is O_h. The optical spectra supported the contention (29, 139) that N_2 is a strong field-ligand having an observed $10Dq$ of 28,800 cm^{-1} in $Cr(N_2)_6$. The optical spectra and assignments had been previously discussed (128).

The reaction of plutonium metal with N_2 in a sputtering device (41) resulted in the observation of matrix-isolated, plutonium nitride species. The species observed were PuN and PuN_2, the latter being a linear species.

This section on N_2 complexes concludes with a brief mention of a Mössbauer study of the Fe/N_2 system (8). The N_2 stoichiometries were not determined, but it was found that N_2 reacts only with the Fe dimer. The ^{57}Fe isomer shift suggested less σ-donation or π-backbonding, or both, in the identified iron–nitrogen complex than in stable, dinitrogen compounds.

D. Miscellaneous Complexes

A number of other metal atom–matrix studies have appeared in the literature, with such typical inorganic ligands as NO and H. In the following Section, we shall briefly summarize some of these results.

A Japanese group reacted iron vapor with nitric oxide at 77 K (6). Two different species were observed, with ν_{NO} at 1800 and 1720 cm^{-1}, that were assigned as NO species adsorbed on oxidized and metallic iron, respectively. Although no evidence was presented as to the nuclearity of the products, the authors considered the species to be models for the chemisorption of NO on iron surfaces.

Timms and Atkins (181) reacted the novel ligand, PN (180) (generated by pyrolyzing P_5N_3), with a number of different metal atoms. With Cu, a bis(PN) complex was formed, although it was not possible to distinguish between the two isomers

$$N \equiv P \rightarrow Cu \leftarrow P \equiv N \quad \text{or} \quad P \equiv N \rightarrow Cu \leftarrow N \equiv P. \tag{24}$$

MO calculations (86) suggested that coordination through phosphorus may be more probable. Au atoms also formed a Au(PN)$_2$ species in Kr. These Cu and Au species are straw-colored; in contrast, Ag atoms and PN form an intensely blue matrix, with IR bands suggesting the presence of two species, assigned as Ag(PN) and Ag$_2$(PN)$_2$, the latter having bridging, PN ligands (suggested by the much lower ν_{PN}, the 150-cm^{-1} shift being similar to the shifts observed between terminal and bridging CO ligands). Reactions with Co, Ni, or Pd atoms also yielded complexes, although the stoichiometries were not elucidated.

Iron atoms have been reacted with various phosphine ligands. Verkade et al. (194) reacted Fe with P(OMe$_3$)$_3$ on a macro-scale, both by cocondensing the two simultaneously, and by depositing Fe onto a cooled solution of the ligand. The product isolated from the reaction mixture was Fe[P(OMe)$_3$]$_5$, which could also be synthesized by the sodium amalgam reduction of FeCl$_2$ in the presence of an excess of the ligand. Another complex, Fe[P(OCH$_2$)$_3$CCH$_2$CH$_3$]$_5$, was synthesized by cocondensing Fe atoms with 1,5-COD in a static reactor precharged with the phosphine ligand. Yields were, however, quite low (<20%) in the metal-atom reactions. King and co-workers (63) reacted Fe vapor with aminodifluorophosphines, to form Fe[(CH$_3$)$_2$NPF$_2$]$_3$ in 14% yield.

In addition, a similar reaction resulted in a 1% yield of Fe[CH$_3$N(PF$_2$)$_2$]$_4$.

The iron complex Fe[P(OC$_6$H$_5$)$_3$]$_2$[(C$_6$H$_4$O)P(OC$_6$H$_5$)$_2$]$_2$ has been synthesized by metal-atom evaporation-techniques (*190*). The complex is, formally, the result of two ortho-oxidative, C–H additions, accompanied by loss of a molecule of H$_2$.

Ytterbium atoms have been reacted with thermally generated hydrogen or deuterium atoms, with the resultant formation of YbH and YbD (*198*). The IR ν_{YbH} stretching-frequency was observed at 1214.9 cm^{-1}. In addition, Yb atom and YbH absorption and emission spectra were observed. The magnetic parameters of YbH were determined from the esr spectra of the $^2\Sigma$ molecules (with Yb nuclear spin I = 0 and I = ½) to be $g_\parallel$ = 1.9953, $g_\perp$ = 1.9402, $A_\parallel$(H) = 226 MHz, $A_\perp$(H) = 224 MHz, $A_\perp[^{171}\text{Yb}$ (I = ½)] = 5.266 GHz, $A_\parallel[^{171}\text{Yb}$ (I = ½)] = 5.724 GHz. The hyperfine parameters indicated that the spin density is less than 20% on the hydrogen, and that the bonding is largely Yb$^+$H$^-$.

Another study (*200*) presented IR data for a number of hydride and deuteride species. Using matrix-isolation spectroscopy in conjunction with a hollow-cathode, sputtering source (the apparatus for which is shown in Fig. 36), the IR-active vibrations of the diatomic hydrides and deuterides of aluminum, copper, and nickel were observed. The vibra-

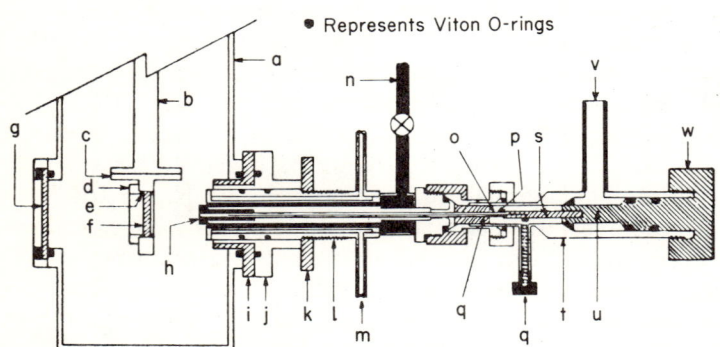

FIG. 36. Cross-sectional view of a matrix-isolation, hollow-cathode, sputtering device. See ref. (*200*) for explanation of components. The letters indicate the following: (a) vacuum chamber; (b) Cryogenic Technology, Inc., Model 21 closed-cycle, helium refrigerator; (c) OFHC-copper-deposition plate-holder; (d) OFHC-copper retaining ring; (e) indium gaskets; (f) KCl or CsI deposition plate; (g) quartz window; (h) metal cathode; (i) Teflon spacer; (j) flange; (k) brass, threaded ring; (l) cathode support; (m) water-cooled tubes; (n) gas-flow inlet; (o) quartz tube; (p) platinum-wire anode; (q) Teflon plug; (r) spring-loaded electrode; (s) brass rod; (t) modified, glass stopcock; (u) stopcock valve: (v) discharge-gas inlet; (w) stopcock knob.

tions at 14 K in an Ar matrix were AlH, 1593; AlD, 1158; CuH, 1882; CuD, 1356; NiH, 1906; and NiD, 1374 cm^{-1}. Interestingly, the compounds were produced only when hydride or deuteride atoms were formed in the discharge; on codeposition of the metals with molecular H_2, hydrides were not observed. Spectra of molecular hydrides and deuterides of Ti, V, Zr, Nb, Mo, and W were also isolated (200), but, because of spectral complications, the compound stoichiometries were not reported.

VI. Organometallic Complexes

It is in the synthesis of organometallic complexes that the metal-atom technique shows its greatest utility. From metal vapors, many complexes may be synthesized on a macroscale that are difficult, if not impossible, to prepare by standard, wet-chemical techniques (64, 65). In this section, we shall illustrate the vast potential that the method has in this area, although, to be sure, it is evident throughout this entire review.

A. Arene Complexes

Chromium atoms have been reacted at 77 K with a mixture of benzene and pentafluorobenzene (84), to give a 22% yield of the sandwich complex [$(C_6H_6)Cr(C_6F_5H)$]; this was the starting point in an interesting series of reactions (85). The complex is readily lithiated at $-78°C$, with subsequent conversion into [$(C_6H_6)Cr(C_6F_5X)$], where X = $SiMe_3$, CO_2Li, CMe_2OH, $(\eta$-$C_5H_5)Fe(CO)_2$, or $(\eta$-$C_5H_5)Fe(\eta$-$C_5H_4CHOH)$. In a similar way, the complexes 1,2,3,5-tetrafluorochromarene, 1,2,4,5-tetrafluorochromarene, 2,3,4,5,6-pentafluoro-1,1′-dimethylchromarene, and 1,1′-dimethylchromarene (85) were obtained. [Chromarene was the authors' name for $(C_6H_6)_2Cr$, the derivatives being named by analogy with ferrocene (85).]

Benzene, benzene-d_6, and fluorobenzene were found (170, 171) to react with chromium, cobalt, iron, and nickel atoms on codeposition in the neat ligand at 77 K, or in argon matrices at 10–12 K. IR studies of the products indicated that the initial reaction of these transition-metal atoms with an aromatic system is π-complex formation. Studies of ligand concentration-effects showed that the chromium-atom reaction is approximately second-order with respect to benzene, yielding the previously known (182) complex $(C_6H_6)_2Cr$, whereas, with the other metals, the reaction is first-order, yielding $M(C_6H_6)$, M = Co, Fe, or Ni. The absence of $Cr(C_6H_6)$ is probably a reflection of (a) the fact that the

M:L ratio in this particular study was not sufficiently low to permit isolation of $Cr(C_6H_6)$, or (b) the exceedingly high lability of the complex. Stoichiometries were determined by mixed C_6H_6/C_6D_6 or C_6H_6/C_6H_5F depositions, an example of which is shown in Fig. 37. The IR shifts of some modes of $M(C_6H_6)$ (M = Co, Fe, or Ni) and $Cr(C_6H_6)_2$, tabulated in Table XV, indicated that the relative bond-strengths of the metal–arene bonds are Cr > Fe > Co > Ni. Evidence for binuclear or higher cluster complexes, or both, was observed, but these species were not characterized. Manganese atoms were found not to react with benzene under the conditions studied, possibly as a result of manganese cluster-formation.

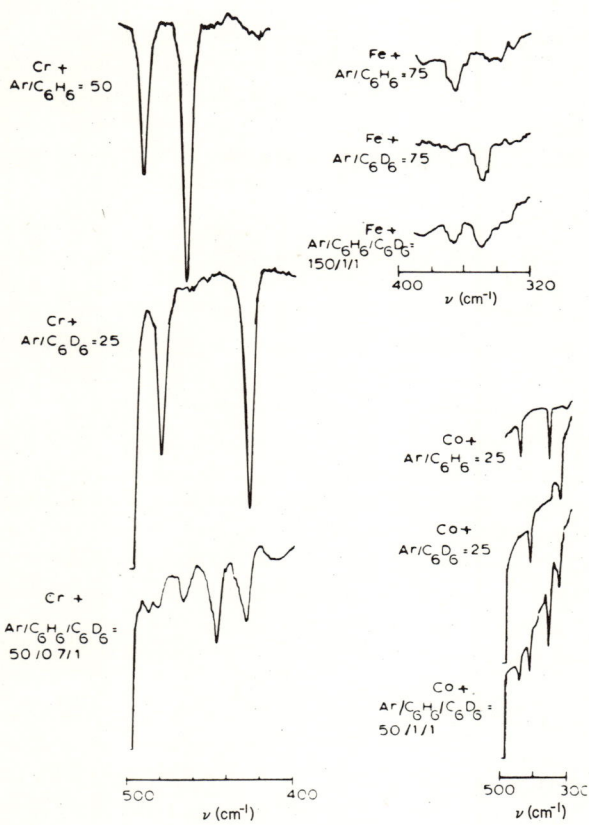

FIG. 37. Far-infrared spectra of chromium, iron, and cobalt atom reactions with benzene, benzene-d_6 and benzene/benzene-d_6 mixtures in argon matrices at 10–12K (*171*).

TABLE XV
Far-Infrared Bands and Shifts of the νC=C Stretching and δC–H o.p. Bending Vibrations of Benzene upon Complexation[a] (171)

Complex	$\Delta\nu$(C=C) stretch[b] (cm^{-1})	δ(C–H) o.p. bend[b] (cm^{-1})	Far-infrared (cm^{-1})
(C$_6$H$_6$)$_2$Cr	−49	+122	492, 466
(C$_6$H$_6$)Fe	−40	+89	485,[c] 366
(C$_6$H$_6$)Co	−39	+85	454, 366
(C$_6$H$_6$)Ni	−31	+75	445,[c] 346

[a] Argon matrices at 10–12K, unless otherwise noted. [b] The symbols + and − indicate shifts to higher and lower wavenumbers, respectively. [c] Benzene matrix at 77 K.

Complexation with polyaromatic systems has also been observed. For instance, M(naphthalene)$_2$, M = Cr (88, 183), Mo (183), V (183), or Ti (183) may be synthesized in a solution reactor with the appropriate, metal vapors at liquid-nitrogen temperature. The Cr/naphthalene complex is less stable (dec. 160°C) than Cr(C$_6$H$_6$)$_2$ (m.p. 283–284° C). In fact, the naphthalene ligand is sufficiently labile to allow reaction under mild conditions, to afford CrL$_6$ (L = CO or Bu^{t}NC), or Cr(naphth)L$_3$ [L = PF$_3$, P(OMe)$_3$, or PMe$_3$]. The Mo, V, and Ti species are equally reactive. Analogous 1-methylnaphthalene complexes were also isolated (183). In addition, the complexes shown in Fig. 38 were synthesized by reaction, at the temperature of liquid nitrogen, of Cr atoms with 1,4-diphenylbutane (35, 201, 202). Analogous complexes were formed with 1,5-diphenylbutane (202).

Lagowski et al. (79) synthesized a very large series of bis(arene) Cr(0) compounds, thereby providing a good example of the use of metal-atom chemistry in synthesizing homologous series of compounds

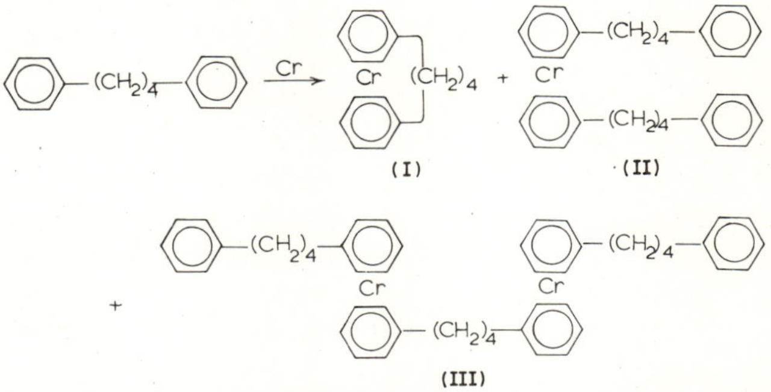

Fig. 38. Complexes synthesized by the reaction of chromium atoms with 1,4-diphenylbutane at the temperature of liquid nitrogen. From ref. (201).

in order to observe the evolution of various properties as a function of ligand and metal. In this case, the ^{13}C-NMR spectra were systematically recorded. For monosubstituted, bis(arene)Cr complexes containing substituents known (78) to perturb the resonance system of the arene (e.g., CH_3, CO_2CH_3, and $N(CH_3)_2$), an analysis of the C-4 chemical-shift indicated that there is no transmission of substituent effects between the complexed rings. The results were interpreted as meaning that a significant diminution of ring aromaticity occurs for complexation to chromium, an effect attributed to the donation of arene, π-electron density into vacant metal-orbitals. This lowering of aromaticity explains the failure of the compounds to undergo electrophilic, aromatic substitution. The availability of complexes having good leaving-groups, such as F and Cl, together with the lessened aromatic character of the ring, suggested that nucleophilic, aromatic substitution-reactions may be feasible.

As was suggested in the preceding discussion, most of the arene complexes isolated by metal-atom techniques are benzene derivatives. However, heterocyclic ligands are also known to act as 5- or 6-electron donors in transition-metal π-complexes (79), and it has proved possible to isolate heterocyclic complexes via the metal-atom route. Bis(2,6-dimethylpyridine)Cr(0) was prepared by cocondensation of Cr atoms with the ligand at 77 K (79). The red-brown product was isolated in only 2% yield; the stoichiometry was confirmed by mass spectrometry, and the structure determined by X-ray crystal-structure analysis, which supported a sandwich formulation.

Using an electron-gun source, tungsten atoms were reacted with benzene, toluene, or mesitylene at 77 K, to form the expected (arene)$_2$W complex (42) in a yield of ~30%, compared with the ~2% yield from the previously published, bis(benzene)W synthesis (32). These arene complexes are reversibly protonated, to give the appropriate [(η-arene)$_2$WH]$^+$ species. By using the same technique, the analogous, niobium complexes were isolated (43).

Metal-σ-aryl complexes may also be generated via the metal-atom route. For example, the cocondensation of cobalt atoms with C_6F_5Br yields $(C_6F_5)_2Co$ and $CoBr_2$ [70], both Co(II) species, although the mechanism for their formation was not elucidated. On adding toluene to the reaction mixture, the complex $(C_6F_5)_2Co(toluene)$ was isolated (70). The X-ray structure of the compound showed a cobalt atom σ-bonded to two F-phenyl ligands and π-bonded to one toluene. According to the authors, this was the first example of an η^6-arene complex of an R_2M compound. An isostructural complex was formed in the reaction of nickel atoms with toluene and bromopentafluorobenzene (71). These materials are proving to be interesting arene hydrogenation catalysts (70, 71). An intermediate species, C_6F_5NiBr, observed to be stable to $-80°C$, can be trapped by addition of PEt_3. On warming this compound in the absence of toluene, decafluorobiphenyl was formed.

B. ALKENE AND ALKYNE COMPLEXES

1. Mono-enes and Mono-ynes

Nickel and palladium react with a number of olefins other than ethylene, to afford a wide range of binary complexes. With styrene (11), Ni atoms react at 77 K to form tris(styrene)Ni(0), a red-brown solid that decomposes at -20 °C. The ability of nickel atoms to coordinate three olefins with a bulky phenyl substituent illustrates that the steric and electronic effects (54, 141) responsible for the stability of a tris (planar) coordination are not sufficiently great to preclude formation of a tris complex rather than a bis (olefin) species as the highest-stoichiometry complex. In contrast to the nickel-atom reaction, chromium atoms react (11) with styrene, to form both polystyrene and an intractable material in which chromium is bonded to polystyrene. It would be interesting to ascertain whether such a polymeric material might have any catalytic activity, in view of the current interest in polymer-supported catalysts (51).

The systematic synthesis and spectral examination of a large series of complexes, $M(ol)_n$, $n = 1-3$, M = Ni or Pd, has been performed (140 –142), with special reference to the optical spectra of the products, again affirming the usefulness of the technique for observing spectral trends as a function of substituent (see later). A number of interesting points emerged from this study, some of which have already been alluded to. The optical data for the nickel and palladium complexes respectively are reported in Tables XVI and XVII.

For all the olefins studied, alkyl-, fluoro-, or chloro-substituted, *three* binary, mononuclear species were observed. It now seems that it is a general property of Ni and Pd atom–olefin reactions at cryogenic temperatures to form complexes that have a maximum coordination of three olefin molecules per metal atom, regardless of the electronic or steric attributes of the substituent(s). As intimated previously, the absence of higher stoichiometry species, even for unsubstituted ethylene, is, most probably, the result of steric interactions (54).

As was the case for the Ni (123) and Pd/C_2H_4 (140) systems, each of the binary olefin complexes isolated has associated with it a moderately intense, UV band, the bands for Pd complexes lying at higher energy than those of the nickel complexes; in addition, for each olefin sys-

TABLE XVI
Optical Spectroscopic Data for $Ni(ol)_n$ (where n = 1, 2, and 3) (141)

	Olefin	n	$\lambda_{max}(cm^{-1})$	$T_{dec.}(K)^a$	$-E_{\pi^*}(eV)$
A	C_2H_4 (ethylene)	1	31 350	35	2.91
		2	35 714	28	
		3	42 373	80	
B	C_3H_6 (propene)	1	30 864	40	2.58
		2	34 843	35	
		3	41 152	67	
C	C_4H_8 (1-butene)	1	30 675	49	2.48
		2	34 722	39	
		3	40 160	77	
D	C_4H_8 (isobutene)	1	30 488	48	2.51
		2	35 088	36	
		3	39 370	80	
E	C_4H_8 (*cis*-2-butene)	1	30 303	105	2.01
		2	34 483	36^b	
		3	41 152	105	
F	C_4H_8 (*trans*-2-butene)	1	30 675	51	2.16
		2	34 722	25^b	
		3	39 682	80	
G	C_6H_{12} (1-hexene)	1	30 650	38	2.29
		2	34 602	28	
		3	39 682	38	
H	CH_2CHCl (vinyl chloride)	1	31 153	65	3.43
		2	35 714	25	
		3	41 841	225	
I	CH_2CHF (vinyl fluoride)	1	31 056	37	3.08
		2	35 336	31	
		3	41 666	65	
J	$CClFCF_2$ (chlorotrifluoroethylene)	1	31 056	25	2.44
		2	34 602	60	
		3	40 000	60	
K	CH_2CHCH_2Cl (allyl chloride)	1	30 864	100	
		2	35 461	100	
		3	43 290	100	
L	C_2F_4 (tetrafluoroethylene)	1	30 675	35	1.68
		2	35 714	37	
		3	41 322	37	

[a] Highest temperature at which a species absorbance was observed. This constitutes a lower limit for complex stability. [b] Estimated value, due to band-overlap difficulties in these spectra.

TABLE XVII
OPTICAL SPECTROSCOPIC DATA FOR $Pd(ol)_n$
(WHERE $n = 1, 2,$ AND 3) (141)

	Olefin (ol)	n	λ_{max} (nm)
A	C_2H_4	1	240
		2	221
		3	204
B	C_3H_6	1	240
		2	221
		3	204
C	C_2H_3Cl	1	239
		2	221
		3	204
D	$CH_2=CHCH_2Cl$	1	238
		2	221
		3	203

tem, the transition energies move to higher energy with increasing olefin stoichiometry. The salient feature of the optical spectra is, however, the insensitivity of the UV-absorption energy as a function of olefin substituent (see Tables XVI and XVII). There is a suggested correlation between the value of λ_{max} for Ni(ol) and the π^* orbital energy of the free olefin, as shown in Fig. 39; however, the fact that the shifts for λ_{max} are so small, whereas the π^* orbital energy varies so much (see Table XVI) is inconsistent with the previous assignment (123) of these bands as metal-to-ligand charge-transfer (MLCT). GVB–CI–MO calculations for $Ni(C_2H_4)$ (39, 191) suggested that these transitions might be better described as metal-localized, $M(d) \rightarrow M(p_y)$, excitations, which would tend to be less sensitive to the olefin coordinated to the metal. Were this the case, the slight sensitivity of the transition to E_{π^*} may be rationalized by observing that the $M(p_y)$ orbital could mix somewhat with the olefin π^* orbital.

One other point to note in regard to this study (141) is that any evidence of oxidative addition, particularly with the chloro-olefins, was absent. The similarity of the spectra, coupled with the nonobservation of any bands in the visible region, as well as the observation of $\nu_{C=C}$ in the region commonly associated with π-complexation of an olefin (141, 142), all argue in favor of normal π-coordination, rather than oxidative insertion of the metal atom into, for example, a C–Cl bond. Oxidative, addition reactions of metal atoms will be discussed subsequently.

The cocondensation of iron atoms with styrene or phenylacetylene at 77 K produces (14) a low yield of polystyrene and a triphenylbenzene,

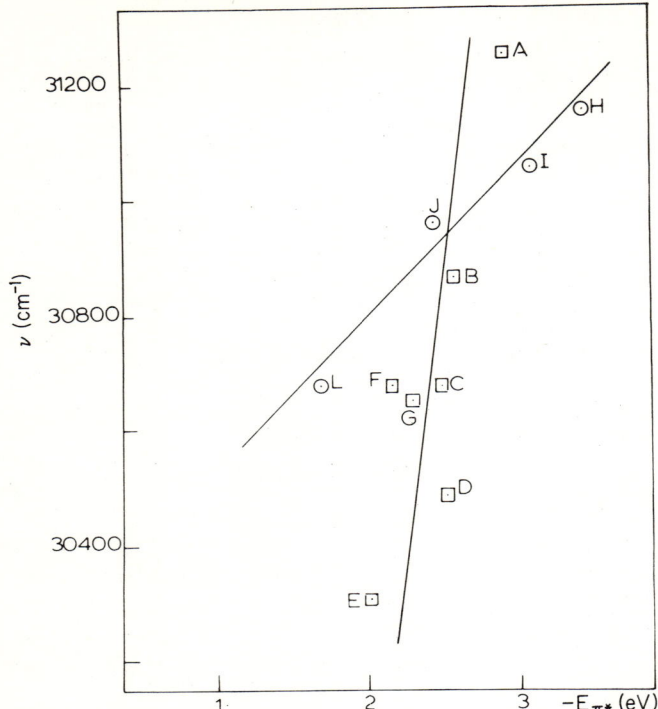

FIG. 39. Graphical representation of the correlation between $\bar{\nu}$ (cm^{-1}) for Ni(ol) and the energy (E_π^*) of the π^* orbital of the free olefin. The olefins are lettered as in Table XVI. Squares indicate hydrocarbon olefins, and circles, fluoro- or chloro-olefins (141).

respectively. However, warm-up in the presence of CO produced (styrene)Fe(CO)$_3$, (styrene)Fe(CO)$_4$, and Fe(CO)$_5$.

The reactions of the copper-triad metals with C$_2$H$_4$ have also been studied, the copper–ethylene system (122, 124) having already been described. Silver and gold atoms behave somewhat differently from copper atoms, forming only M(C$_2$H$_4$). The Ag(C$_2$H$_4$) species was first reported by Kasai and McLeod (59) in an esr study of the reaction of Ag atoms with ethylene. Subsequent IR and UV–visible studies (143) confirmed the authenticity of this purple complex. The existence of Au(C$_2$H$_4$) has also been demonstrated by the metal-atom route (143), this synthesis constituting the first example of a zero-valent, gold–olefin complex. A comparison of the gold and silver optical spectra is made in Fig. 40.

Moving from alkenes to alkynes, it was found that a variety of transition-metal atoms react with hexafluoro-2-butyne (HFB) to form new

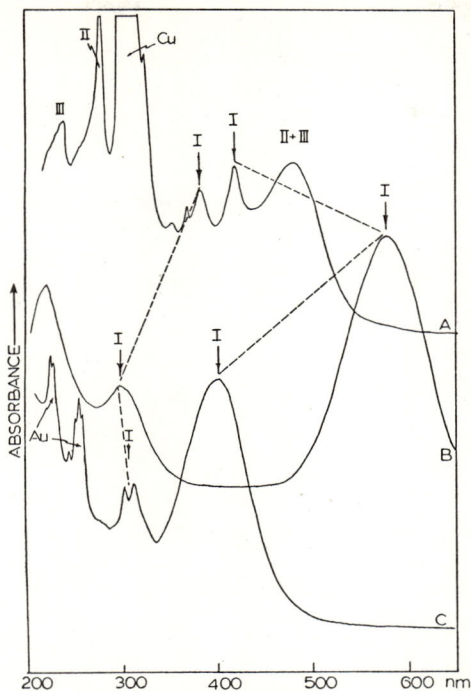

FIG. 40. Optical spectra of matrix-entrapped (A) $Cu(C_2H_4)_{1,2,3}$ in Ar, (B) $Ag(C_2H_4)$ in C_2H_4, and (C) $Au(C_2H_4)$ in C_2H_4 at 10–12K where dotted lines indicate correlations between corresponding "low-energy" $6a_1 \rightarrow 3b_2$ and "high-energy" $5a_1 \rightarrow 6a_1$ excitations of $M(C_2H_4)$, where M = Cu, Ag, Au. Lines ascribed to free Cu and Au atoms in the matrix are indicated with arrows (143, 148).

chemical species (72). Nickel or palladium atoms dispersed in an excess of HFB and subsequently treated with CO at low temperatures yield labile $M(CO)_2(HFB)$ complexes that spontaneously form cluster systems on warming to near room temperature. The clusters formed are $Ni_4(CO)_4(HFB)$, which was already known (178), and the new species $Pd_4(CO)_4(HFB)_3$, in which the alkyne is very labile. The chemistry of these species, from metal-atom generation to ultimate decomposition, is illustrated in Scheme 5. Support for the postulated, binary, mononuclear, metal–alkyne species stems from Kasai and McLeod's esr characterization of $Cu(C_2H_2)_{1,2}$ (60), and unpublished, IR/UV–visible spectral results of Ozin and Power (144) which showed the formation of the species $Ni(C_2H_2)_n$, $n = 1$ or 2, as well as a species $Ni_2(C_2H_2)_m$ (where m is, most probably, unity). Attempts to make the analogous Au, Co, and Pt clusters in a similar way failed (72). However, it was

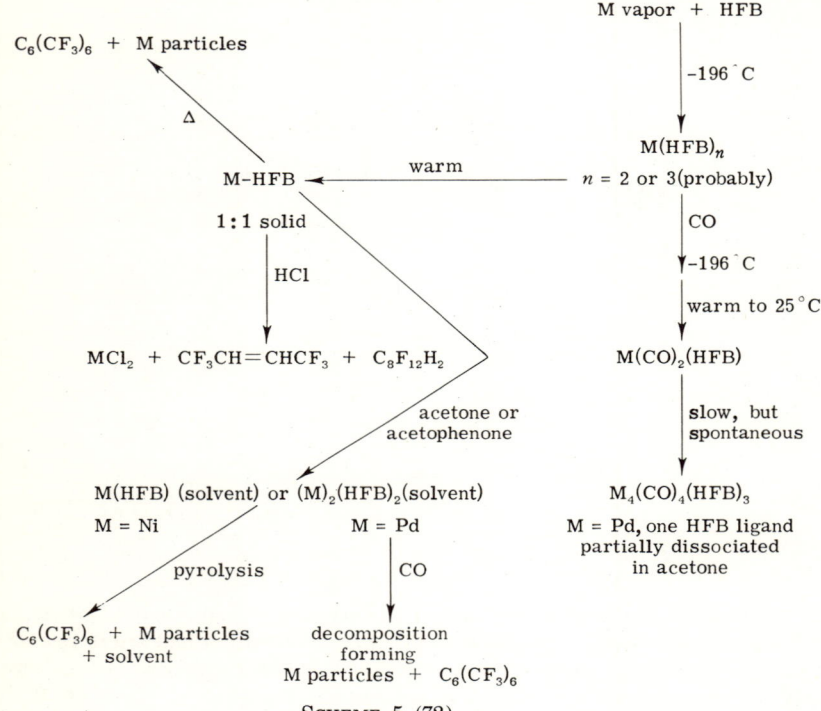

SCHEME 5 (72)

found that Co, Ni, Pd, Pt, Cu, and Au–HFB complexes lacking CO are stable (although the stoichiometries were not reported), and Ni–HFB(solvent) and $Pd_2(HFB)_2$(solvent) were prepared as thermally stable, soluble complexes that have a high propensity for the formation of hexa(trifluoromethyl)benzene, $C_6(CF_3)_6$ (a metal-atom-induced, trimerization reaction). In comparison, the main-group elements Ge and Sn interact with HFB (72) to form low-molecular-weight polymers incorporating a great deal of metal, quite reminiscent of the Cr/styrene reaction (11) already mentioned. The alkynes, 2-butyne and 1-propyne, form high polymers, and these reactions are, apparently, radical processes.

The copper atom–acetylene matrix-reaction, monitored originally by esr spectroscopy (60) has now been investigated by IR/UV–visible spectroscopy (144), and there is general agreement on the identification of two mononuclear species, $Cu(C_2H_2)_{1,2}$. The esr/IR/UV–visible data are consistent with a π-complex formulation for both species, which is in sharp contrast with the Al atom–acetylene, cocondensation product (61), formulated as the radical species

$$\begin{array}{c} H \\ \diagdown \\ C = \dot{C}H \\ \diagup \\ Al \end{array} \qquad (25)$$

which is also distinct from the Al atom–ethylene, cocondensation product formulated as π-bonded $Al(C_2H_4)$ (62).

2. Dienes, Conjugated Dienes, and Cyclic Species

Cyclic dienes and conjugated dienes have also been fashionable systems in the metal-atom synthesis of organometallic complexes. Chromium atoms have been reacted with 1,3-cyclohexadiene at 77 K (12) and, upon subsequent treatment with an excess of PF_3, the complex bis(η^4-cyclohexadiene)–bis(trifluorophosphine)chromium was isolated. Reaction of chromium atoms with 1,4-cyclohexadiene did not afford the analogous complex (12); instead, a η^5-cyclohexadienylhydridotris(trifluorophosphine)chromium(II) species, stable in solution to 50 °C in the absence of air, was formed. Chromium atoms react (13) at 77 K with cycloheptatriene in a hexane matrix to form a brown material which, upon warming to -20 °C, forms the known (33) species $[Cr(C_7H_7)(C_7H_{10})]$. That the low-temperature species contains only coordinated cycloheptatriene was confirmed by warming this species in the presence of PF_3, which resulted in the formation of $[Cr(C_7H_8)(PF_3)_3]$.

Timms and Turney (184) studied the reactions at low temperatures of a wide range of transition-metal atoms with cycloheptatriene and cyclo-octatetraene. With C_7H_8, the reactions are accompanied by extensive hydrogen-migration. Thus, cocondensation of C_7H_8 with the vapors of Ti, V, Cr, Fe, and Co affords $[Ti(\eta-C_7H_7)(\eta^5-C_7H_9)]$, $V(C_{14}H_{16})$, $[Cr(\eta-C_7H_7)(\eta^4-C_7H_{10})]$, $[Fe(\eta^5-C_7H_7)(\eta^5-C_7H_9)]$ or $Fe(C_{14}H_{18})$, and $Co(C_{14}H_n)$, $n = 15$, 17, or 19, respectively. Where the mode of coordination of the cyclic triene is not specified, the data were not sufficiently conclusive to permit determining it. No reaction products were isolated with Mn, Ni, or Pd atoms. However, cocondensation of PF_3/C_7H_8 mixtures with vapors of Cr or Co affords $[Cr(C_7H_8)(PF_3)_3]$ or $[Co(C_7H_8)(PF_3)_3]$, respectively. In general, the complexes formed with C_7H_8 are new examples of sandwich complexes. Reaction of C_8H_8 with Ti, Fe, and Co atoms resulted in the formation of unidentified, intractable polymers; with chromium atoms, the complex $[Cr_2(C_8H_8)_3]$

was obtained, an interesting observation in view of the fact that C_8H_8 complexes of these metals are well established (184).

In contrast, Skell and co-workers (169) demonstrated that there could be prepared, by the metal atom method, a reasonably well-defined, paramagnetic, yellow $Ti_2(C_8H_8)_3$ compound which, in THF, is rapidly reduced with potassium to yield a fairly stable, green solution of the diamagnetic dianion. The ^{1}H-NMR spectrum and the analytical data were all consistent with the formulation of the green dianion shown, which appears to be the

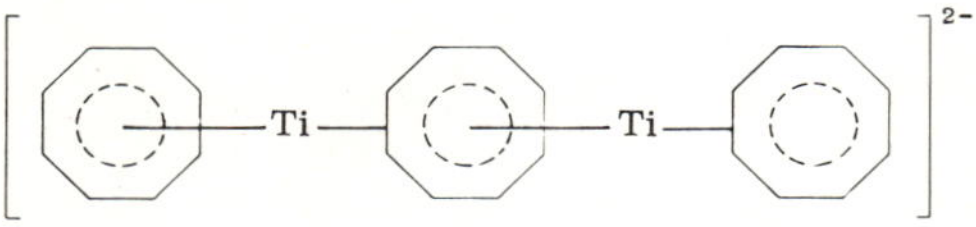

$$\left[\bigcirc \!\!-\! Ti \!-\! \bigcirc \!\!-\! Ti \!-\! \bigcirc \right]^{2-} \quad (26)$$

first instance of such a "triple-decker" sandwich complex.

Reaction of iron atoms with cycloheptatriene to form $[Fe(\eta^5\text{-}C_7H_7)(\eta^5\text{-}C_7H_9)]$ was confirmed by another group (15); these workers determined the crystal structure of the species, demonstrating a sandwich structure with the open faces of the two η^5-systems skewed to each other. The temperature-dependent NMR spectrum of this species (16) indicated two types of fluxional behavior in solution. Evidence for a 1,2-shift mechanism of the 1-5-η-cycloheptatrienyl moiety in the structure shown,

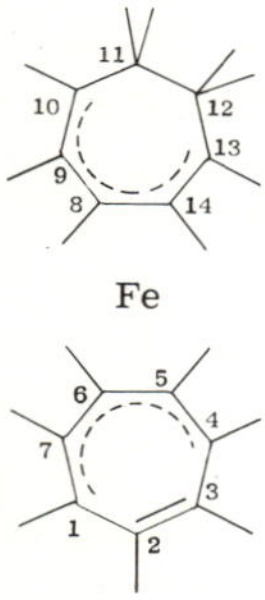

(27)

with respect to the central, iron atom, as well as a low-temperature, rocking motion of both rings, was observed.

Fe atoms have been reacted with butadiene at liquid-nitrogen temperature (14). Upon warm-up in an atmosphere of CO or PF_3, only bis(butadiene)Fe(CO) or bis(butadiene)Fe(PF_3) was isolated. One of the butadienes could be replaced by warming the species in $P(OMe)_3$, to form (butadiene)Fe[$P(OMe)_3$]$_3$. A similar reaction led to the formation of the analogous 2,3-dimethylbutadiene species. In addition, Fe atoms react with 1,5-cyclooctadiene to form (1,5-COD)$_2$Fe (185, 189) which, upon warming in the presence of phosphine (189), yields (1,3-COD)FeP$_3$, P = P(OMe)$_3$, P(OEt)$_3$, or P(O-i-Pr)$_3$. Again, hydrogen migration is evident in the isomerization of the COD ligand.

Lanthanide metal-atom chemistry has also become popular. Reaction of various lanthanide-metal atoms at 77 K with cyclo-octatetraene, with subsequent extraction of the reaction mixture with THF, resulted in a series of lanthanide complexes of formula [Ln(C$_8$H$_8$)(THF)$_2$][Ln(C$_8$H$_8$)$_2$], where Ln is La, Ce, Nd, or Er (24, 25). Reaction with Yb atoms yielded the known compound (51) Yb(C$_8$H$_8$). A crystal-structure determination of the neodymium compound was completed (25), and it showed that all of the C$_8$H$_8$ rings are present as the ten-π-electron, aromatic dianion. The structure was strikingly asymmetrical, in comparison to other lanthanide and actinide C$_8$H$_8$ complexes (25), as is illustrated in Fig. 41. All of these compounds are extremely sensitive to air and moisture, with traces of O$_2$ and H$_2$O causing decomposition. On exposure to air, they inflame spontaneously, often leaving a colored powder of Ln$_2$O$_3$.

A new class of organolanthanide complex has been reported from the metal-atom reaction of lanthanide atoms and butadiene (BD) or 2,-

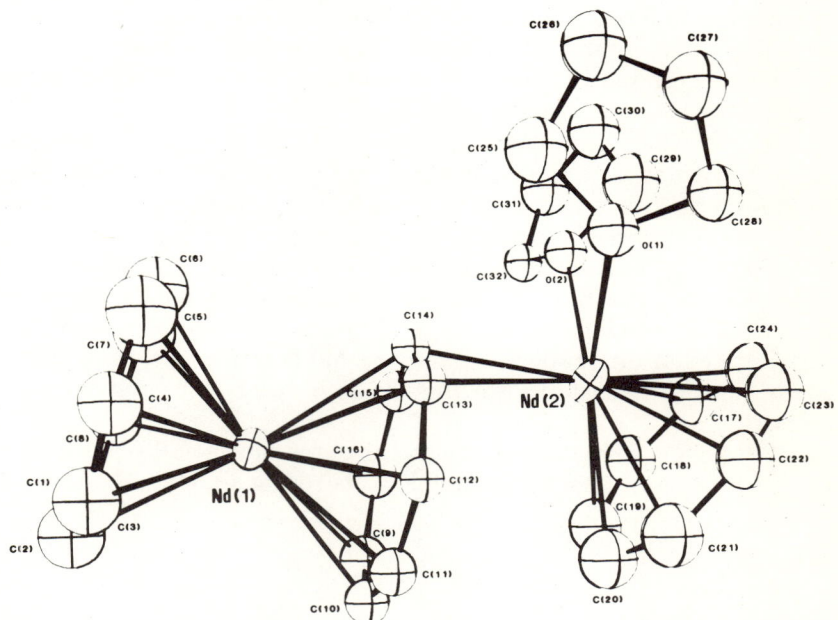

FIG. 41. An ORTEP drawing of [Nd(COT)(THF)$_2$][Nd(COT)$_2$], approximately along the c axis (25).

3-dimethyl-1,3-butadiene (DMBD) (*31*), with M(BD)$_3$ resulting for M = Er, Ne, or Sm, and M(DMBD)$_2$ for M = La or Er. Consideration of the spectral and analytical data suggested that the bonding in these species is best represented as intermediate between structures A and B.

$$\underset{A}{\underset{M}{C{\overset{C-C}{\diagup\diagdown}}C}} \qquad \underset{B}{\underset{M}{C{\overset{C=C}{\diagup\diagdown}}C}} \qquad (28)$$

C. Oxidative Addition Reactions

Although, as has already been mentioned, under matrix conditions between 10 and 77 K, there is no oxidative addition of a chloroolefin to nickel or palladium atoms (*141*), it is evident that this is simply a function of reaction and processing conditions, as it has been shown (*68*) that oxidative addition to C–C or C–H bonds by nickel atoms leads to "pseudocomplexes" having Ni:C:H ratios of 2–5:1:2. Klabunde and co-workers investigated the oxidative addition-reactions of palladium atoms with alkyl halides (*73*) and benzyl chlorides (*74*).

In the former (*73*), oxidative addition of palladium atoms to RX species resulted in formation of RMX compounds. Trapping experiments, free-radical scavenging experiments, and decomposition-product distributions suggested that the metal atom, C–X bond insertion occurs directly, via a caged, radical pair. A number of these RMX species, such as (CF$_3$PdI)$_n$, (C$_2$F$_5$PdI)$_n$, and (*n*-C$_3$F$_7$PdI)$_n$, were isolable, whereas, with such organics as CH$_3$I, the products were not isolable. Without going into too much detail, a number of points had to be reconciled in order to predict a mechanistic pathway.

(1) Coupling and disproportionation reactions of R radicals are a minor process, usually not observed. Hence, these radicals, if formed, are not mobile in the matrix.

(2) Related to this, radical scavengers did not affect product yield or distribution, therefore minimizing the importance of a radical chain process.

(3) Photolysis from the vaporization source had no effect. [Interestingly, it is possible that such photolysis is responsible for the reaction (*186*)

$$\text{Ni} + \text{CH}_2\text{CHCH}_2\text{Cl} \longrightarrow \text{Ni}\underset{\text{Cl}}{\overset{\text{Cl}}{\diagup\diagdown}}\text{Ni} \qquad (29)$$

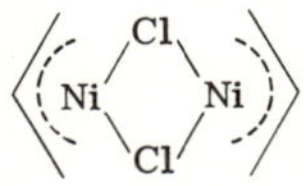

or, at least, the first step (*141*).]

(4) *tert*-Alkyl halides react as effectively as primary halides, indicating a nonbackside-attack mechanism, such as SN2.

(5) Vibrationally excited radicals are probably formed.

(6) HPdX is an intermediate that can reduce R–X to R–H.

(7) Gaseous products arise from the decomposition of RPdX, rather than from the process of formation of RPdX. Hence, if RPdX is stable, it is the only organic product formed.

(8) Unsaturated alkyl halides react first by π-complexation (*141*), followed by C–X oxidative addition, probably on matrix warm-up [but see the preceding point 3, and see ref. (*81*), which suggests that pyrolysis and radical production can occur on the crucible insulating material to cause reaction].

(9) It is likely that complexation to form RX $\cdots$ M occurs at 77 K, with oxidative addition occurring on warm-up (see point 8).

The mechanism shown in Scheme 6 is, for the most part, consistent with points (1) to (9). Thus, initially formed is a σ-complex that is stable only at low temperatures. Upon matrix warm-up, a caged, radical pair forms and, if the R portion possesses a sufficient excess of vibrational energy, decomposition processes may occur. The radicals combine to form RPdX, which may, or may not, be isolated.

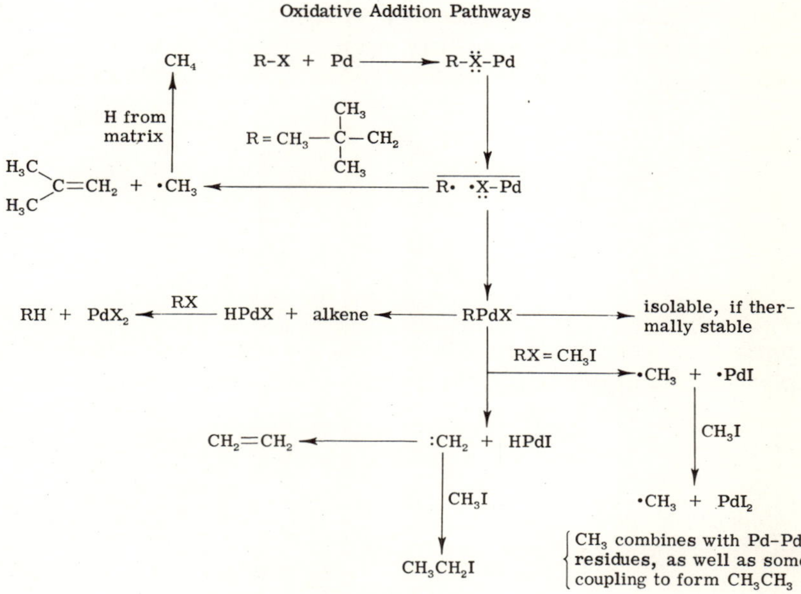

Oxidative Addition Pathways

$$R = CH_3-\underset{\underset{CH_3}{|}}{\overset{\overset{CH_3}{|}}{C}}-CH_2$$

SCHEME 6 (*73*)

As far as the reactions with benzyl chlorides are concerned (74), the oxidative addition of benzyl chloride and substituted benzyl chlorides to palladium atoms yields η^3-benzylpalladium chloride dimers. The parent compound, bis(1,2,3-η^3-benzyl)di-μ-chloro-palladium(II), quantitatively adds four molecules of PEt$_3$ by first forcing the η^3-benzyl–η^1-benzyl transformation, with subsequent breakage of the Pd–Cl bridges to form trans-bis(PEt$_3$)(benzyl)chloroPd(II). The spectral characteristics of the parent molecule are indicative of the allylic type of bonding. Similar η^3-benzyl compounds were formed from 4-methylbenzyl chloride, 2-chloro-1,1,1-trifluoro-2-phenylethane, and 3,4-dimethylbenzyl chloride.

D. Organic Reactions

In addition to complex-formation, the interaction of transition-metal atoms with organic substrates at low temperatures can result in rearrangement of the organic moiety without complexation. Two such reactions have already been briefly mentioned, namely, the polymerization of hexafluoro-2-butyne by Ge and Sn atoms (72) and the polymerization of styrene by Cr atoms (11). In this section we shall briefly summarize some of these transition-metal-atom-promoted, organic rearrangements.

Chromium atoms were cocondensed with benzyl sulfide at 77 K (35), the primary result being desulfurization to form bibenzyl and trans-stilbene. Coordination compounds were not characterized in this system.

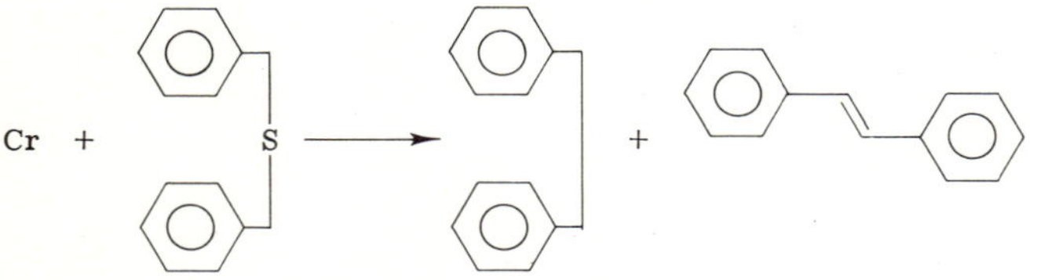

$$\text{Cr} + \text{(PhCH}_2\text{)}_2\text{S} \longrightarrow \text{PhCH}_2\text{CH}_2\text{Ph} + \text{PhCH=CHPh} \quad (30)$$

In contrast, when chromium atoms were reacted with benzyl ether, complexation to only one phenyl group of the ether was observed.

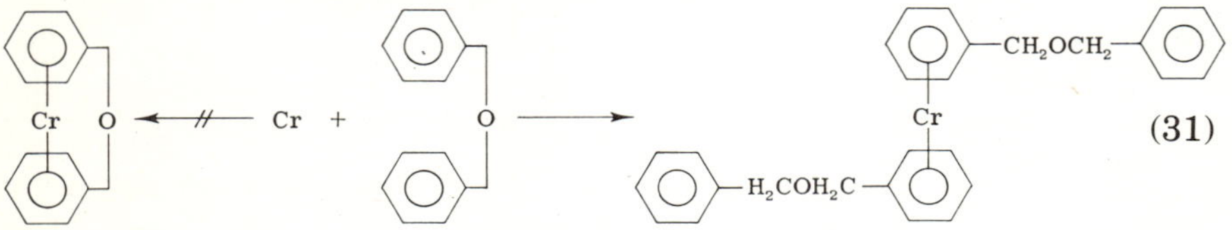

$$\text{(Ph)Cr-O-CH}_2\text{Ph} \;\not\leftarrow\; \text{Cr} + (\text{PhCH}_2)_2\text{O} \longrightarrow (\text{PhCH}_2\text{OCH}_2\text{Ph})\text{Cr} \quad (31)$$

It has, however, proved possible to synthesize the following chromium complexes via a 77 K, Cr-atom, cocondensation reaction with 2,-

2-paracyclophane (*202a*). This is a significant discovery, in view of the fact that

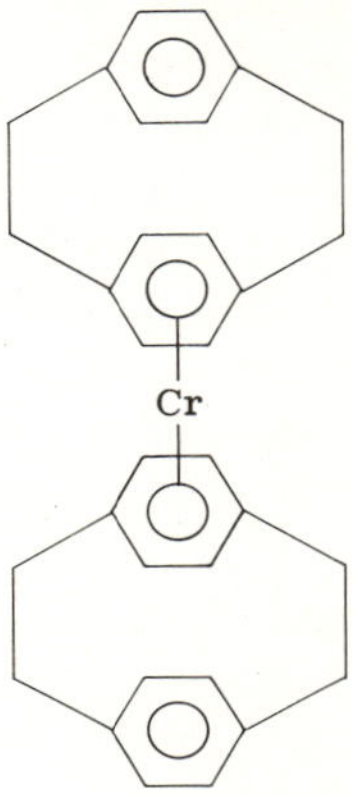

(1) 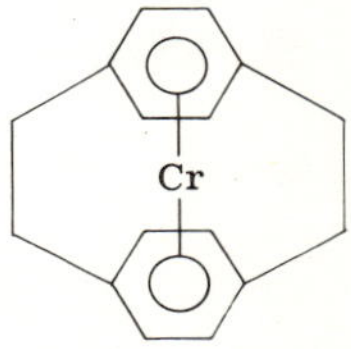   (2)

cyclophanes could hitherto be coordinated to transition metals only in combination with carbonyl ligands, an example (3) being shown.

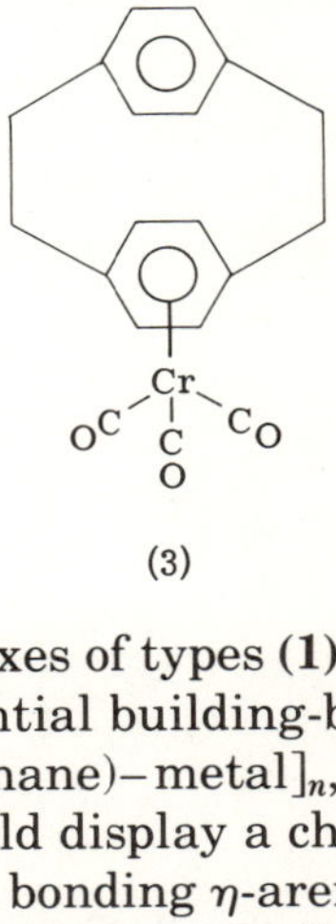

(3) 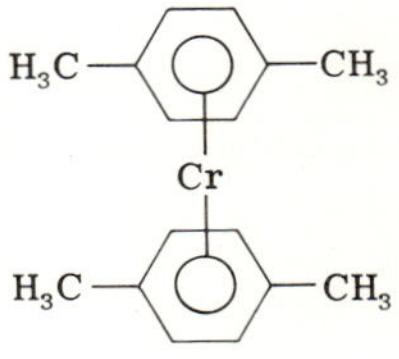   (4)

Binary complexes of types (1) and (2) are of special interest, as (1) represents a potential building-block for columnar structures of composition $[(\eta\text{-cyclophane})-\text{metal}]_n$, and (2) is expected to be kinetically very inert and should display a change from intramolecular, π-π repulsion (free ligand) to bonding η-arene–metal interaction (complex). Interestingly, complex (2) does, in fact, display compressed-sandwich properties, as seen from the esr, hyperfine coupling-constants for the ring protons and the ^{53}Cr site of the monocation of (2), which are increased and decreased, respectively, compared to the corresponding values for the monocation of (1) and (4). This finding pointed to a slightly more extensive $Cr(3d_z^2) \rightarrow$ arene $(\sigma_{a_{1g}})$ spin-delocalization in the monocation of (2), which is probably caused by a shorter metal–ligand distance than in the cation of (1) and (4). The high stability of the cation of (2) [compared to either (1) or (4)] to solvolytic, metal–ligand cleavage has been explained in terms of the "concave sides" of the nonplanar, benzene rings of 2,2-paracyclophane functioning as a rigid, chelate ligand, oriented towards the central metal, thereby shielding the latter from solvolytic attack (202a). For the cation of (1), however, the cleavage can take place in two steps, via the half-sandwich complex (η^6-2,2-paracyclophane)Cr$^+$, whereby the initial attack is easier than in, for example, the cation of (4), because coordination via the "convex sides" of the nonplanar, benzene rings leaves the central metal-atom exposed to attack by solvent. Thus, the observed order of reactivity $(1)^+ > (4)^+ > (2)^+$ may be understood (202a).

Transition-metal atoms have been shown to deoxygenate epoxides to alkenes (36). Chromium and titanium atoms emerged as the most effective species in this regard, abstracting over two equivalents of oxygen. By studying the reaction of a wide range of epoxides with chromium atoms, the reaction

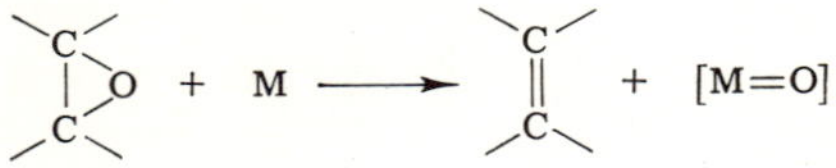

$$\begin{matrix} \diagdown C \diagup \\ | \diagdown O \\ \diagup C \diagdown \end{matrix} \quad + \quad M \quad \longrightarrow \quad \begin{matrix} \diagdown C \diagup \\ || \\ \diagup C \diagdown \end{matrix} \quad + \quad [M=O] \tag{32}$$

was found to be general. In addition, metal atoms also abstract oxygen from compounds containing nitrogen–oxygen multiple bonds (37). On cocondensing chromium atoms with nitrobenzene, a number of interesting organic moieties were observed.

(33)

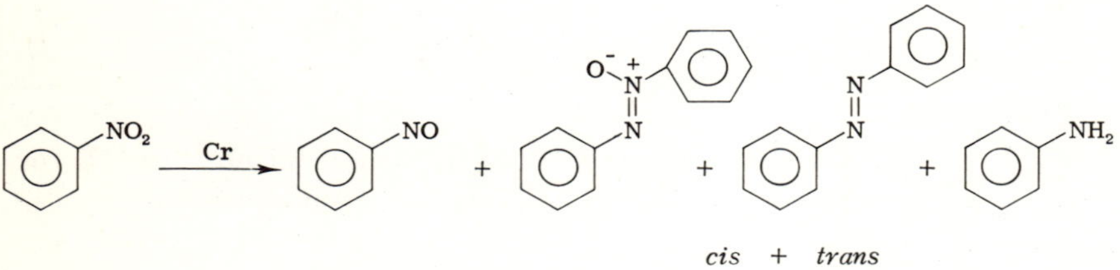

PhNO$_2$ →[Cr] PhNO + Ph–N(O$^-$)=N–Ph + Ph–N=N–Ph + PhNH$_2$

cis + trans

The yields of the products were dependent on the metal:organic ratio. Similar reactions do not occur on preformed, chromium surfaces.

When chromium atoms were cocondensed at 77 K with 1,7-cyclodecadiyne (38), complexation was not observed; however, an organic trimer of the starting material was formed. Standard, organic characterization-techniques showed that this trimer is the one depicted, rather

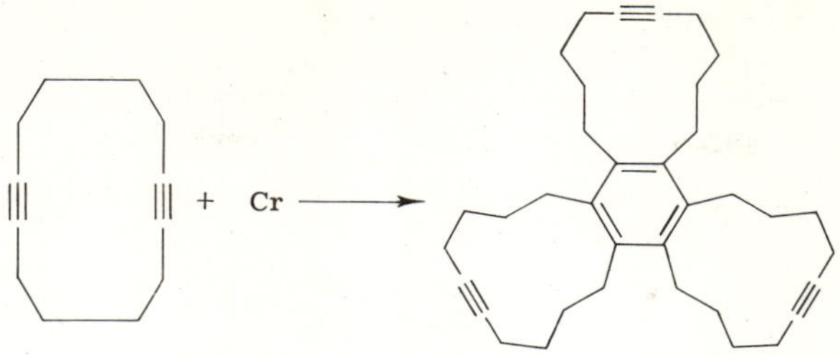

$$\text{[cyclic diyne]} + \text{[cyclic triyne]} + \text{Cr} \longrightarrow \text{[arene complex]} \tag{34}$$

than a cage species [which, it had been supposed, a compound having identical physical properties might be (*176*)].

In addition, Lagowski and Simons showed (*80*) that the black, nickel-containing substances produced by the cocondensation of nickel atoms and alkynes are active, homogeneous catalysts for the oligomerization of terminal acetylenes under mild conditions. Table XVIII shows the yields of the oligomerization of propylene by these catalysts.

E. Miscellaneous Organometallic Species

Somewhat related to the desulfurization reaction already discussed (*35*), the cocondensation of Cr and Fe atoms with thiophenes at 77 K leads to desulfurization of the thiophene (*22, 187*). Warm-up of the iron–thiophene cocondensate in a CO atmosphere produces tricarbonylferrocyclopentadiene–tricarbonyliron.

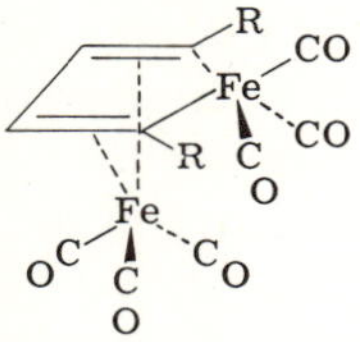

$$\text{[Fe}_2\text{(CO)}_6\text{ diene complex]} \tag{35}$$

The cocondensation of nickel atoms and CS_2 at 12 K resulted in the formation of *three* binary, mononuclear, nickel/CS_2 complexes, $Ni(CS_2)_n$, $n = 1-3$ (*145*). Mixed $^{12}CS_2/^{13}CS_2$ isotopes were used to identify the lowest stoichiometry species. An interpretation of the IR and UV–visible spectra, as well as normal-coordinate analyses (*144*), suggested that these species are best considered as normal π-complexes, with the nickel atom coordinated to the C=S bond in a manner analogous to C=C bond coordination (*123*).

An interesting reaction involves the cocondensation of transition-metal atoms with liquid methylphenylsiloxane polymers at -20 to $0°C$

TABLE XVIII
Oligomers of Propyne (80)

Oligomer	Mol % of oligomer produced					
	10^a	30^a	15^a	34^b	6^b	5^b
	I	II	III	IV		

[a] Integration of the ^{1}H-NMR spectrum shows 55% of tetramer; proportions of individual isomers were estimated from GC/MS data. Tetramer mixtures show an asymmetrical, vinyl resonance centered at δ 5.40, and asymmetrical, methyl resonances centered at δ 1.70 (vs. Me$_4$Si in CS$_2$ solution). [b] Integration of the ^{1}H-NMR spectrum.

(188), a much higher temperature than is normally used in metal-atom work. Polymeric compounds (liquid at room temperature) in which two phenyl rings in the polymer are coordinated to each metal atom were formed with Ti, V, Cr, Mo, and W vapors. With chromium atoms, for example, reaction proceeded until about 50% of the available phenyl groups became coordinated to the metal.

An interesting study combining metal-atom chemistry with previously synthesized gold complexes resulted in the novel synthesis, and subsequent Mössbauer investigation, of gold cluster species (195). A typical experiment involved the evaporation of gold metal into an ethanol film (at $-100°C$) containing Au(PAr$_3$)X with the ratios of reactants being 8:3:4. The Mössbauer spectra of the products, Au$_{11}$(PAr$_3$)$_7$X$_3$ (Ar = C$_6$H$_5$, p-ClC$_6$H$_4$, X = SCN,I, or CN), showed different gold sites in the cluster. An interpretation involving five different gold sites (according to the crystal structure) was possible. The yield from the metal-atom route to these gold cluster complexes was 70%.

To conclude, we shall mention some metal-atom reactions with boranes (172) and carboranes (173). When cobalt atoms reacted with pentaborane(9) and cyclopentadiene, a number of new metalloborane clusters were formed (172), two of which were B$_5$H$_5$Co$_3$(η-C$_5$H$_5$)$_3$ and cyclopentyl-B$_5$H$_4$Co$_2$(η-C$_5$H$_5$)$_3$. Possible structures for the former are shown in Fig. 42. The reaction of cyclopentadiene, pentaborane(9), and 2-butyne with cobalt atoms yielded the metallocarborane species illustrated in Fig. 43 (173).

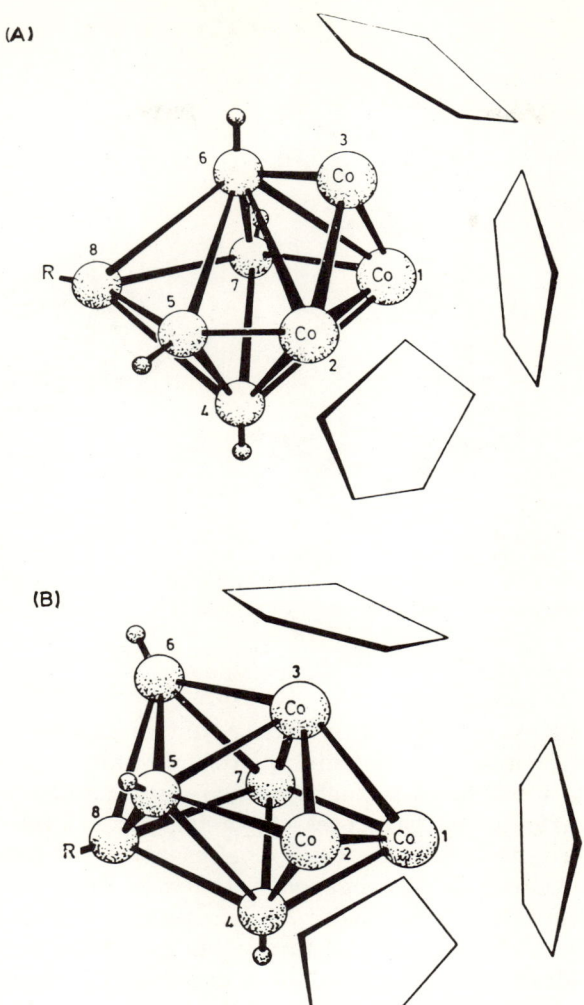

FIG. 42. Possible capped (A) and dodecahedral (B) structures for compound **(I)** (R = H) and compound **(II)** (R = cyclopentyl) (*172*).

$$B_5H_9 + C_5H_6 + CH_3C{\equiv}CCH_3 + Co\ (atoms) \longrightarrow$$

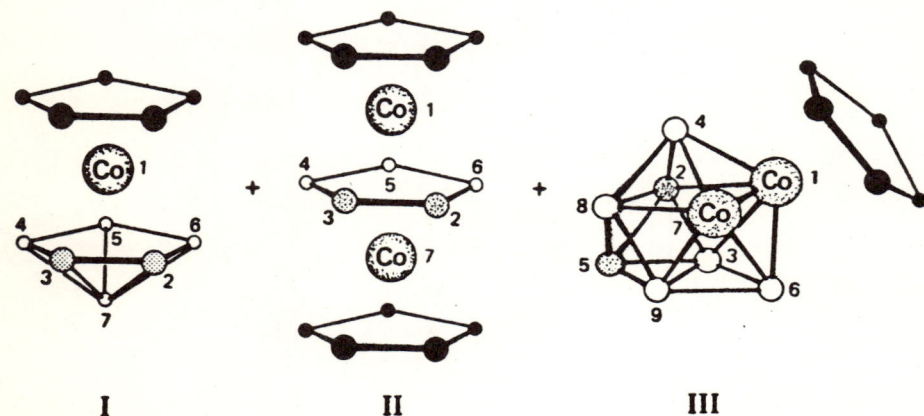

FIG. 43. Reaction of cyclopentadiene, pentaborane(9), and 2-butyne with cobalt atoms. Open circles, BH units; shaded circles, C–CH$_3$ units; and dark circles, C–H units. One cyclopentadienyl ring has been omitted on compound **III**, for clarity (*173*).

VII. Conclusion

It should be evident from this review that the fields of macro- and matrix-scale, metal-vapor chemistry are flourishing and are finding many unique applications in such areas as organic and organometallic synthesis, metal nucleation processes, photography, cluster model theory, surface-reaction intermediates, chemisorption models, alloy and bimetallic clusters, and heterogeneous catalysis. The interdisciplinary nature of the field is now apparent, and many new directions are expected to emerge in the near future. A logical move at this time would be a closer interaction between metal-vapor chemists and researchers in the fields of organometallic synthesis, homogeneous and heterogeneous catalysis, surface chemistry, and physics.

Addendum

The purpose of this section, added in proof, is to include certain references inadvertently omitted in the initial preparation of this article, as well as to update the review. We consider the literature coverage to be complete to about July 31, 1979. The papers referred to in this section will not be discussed in detail.

It is worth noting, prior to citing actual metal atom studies, the recent secondary ion mass spectrometry (SIMS)[1] on an argon matrix-isolated propene sample, demonstrating the applicability of SIMS analysis to the characterization of matrix-isolated species. The same group has reported the first ^{13}C NMR spectra of organic molecules trapped in an argon matrix.[1]

Further investigation of the optical spectra of Cu, Ag, and Au atoms isolated in H_2, O_2, N_2, and CH_4 matrices has been reported.[2]

Mössbauer spectra for the species Fe, Fe_2, Fe_2Mn, and Fe_3 have been investigated.[3]

The absorption spectra of Zr atoms isolated in a variety of matrices have been reported.[4] In addition, the diatomic molecule ZrN, prepared using a hollow cathode source and N_2, was observed.[4] Other work involving N_2 included the identification of ThN and Th(N_2),[5] and TaN[6] in various matrices.

Weltner et al.[7] have performed an ESR and optical study of the high spin molecules MnH and MnH_2 (and possibly MnH_3) in Ar and Ne at 4K.

Three separate groups have recently published metal vapor–phosphine studies. Bowmaker[8] isolated the series of complexes Ni(PH_3)$_n$ ($n = 1 \to 4$) and Cu(PH_3)$_n$ ($n = 1 \to 3$) via matrix isolation, the former series also having been investigated by Trabelsi and Loutellier.[9] Co-condensation of $(CH_3)_2NPF_2$ with Cr, Fe, and Ni vapor at 77K gave the appropriate M($PF_2N(CH_3)_2$)$_n$ complexes,[10] where M = Cr, $n = 6$; M = Fe, $n = 5$; and M = Ni, $n = 4$. Co-condensation of Cr, Fe, Co, and Ni vapors with $CH_3N(PF_2)_2$[10] gave Cr[$(PF_2)_2NCH_3$]$_2$, Fe[$(PF_2)_2NCH_3$]$_4$, Co_2[$(PF_2)_2NCH_3$]$_5$, and {Ni[$(PF_2)_2NCH_3$]$_2$}$_n$. Mixed ligand co-condensations were also studied.[10–12]

With respect to CO complexes, the luminescence spectra of a series of Group VI metal carbonyls and substituted carbonyls were obtained in frozen gas matrices at 12K.[13] In addition, the IR spectra of HCo(CO)$_4$ and HCo(CO)$_3$ (proposed as an intermediate in hydroformylation) were observed in an argon matrix.[14]

Recent organometallic studies have been reported. Mössbauer spectra have shown that the Fe_2 dimer reacts at 4K with CH_4 to oxidatively cleave the C—H bond to form HFe_2CH_3.[15] Zirconium atoms have also been shown to oxidatively cleave the C—H and C—C bonds of alkanes to form discrete organometallic species.[16]

Co-condensation of Hf and Zr atoms from an electron-gun evaporation device, with P(Me)$_3$ and arenes at 77K gave good yields of the species [M(arene)$_2$P(Me$_3$)].[17] Metal vapor synthesis led to Fe(η^6-arene)L$_2$ and Fe(η^6-arene)-(η^4-diene),[18] where L is a phosphorus ligand. In addition, complexes of stoichiometry Fe(η^4-diene)L$_3$ (where L is again a phosphorus ligand) were prepared for a variety of simple cyclic and acyclic dienes,[19] and their NMR temperature dependence studied.

Ruthenium vapor, obtained by heating tungsten filaments coated with a mixture of ruthenium powder and epoxy resin, yielded $Ru(Me_2NPF_2)_5$ and $Ru(C_6H_6)_2$ on reaction with the appropriate ligand.[20a] The atomic nature and purity of Ru vapor generated in this way were subsequently confirmed by matrix isolation optical spectroscopy.[20b] In this study, other refractory metal atoms and clusters ($Zr_{2,3}$, $Pd_{1,2}$, $Au_{2,3}$, Hf, NbMo) were also investigated by matrix UV–visible methods. Bis(arene) complexes, including such species as bis(toluene)-molybdenum and bis(1-methylnaphthalene)molybdenum, can also be synthesized by condensing potassium atoms into THF solutions of the metal halides and arenes at 173K.[21]

Organic rearrangements have also been observed in metal atom chemistry. When nickel atoms were condensed with dimethylfulvene at 77K,[22] the only observed product was a dimer of the starting organic, resulting from a formal [6 + 6] cycloaddition. Nickel atoms were also suggested to cyclodimerize norbornadiene via nickel–cyclopentane intermediates.[23] Metal atoms were also reacted with organocyclopropanes and spirocycles,[24] with the result that metal atoms reaggregated at a faster rate than they underwent reaction with the organic molecule, unless a functional group was present with which the metal atom could strongly interact.

A method has recently been described for wrapping polymers around metal atoms and very small metal clusters using both matrix and macroscale metal vapor–fluid polymer synthetic techniques.[25] Significant early observations are that (*i*) the experiments can be entirely conducted at, or close to room temperature, (*ii*) the resulting "polymer stabilized metal cluster" combinations are homogeneous liquids which are stable at or near room temperature, and (*iii*) the methodology is easily extended to bimetallic and trimetallic polymer combinations.[25]

Addendum References

1. H.T. Jonkman and J. Michl, *J. Chem. Soc. Chem. Commun.* 751 (1978); K.W. Zilm, R.T. Colin, D.M. Grant, and J. Michl, *J. Am. Chem. Soc.* **100**, 8038 (1978).
2. H. Abe, W. Schulze, and D.M. Kolb, *Chem. Phys. Lett.* **60**, 208 (1979).
3. W. Dyson and P.A. Montano, *J. Am. Chem. Soc.* **100**, 7439 (1979).
4. J.K. Bates and D.M. Gruen, *High Temp. Sci.* **10**, 27 (1978).
5. D.M. Gruen and G.T. Reedy, *J. Mol. Spec.* **74**, 423 (1979).
6. J.K. Bates and D.M. Gruen, *J. Chem. Phys.* **70**, 4428 (1979).
7. R.J. Van Zee, T.C. DeVore, J.L. Wilkerson, and W. Weltner, Jr., *J. Chem. Phys.* **69**, 1869 (1978).

8. G.A. Bowmaker, *Aust. J. Chem.* **31**, 2549 (1978).
9. M. Trabelsi and A. Loutellier, *J. Mol. Struct.* **43**, 151 (1978).
10. R.B. King and M. Chang, *Inorg. Chem.* **18**, 364 (1979).
11. M. Chang, M.G. Newton, and R.B. King, *Inorg. Chim. Acta* **30**, L341 (1978).
12. M. Chang, M.G. Newton, R.B. King, and T.J. Lotz, *Inorg. Chim. Acta* **28**, L153, (1978).
13. T.M. McHugh, R. Neurayanaswamy, A.J. Rest, and K. Salisbury, *J. Chem. Soc. Chem. Commun.* 208 (1979).
14. P. Wermer, B.S. Ault, and M. Orchin, *J. Organomet. Chem.* **162**, 189 (1978).
15. P.H. Barrett, M. Pasternak, and R.G. Pearson, *J. Am. Chem. Soc.* **101**, 222 (1979).
16. R.J. Remick, T.A. Asunta, and P.S. Skell, *J. Am. Chem. Soc.* **101**, 1320 (1979).
17. F.G.N. Cloke and M.L.H. Green, *J. Chem. Soc. Chem. Commun.* 127 (1979).
18. S.D. Ittel and C.A. Tolman, *J. Organomet. Chem.* **172**, C47 (1979).
19. S.D. Ittel, F.A. Van-Catledge, and J.P. Jesson, *J. Am. Chem. Soc.* **101**, 3874 (1979).
20a. P.L. Timms and R.B. King, *J. Chem. Soc. Chem. Commun.* 898 (1978).
20b. W. Klotzbücher and G.A. Ozin, *Inorg. Chem.* (in press).
21. P.N. Hawker, E.P. Kündig, and P.L. Timms, *J. Chem. Soc. Chem. Commun.* 370 (1978).
22. N. Hao, J.F. Sawyer, B.G. Sayer, and M.J. McGlinchey, *J. Am. Chem. Soc.* **101**, 2203 (1979).
23. J.R. Blackborow, V. Feldhoff, F-W. Grevels, R.H. Grubbs, and A. Mingashita, *J. Organomet. Chem.* **173**, 253 (1979).
24. J.A. Gladysz, J.G. Fulcher, R.C. Ugolick, A.J.L. Hanlan, and A.B. Bocarsly, *J. Am. Chem. Soc.* **101**, 3388 (1979).
25. G.A. Ozin and C.G. Francis, *Proc. EUCMOS Conf., Frankfurt, Sept. 1979*, in "Spectroscopy in Chemistry and Physics," Elsevier, New York, 1979, and *J. Mol. Struct.* (1979) (in press); C.G. Francis, G.A. Ozin, and H. Huber, *J. Am. Chem. Soc.* **101**, 6250 (1979); *Inorg. Chem.* **19**, 219 (1980); *Angew. Chem. Int. Ed. Engl.* (in press); C.G. Francis and G.A. Ozin, in "Organometallic Polymers" (E. Carraher, Jr., ed.). Academic Press, New York, 1980.

Acknowledgments

G.A.O. acknowledges the invaluable collaboration of his students and colleagues whose names appear in the cited articles, and expresses his special indebtedness to Mr. Ted Huber for his invaluable technical assistance, Mr. Alex Campbell, Mr. Karl Molnar, Mr. Martin Mittelstadt, and Mr. Bob Torbet for their expert, machine-shop contributions, and Mrs. Elinor Foden for typing this manuscript. The financial assistance of the National Research Council of Canada New Ideas, Strategic Energy, and Operating Grant programmes, Imperial Oil of Canada, Atkinson Foundation, Connaught Fund, Lash Miller Chemical Laboratories, and Erindale College is all gratefully acknowledged. W.J.P. is indebted for an NRCC graduate scholarship throughout the tenure of his research.

References

1. Abramowitz, S., Acquista, N., and Legin, I. W., *Chem. Phys. Lett.* **50**, 423 (1977).
2. Anderson, A. B., *J. Chem. Phys.* **66**, 5108 (1977).

3. Andrews, L., and Ozin, G. A. (Coblentz Award); Timms, P. L., Green, M. L. H., and Turner, J. J. (Corday–Morgan Award); Ozin, G. A., Burdett, J., and Poliakoff, M. (Meldola Award).
3a. Andrews, L., and Smardzewski, R. R., *J. Phys. Chem.* **77,** 801 (1973).
3b. Andrews, L., Hwang, J. T., and Trindle, C. T., *J. Phys. Chem.* **77,** 1065 (1973).
4. Arai, H., and Tominaga, H., *J. Catal.* **43,** 131 (1976).
5. Baetzold, R. C., *J. Chem. Phys.* **55,** 4363 (1971).
6. Bandow, H., Onishi, T., and Tamaru, K., *Chem. Lett.* p. 83 (1978).
7. Barrett, P. H., and Montano, P. A., *J. Chem. Soc., Faraday Trans. 2* p. 378 (1977).
8. Barrett, P., and Montano, P., *Ber. Bunsenges. Phys. Chem.* **82,** 30 (1978).
9. Barrow, R. F., and Griffiths, M. J., *J. Chem. Soc., Faraday Trans. 2* p. 943 (1977).
10. Bates, J. K., and Gruen, D. M., *Inorg. Chem.* **16,** 2450 (1977).
11. Blackborow, J. R., Grubbs, R., Miyashita, A., and Scrivanti, A., *J. Organomet. Chem.* **120,** c49 (1976).
12. Blackborow, J. K., Grubbs, R. H., Miyashita, A., Scrivanti, A., and von Gustorf, E. A. K., *J. Organomet. Chem.* **122,** c6 (1976).
13. Blackborow, J. R., Eady, C. R., von Gustorf, E. A. K., Scrivanti, A., and Wolfbeis, O., *J. Organomet. Chem.* **108,** c32 (1976).
14. Blackborow, J. R., Eady, C. R., von Gustorf, E. A. K., Scrivanti, A., and Wolfbeis, O., *J. Organomet. Chem.* **111,** c3 (1976).
15. Blackborow, J. R., Hildenbrand, K., von Gustorf, E. A. K., Scrivanti, A., Eady, C. R., Ehntolt, D., and Krüger, C., *J. Chem. Soc., Chem. Commun.* p. 16 (1976).
16. Blackborow, J. R., Grubbs, R. H., Hildenbrand, K., von Gustorf, E. A. K., Miyashita, A., and Scrivanti, A., *J. Chem. Soc., Dalton Trans.* p. 2205 (1977).
17. Bor, G., Dietler, U. K., and Noack, K., *J. Chem. Soc., Chem. Commun.* p. 914 (1976).
18. Brown, C. M., and Ginter, M. L., *J. Mol. Spectrosc.* **69,** 25 (1978).
19. Brown, T. L., and Sweany, R. L., *Inorg. Chem.* **16,** 415 (1977).
20. Brown, T. L., and Sweany, R. L., *Inorg. Chem.* **16,** 421 (1977).
21. Burdett, J. K., *Coord. Chem. Rev.* **27,** 1 (1978).
21a. Burdett, J. K., *J. Chem. Soc., Faraday Trans. 2* p. 1599 (1974).
22. Chivers, T., and Timms, P. L., *Can. J. Chem.* **55,** 3509 (1977).
23. Craddock, S., and Hinchcliffe, A. J., "Matrix Isolation." Cambridge Univ. Press, London and New York, 1975.
24. DeKock, C. W., Ely, S. R., and Hopkins, T. E., *J. Am. Chem. Soc.* **98,** 1624 (1976).
25. DeKock, C. W., Ely, S. R., Hopkins, T. E., and Brault, M. A., *Inorg. Chem.* **17,** 625 (1978).
26. DeKock, R. L., *Inorg. Chem.* **10,** 1205 (1971).
27. Demuth, J., Ibach, H., and Lehwald, S., *Surf. Sci.* (1978; in press), and personal communication.
28. DeVore, T. C., and Franzen, H. T., *Inorg. Chem.* **15,** 1318 (1976).
29. DeVore, T. C., *Inorg. Chem.* **15,** 1315 (1976).
30. Efremov, Y. M., Samoilova, A. N., and Gurvich, L. V., *Chem. Phys. Lett.* **44,** 108 (1976).
31. Evans, W. J., Engerer, S. C., and Neville, A. C., *J. Am. Chem. Soc.* **100,** 332 (1978).
32. Fischer, E. O., Scherer, F., and Stahl, H. O., *Chem. Ber.* **93,** 2065 (1960).
33. Fischer, E. O., Reckziegel, A., Muller, J., and Goser, P., *J. Organomet. Chem.* **11,** 13 (1968).
34. (a) Garland, C. W., and Yang, A. L., *J. Phys. Chem.* **61,** 1504 (1975); (b) Garland, C. W., Lord, R. C., and Troiano, P. F., *ibid.* **69,** 1188 (1965).

35. Gladysz, J. A., Fulcher, J. G., and Bocarsly, A. B., *Tetrahedron Lett.* p. 1725 (1978).
36. Gladysz, J. A., Fulcher, J. G., and Togashi, S., *J. Org. Chem.* **41,** 3648 (1976).
37. Gladysz, J. A., Fulcher, J. G., and Togashi, S., *Tetrahedron Lett.* p. 521 (1977).
38. Gladysz, J. A., Fulcher, J. G., Lee, S. J., and Bocarsly, A. B., *Tetrahedron Lett.* p. 3421 (1977).
39. Goddard, III, W. A., and Upton, T. H., *J. Am. Chem. Soc.* **100,** 321 (1978).
40. Green, D. W., and Reedy, G. T., *J. Chem. Phys.* **69,** 544 (1978).
41. Green, D. W., and Reedy, G. T., *J. Chem. Phys.* **69,** 552 (1978).
42. Green, M. L. H., Cloke, F. G. N., and Morris, G. E., *J. Chem. Soc., Chem. Commun.* p. 72 (1978).
43. Green, M. L. H., Cloke, F. G. N., and Price, D. H., *J. Chem. Soc., Chem. Commun.* p. 431 (1978).
44. Grinter, R., Barton, T. J., and Thomson, A. J., *J. Chem. Soc. Dalton Trans.* p. 608 (1978).
45. Grinter, R., Barton, T. J., and Thomson, A. J., *Ber. Bunsenges. Phys. Chem.* **82,** 131 (1978).
46. Grinter, R., Barton, T. J., and Thomson, A. J., *Chem. Phys. Lett.* **40,** 399 (1976).
47. Gruen, D. M., Green, D. W., and Hodges, R. V., *Inorg. Chem.* **15,** 970 (1976).
48. (a) Gruen, D., Carstens, D. H. W., Brashear, W., and Eslinger, D. E., *Appl. Spectrosc.* **26,** 185 (1972); (b) Guerra, C. R., and Schulman, J. H., *Surf. Sci.* **7,** 229 (1967).
49. Hanlan, A. J. L., Ph.D. Thesis, University of Toronto, 1978; and unpublished work.
50. Harrod, J. F., Roberts, R. W., and Rissman, E. F., *J. Phys. Chem.* **71,** 343 (1967).
51. Hartley, F. R., and Vezey, P. N., *Adv. Organomet. Chem.* **15,** 189 (1977).
52. Hayes, L. G., and Thomas, J. L., *J. Am. Chem. Soc.* **91,** 6876 (1969).
53. Hoffman, R., and Elian, M., *Inorg. Chem.* **14,** 1058 (1975).
54. Hoffman, R., and Elian, M., *Inorg. Chem.* **14,** 375 (1975).
55. Ibers, J. A., Franz, B. A., and Enemark, J. H., *Inorg. Chem.* **8,** 1288 (1969).
56. Kasai, P. H., and McLeod, Jr., D., *J. Chem. Phys.* **55,** 1566 (1971).
57. Kasai, P. H., and McLeod, Jr., D., *Ber. Bunsenges. Phys. Chem.* **82,** 103 (1978).
58. Kasai, P. H., and McLeod, Jr., D., *J. Phys. Chem.* **82,** 1554 (1978).
59. Kasai, P. H., and McLeod, Jr., D., *J. Am. Chem. Soc.* **97,** 6602 (1975).
60. Kasai, P. H., and McLeod, Jr., D., *J. Am. Chem. Soc.* **100,** 625 (1978).
61. Kasai, P. H., McLeod, Jr., D., and Watanabe, T., *J. Am. Chem. Soc.* **99,** 3521 (1977).
62. Kasai, P. H., and McLeod, Jr., D., *J. Am. Chem. Soc.* **97,** 5610 (1975).
63. King, R. B., Chang, M., and Newton, M. G., *J. Am. Chem. Soc.* **100,** 998 (1978).
64. Klabunde, K. J., *Acc. Chem. Res.* **8,** 393 (1975).
65. Klabunde, K. J., and Murdock, T. O., *Chem. Tech. (Leipzig)* p. 624 (1975).
66. Klabunde, K. J., Efner, H. F., Murdock, T. O., and Ropple, R., *J. Am. Chem. Soc.* **98,** 1021 (1976).
67. Klabunde, K. J., Efner, H. F., Satek, L., and Donley, W., *J. Organomet. Chem.* **71,** 309 (1974).
68. Klabunde, K. J., and Davis, S. C., *J. Am. Chem. Soc.* **100,** 5975 (1978).
69. Klabunde, K. J., and Davis, S. C., *J. Catal.* **54,** 254 (1978).
70. Klabunde, K. J., Anderson, B. B., Behrens, C. L., and Radonovich, L. J., *J. Am. Chem. Soc.* **98,** 5390 (1976).
71. Klabunde, K. J., Anderson, B. B., Bader, M., and Radonovich, L. J., *J. Am. Chem. Soc.* **100,** 1313 (1978); Klabunde, K.J., *EUCHEM Conf., Venice, Nov. 1979*.
72. Klabunde, K. J., Groshens, T., Brezinski, M., and Kennelly, W., *J. Am. Chem. Soc.* **100,** 4437 (1978).
73. Klabunde, K. J., and Roberts, J. S., *J. Organomet. Chem.* **137,** 113 (1977).

74. Klabunde, K. J., and Roberts, J. S., *J. Am. Chem. Soc.* **99,** 2509 (1977).
75. Kolb, D. M., Forstmann, F., Leutloff, D., and Schulze, W., *J. Chem. Phys.* **66,** 2806 (1977).
76. Kolb, D. M., and Leutloff, D., *Chem. Phys. Lett.* **55,** 264 (1978).
77. Kolb, D. M., and Forstmann, F., *Ber. Bunsenges. Phys. Chem.* **82,** 30 (1978).
78. Lagowski, J. J., and Graves, V., *Inorg. Chem.* **15,** 577 (1976).
79. (a) Lagowski, J. J., Simons, L. H., Riley, P. E., and Davis, R. E., *J. Am. Chem. Soc.* **98,** 1044 (1976); (b) Lagowski, J. J., and Peer, W. J., *ibid.* **100,** 6262, (1978).
80. Lagowski, J. J., and Simons, L. H., *J. Org. Chem.* **16,** 3247 (1978).
81. Klabunde, K. J., Groshens, T., Efner, W. F., and Kramer, M., *J. Organomet. Chem.* **157,** 91 (1978).
82. Mann, D. M., and Broida, H. P., *J. Chem. Phys.* **55,** 84 (1971).
83. Maslowski, E., "Vibrational Spectra of Organometallic Compounds," Wiley-Interscience, New York, 1977.
84. McGlinchey, M. J., and Tan, T.-S., *J. Chem. Soc., Chem. Commun.* p. 155 (1976).
85. McGlinchey, M. J., Agarwal, A., and Tan, T.-S., *J. Organomet. Chem.* **141,** 85 (1977).
86. McLean, A. D., and Yoshimine, M., *IBM J. Res. Dev. Suppl.* **12,** 206 (1968).
87. Messmer, R. P., *in* "The Theoretical Basis for Heterogeneous Catalysis" (E. Drauglis and R. K. Jaffee, eds.), Plenum, New York, 1975.
88. Möckel, R., and Elschenbroich, C., *Angew. Chem. Int. Ed. Engl.* **16,** 870 (1977).
89. Montano, P. A., *J. Appl. Phys.* **49,** 1561 (1978); **49,** 4612 (1978); W. Dyson and P. A. Montano, *J. Am. Chem. Soc.* **100,** 7439 (1978).
90. Moore, C., *Natl. Bur. Stand. (U.S.) Circ.* 467; Vol. I, 1949; Vol. II, 1952; and Vol. III, 1958.
91. Moskovits, M., and Ozin, G. A., "Cryochemistry," Wiley-Interscience, New York, 1976.
92. Moskovits, M., and Hulse, J. E., *J. Chem. Soc. Faraday Trans.* 2 p. 471 (1977).
93. Moskovits, M., and Hulse, J. E., *J. Chem. Phys.* **66,** 3988 (1977).
94. Moskovits, M., and Hulse, J. E., *J. Chem. Phys.* **67,** 4271 (1977).
95. Moskovits, M., and Hulse, J. E., *Surf. Sci.* **57,** 125 (1976).
96. (a) Moskovits, M., and Hulse, J. E., *Surf. Sci.* **61,** 302 (1976); (b) M. Moskovits, *Acc. Chem. Res.* **12,** 229 (1979).
97. Nibler, J., Rojhantalab, H. M., and Allamandola, L., *Ber. Bunsenges. Phys. Chem.* **82,** 107 (1978).
98. Ogden, J. S., Darling, J. H., and Garton-Sprenger, M. B., *Faraday Symp. Chem. Soc.* **8,** 75 (1973).
99. Ozin, G. A., *Appl. Spectrosc.* **30,** 573 (1976).
100. Ozin, G. A., *Catal. Rev. Sci. Eng.* **16,** 191 (1977); *Coord. Chem. Rev.* **28,** 117 (1979); *Faraday Soc. Disc., Chem. Soc.* **14,** 1 (1980).
101. Ozin, G. A., Power, W. J., Upton, T., and Goddard, III, W. A., *J. Am. Chem. Soc.* **100,** 4750 (1978).
102. Ozin, G. A., Huber, H., McIntosh, D. F., Mitchell, S. A., Norman, Jr., J. G., and Noodleman, L., *J. Am. Chem. Soc.* **101,** 3504 (1979).
103. Ozin, G. A., *Acc. Chem. Res.* **10,** 21 (1977).
104. Ozin, G. A., and Power, W. J., *Inorg. Chem.* **16,** 212 (1977).
105. Ozin, G. A., and Power, W. J., *Ber. Bunsenges. Phys. Chem.* **82,** 93 (1978).
106. Ozin, G. A., Kündig, E. P., and Moskovits, M., *Nature* **254,** 503 (1975).
107. Ozin, G. A., Klotzbücher, W., and Busby, R., *J. Am. Chem. Soc.* **98,** 4013 (1976).
108. Ozin, G. A., Ford, T. A., Huber, H., Klotzbücher, W., Kündig, E. P., and Moskovits, M., *J. Chem. Phys.* **66,** 425 (1977).

109. Ozin, G. A., Klotzbücher, W. E., and Mitchell, S. A., *Inorg. Chem.* **16,** 3063 (1977).
110. Ozin, G. A., and Klotzbücher, W. E., *Inorg. Chem.* **16,** 984 (1977).
111. Ozin, G. A., and Klotzbücher, W. E., *Inorg. Chem.* **15,** 292 (1976).
112. Ozin, G. A., and Huber, H., *Inorg. Chem.* **17,** 155 (1978).
113. Ozin, G. A., Klotzbücher, W. E., Norman, Jr., J. G., and Kolari, H. J., *Inorg. Chem.* **16,** 2871 (1977).
114. Ozin, G. A., and Klotzbücher, W. E., *J. Mol. Catal.* **3,** 195 (1977/78).
115. Ozin, G. A., and Klotzbücher, W. E., *J. Am. Chem. Soc.* **100,** 2262 (1978).
116. Ozin, G. A., and Klotzbücher, W. E., *Inorg. Chem.* **18,** 2101 (1979).
117. Ozin, G. A., Huber, H., Kündig, E. P., and Moskovits, M., *J. Am. Chem. Soc.* **97,** 2097 (1975).
118. Ozin, G. A., McIntosh, D., and Moskovits, M., *Inorg. Chem.* **15,** 1669 (1976).
119. Ozin, G. A., and Hanlan, A. J. L., *Inorg. Chem.* **16,** 2848 (1977).
120. Ozin, G. A., and Hanlan, A. J. L., *Inorg. Chem.* **16,** 2857 (1977).
121. Ozin, G. A., Hanlan, A. J. L., and Power, W. J., *Inorg. Chem.* **17,** 3648 (1978).
122. Ozin, G. A., Huber, H., and McIntosh, D., *Inorg. Chem.* **16,** 3070 (1977).
123. Ozin, G. A., Huber, H., and Power, W. J., *J. Am. Chem. Soc.* **98,** 6508 (1976).
124. Ozin, G. A., Huber, H., and McIntosh, D., *J. Organomet. Chem.* **112,** c50 (1976).
125. Ozin, G. A., Ford, T. A., Huber, H., Klotzbücher, W., and Moskovits, M., *Inorg. Chem.* **15,** 1666 (1976).
126. Ozin, G. A., Hanlan, L., and Huber, H., *Inorg. Chem.* **15,** 2592 (1976).
127. Ozin, G. A., and Moskovits, M., in "Vibrational Spectra and Structure" (J. Durig, ed.), Elsevier, Amsterdam, 1975.
128. Ozin, G. A., and Lever, A. B. P., *Inorg. Chem.* **16,** 2012 (1977).
129. Ozin, G. A., Hanlan, L. A., Huber, H., Kündig, E. P., and McGarvey, B. R., *J. Am. Chem. Soc.* **97,** 7054 (1975).
130. Ozin, G. A., and McIntosh, D., *J. Am. Chem. Soc.* **98,** 3167 (1976).
131. Ozin, G. A., and McIntosh, D., *Inorg. Chem.* **16,** 51 (1977).
132. Ozin, G. A., Huber, H., and McIntosh, D., *Inorg. Chem.* **16,** 975 (1977).
133. Ozin, G. A., Huber, H., and McIntosh, D., *Inorg. Chem.* **17,** 1472 (1978).
134. Ozin, G. A., and McIntosh, D., *Inorg. Chem.* **16,** 59 (1977).
135. Ozin, G. A., and McIntosh, D., *Inorg. Chem.* **15,** 2869 (1976).
136. Ozin, G. A., and Huber, H., *Inorg. Chem.* **16,** 64 (1977).
137. Ozin, G. A., Huber, H., Klotzbücher, W., and Vander Voet, A., *Can. J. Chem.* **51,** 2722 (1973).
138. Ozin, G. A., Busby, R., and Klotzbücher, W., *Inorg. Chem.* **16,** 822 (1977).
139. Ozin, G. A., Huber, H., Ford, T. A., and Klotzbücher, W., *J. Am. Chem. Soc.* **98,** 3176 (1976).
140. Ozin, G. A., Huber, H., and Power, W. J., *Inorg. Chem.* **16,** 979 (1977).
141. Ozin, G. A., and Power, W. J., *Inorg. Chem.* **17,** 2836 (1978).
142. Ozin, G. A., and Power, W. J., *Inorg. Chem.* **16,** 2864 (1977).
143. Ozin, G. A., and McIntosh, D., *J. Organomet. Chem.* **121,** 127 (1976).
144. Ozin, G. A., and Power, W. J., unpublished data.
145. Ozin, G. A., Huber, H., and Power, W. J., *Inorg. Chem.* **16,** 2234 (1977).
146. Ozin, G. A., Huber, H., and McKenzie, P., *J. Am. Chem. Soc.* (in press).
147. Ozin, G. A., Kenney-Wallace, G., Farrel, J., Mitchell, S., and Huber, H., *J. Am. Chem. Soc.* (submitted).
148. Ozin, G. A., McIntosh, D., Mitchell, S., and Messmer, R. P. (to be published).
149. Ozin, G. A., and Mitchell, S., *J. Am. Chem. Soc.* **100,** 6776 (1978).
150. Ozin, G. A., Mitchell, S., Huber, H., Norman, J. G., and Noodleman, L., *J. Am. Chem. Soc.* **101,** 3504 (1979).

151. Ozin, G. A., and Mitchell, S., paper presented at the "Cluster Symposium," *Am. Chem. Soc. Meet.*, Anaheim, March 1977; and *Inorg. Chem.* **18**, 2932 (1979).
152. Ozin, G. A., Huber, H., and Mitchell, S. (to be published).
153. Ozin, G. A., and Hanlan, A. J. L., *J. Am. Chem. Soc.* **96**, 6324 (1974).
154. Ozin, G. A., and Hanlan, A. J. L., *Inorg. Chem.* **18**, 2091 (1979); **18**, 1781 (1979).
155. Pritchard, J., and Sims, M. L., *Trans. Faraday Soc.* **70**, 427 (1969).
156. Rehder, D., *J. Organomet. Chem.* **37**, 303 (1972).
157. Rest, A. J., and Taylor, D. L., *J. Chem. Soc., Chem. Commun.* p. 717 (1977).
158. Rieke, R. D., *Acc. Chem. Res.* **10**, 301 (1977).
159. Rieke, R. D., Wolf, W. J., Kujundžić, N., and Kavoluinas, A. V., *J. Am. Chem. Soc.* **99**, 4159 (1977).
160. Roberts, M. W., *Chem. Soc. Rev.* **6**, 373 (1977).
161. Roberts, M. W., and Kishi, K., *Surf. Sci.* **62**, 252 (1977).
162. Rothschild, W. G., and Yao, H. C., *J. Chem. Phys.* **68**, 4774 (1978).
163. Schulze, W., Becker, H. V., Minkwitz, R., and Marzel, K., *Chem. Phys. Lett.* **55**, 59 (1978).
164. Schulze, W., *Faraday Soc. Disc. Chem. Soc.* **14** (1980).
165. Seff, K., and Kim, Y., *J. Am. Chem. Soc.* **99**, 7055 (1977).
166. Sheppard, N., Prentice, J. D., and Lesuinans, A., *J. Chem. Soc. Chem. Commun.* p. 76 (1976).
167. Sinfelt, J., *Acc. Chem. Res.* **10**, 15 (1977); Sinfelt, J. H., Cusumano, J. A., Burton, J. J., and Garten, R. L., *in* "Advanced Material in Catalysis" (Burton, J. J., and Garten, R. L., eds.), Academic Press, New York, 1977.
168. Skell, P. S., Wescott, Jr., L. D., Goldstein, J. P., and Engerl, R. R., *J. Am. Chem. Soc.* **87**, 2829 (1965).
169. Skell, P. S., Kolesnikov, S. P., and Dobson, J. E., *J. Am. Chem. Soc.* **100**, 999 (1978).
170. Smardzewski, R. R., Efner, H. F., Fox, W. B., and Tevault, D. E., *Inorg. Chim. Acta* **24**, L93 (1977).
171. Smardzewski, R. R., Efner, H. F., Fox, W. B., and Tevault, D. E., *J. Organomet. Chem.* **146**, 45 (1978).
172. Sneddon, L. G., Hall, L. W., and Zimmerman, G. J., *J. Chem. Soc. Chem. Commun.* p. 45, 1977.
173. Sneddon, L. G., Zimmerman, G. J., and Wilczynski, R., *J. Organomet. Chem.* **154**, c29 (1978).
174. Somorjai, G., *Angew. Chem. Int. Ed. Engl.* **16**, 92 (1977).
175. Somorjai, G., *Acc. Chem. Res.* **9**, 248 (1976).
176. Stevens, R. D., *J. Org. Chem.* **38**, 2260 (1973).
177. Stone, F. G. A., Treichel, P. M., Pitcher, E., and King, R. B., *J. Am. Chem. Soc.* **83**, 2593 (1961).
178. Stone, F. G. A., King, R. B., Bruce, M. I., and Phillips, J. R., *Inorg. Chem.* **5**, 684 (1966).
179. Timms, P. L., and Turney, T. W., *Adv. Organomet. Chem.* **15**, 53 (1977).
180. Timms, P. L., and Atkins, R. M., *Spectrochim. Acta Part A* **33**, 853 (1977).
181. Timms, P. L., and Atkins, R. M., *Inorg. Nucl. Chem. Lett.* **14**, 113 (1978).
182. Timms, P. L., *J. Chem. Soc. Chem. Commun.* p. 1033 (1969).
183. Timms, P. L., and Kündig, E. P., *J. Chem. Soc. Chem. Commun.* p. 912 (1977).
184. Timms, P. L., and Turney, T. W., *J. Chem. Soc. Dalton Trans.* p. 2021 (1976).
185. Timms, P. L., and Mackenzie, R., *J. Chem. Soc. Chem. Commun.* p. 650 (1974).
186. Timms, P. L., and Piper, M., *J. Chem. Soc. Chem. Commun.* p. 52 (1972).
187. Timms, P. L., and Chivers, T., *J. Organomet. Chem.* **118**, c37 (1976).

188. Timms, P. L., and Francis, C. G., *J. Chem. Soc. Chem. Commun.* p. 466 (1977).
189. Tolman, C. A., English, A. D., and Jesson, J. P., *Inorg. Chem.* **15,** 1730 (1976).
190. Tolman, C. A., English, A. D. Ittel, S. D., and Jesson, J. P., *Inorg. Chem.* **17,** 2374 (1978).
191. Turner, J. J., Burdett, J. K., Perutz, R. N., and Poliakoff, M., *Pure Appl. Chem.* **49,** 271 (1977).
192. Turner, J. J., and Perutz, R. N., *J. Am. Chem. Soc.* **97,** 4791 (1975).
193. Turner, J. J., Burdett, J. K., Grzybowski, J. M., and Poliakoff, M., *J. Am. Chem. Soc.* **98,** 5728 (1976).
194. Verkade, J. G., Tolman, C. A., and Yarbrough, II, L. W., *Inorg. Chem.* **16,** 479 (1977).
195. Vollenbroek, F. A., Bouten, P. C. P., Trooster, J. M., van der Berg, J. P., and Bons, J. J., *Inorg. Chem.* **17,** 1345 (1978).
196. Wada, N., *J. Phys. (Paris)* **C2,** 219 (1977), and references cited therein.
197. Weltner, Jr., W., Ferrante, R. F., Wilkerson, J. L., and Graham, W. R. M., *J. Chem. Phys.* **67,** 5904 (1977).
198. Weltner Jr., W., Van Zee, R. J., and Seely, M. L., *J. Chem. Phys.* **67,** 861 (1977).
199. Wilke, G., Fischer, K., and Jonas, K., *Angew. Chem.* **85,** 620 (1973); *Angew. Chem. Int. Ed. Engl.* **12,** 565 (1973).
200. Wright, R. B., Bates, J. K., and Gruen, D. M., *Inorg. Chem.* **17,** 2275 (1978).
201. Zaitseva, N. N., Nesmeyanov, A. N., Domracher, G. A., Zinov'ev, V. D., Yur'eva, L. P., and Tverdokhlebova, I. I., *J. Organomet. Chem.* **121,** c52 (1976).
202. Zaitseva, N. N., Nesmeyanov, A. N., Yur'eva, L. P., Domracher, G. A., and Zinov'ev, V. D., *J. Organomet. Chem.* **153,** 181 (1978).
202a. Zenneck, U., Elschenbroich, C., and Möckel, R., *Angew. Chem. Int. Ed. Engl.* **17,** 531 (1978).
203. Zmbova, B., Ihle, H. R., and Langenscheidt, E., *J. Chem. Phys.* **66,** 5105 (1977).

NEW METHODS FOR THE SYNTHESIS OF TRIFLUOROMETHYL ORGANOMETALLIC COMPOUNDS

RICHARD J. LAGOW

Department of Chemistry, University of Texas at Austin, Austin, Texas

and
JOHN A. MORRISON

Department of Chemistry, University of Illinois, Chicago Circle, Chicago, Illinois

I. Introduction	178
II. Plasma-Generated Trifluoromethyl Radicals as a Synthetic Reagent	181
A. Reaction with Mercuric Halides	183
B. Reaction with Germanium Tetrabromide	184
C. Reaction with Tin Tetraiodide	186
D. Reaction with Bismuth Triiodide	187
E. Reaction with Tellurium Tetrabromide	187
F. Reactions of Trifluoromethyl Radicals with Sulfur Vapor	188
G. Reactions of Trifluoromethyl Radicals with Organic Halides	189
H. Summary	189
III. Bis(trifluoromethyl)mercury as a Synthetic Reagent	192
A. Reaction with Germanium Tetraiodide	193
B. Reaction with Tin Tetrabromide	194
C. Reaction with Silicon Tetrahalides	196
D. Chemical Integrity of the CF_3Ge and CF_3Sn Bonds	196
E. Summary	197
IV. Synthesis of Trifluoromethyl Organometallic Compounds by Direct Fluorination	197
A. Experimental Methods	197
B. The Controlled Reaction of Metal Alkyls with Elemental Fluorine	198
V. A New General Synthesis for Trifluoromethyl Organometallic Compounds and Other Sigma-Bonded Metal Compounds Based on Metal Vapor as a Reagent	203
A. Introduction	203
B. Apparatus	204
C. Synthesis of Trifluoromethyl Organometallic Compounds by Cocondensation of Trifluoromethyl Radicals and Metals	204

D. Synthesis of Methyl Organometallic Compounds by Cocondensation of Methyl Radicals and Metals 205
E. Synthesis of Bis(trifluorosilyl)mercury by Cocondensation of Mercury and Trifluorosilyl Radicals 207
F. Prospects and Future Applications of This Technique 207
References 208

I. Introduction

As most readers will undoubtedly be aware, the study of trifluoromethyl organometallic compounds, and the initiation of this area as a field of research, began with the discovery of bis(trifluoromethyl)mercury by Eméleus and Haszeldine (1) in 1949 and continued with their associated research programs. Since that time, a steady, and sometimes spectacular, international research effort has continued over the past twenty-five years that has involved many laboratories (2).

Nevertheless, it could be fairly stated that, in the next 20-year period through the end of the sixties, preparation of even the initial bis(trifluoromethyl)mercury compound in quantity required considerable time and effort and was, in general, painstakingly laborious. Furthermore, it is quite clear from the literature that no truly general methods were known for the synthesis of trifluoromethyl organometallic compounds. Indeed, the literature contains many rationalizations, based on suppositions of instability of compounds, for the failure of certain synthetic methods to afford highly substituted trifluoromethyl compounds that are now known to be stable, in many cases to temperatures of over 100°C. Many of the more established synthetic methods in the area also have associated activation-energy problems, making it difficult or impossible to prepare trifluoromethyl organometallic compounds of marginal stability.

The state of the synthetic art in this area, in 1979, is much more satisfactory. During the past decade, several new synthetic developments have occurred such that we are closer to the point where the limitations upon synthesis of trifluoromethyl compounds are related more to stability problems in isolated cases, and are not nearly so much due to lack of widely applicable synthetic techniques. We find ourselves, for example, in a position in 1979 where the germanium compound, $Ge(CF_3)_4$, which in the past decade, was considered by many workers to be of insufficient stability to permit isolation, has been prepared by four independent methods and is known to be stable to over 100°C. Many of these new synthetic techniques have emerged from studies conducted in our laboratory at the University of Texas and previously at the Massachusetts Institute of Technology during the past several years.

The four general areas of research discussed in this article are in some cases only vaguely interrelated with respect to methodology, even when they are capable of producing the same compounds. They range from such rather exotic approaches as elemental fluorination of alkyls, which has, surprisingly, proved to be a practical synthetic method for several trifluoromethyl organometallic compounds, to the development of an extremely general synthesis involving metal atoms and free radicals as precursors, a method that impacts not only on the area of trifluoromethyl organometallic chemistry, but on such diverse areas as the synthesis of new methyl alkyls and new classes of sigma-bonded, alkyl-like compounds both within and outside the realm of fluorine chemistry.

Perfluoroalkyl derivatives of the inorganic elements have long been known to possess properties quite different from those of their perhydrogenated analogs. Although the properties of the alkyl and perfluoroalkyl derivatives of many elements could be contrasted, one example will suffice. Bis(trifluoromethyl)mercury, the first perfluoroalkyl organometallic compound prepared (1), is a sublimable solid, having no known liquid phase, that is soluble in and easily recoverable from water, and *appears* to be physiologically inert, as it has been handled routinely in the open laboratory atmosphere many times without adverse effect (3). Additionally, it has been reported that this perfluoromethyl mercurial does not exchange ligands with other inorganic halides to yield trifluoromethyl inorganic compounds and mercuric halides (2). All of these properties are in marked contrast to the properties of the perhydrogen analog, dimethylmercury. They are also among the first indications that the properties and reactivities of perfluorinated organometallic compounds may be strikingly different from those of the more-usual, hydrogen-containing, organometallic compounds.

The trifluoromethyl ligand has the characteristics both of a pseudohalogen and an alkyl group. For example, the trifluoromethyl group stabilizes the highest-valence states of such elements as arsenic, e.g., $(CF_3)_3AsCl_2$, (4), for which even the pentachloride is unstable. One especially interesting example of the interplay of the various properties of the trifluoromethyl group concerns the ground states of pentavalent phosphoranes where, originally, it had been proposed that the site preference of different ligands is related solely to the electronegativity of the ligand (5). Results have, however, now been obtained that indicate that other factors predominate, and the CF_3 ligand is considered to be located equatorially in mixed chloro(trifluoromethyl)phosphoranes (6).

Although a great number of trifluoromethyl-containing compounds have been prepared, one curious feature is that, until recently, almost all of these compounds had been prepared either directly from, or through, the intermediacy of only one precursor, trifluoromethyl iodide. This reagent has been found to oxidize a few elements directly, and to add oxidatively to low-valent complexes of a large number of other main-group and transition-metal elements, generating compounds containing from one to as many as three trifluoromethyl groups. The most thermally stable of these compounds are formed directly by the lighter elements of Groups 5 and 6A. Antimony, arsenic, phosphorus, and selenium all form the fully substituted, trifluoromethyl derivatives simply upon heating with trifluoromethyl iodide; elemental sulfur forms $(CF_3)_2S$ upon UV irradiation of the $(CF_3)_2S_2$ that is initially formed by reaction of trifluoromethyl iodide with the element. The mercurial, $(CF_3)_2Hg$, is, however, formed only if the element has previously been amalgamated.

Monosubstituted trifluoromethyl derivatives of the Group 4A elements have also been prepared, but by very specific reactions. The divalent halides SiF_2 and GeI_2 react with CF_3I to form CF_3SiF_2I and CF_3GeI_3 and a very small proportion of $(CF_3)_2GeI_2$, but the more stable SnI_2 is unaffected. The metal–metal bond in hexamethyldistannane is cleaved by CF_3I to form CF_3SnMe_3, but the stronger bonds in hexamethyldisilane and hexamethyldigermane are unreactive. Additionally, the reaction of $CF_3Sn(CH_3)_3$ with BF_3 has been shown (7) to yield $CF_3BF_3^-$. Trifluoromethyl iodide has thus proved to be an extremely versatile reagent, but it seems likely that other reagents and conditions could be found that would prove to be superior for the formation of currently unknown, per(trifluoromethyl) compounds of the main-group elements. Thus, a number of investigations have been undertaken to assess the potential of several, alternative synthetic techniques for the preparation of trifluoromethyl-containing organometallic compounds. One possible route to these compounds was the discharge reaction of hexafluoroethane to produce very reactive trifluoromethyl radicals that could react with metal halides. Another involved reinvestigation of the claim that the trifluoromethyl mercurial was much less inclined to exchange ligands than alkyl mercurials. A third possibility concerned the reaction of elemental fluorine with alkyl compounds, a technique described (8) as "... not suitable for the preparation of perfluoroalkyl-metallic derivatives." Finally, the reaction of atomic metal vapors with trifluoromethyl radicals has been preliminarily surveyed.

As a test of the suitability of these potential, synthetic routes, the preparation of several molecules that had not previously been synthesized was attempted. They included the compounds $(CF_3)_2Te$ and $(CF_3)_3Bi$, the fully substituted derivatives of the lower elements in Groups 5 and 6A (which had previously been sought, but not isolated), and the tetrasubstituted compounds of Group 4A elements, e.g., $(CF_3)_4Ge$, as, with the trivial exception of perfluoroneopentane, no compound containing more than three trifluoromethyl groups attached to any element was known. Because relatively little information was available as to the chemical properties of Group 4A compounds in which one or more of the ligands was a CF_3 group, the chemical stabilities of the CF_3Ge and CF_3Sn linkages were tested by reaction of the trifluoromethylgermanium or trifluoromethyltin halides with a variety of standard reagents to determine whether the CF_3–metal bonds are reactive.

II. Plasma-Generated Trifluoromethyl Radicals as a Synthetic Reagent

Hexafluoroethane is a particularly promising precursor for trifluoromethyl-containing species, as it is relatively inexpensive, readily available, and easy to manipulate. Conceptually, if a process could be found that would preferentially cleave the carbon–carbon bond in this molecule to generate, e.g., trifluoromethyl radicals, these could then react with another substrate, such as a metal halide, to form metal or metalloid compounds highly substituted with CF_3 groups. Although the bond strengths in C_2F_6 are controversial (9), the best estimates currently available indicate that the carbon–fluorine bond is ~40 kcal/mole stronger than the carbon–carbon bond (10) (see Fig. 1). Thus, were hexafluoroethane dissociated under relatively mild conditions, it would be expected that cleavage of the weaker carbon–carbon bond to produce trifluoromethyl radicals (or ions) would predominate, and that cleavage of the carbon–fluorine bond, to produce, e.g., pentafluoroethyl radicals, would be a very minor reaction.

Low-temperature glow-discharges were utilized to cause bond rupture in hexafluoroethane (11). In these experiments, the power to support the discharge was supplied by a radio-frequency discharge that delivered ~25 W of power, at a frequency of 8.6 MHz, to the copper coil surrounding the Pyrex reactor (see Fig. 2). The load coil was induc-

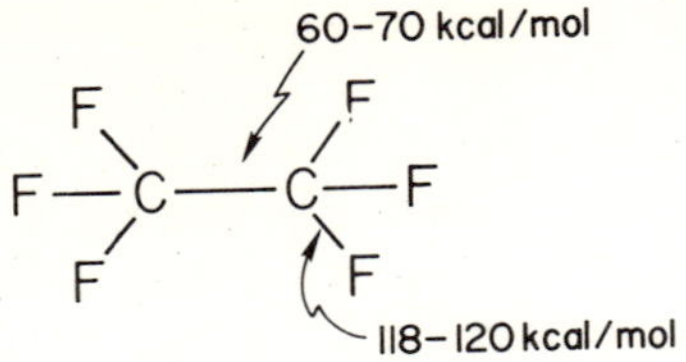

F—C(F)(F)—C(F)(F)—F   60–70 kcal/mol (C–C)
118–120 kcal/mol (C–F)

$CF_3^* + MX_n \longrightarrow M(CF_3)_n + X_2$

$CF_3^* + CX_n \longrightarrow C(CF_3)_n + X_2$

(X = Br, Cl, or I)

FIG. 1. Estimated strengths of the carbon–carbon and carbon–fluorine bonds in hexafluoroethane.

tively coupled to the plasma within the reactor, and any free electrons in the discharge region were accelerated by the rapidly fluctuating magnetic field. Other ions present were too massive to gain significant kinetic energy before the polarity of the field reversed. The critical difference between this type of discharge and dc and 60-Hz discharges is that only the electrons gain appreciable translational energy from the magnetic field. As the pressure is low, little kinetic energy is transferred from the electrons to the larger species, and thus, the temperature of the gaseous molecules, T_g, is low. This is, perhaps, the key feature for the success of this synthetic method. The higher translational energies, which are associated with the generation of such high-energy reactants by thermal means, would result in pyrolysis of many marginally stable reactants, such as hydrocarbon compounds. When the apparatus is in operation, the temperatures observed immediately ad-

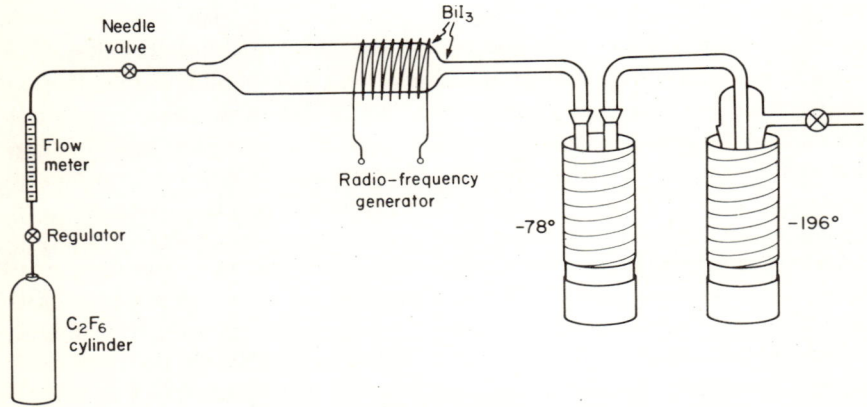

FIG. 2. Diagram of plasma apparatus.

jacent to the discharge region are ~40–50°C. In other types of discharges, quartz reactors have frequently been needed, in order to withstand the severe heat-strains encountered (12).

In operation, the gaseous precursor, hexafluoroethane, at a pressure of 0.1 to 1 mm Hg, flows into the discharge region (see Fig. 2), where the radicals and ions are formed. As the gas flows out of the discharge region, abstractions and recombinations quickly lower the radical concentrations to zero. In attempts to utilize these reactive species, halides of metals or metalloids were placed at the tail of the plasma, to react and afford volatile, trifluoromethyl-substituted compounds which then would be swept away from the discharge region by the gas flow and into the traps, later to be isolated, identified, and characterized by the usual techniques. Thus, the objective of the experiments was to promote such reactions as those in Eq. (1)–(3) at the expense of Eq. (4). Equations (5) and (6) indicate the observed fate of many of the halogen atoms.

$$C_2F_6 \xrightarrow{\text{discharge}} 2 \cdot CF_3 \tag{1}$$

$$\cdot CF_3 + MX_n \rightarrow CF_3MX_{n-1} + X \cdot \tag{2}$$

$$\cdot CF_3 + CF_3MX_{n-1} \rightarrow (CF_3)_2MX_{n-2} + X \cdot \tag{3}$$

$$2 \cdot CF_3 \rightarrow C_2F_6 \tag{4}$$

$$\cdot CF_3 + X \cdot \rightarrow CF_3X \tag{5}$$

$$2X \cdot + M \rightarrow X_2 + M^* \tag{6}$$

A. Reaction with Mercuric Halides

The reactions of mercuric iodide, mercuric bromide, and mercuric chloride with the excited species produced in the hexafluoroethane plasma were examined first, as the expected products were known to be stable and had been well characterized (13). Thus, these reactions constituted a "calibration" of the system. Bis(trifluoromethyl)mercury was obtained from the reaction of all of the mercuric halides, but the highest yield (95%, based on the amount of metal halide consumed) was obtained with mercuric iodide. The mole ratios of bis(trifluoromethyl)mercury to (trifluoromethyl)mercuric halides formed by the respective halides is presented in Table I, along with the weight in grams of the trifluoromethyl mercurials recovered from a typical, five-hour run.

As might have been expected, the highest ratio of disubstituted compound, as well as the largest total amount of trifluoromethyl mercurials, was formed by the reagent having the weakest metal–halogen

TABLE I

MOLE RATIOS AND WEIGHTS OF TRIFLUOROMETHYL MERCURIALS FORMED

Mercuric halide	Mole ratio $(CF_3)_2Hg/CF_3HgX$	Mercurials recovered (g) $(CF_3)_2Hg + CF_3HgX$
HgI_2	43.0	0.82
$HgBr_2$	0.85	0.55
$HgCl_2$	0.11	0.44

bond, namely, mercuric iodide. This is illustrated very clearly in Figs. 3–5, which show the ^{19}F-NMR spectra of the unseparated mixtures of products produced by the reactions of CF_3 radicals with HgI_2, $HgBr_2$, and $HgCl_2$, respectively. In Fig. 3, the small peak to the left of the central, $Hg(CF_3)_2$ resonance is due to CF_3HgI. By varying the experimental parameters, the quantity of $(CF_3)_2Hg$ produced from HgI_2 was increased to 8 g/day. These reactions indicated that plasmas could be successfully used to generate trifluoromethyl organometallic compounds in relatively large amounts. An attractive feature of the technique is that very little of the operator's time is required, as the reactions, once started, proceed virtually unattended.

B. REACTION WITH GERMANIUM TETRABROMIDE

At a point near the tail of the plasma, where the blue glow was near extinction, gaseous germanium tetrabromide was slowly admitted to the reactor from a side arm. Tetrakis(trifluoromethyl)germane, previously unknown, was prepared in 64% yield, along with smaller proportions of $(CF_3)_3GeBr$, also new, and $(CF_3)_2GeBr_2$. These species were isolated and identified by the usual techniques (11), and, during the course of the characterization, two noteworthy observations were made. The first was that, during the isolation of $(CF_3)_3GeBr$, or especially $(CF_3)_3GeI$ (if GeI_4 had been used as the reactant), as the compound was concentrated, all of the $(CF_3)_4Ge$, as well as the material of the formula $(CF_3)_2GeX_2$, was removed. However, if the sample containing, e.g., $(CF_3)_3GeI$, was kept at room temperature for a few days and then examined, it was found that more $(CF_3)_4Ge$ and $(CF_3)_2GeI_2$ could be isolated. Clearly, the tris(trifluoromethyl)germanes were undergoing ligand redistribution-reactions in which the CF_3 group as a unit was exchanging metal centers, a type of reaction previously reported (14) not to proceed, even at 180°C. Later studies (15) showed that neat $(CF_3)_3GeI$ does, indeed, exchange ligands, to form $(CF_3)_4Ge$ and $(CF_3)_2GeI_2$, as shown in Table II. The second observation, as yet un-

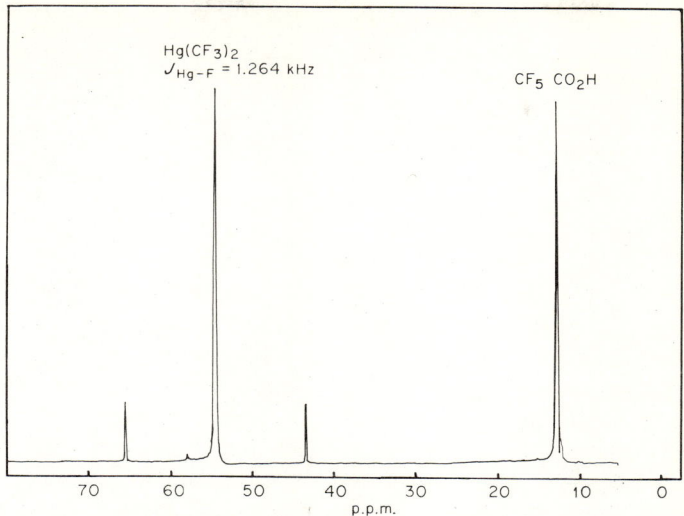

FIG. 3. ^{19}F-NMR spectrum of the mixture from the reaction of CF_3 radicals with HgI_2.

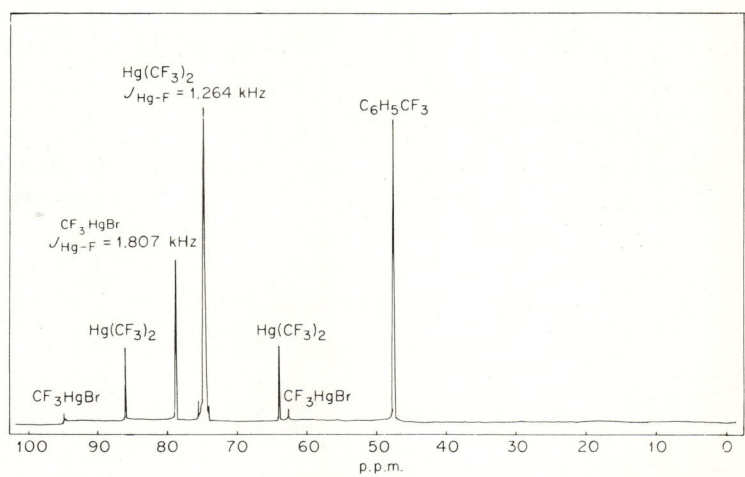

FIG. 4. ^{19}F-NMR spectrum of the mixture from the reaction of CF_3 radicals with $HgBr_2$.

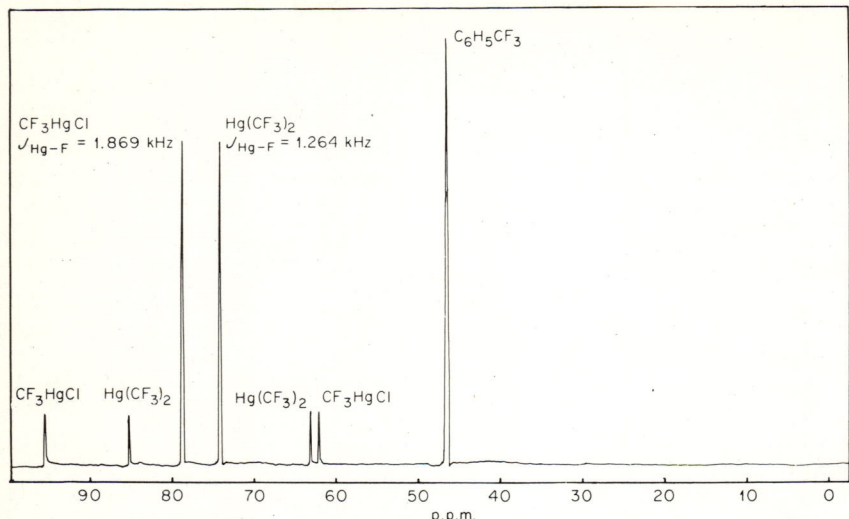

FIG. 5. ^{19}F-NMR spectrum of the mixture from the reaction of CF_3 radicals with $HgCl_2$.

exploited, was that a strong metastable ion observed in the mass spectrum of $(CF_3)_4Ge$ could be ascribed to the following decomposition.

$$Ge(CF_3)_3CF_2^+ \rightarrow {:}Ge(CF_3)_2 + C_2F_5^+$$

The facile elimination of the methylene-like species $(CF_3)_2Ge$: indicates that, under the appropriate conditions, $(CF_3)_4Ge$ may well serve as a useful laboratory source of trifluoromethyl-substituted Ge(II).

C. Reaction with Tin Tetraiodide

Tin tetraiodide was placed in a quartz boat which was then positioned at the tail of the plasma. Tetrakis(trifluoromethyl)tin was syn-

TABLE II

Redistribution Reaction of $(CF_3)_3GeI$ at 155°

Time (h)	Amount of trifluoromethylgermane[a] (mole %)		
	$(CF_3)_2GeI_2$	$(CF_3)_3GeI$	$(CF_3)_4Ge$
0	0	100	0
15	8	86	6
30	12	77	11
90	19	71	19

[a] Sealed, 4-mm tubes; monitored by ^{19}F-NMR.

thesized in 90% yield, along with $(CF_3)_3SnI$ and $(CF_3)_2SnI_2$ which were formed in small proportions (11). Tetrakis(trifluoromethyl)germane and tetrakis(trifluoromethyl)tin were the first organometallic compounds prepared in which an atom of any element had more than three trifluoromethyl ligands. The preparation of $(CF_3)_4Sn$ is also an indication that this type of discharge is capable of synthesizing compounds that have only marginal thermal stability, as $(CF_3)_4Sn$ was shown to decompose totally in 24 hours at 100°C. Even at temperatures as low as 66°C, the majority of the sample decomposes within 24 hours (15, 16).

D. Reaction with Bismuth Triiodide

Bismuth triiodide (18 g) was introduced into the reactor, and spread out on the walls of the Pyrex vessel shown in Fig. 2. After evacuation of the reactor and introduction of hexafluoroethane at a pressure of ~1 torr, the discharge was initiated and maintained for 100 hours. Tris(trifluoromethyl)bismuthine, 0.8 g, was isolated in 32% yield (17), along with the previously prepared $(CF_3)_2BiI$. This reaction again demonstrates the utility of this synthetic route as a preparative tool, as several early attempts to prepare $(CF_3)_3Bi$ by more traditional routes had proved unsuccessful (18). In one closely related experiment, $\cdot CF_3$ radicals, generated by the pyrolysis of hexafluoroacetone, were reacted with a bismuth mirror in a Paneth type of reaction, but no trifluoromethyl-substituted bismuthines were isolated (19). In each of the earlier studies, temperatures over 200° were needed (18, 19), and yet tris(trifluoromethyl)bismuth has been shown to decompose (17) within a few minutes at 100°C. Thus, if $(CF_3)_3Bi$ had been prepared by, e.g., ligand exchange-reactions (18), it would have immediately decomposed.

E. Reaction with Tellurium Tetrabromide

Tellurium tetrabromide (2.6 g) was placed in a Vycor sample-boat which was then positioned in the tail of the plasma. After exposure to the discharge for 46 hours, 1.5 g of the $TeBr_4$ had been consumed, and three major, tellurium-containing products had been formed (11). Bis(trifluoromethyl)tellurium, $(CF_3)_2Te$, a new compound, was isolated in 20% yield, and later found to react with gaseous bromine to give $(CF_3)_2TeBr_2$. When $(CF_3)_2Te$ in a sealed tube was gently heated with a flame, the compound decomposed, with the evolution of fluorocarbons and $(CF_3)_2Te_2$. The second compound synthesized was bis(trifluoromethyl)ditelluride, $(CF_3)_2Te_2$, a previously known compound, which was formed in 33% yield, based on the $TeBr_4$ consumed. The final product, formed in 36% yield, was a solid of low volatility and solubility. The mass-spectral and ^{19}F-NMR data indicated a mixed (trifluoromethyl)tellurium bromide.

This reaction provides a third indication of the usefulness of a radiofrequency discharge in the synthesis of compounds of low thermal stability. The more-stable $(CF_3)_2Te_2$ had been prepared by the interaction of CF_3 radicals, formed in the pyrolysis of $(CF_3)_2CO$, with a tellurium mirror (19). The less-stable $(CF_3)_2Te$ was not, however, observed in that experiment.

F. REACTIONS OF TRIFLUOROMETHYL RADICALS WITH SULFUR VAPOR

Although belonging to a slightly different class of reactions, the reaction of trifluoromethyl radicals with sulfur vapor has been shown to provide a route to trifluoromethyl polysulfide compounds (20). Instead of using sulfur halides, which undoubtedly would also give positive results, elemental sulfur (S_8) was vaporized and dissociated into atomic and polyatomic sulfur species.

$$S_{8(g)} \xrightarrow{rf} S + S_2 + S_3 + S_4 + \ldots \ldots$$

The apparatus used is shown in Fig. 6. The reaction proceeds according to the equation

$$S_m + CF_3 \cdot \rightarrow CF_3S_nCF_3 + C_2F_5S_nCF_3 + C_2F_5S_nC_2F_3,$$

where $m = 1-8$, and $n = 1-4$.

The primary products are the trifluoromethyl compounds. The perfluoroethyl species are considered to arise through an unusual mechanistic process (20), instead of being generated directly, as penta-

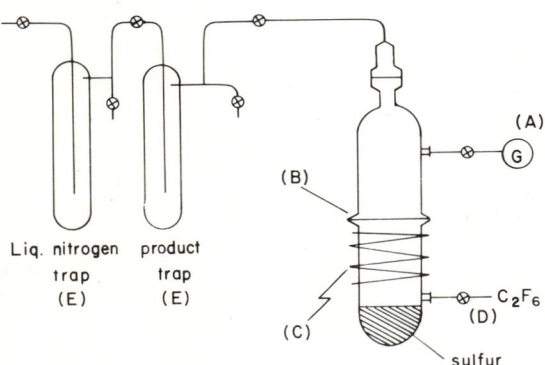

FIG. 6. Plasma apparatus.

fluoroethyl radicals, from C_2F_6. Although the majority of these species had been reported previously, several novel compounds resulted from these synthetic efforts.

G. Reactions of Trifluoromethyl Radicals with Organic Halides

For completeness, it should be mentioned that the reaction of trifluoromethyl radicals to replace halogens is extremely general, and not confined solely to metal species. Plasma-generated trifluoromethyl radicals will react with halocarbons according to the reaction (21)

$$CF_3\cdot + CX_n \rightarrow C(CF_3)_n + n/2\ X_2,$$

where X = Br, Cl, or I.

Specifically, the reaction of trifluoromethyl radicals with carbon tetraiodide produces perfluoro-*tert*-butyl iodide and perfluoroneopentane in the ratio of 3:1. Incomplete substitution is presumably due to steric factors around the crowded, central carbon atom.

$$CF_3\cdot + CI_4 \rightarrow \underset{(24\%)}{C(CF_3)_4} + \underset{(73\%)}{(CF_3)_3CI} + I_2$$

Such reactions appear to be rather general in scope, since an aryl example, the substitution for bromine in bromobenzene, and for halogen in a normal alkyl halide has also been observed (21).

$$PhBr + CF_3\cdot \rightarrow PhCF_3 + \tfrac{1}{2}Br_2$$

$$C_7H_{15}I + CF_3\cdot \rightarrow C_7H_{15}CF_3 + \tfrac{1}{2}I_2$$

H. Summary

Hexafluoroethane has been demonstrated to be a very useful precursor to trifluoromethyl-containing compounds of the main-group elements. Previously unknown compounds in which the ligands were solely trifluoromethyl groups have been prepared in good to excellent yield by the interaction of the halides of metal or metalloid elements with the reactive intermediates produced in the radio-frequency discharge of hexafluoroethane. In several instances, the compounds formed possess very limited thermal stability, and yet they could be synthesized readily. Because the emphasis in these experiments was development of a new and more useful preparative tool, the formation of known trifluoromethyl-containing species was not attempted. For example, $(CF_3)_3$As could be readily prepared from AsI_3 by the preceding method.

The various spectral and physical properties of the compounds prepared, including their elemental analysis, and IR, NMR, and mass spectra (which contained the appropriate ions, each of the intensity demanded by the isotopic composition of the ion), all fully supported the formulation of the species as reported. With two exceptions, all of the new compounds were found to be colorless liquids, typically having a relatively short liquid range, and they are usually very volatile for their molecular weight. The two exceptions are $(CF_3)_2Te$, which is yellow-green, and $(CF_3)_2Te_2$, which is red-brown (11).

Experimentally, there has been but little direct observation of the reactive intermediates present in this, or any other, type of discharge, as it is extraordinarily difficult to analyze a reactive plasma without perturbing the system.

Ideally, it would be desirable to determine many parameters in order to characterize and mechanistically define these unusual reactions. This has been an important objective that has often been considered in the course of these studies. It would be helpful to know, as a function of such parameters of the plasma as the radio-frequency power, pressure, and rate of admission of reactants, (1) the identity and concentrations of all species, including trifluoromethyl radicals, (2) the electronic states of each species, (3) the vibrational states of each species, and (4) both the rotational states of each species and the average, translational energies of, at least, the trifluoromethyl radicals.

Objective (1) has been the impetus for a considerable amount of contemporary research, but it has proved most difficult to design a mass-spectrometric sampling-system that does not perturb the species when they are admitted to the ionization chamber (which must necessarily be at a pressure lower than the 1 mm Hg required for plasma conditions). Only recently has convincing evidence been adduced that nonspurious results might be achieved by use of slit sampling-devices and quadrupole, mass spectrometers. To observe the excited electronic states, one option would, of course, be to examine the vacuum-ultraviolet spectrum of such species, but this has been found to be exceedingly complex because many lines arise from the multiple species. Attachment of a vacuum-ultraviolet spectrometer to a plasma device is also rather awkward. Gas-phase photoelectron spectroscopy offers some possibilities here but the attendant radio frequency and microwave generated field cause instrumental problems with currently available spectrometers. Vibrationally, the multiplicity of species and broadening of lines observed in plasmas constitute a source of some difficulty with infrared or Raman spectroscopy. In studies of rotational states by microwave spectroscopy, the microwave radiation would be likely to cause fluctuations and variations in the plasma, since microwave sources are also capable of producing discharges.

Another technique that should eventually prove fruitful in the investigation of plasma reactions is EPR spectroscopy, by which each radical species could be observed, but not in a strictly quantitative way, and could be characterized from its EPR spectrum. Again, in this instance, there are permutations of the plasma conditions caused by the EPR klystron sources. Thus, to characterize such a seemingly simple reaction adequately, a tremendous amount of spectroscopic data should be collected simultaneously. After all of this instrumentation had been assembled, perhaps the most perplexing problem would be the fact that the radio-frequency- or microwave-generated plasma is not uniform in intensity, or with respect to any other parameter, but changes continually, from the central part of the coils towards the edge, or "tail," of the plasma. Thus, the fact that certain energy states and species were observed at one particular point would not necessarily mean that they were uniform throughout and, in fact, to characterize the plasma completely, a number of such points of observation would have to be considered. The solution to this multiple dilemma appears to lie in various types of laser spectrometers that can be focused on precise points in a plasma apparatus. It would be expected that spectroscopic information of this type would be useful in increasing the yields of known processes and in developing new syntheses.

Operationally, however, the vast majority of the metal-containing products are those expected from the reactions already presented, namely, successive exchange of trifluoromethyl for halogen. Equations (7) and (8), for example, represent the last two steps of the formation of $(CF_3)_4Ge$.

$$\cdot CF_3 + (CF_3)_2GeBr_2 \rightarrow (CF_3)_3GeBr + \cdot Br \tag{7}$$

$$\cdot CF_3 + (CF_3)_3GeBr \rightarrow (CF_3)_4Ge + \cdot Br \tag{8}$$

With the exception of the reactions of trifluoromethyl radicals with sulfur vapor, which is really a separate class of reactions, if the power supplied to the load coil surrounding the reactor (see Fig. 2) was maintained at, or near, the minimum amount needed to support the discharge, in only two cases were compounds found that clearly resulted from reactions other than replacement of halogen by trifluoromethyl. The reaction of tellurium tetrabromide (or the chloride) gave, in addition to the products just reported, very small proportions of such species as $BrCF_2TeCF_2Br$ and $(C_2F_5)_2Te$, which were isolated in *yields of much less than 1%*. The isolation of these products was made possible by the fact that they are highly colored. Presumably, similar types of products were formed in comparable yields during the reaction of the other metal halides. The isolation of these products indicates that other reactive intermediates, e.g., $:CF_2$ or $\cdot C_2F_5$, may well be present, but, as expected, in very low concentration.

III. Bis(trifluoromethyl)mercury as a Synthetic Reagent

As noted in the Introduction, the trifluoromethyl group is a substituted alkyl species that appears to react as though it were a pseudohalogen. Somewhat surprisingly, however, ligand-exchange reactions that result in the transfer of a trifluoromethyl group from one metallic element to another appear to have been but little investigated as a synthetic procedure. The results of several earlier experiments had been summarized (8): "All attempts to prepare new perfluoroalkyl organometallics from the mercury compounds have been unsuccessful." In only one instance had more than one trifluoromethyl ligand been attached to another element by an exchange reaction, and, in that study, no compounds could be isolated, but the appearance of new resonances in the fluorine-NMR spectrum was postulated (22) to be due to the formation of $(CF_3)_2Cd$.

However, the formation of $(CF_3)_4Ge$ and $(CF_3)_2GeI_2$ in aged samples that had originally contained only $(CF_3)_3GeI$ indicated that the transfer of trifluoromethyl ligands, at least among germanium centers, must take place fairly easily. To learn more concerning the potential of ligand-exchange reactions for the preparation of trifluoromethyl-containing compounds, the interaction of bis(trifluoromethyl)mercury with the halides of several Group 4A elements was examined, as this synthetic route might nicely complement the discharge synthesis. In the discharge reaction, the yields of the compounds containing only CF_3 groups as ligands was rather high, and, consequently, the yields of such partially substituted species as $(CF_3)_2SnBr_2$ was necessarily low. Should the ligand-exchange reaction between $(CF_3)_2Hg$ and the Group 4A tetrahalides proceed to afford the (trifluoromethyl)Group 4A halides, suitable control of the reaction conditions might well result in a synthesis that preferentially resulted in compounds containing only two, or three, trifluoromethyl ligands. As little was known of the chemical stability of a CF_3 group bound to a Group 4A element, the reactivity of this linkage was assessed by exposing representative (trifluoromethyl)germanium and (trifluoromethyl)tin halides to a spectrum of common reagents, in order to determine whether the CF_3Ge or

CF_3Sn bond is stable to reaction conditions that cleave the metal–halogen bond.

The preparative reactions were conducted in sealed tubes in which ~1–3 g of the reagents had been placed. After the vessels had been maintained at the indicated temperatures for the designated times, the contents were removed, to be separated by fractional condensation and GLC. In addition to the (trifluoromethyl)Group 4A halides reported next, each sample contained unreacted $(CF_3)_2Hg$, the expected (trifluoromethyl)mercuric halide, and the mercuric halide, identified by fluorine-NMR spectroscopy and mass spectrometry.

In our laboratory, we find that the plasma reaction of trifluoromethyl radicals with mercuric iodide is an excellent source of bis(trifluoromethyl)mercury. For those laboratories that lack access to radiofrequency (rf) equipment (a 100-W, rf source can at present be purchased for less than $1,000), synthesis of bis(trifluoromethyl)mercury by the thermal decarboxylation of $(CF_3CO_2)_2Hg$ is also a functional, and quite convenient, source of bis(trifluoromethyl)mercury (23).

A. Reaction with Germanium Tetraiodide

Germanium tetraiodide (3 g) and sufficient $(CF_3)_2Hg$ to provide molar ratios (of mercurial to germane) of 0.55, 0.98, 1.72, and 2.00 were placed in tubes (10-mm diam.) that were then degassed, sealed, and placed in an oven held at 120°. After 120 hours, the tubes were opened, and the contents separated. The yields of the (trifluoromethyl)germanium halides formed (16), based on GeI_4, are presented in Table III.

As these results indicate, the reaction of $(CF_3)_2Hg$ with germanium tetraiodide provides a convenient source of the (trifluoromethyl)germanium iodides in good to excellent yields, and these yields can be varied, to increase the proportion of a particular (trifluoromethyl) germa-

TABLE III

Product Distribution for the Reaction[a] of $(CF_3)_2Hg$ with GeI_4

Molar ratio $(CF_3)_2Hg/GeI_4$	CF_3GeI_3	$(CF_3)_2GeI_2$	$(CF_3)_3GeI$	$(CF_3)_4Ge$	$(CF_3)_3GeF$
0.55	90	5			
0.98	13	53	16		
1.72			72	22	
2.00			72	15	11

[a] During 120 h at 120°.

nium halide, by appropriate selection of the proportions of the reagents employed. Presumably, the formation of the fluoride $(CF_3)_3GeF$ is acid-catalyzed, as both the mercurial $(CF_3)_2Hg$[1] and the germanes (see Table II) are stable at the temperatures employed. The fluoride is "anomalous," in that the sublimation point of this compound is normal when compared to the boiling point of the other halides, $(CF_3)_3GeX$ (X =Cl, Br, or I), but the melting point is quite abnormal, being ~125° above the melting point of the chloride; this is an indication of association in the solid state. The fluorine-NMR spectrum of the (heated) neat liquid consists of two broad singlets. In solution in diethyl ether, however, the resonances become resolved into the expected doublet and decet, again demonstrating association in the condensed phase that is broken up (on the NMR time-scale) by the intervention of solvent molecules (15). Similar effects have been reported for related compounds, and constitute indications that the germanium fluorides occupy a middle ground between those of carbon and silicon, which are very volatile, nonassociated fluorides, and those of tin and lead, which are essentially nonvolatile at lower temperatures and are strongly associated in the solid phase.

In order to test the generality of the synthesis, the reaction of $(C_2F_5)_2Hg$ with GeI_4 was assessed. This reaction required a higher temperature, 135°, before it proceeded measurably. (Pentafluoroethyl)germanium triiodide was the only (perfluoroethyl)germanium halide produced, in 54% yield (15). From this reaction, perfluoroethyl iodide was also isolated (in 30% yield). The reactions of $GeBr_4$ with $(CF_3)_2Hg$ and $(C_2F_5)_2Hg$ were similar in nature (16).

B. Reaction with Tin Tetrabromide

Under a variety of conditions, the reaction of $SnBr_4$ with $(CF_3)_2Hg$ results in the formation of CF_3SnBr_3 and $(CF_3)_2SnBr_2$ as the only volatile, tin-containing products. All attempts to produce more fully substituted (trifluoromethyl)tin bromides by further reaction of $(CF_3)_2SnBr_2$ with more $(CF_3)_2Hg$ failed (16, 17). At 80°, there was no reaction; at 100°, the $(CF_3)_2SnBr_2$ decomposed. The nature of this reaction was further studied by determining the proportions of CF_3SnBr_3 and $(CF_3)_2SnBr_2$ present during the course of the reaction at 112, 121, and 130°. The proportions of these compounds formed from a 2:1 molar ratio of mercurial to stannane, at the temperatures reported (24), are shown in Fig. 7. Although sealed-tube reactions tend to be irregular, the trends in Fig. 7 are clear. The first step is the formation of an equilibrium mixture containing CF_3SnBr_3 and CF_3HgBr in approximately

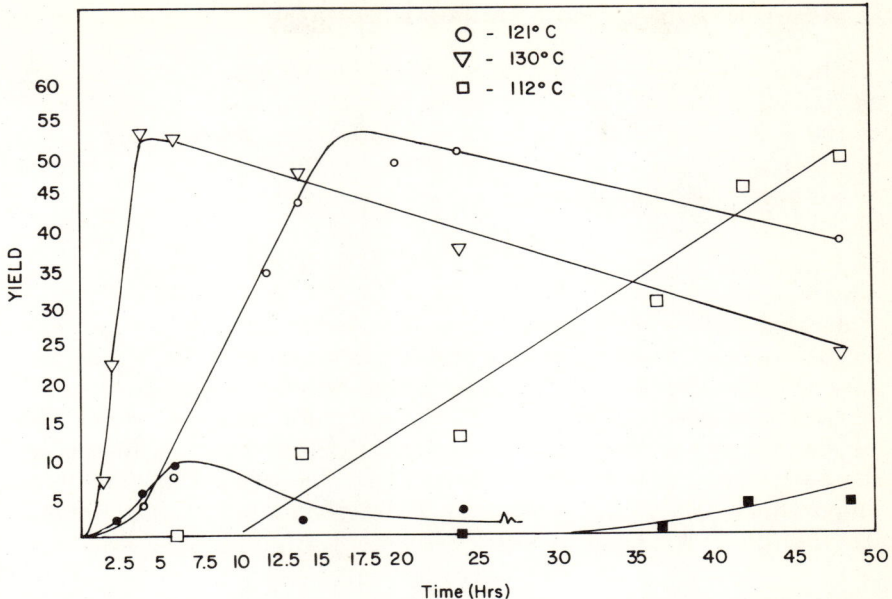

Fig. 7. The proportions of CF_3SnBr_3 formed at 112 (□), 121 (○), and 130 (▽), and of $(CF_3)_2SnBr_2$ formed at 112 (■) and 130° (●), in sealed-tube reactions.

equal proportions. The second step, the production of $(CF_3)_2SnBr_2$, results in the formation of a thermally unstable product. Thus, the formation of $(CF_3)_2SnBr_2$ is initially faster than the decomposition, but, in the later stages of the reaction, the decomposition predominates, and only small, "steady-state" concentrations of $(CF_3)_2SnBr_2$ are observed (24). These reactions are shown in Eqs. 9–11.

$$SnBr_r + (CF_3)_2Hg \rightarrow CF_3SnBr_3 + CF_3HgBr \qquad (9)$$

$$CF_3SnBr_3 + (CF_3)_2Hg \rightarrow (CF_3)_2SnBr_2 + CF_3HgBr \qquad (10)$$

$$3\ (CF_3)_2SnBr_2 \xrightarrow{\Delta} 3\ CF_3SnBr_2F + C_3F_6 \qquad (11)$$

The temperature dependence of the reaction is remarkable. For a 5-h reaction, a temperature change of 18° results in a change in the yield of CF_3SnBr_3 from 0 at 112° to 55% at 130°. Alternatively, for a 35-h reaction, the yield of $(CF_3)_2SnBr_2$ is very low at both 112 and 130°; in the former, because the $(CF_3)_2SnBr_2$ has yet to be formed, whereas, in the latter, most of the $(CF_3)_2SnBr_2$ has decomposed. These results provide one explanation of the cause of the failure of the earlier study (8) to be productive: the temperature regime is critical.

C. Reaction with Silicon Tetrahalides

Silicon tetra-chloride, -bromide, or -iodide (~5 mmol) was similarly treated with $(CF_3)_2Hg$ (1-20 mmol) at temperatures that varied from 0 to 100°C. Although C_3F_6, SiF_4, and e.g., HgI_2 were produced, and identified by mass spectrometry, in no case were trifluoromethyl-substituted silanes discerned *under the conditions employed*.

D. Chemical Integrity of the CF_3Ge and CF_3Sn Bonds

Because the thermal stabilities of the (trifluoromethyl)germanium halides and the (trifluoromethyl)tin halides appeared to vary considerably, the chemical stability of these linkages was assessed by treating $(CF_3)_3GeI$ and, e.g., CF_3SnBr_3 with a variety of reagents *(15, 24)*. The germane is remarkably resistant to chemical reaction, as, in each of the reactions reported in Equations 12-15, the new compounds were prepared in good yield by reaction of the germanium–halogen bond, leaving the CF_3Ge bond unscathed.

$$(CF_3)_3GeI + AgX \rightarrow (CF_3)_3GeX \quad (12)$$
$$(X = F, Cl, or Br) \quad 72, 92, or 95\%$$

$$(CF_3)_3GeI + HgO \rightarrow (CF_3)_3GeOGe(CF_3)_3 \; (100\%) \quad (13)$$

$$(CF_3)_3GeI + Na/Hg \rightarrow (CF_3)_3GeGe(CF_3)_3 \; (60\%) \quad (14)$$

$$(CF_3)_3GeI + BH_4^- \rightarrow (CF_3)_3GeH \; (93\%) \quad (15)$$

Tris(trifluoromethyl)germanium iodide is unstable in $3M$ base, however, and yields fluoroform quantitatively. All of the compounds showed good thermal stability *(15)*.

The trifluoromethyl–tin bond is, however, much less stable chemically *(24)*. Reaction of, e.g., $(CF_3)_2SnBr_2$ with an excess of the relatively covalent, methylating agent $(CH_3)_2Cd$ results in the very slow substitution for one of the Sn–CF_3 bonds, but the reaction of CF_3SnBr_3 with an excess of the more powerful, more ionic reagent methyllithium results in the displacement of all of the ligands, and the formation of $(CH_3)_4Sn$ as shown in Equations 16 and 17.

$$(CF_3)_2SnBr_2 + (CH_3)_2Cd \rightarrow CF_3Sn(CH_3)_3 \; (50\%) \quad (16)$$

$$CF_3SnBr_3 + CH_3Li \rightarrow (CH_3)_4Sn \quad (95\%) \quad (17)$$

A variety of reductions designed to prepare the stannane CF_3SnH_3 by reaction with, e.g., $LiAlH_4$, failed *(24)*. In each instance, SnH_4 was produced, but little CF_3SnH_3 was isolated.

E. Summary

These studies have shown that $(CF_3)_2Hg$ and GeI_4, or $SnBr_4$, do exchange ligands to provide the compounds $(CF_3)_nGeI_{4-n}$, $n = 1-4$, or $(CF_3)_nSnBr_{4-n}$, $n = 1$ or 2, and that these reactions constitute convenient syntheses of trifluoromethyl-containing germanium and tin compounds (16, 17, 22). The method is especially effective for the more-electronegative element germanium, where the trifluoromethyl–germanium linkage is stable both thermally and chemically, a stability akin to the stabilities found for other substituted alkyl ligands (16). The trifluoromethyl–tin bond, a bond to a more electropositive element, is much less stable, both thermally and chemically, and is readily displaced in reactions of the (trifluoromethyl)tin halides with more powerful reagents, such as CH_3Li or $LiAlH_4$, with the CF_3 ligand reacting somewhat like a pseudohalogen (24). Towards more covalent, more discriminating reagents, such as $(CH_3)_2Cd$, however, the distinction between the CF_3–Sn and the Br–Sn bond is differentiable, as $(CF_3)_2Sn(CH_3)_2$ may be prepared from $(CF_3)_2SnBr_2$ in good yield. These methods, coupled with the plasma technique, have been so effective synthetically that all $Ge(CF_3)_n(X)_{4-n}$ (X = I, Br, Cl, F) compounds are now known due to a research effort by a former member of our research team (16b).

IV. Synthesis of Trifluoromethyl Organometallic Compounds by Direct Fluorination

A quite surprising development, even to experienced workers in elemental-fluorine chemistry, has been the synthesis of trifluoromethyl organometallic compounds by direct fluorination of metal alkyls (25). Even more surprising is the fact that, for certain metal and metalloid systems, such as the reaction of elemental fluorine with tetramethylgermane, this type of low-temperature synthesis is a practical method (26) for the laboratory preparation of the perfluoro analog.

A. Experimental Methods

This method of low-temperature, direct fluorination involves very precise control of fluorine concentrations during the reaction, and initial high dilution of the fluorine with helium. The reaction of elemental fluorine with organometallic compounds is conducted (27) in a cryogenic-zone reactor (see Fig. 8) at temperatures in the range of -78 to

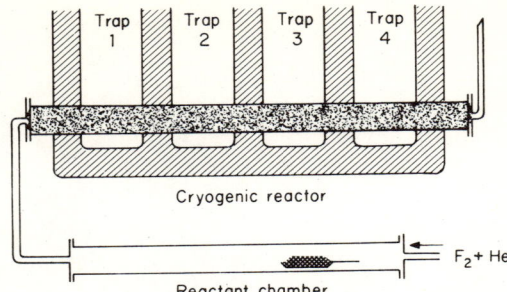

Fig. 8. Cryogenic reactor system.

−100°C. Details on methodology, and applications of these synthetic capabilities, are provided in a review article (28). The preservation of metal–carbon and metalloid–carbon bonds during direct fluorination is one of the most impressive cases of successful, direct fluorination described to date.

B. The Controlled Reaction of Metal Alkyls with Elemental Fluorine

Under the appropriate conditions, $Ge(CF_3)_4$ is obtained from $Ge(CH_3)_4$ in 63% yield, on a three-gram scale.

$$Ge(CH_3)_4 + F_2\text{-He} \rightarrow Ge(CF_3)_4 + \text{polyfluorotetramethylgermanes} + \text{fluorocarbons}$$
$$(63\%)$$

That the carbon–metal or carbon–metalloid bonds are preserved at all in these reactions is quite surprising. With tetramethylgermanes, for example, this free-radical reaction must be a 24-step process. The success in preserving carbon–germanium bonds must arise from very rapid, molecular-vibrational, rotational, and translational relaxation-processes occurring on the cryogenically cooled surfaces, such that the energy from the extremely exothermic reaction is smoothly dissipated. Under milder conditions of fluorination, a range of partially fluorinated methylgermanium compounds is produced (30) (see Tables IV–VI).

The reaction of tetramethylsilane with fluorine led to the isolation of several, partially fluorine-substituted tetramethylsilanes (see Tables VII–IX), and preservation of over 80% of the silicon–carbon bonds in the initial, tetramethylsilane reactant. The stability of many of the partially fluorinated germanes and silanes (some are stable to over 100°C) is very surprising, for the possibility of elimination of hydrogen fluoride is obvious. Indeed, before the first reported synthesis (12) of

TABLE IV
Fluorine-NMR Spectra of Polyfluorotetramethylgermanes

Compound	CF_3*	J_{FF}	CHF_2**	J_{HF}	J_{FF}	CH_2F***	J_{HF}	J_{FF}
$Ge(CF_3)_4$	−27.0							
$Ge(CF_3)_3(C_2F_5)$	−27.7	2.7[a]						
$Ge(CF_3)_3(CF_2H)$	−27.2	3.0[b]	49.0	45.5	3.1[c]			
$Ge(CF_3)_3(CFH_2)$	−25.8							
$Ge(CF_3)_2(CF_2H)_2$	−27.6	3.2[d]	49.4	46.0	3.1[e]			
$Ge(CF_3)_2(CF_2H)(CFH_2)$	−27.6	3.3[f]	50.6	45.5	3.0[g]	193.2	46.5	3.3[h]
$Ge(CF_3)(CF_2H)_3$	−27.9	3.2[i]	49.7	46.0	3.1[j]			
$Ge(CF_3)_2(CFH_2)_2$	−24.8	3.4[k]				193.0	47.0	3.5[l]
$Ge(CF_3)(CF_2H)_2(CFH_2)$	−26.8	3.2[m]	50.5	45.6	3.0[n]	193.0	46.0	2.9[o]
$Ge(CF_3)(CF_2H)_2(CH_3)$	−23.5	3.4[p]	53.0	46.5	3.2[q]			
$Ge(CF_3)(CF_2H)(CFH_2)_2$	−25.2	3.2[r]	51.8	45.5	3.0[s]	192.0	46.0	2.7[t]
$Ge(CF_2H)_3(CFH_2)$			50.5	46.2	2.2[u]	192.9	46.0	2.4[v]
$Ge(CF_2H)_2(CFH_2)_2$			51.4	46.0	2.5[w]	192.5	46.6	2.5[x]
$Ge(CF_3)(CF_2H)(CFH_2)(CH_3)$	−22.3	3.4[y]	53.7	46.0	3.0[z]	192.5	46.0	3.0
$Ge(CF_2H)(CFH_2)_3$			52.0	46.0	2.1[aa]	191.9	46.6	1.8[bb]
$Ge(CF_2H)(CFH_2)_2(CH_3)$			53.9	46.2	2.1[cc]	192.5	46.5	2.0[dd]
$Ge(CFH_2)_4$						191.4	46.7	
$Ge(CF_3)_3(OH)$	−21.6							

* Singlet. ** Doublet. *** Triplet. Shifts in p.p.m. from external TFA + upfield from TFA. Coupling constants in Hertz. [a] Basic quartet, C_2F_5 group: CF_3, 6.76, multiplet; CF_2, 38.4, J_{FF} = 3.3, septet. [b] Triplet. [c] 4 of 10 lines. [d] Pentet. [e] Septet. [f] Quartet. [g] 6 of 8 lines. [h] 7 of 9 lines. [i] Septet. [j] Quartet. [k] Triplet. [l] 5 of 7 lines. [m] Sextet. [n] Pentet. [o] 6 of 8 lines. [p] Pentet. [q] Quartet. [r] Pentet. [s] Sextet. [t] Sextet. [u] Doublet. [v] 5 of 7 lines. [w] Pentet. [x] Pentet. [y] Quartet. [z] Pentet. [aa] Quartet. [bb] Triplet. [cc] Triplet. [dd] Triplet.

$Ge(CF_3)_4$ and $Sn(CF_3)_4$ in 1975, many authors had given detailed reasons for the supposed instability of these species as a rationale for the failure of conventional syntheses to produce them.

The first successful, fluorine reaction observed with organometallics was the conversion of dimethylmercury into bis(trifluoromethyl)mercury (25).

$$Hg(CH_3)_2 \xrightarrow[-110°C]{He-F_2} Hg(CF_3)_2$$
$$(6.8\%)$$

It is quite probable that, were this reaction to be repeated, yields much higher than the 6.8% yield reported (25) could be obtained. In fact, it is even possible that this method, too, could eventually be developed into a practical synthesis for bis(trifluoromethyl)mercury.

The reaction of elemental fluorine with tetramethyltin is very different, and even more unusual (31). Under the conditions studied, no volatile compounds, such as $Sn(CF_3)_4$, were obtained. If fact, it appears

TABLE V

PROTON-NMR SPECTRA OF POLYFLUOROTETRAMETHYLGERMANES

Compound	CH$_3$*	CH$_2$F**	J_{HF}	CHF$_2$***	J_{HF}
Ge(CF$_3$)$_3$(CF$_2$H)				6.10^a	45.0^b
Ge(CF$_3$)$_3$(CFH$_2$)					
Ge(CF$_3$)$_2$(CF$_2$H)$_2$				6.23	45.5
Ge(CF$_3$)$_2$(CF$_2$H)(CFH$_2$)		4.98	46.5	6.24	45.7
Ge(CF$_3$)(CF$_2$H)$_3$				6.25	45.5
Ge(CF$_3$)$_2$CFH$_2$)$_2$					
Ge(CF$_3$)(CF$_2$H)$_2$(CFH$_2$)		4.89	46.0	6.15	45.5
Ge(CF$_3$)(CF$_2$H)$_2$(CH$_3$)	0.51			6.10	45.6
Ge(CF$_3$)(CF$_2$H)(CFH$_2$)$_2$		4.90	46.0	6.25	45.6
Ge(CF$_2$H)$_3$(CFH$_2$)		5.02	45.7	6.28	45.6
Ge(CF$_2$H)$_2$(CFH$_2$)$_2$		4.97	46.0	6.26	45.2
Ge(CF$_3$)(CF$_2$H)(CFH$_2$)(CH$_3$)	0.47	4.79	46.0	6.08	46.0
Ge(CF$_2$H)(CFH$_2$)$_3$		4.97	46.5	6.29	46.0
Ge(CF$_2$H)(CFH$_2$)$_2$(CH$_3$)	0.34	4.78	46.0	6.10	46.0
Ge(CFH$_2$)$_4$		4.87	47.0		
Ge(CF$_3$)$_3$(OH)^c					

* Singlet. ** Doublet. *** Triplet. ^{a} Chemical shifts in p.p.m. + downfield from external Me$_4$Si. ^{b} Coupling constants in Hertz. ^{c} OH (2.43), singlet.

TABLE VI

WEIGHT PERCENTAGE YIELDS^a OF POLYFLUOROTETRAMETHYLGERMANIUM

Compound	Yield
Ge(CF$_3$)$_4$	63.5^b
Ge(CF$_3$)$_3$(C$_2$F$_5$)	0.25
Ge(CF$_3$)$_3$(CF$_2$H)	0.38
Ge(CF$_3$)$_3$(CFH$_2$)	0.06
Ge(CF$_3$)$_2$(CF$_2$H)$_2$	6.20
Ge(CF$_3$)$_2$(CF$_2$H)(CFH$_2$)	4.16
Ge(CF$_3$)(CF$_2$H)$_3$/Ge(CF$_3$)$_2$(CFH$_2$)$_2$/Ge(CF$_3$)(CF$_2$H)$_2$(CH$_3$)	13.35^c
Ge(CF$_3$)(CF$_2$H)$_2$(CFH$_2$)/Ge(CF$_3$)(CF$_2$H)(CFH$_2$)(CH$_3$)	13.94^d
Ge(CF$_3$)(CF$_2$H)(CFH$_2$)$_2$	28.51
Ge(CF$_2$H)$_3$(CFH$_2$)	15.47
Ge(CF$_2$H)$_2$(CFH$_2$)$_2$/Ge(CF$_2$H)(CFH$_2$)$_2$(CH$_3$)	11.87^e
Ge(CF$_2$H)(CFH$_2$)$_3$	4.18
Ge(CF$_3$)$_3$(OH)	1.40
Ge(CFH$_2$)$_4$	0.22

^{a} Yield for Ge(CF$_3$)$_4$ calculated from reaction conditions designed to maximize Ge(CF$_3$)$_4$. When yield of Ge(CF$_3$)$_4$ is high, only traces of two other compounds, Ge(CF$_3$)$_3$(C$_2$F$_5$) and Ge(CF$_3$)$_3$(OH), appear. ^{b} Based on 0.87 g of Ge(CH$_3$)$_4$ as starting material. ^{c} Ratio of three compounds, 6.1:1.4:1. ^{d} Ratio of two compounds, 10:1. ^{e} Ratio of two compounds, 14.3:1.

TABLE VII
Melting Points of Polyfluorotetramethylsilanes

Compound	Melting point (°C)
$Si(CH_3)_3(CH_2F)$	−86.5 to −85.0
$Si(CH_3)(CH_2F)_3$	−89.5 to −84.5
$Si(CH_3)(CH_2F_2)(CHF_2)$	−63.5 to −62.0
$Si(CH_2F)_4$	−18.0 to −16.6
$Si(CH_3)(CH_2F)(CHF_2)_2$	−58.7 to −56.8
$Si(CH_3)(CH_2F)_2(CF_3)$	−144 to −142.7
$Si(CH_2F)_3(CHF_2)$	−54.0 to −53.2
$Si(CH_3)(CH_2F)(CHF_2)(CF_3)$	−144 to −138
$Si(CH_2F)_2(CHF_2)_2$	−68.2 to −66.8
$Si(CH_2F)(CHF_2)_3$	−72.6 to −71.0
$Si(CH_3)_4$	−91.1

that the initial step in the reaction is cleavage of some of the tin–carbon bonds, to produce involatile, partially fluorinated, organometallic tin fluorides.

$$Sn(CH_3)_4 \xrightarrow[-100°C]{F_2-He}$$
$$Sn(CF_3)_2F_2 + Sn(CF_3)F_3 + Sn(CH_2F)F_2 + Sn(CHF_2)_2F_2 + Sn(CHF_2)F_3 + SnF_4$$
$$\text{I} \qquad \text{II} \qquad \text{III} \qquad \text{IV} \qquad \text{V} \qquad \text{VI}$$

TABLE VIII
Proton-NMR Spectra of Substituted Silanes

Compound	CH_3	CH_2F*	J_{HF}	CHF_2**	J_{HF}
$Si(CH_3)_3(CH_2F)^a$	0.06	4.30	46.9		
$Si(CH_3)_2(CH_2F)_2$	−0.18	4.16	47.5		
$Si(CH_3)_3(CHF_2)$	0.07			6.43	52.0
$Si(CH_3)(CH_2F)_3$	−0.23	4.17	47.0		
$Si(CH_3)_2(CH_2F)(CHF_2)$	−0.11	4.22	47.4	5.62	46.2
$Si(CH_3)_2(CHF_2)_2$	0.31			5.91	45.6
$Si(CH_3)(CH_2F)_2(CHF_2)$	−0.06	4.32	47.3	5.68	45.9
$Si(CH_2F)_4$		4.24	47.0		
$Si(CH_3)(CH_2F)(CHF_2)_2$	0.04	4.39	46.8	5.74	45.4
$Si(CH_3)(CH_2F)_2(CF_3)$	0.12	4.42	47.1		
$Si(CH_2F)_3(CHF_2)$		4.54	46.6	5.87	45.4
$Si(CH_3)(CH_2F)(CHF_2)(CF_3)$	0.08	4.36	46.3	5.69	45.4
$Si(CH_2F)_2(CHF_2)_2$		4.66	46.6	5.91	45.4
$Si(CH_2F)(CHF_2)_3$		4.68	46.2	5.82	45.0

* Doublet. ** Triplet. Shifts in p.p.m. from external Me_4Si, + downfield from Me_4Si. Coupling constants in Hertz. *a* Lit. (29) CH_3 (0.31), CH_2F (4.53), J_{HF} (46.8).

TABLE IX

FLUORINE-NMR SPECTRA OF SUBSTITUTED SILANES

Compound	CF_3	CHF_2*	J_{HF}	CH_2F**	J_{HF}
$Si(CH_3)_3(CH_2F)$ (29)				196	46.8
$Si(CH_3)_2(CH_2F)_2$				195.14	47.1
$Si(CH_3)(CH_2F)_3$				197.73	46.8
$Si(CH_3)_2(CH_2F)(CHF_2)$		61.17	45.6	196.61	46.4
$Si(CH_3)_2(CHF_2)_2$		59.22	46.0		
$Si(CH_3)(CH_2F)_2(CHF_2)$		61.08	45.8[a]	199.60	47.0[b]
$Si(CH_2F)_4$				200.13	46.9
$Si(CH_3)(CH_2F)(CHF_2)_2$		60.85	45.7[c]	201.00	46.7[d]
$Si(CH_3)(CH_2F)_2(CF_3)$	−15.48[e]			199.92	46.3
$Si(CH_2F)_3(CHF_2)$		60.03	45.4[f]	201.41	46.6[g]
$Si(CH_3)(CH_2F)(CHF_2)(CF_3)$	−16.53[h]	61.15	45.0	201.19	+
$Si(CH_2F)_2(CHF_2)_2$		58.88	45.4[i]	202.87	46.4[j]
$Si(CH_2F)(CHF_2)_3$		59.57	45.0[k]	204.39	46.6[l]

* Doublet. ** Triplet. + Not recorded. Shifts in p.p.m. from external TFA, + upfield from TFA. Coupling constants in Hertz. [a] $J_{FF} = 1.4$ (triplet). [b] $J_{FF} = 1.5$ (multiplet), $J_{HHF} \approx 1.0$. [c] $J_{FF} = 1.8$ (doublet). [d] $J_{FF} = 1.4$ (multiplet), $J_{HHF} \approx 0.5$. [e] $J_{FF} = 3.2$ (quartet of triplets), $J_{HHF} \approx 1.0$. [f] $J_{FF} = 1.8$ (quartet). [g] $J_{FF} = 1.6$ (triplet). [h] $J_{FF} = 3.0$ (quartet or doublet of triplets). [i] $J_{FF} = 2.0$ (triplet). [j] $J_{FF} = 1.9$ (pentet). [k] $J_{FF} = 2.1$. [l] $J_{FF} = 2.0$, $J_{HHF} \approx 0.4$.

Subsequently, exchange with dimethylcadmium produces mixed fluoromethyl compounds, according to the following reaction (31).

$$\text{I–VI} \xrightarrow{Cd(CH_3)_2} Sn(CF_3)_2(CH_3)_2 + Sn(CF_3)(CH_3)_3 + Sn(CH_2F)_2(CH_3)_2 + Sn(CHF_2)_2(CH_3)_2 + Sn(CHF_2)(CH_3)_3 + Sn(CH_3)_4$$

The properties of these compounds, and the relative composition of the products are shown in Table X.

Two additional areas are under study at this time. The reaction of fluorine with transition-metal alkyls appears to be quite promising.

An attempt is also being made to preserve metal–metal bonds during direct fluorination. It has been found that the reaction of fluorine with hexamethyldigermane leads primarily to tris(trisfluoromethyl)-germanium fluoride (34).

$$(CH_3)_3Ge\text{-}Ge(CH_3)_3 \xrightarrow[-100°C]{F_2\text{-}He} (CF_3)_3GeF$$
$$70\% \text{ (based on Ge)}$$

Other studies are under way, with other metal–metal bonded compounds, such as the tin and silicon alkyls.

TABLE X

COMPOSITION AND MELTING POINTS OF FLUORINATION PRODUCTS OF THE REACTION OF FLUORINE WITH $Sn(CH_3)_4$

Compound	Weight % of products	Melting point (°C)
$Sn(CH_3)_4$	48.4	−54.2 to −53.5[a]
$Sn(CH_3)_3(CF_3)$[b]	12.9	−57.0 to −53.2
$Sn(CH_3)_3(CH_2F)$	12.9	−62.5 to −59.0
$Sn(CH_3)_3(CHF_2)$[c]	3.2	−70.5 to −66.4
$Sn(CH_3)_2(CF_3)_2$	12.9	−34.5 to −32.0
$Sn(CH_3)_2(CH_2F)_2$	9.7	>80

* Cloudy until −55°. ** Cloudy until −63°. [a] Lit. m.p. −54.8°. [b] Ref. (32), b.p. 97–101° (745 mm Hg). [c] Ref. (33), b.p. 111.5°.

V. A New General Synthesis for Trifluoromethyl Organometallic Compounds and Other Sigma-Bonded Metal Compounds Based on Metal Vapor as a Reagent

A. INTRODUCTION

Just beginning to unfold in the literature is a new synthesis for metal alkyls $M(R)_n$ and sigma-bonded metal compounds that appears to be of an extremely general nature (35). This new method is, perhaps, the most promising development reported in the present review. It appears, at this writing, to provide an absolutely general synthesis both of trifluoromethyl, organometallic compounds and many new classes of alkyl-like, sigma-bonded, metal compounds. It employs the cocondensation of metal vapor of almost any type with free radicals under conditions where the metal is oxidized to the most stable (usually, the highest) oxidation state.

In the field of metal-vapor chemistry, since its inception by the initial papers of Timms (36) and Skell and Engel (37), the major thrust has been towards syntheses involving cocondensation of π-bonding ligands with metal vapor, with the most notable exceptions being the use of metal atoms as dehalogenation agents or in Grignard-like reactions. Most of the products formed are those in which the metal vapor is in a low-valence state, π-bonded or complexed to various ligands. Such reactions have constituted an important, major research-effort in inorganic chemistry during the past ten years (38).

$$M(g) + nL \xrightarrow[-196°C]{} ML_n$$

A new synthesis for sigma-bonded, metal alkyls and similar compounds, involving a reaction between metal vapor and free radicals generated in a radio-frequency glow-discharge, has been reported (35).

$$M(g) + n\,R\cdot \xrightarrow[-196°C]{} MR_n$$

where $R = CH_3$, CF_3, or SiF_3.

The radicals, such as methyl, trifluoromethyl, and trifluorosilyl, used in this work have been found to oxidize zero-valent metals to their highest oxidation-state upon cocondensation with these metals on a cold surface at $-196°C$.

It has been experimentally determined that a number of sources of radicals, including some generated by pyrolysis, may be used for this technique. However, the low-temperature glow-discharge is a convenient source of radicals for synthetic work.

B. Apparatus

The reactor in which the syntheses were accomplished is shown in Fig. 9. The cold finger (A) and the reactor are connected with a glass, O-ring connector (B). The base (C) is made of brass, with water-cooled, electrical feed-throughs (D) for the tungsten basket or crucible heater (E). The base and reactor are connected with an O-ring junction (F). The Pyrex reactor consists of halves separated by an O-ring (G). The radio-frequency power is supplied by a Tegal Corp., 100-W, 13.56-MHz, radio-frequency generator and matching network capacitively coupled to the reactor by two metal rings (H and I). The plasma gas (i.e., hexafluoroethane) enters at the outer jacket of the reactor (J) into a ballast volume, and proceeds through a slit (~ 0.5 in. wide) in the inner tubing and between the metal rings (H and I). The tungsten basket or crucible heater are resistively heated by passing current through them. Metals more volatile than bismuth are evaporated from a quartz crucible heated by tungsten wire around the outside of the crucible. Bismuth and less volatile metals are placed in a tungsten-wire basket which is coated with aluminum oxide cement.

C. Synthesis of Trifluoromethyl Organometallic Compounds by Cocondensation of Trifluoromethyl Radicals and Metals

Using this technique, it is extremely easy to prepare almost all of the trifluoromethyl organometallic compounds that have been discussed

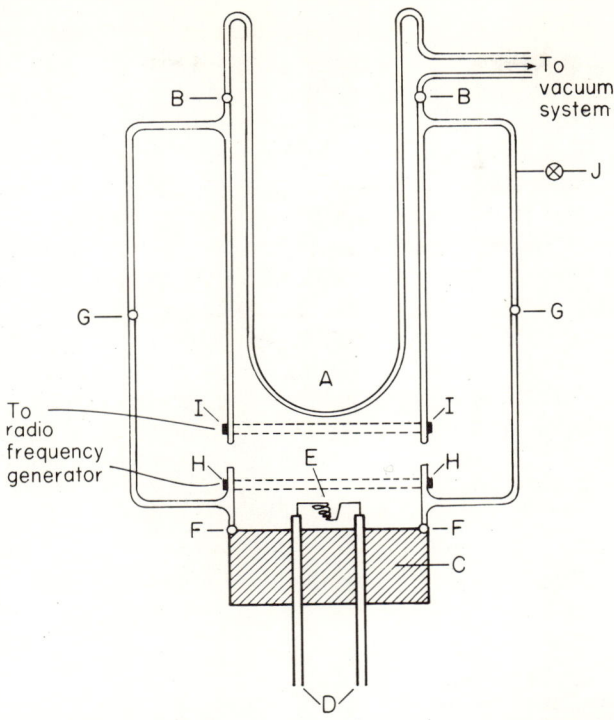

FIG. 9. Reactor for the new general metal vapor synthetic method.

in previous Sections. In each case, at least 1.5 g, and often as much as 4 g, of metal was evaporated during 2 h, and cocondensed with trifluoromethyl radicals. Although the prospects for this type of reaction appear limitless, the examples that have thus far been reported, all at $-196°C$, are as follows.

$$Hg(g) + n\ CF_3 \cdot \rightarrow Hg(CF_3)_2\ (89\%)$$
$$Te(g) + n\ CF_3 \cdot \rightarrow Te(CF_3)_2 + Te_2(CF_3)_2$$
$$\qquad\qquad\qquad\qquad (10\%) \qquad (20\%)$$
$$Bi(g) + n\ CF_3 \cdot \rightarrow Bi(CF_3)_3\ (31\%)$$
$$Sn(g) + n\ CF_3 \cdot \rightarrow Sn(CF_3)_4\ (8\%)$$
$$Ge(g) + n\ CF_3 \cdot \rightarrow Ge(CF_3)_4\ (50\%)$$

D. Synthesis of Methyl Organometallic Compounds by Cocondensation of Methyl Radicals and Metals

This new method appears to be extremely general, and it has applications extending far outside the bounds of fluorine chemistry. With the newer technique, it is possible to use other symmetrical molecules, such as ethane, having a relatively weak central bond to produce methyl radicals, and Si_2F_6 to produce trifluorosilyl radicals. The carbon–carbon bond in ethane has a bond strength of ~83–84 kcal/mole, and the carbon–hydrogen bond-strength (39) is 98–99 kcal/mole. Although this difference in bond strengths of ~15 kcal/mole is substantially smaller than with hexafluoroethane, where it is at least 40 kcal/mole, no substantial amounts (none were observed in NMR spectra) of the metal alkyls produced by the reaction of metal atoms with an ethane discharge contained ethyl groups. This result was somewhat surprising, and, initially, some experiments were conducted that employed sources (of methyl radicals) other than ethane.

A number of methyl-radical reactions with various metals have also been observed to occur in good yield on about the same, 1–4-g scale at $-196°C$.

$$Hg(g) + n\ CH_3\cdot \rightarrow Hg(CH_3)_2\ (9\%)$$
$$Cd(g) + n\ CH_3\cdot \rightarrow Cd(CH_3)_2\ (31\%)$$
$$Bi(g) + n\ CH_3\cdot \rightarrow Bi(CH_3)_3\ (13\%)$$
$$Sn(g) + n\ CH_3\cdot \rightarrow Sn(CH_3)_4\ (87\%)$$
$$Ge(g) + n\ CH_3\cdot \rightarrow Ge(CH_3)_4\ (16\%)$$

It is clear that the "oxidation" of metal atoms to their most stable (and, usually, highest) coordination state by radicals on a cold surface or in the gas phase should be possible for radicals of almost any type. Although it might have been predicted that such highly electronegative radicals as $CF_3\cdot$ and $SF_5\cdot$ would accomplish this oxidation, the fact that it occurs with such facility for the methyl radical (which has no particularly great oxidizing power) is an encouraging sign. Certainly, new organometallic compounds from ethyl, phenyl, and other organic radicals of most types are plausible. The fact that the reaction appears to occur primarily on the cold finger of the reactor, or in the very short time preceding condensation, offers the possibility that very unstable compounds (i.e., compounds that decompose at $-50°C$) might be isolated by extracting the matrix with a cold solvent or by adding other stabilizing ligands, as has become common practice in the previous metal-vapor–ligand reactions. There is a distinct possibility that the work on cocondensation of transition-metal vapors and methyl radicals that is currently in progress could establish the existence of methyl transition-metal alkyl species too unstable to be produced by the conventional routes (which often require a much higher activation-energy).

E. Synthesis of Bis(trifluorosilyl)mercury by Cocondensation of Mercury and Trifluorosilyl Radicals

The first trifluorosilyl "organometallic" compound, bis(trifluorosilyl)mercury, has been prepared by using the new, metal-vapor technique (35).

$$Si_2F_6 \xrightarrow{rf} 2\ SiF_3\cdot$$

$$Hg(g) + n\ SiF_3\cdot \xrightarrow{-196°C} Hg(SiF_3)_2\ (26\%)$$

Previously, trifluorosilyl groups have been bound to phosphorus (40) and silicon via the $SiF_2(g)$, fluorine-bond insertion-mechanism (41). The new compound $Hg(SiF_3)_2$ is readily hydrolyzed, but it can be stored for long periods of time in an inert atmosphere. It is a volatile, white solid that is stable up to at least 80°C. The preparation of bis(trifluorosilyl)mercury, of course, raises the possibility of (a) synthesis of the complete series of trifluorosilyl, "silametallic" compounds, as had previously been done for bis(trifluoromethyl)mercury by using conventional syntheses, and (b) transfer reactions similar to those in Section II, as well as (c) further exploration of the metal-vapor approach. The compound $Hg(SiF_3)_2$ appears also to be a convenient source of difluorosilane upon thermal decomposition, analogous to bis(trifluoromethyl)mercury:

$$Hg(SiF_3)_2 \xrightarrow{\Delta} HgF_2 + 2SiF_2$$

This route could prove to be a very productive condensed phase (solution) method of generating a high concentration of SiF_2. An investigation to determine the reaction differences and similarities to the well known difluorosilane work of Margrave (42) is contemplated.

F. Prospects and Future Applications of This Technique

There are prospects for this type of reaction which are likely to have sweeping applications. There are a number of areas in which significant developments in the near future may be forecast. (1) It is quite clear that there are extensive, experimental applications of this type of metal-vapor reaction in transition-metal chemistry. (2) Currently, work has been successful in generating several new $M(SCF_3)_n$ compounds (43) from CF_3SSCF_3, and $M(SF_5)_n$ compounds (44) from S_2F_{10}. Work is under way on the generation of sigma-bonded-BF_2, organometallic compounds, using B_2F_4 as a source of BF_2 radicals, and with

SF$_5$ radicals generated from S$_2$F$_{10}$ directed toward the preparation of sigma-bonded M(SF$_5$)$_n$. Also, PF$_2$·, SiCl$_3$·, and SiH$_3$· reactions are under investigation. (3) Furthermore, it is clear that the oxidation of many species other than pure metals is possible on cold surfaces with metal radicals. In particular, such metal halides as SnCl$_2$ are converted primarily (45) into Sn(CF$_3$)$_2$Cl$_2$ while PbX$_2$ compounds have yielded the new compound Pb(CF$_3$)$_4$, which is surprisingly stable. Other metal subfluorides or metal oxides that are either stable at ambient temperature, or generated solely as high-temperature species, are also oxidized. Here, a possible criterion for reaction is simply that the ionization potentials of the metal or non-metal compounds must be comparable to those of the corresponding main-group or transition metals, so that oxidation may take place. (4) A further ramification of the present work is the possibility of cocondensing metals and radicals in argon and other types of matrices for spectroscopic study. It would appear that an apparatus as simple as a diathermy plasma-source and metal atoms could be used to prepare, by cocondensation, many new species in matrices. Of particular interest should be the cocondensation of transition metals with such species as methyl radicals, to produce the unsaturated species (i.e., of valence lower than normal) M(R)$_{n-1}$, M(R)$_{n-2}$, etc., where the metal is not coordinatively saturated. The structures of such species would be of unusual interest in view of the fact that unsaturated metal alkyls have been proposed by many workers as being active species in homogeneous catalysis. Such spectroscopic work on the unsaturated metal carbonyls M(CO)$_{n-1}$ and M(CO)$_{n-2}$ had been performed by Ozin, Turner, and others (46).

ACKNOWLEDGMENTS

The financial contributions of the National Science Foundation, the Air Force Office of Scientific Research, The Office of Naval Research, The Research Corporation, and the Research Board of UICC are gratefully acknowledged.

REFERENCES

1. H. J. Emeléus and R. N. Haszeldine, *J. Chem. Soc.*, p. 2953 (1949).
2. See, for example, J. J. Lagowski, *Q. Rev. Chem. Soc.* **13,** 233 (1959); H. J. Emeléus, "The Chemistry of Fluorine and Its Compounds," pp. 45–105. Academic Press, New York, 1969; R. E. Banks, "Fluorocarbons and Their Derivatives," pp. 102–203. McDonald, 1970; R. D. Chambers, "Fluorine in Organic Chemistry," pp. 344–378 Wiley, New York, 1973.
3. This statement does *not* constitute an endorsement of the procedure, as no volatile mercury compound can be considered innocuous.
4. E. G. Walashewski, *Chem. Ber.* **86,** 272 (1953).

5. E. L. Muetterties, W. Mahler, and R. Schmutzler, *Inorg. Chem.* **2**, 613 (1963).
6. K. I. The and R. G. Cavell, *J. Am. Chem. Soc.* **99**, 7841 (1977); K. I. The, J. A. Gibson, and R. G. Cavell, *Inorg. Chem.* **16**, 2887 (1977).
7. For a comprehensive review of this earlier work, see R. E. Banks, "Fluorocarbons and Their Derivatives." Oldbourne Press, London, 1970.
8. H. C. Clark, *Adv. Fluorine Chem.* **3**, 19 (1963).
9. V. H. Dibeler, R. M. Riese, and F. L. Mohler, *J. Chem. Phys.* **20**, 761 (1952); J. B. Farmer, I. H. S. Henderson, F. P. Lossing, and D. B. H. Marsden, *J. Chem. Phys.* **24**, 342 (1956); J. W. Coomber and E. Whittle, *Trans. Faraday Soc.* **63**, 1394 (1967).
10. "Janaf Thermochemical Tables." Dow Chemical Co., Midland, Michigan, 1963.
11. (a) R. J. Lagow, L. L. Gerchman, R. A. Jacob, and J. A. Morrison, *J. Am. Chem. Soc.* **97**, 518 (1975); (b) R. J. Lagow and R. A. Jacob, *J. Chem. Soc., Chem. Commun.* **4**, 104 (1973); (c) R. J. Lagow and R. Eujen, *Inorg. Chem.* **14**, 3128 (1975).
12. A. G. Massey, D. S. Urch, and A. K. Holliday, *J. Inorg. Nucl. Chem.* **28**, 365 (1966).
13. M. D. Rausch and J. R. Van Wazer, *Inorg. Chem.* **3**, 761 (1964).
14. H. C. Clark and C. J. Willis, *J. Am. Chem. Soc.* **84**, 898 (1962).
15. R. J. Lagow, R. Eujen, L. L. Gerchman, and J. A. Morrison, *J. Am. Chem. Soc.* **100**, 1722 (1978).
16. (a) J. A. Morrison, L. L. Gerchman, R. Eujen, and R. J. Lagow, *J. Fluorine Chem.* **10**, 333 (1977); (b) R. Eujen, private communication.
17. J. A. Morrison and R. J. Lagow, *Inorg. Chem.* **16**, 1823 (1977).
18. T. N. Bell, B. J. Pullman, and B. O. West, *Aust. J. Chem.* **16**, 636 (1963).
19. T. N. Bell, B. J. Pullman, and B. O. West, *Aust. J. Chem.* **16**, 722 (1963).
20. R. J. Lagow and T. Yasumura, *Inorg. Chem.* **17**, 3108 (1978).
21. R. J. Lagow, R. A. Jacob, L. L. Gerchman, and T. J. Juhlke, *J. Chem. Soc., Chem. Commun.* **3**, 128 (1979).
22. B. L. Dyatkin, B. I. Martynov, I. L. Knunyants, S. R. Sterlin, L. A. Federof, and Z. A. Stumbrevichute, *Tetrahedron Lett.* 1345 (1971).
23. L. L. Knunyants, Y. F. Komissarov, B. L. Dayatkin, and L. T. Lantseva, *Izv. Akad. Nauk SSR, Ser. Khim.* p. 943 (1973).
24. L. J. Krause and J. A. Morrison, *Inorg. Chem.* **19** (1980).
25. R. J. Lagow and E. K. S. Liu, *J. Am. Chem. Soc.* **98**, 8270 (1976).
26. R. J. Lagow and E. K. S. Liu, *J. Chem. Soc., Chem. Commun.* p. 450 (1977).
27. (a) R. J. Lagow and N. J. Maraschin, *Inorg. Chem.* **12**, 1459 (1973); (b) R. J. Lagow, N. J. Maraschin, B. D. Catsikis, L. H. Davis, and G. Jarvinen, *J. Am. Chem. Soc.* **97**, 513 (1975); (c) R. J. Lagow, J. L. Adcock, and R. A. Beh, *J. Org. Chem.* **40**, 3271 (1975).
28. R. J. Lagow and J. L. Margrave, "Direct Fluorination: A 'New' Approach to Fluorine Chemistry." *Prog. Inorg. Chem.* Vol. 26 (1979).
29. E. S. Alexander, R. N. Haszeldine, M. J. Newlands, and A. E. Tipping, *J. Chem. Soc. A* p. 2285 (1970).
30. R. J. Lagow, and E. K. S. Liu, *J. Organometal. Chem.* **145**, 167 (1978).
31. R. J. Lagow and E. K. S. Liu, *Inorg. Chem.* **17**, 618 (1978).
32. V. V. Khrapov, U. I. Gol'danskii, A. K. Prokof'ev, and R. G. Kostyanovskii, *Zh. Obshch. Khim.* **37**, 3 (1967).
33. Cullen, W. R., Sams, J. R., and Waldman, M. C., *Inorg. Chem.* **9**, 1682 (1970).
34. R. J. Lagow and R. E. Aikman, to be published.
35. R. J. Lagow, T. J. Juhlke, R. W. Braun and T. R. Bierschenk, "The Second Dimension in Metal Vapor Chemistry; A New General Synthesis for Metal Alkyls." *J. Am. Chem. Soc.* **101**, 12, 3229 (1979).
36. P. L. Timms, *J. Chem. Soc., Chem. Commun.* p. 1525 (1968).

37. P. S. Skell and R. R. Engel, *J. Am. Chem. Soc.* **88,** 3749 (1966).
38. For recent reviews, see (a) K. J. Klabunde, *Acc. Chem. Res.* **8,** 393 (1975); (b) P. L. Timms, *Adv. Inorg. Radiochem.* **14,** 121 (1972).
39. R. T. Sanderson, "Chemical Bonds and Bond Energy," 2nd ed., p. 192. Academic Press, New York, 1976.
40. K. G. Sharp, *J. Chem. Soc., Chem. Commun.* p. 564 (1977).
41. P. L. Timms, *Adv. Inorg. Radiochem.* **14,** 121 (1972).
42. J. L. Margrave and P. W. Wilson, *Acc. Chem. Res.* **4,** 145 (1971).
43. T. R. Biersbank and R. J. Lagow, to be published.
44. T. R. Biersbank and R. J. Lagow, to be published.
45. T. Juhlke and R. J. Lagow, to be published.
46. For recent reviews of matrix isolation of coordinatively unsaturated species, see (a) G. A. Ozin and A. VanderVoet, *Prog. Inorg. Chem.* **19,** 105–72 (1975); (b) J. J. Turner, *New Synth. Methods* **3,** 187–201 (1975).

1,1-DITHIOLATO COMPLEXES OF THE TRANSITION ELEMENTS

R. P. BURNS, F. P. McCULLOUGH,* and C. A. McAULIFFE

Department of Chemistry, University of Manchester Institute of Science and Technology, Manchester, England

I. Introduction	211
II. The Ligands	212
III. Transition-Metal Complexes	215
A. Titanium, Zirconium, and Hafnium	215
B. Vanadium, Niobium, and Tantalum	218
C. Chromium, Molybdenum, and Tungsten	221
D. Manganese, Technetium, and Rhenium	232
E. Iron, Ruthenium, and Osmium	236
F. Cobalt, Rhodium, and Iridium	248
G. Nickel, Palladium, and Platinum	254
H. Copper, Silver, and Gold	265
References	269

I. Introduction

We have recently published an article[1] on 1,2-dithio ligands, and have stressed the versatility of these ligands and the academic and industrial importance of their metal complexes.

Dithio acids and dithiols are formed by reaction of carbon disulfide with various nucleophiles (Z^- or Z^{2-}), as follows.

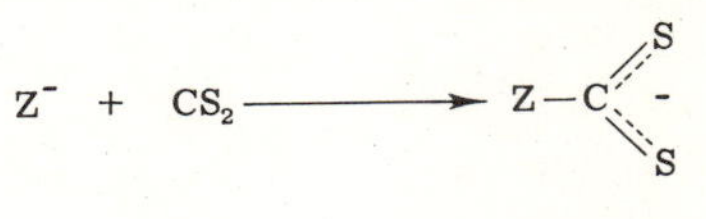

$$Z^- + CS_2 \longrightarrow Z-C(S)(S^-) \qquad Z^{2-} + CS_2 \longrightarrow Z=C(S^-)(S^-)$$

Dithio acid (I)   Dithiol (II)

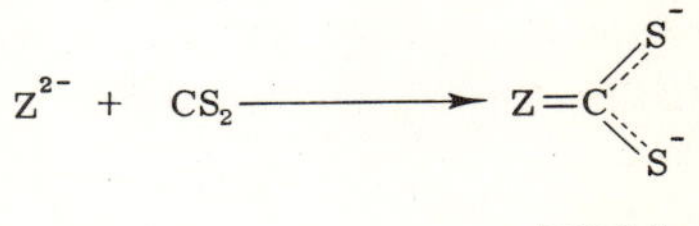

Metal ions react readily with (I) and (II) to yield complexes in which the two sulfur atoms are bound to the same metal, thus forming a four-

---

* Present address: BOC Limited, TechSep, London, England.
[1] Burns, R. P., and McAuliffe, C. A., *Adv. Inorg. Chem. Radiochem.* **22**, (1979).

membered, chelate ring. A wide variety of ligands is thus available by

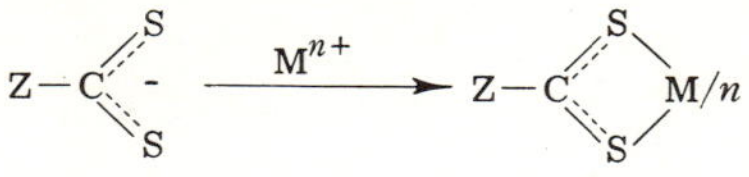

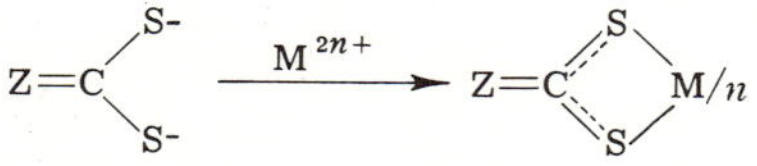

$$Z-C\underset{S}{\overset{S}{\diagup}} \xrightarrow{M^{n+}} Z-C\underset{S}{\overset{S}{\diagup}}M/n$$

$$Z=C\underset{S^-}{\overset{S^-}{\diagup}} \xrightarrow{M^{2n+}} Z=C\underset{S}{\overset{S}{\diagup}}M/n$$

merely varying Z as shown in Table I.

Metal complexes of 1,1-dithiolates have been reviewed by Coucouvanis (1); Eisenberg (2) presented a systematic, structural review of dithiolato chelates, and Stokolosa et al. (3) reviewed dithiophosphate complexes in detail. Earlier reviews (4–8) covered less recent work in greater detail. Following initial work by Delepine (9), 1,1-dithiolato complexes were more intensively studied between 1930 and 1941 (10–16). There is, however, continuous interest in the synthesis, characterization, electronic structures, and bonding of these complexes.

II. The Ligands

Table I outlines the major types of 1,1-dithiolato ligands. Table IA shows the related dithio acid complexes derived from dithiophosphinic, dithiocacodylic (dithioarsinic), dithioarsenate, and dithiophosphoric acids. The complexes most intensively studied to date are those of the dithiocarbamates and dithiophosphates.

(a) N,N'-Disubstituted Dithiocarbamates

When carbon disulfide reacts with either aliphatic or aromatic, primary or secondary amines, dithiocarbamate salts are formed (8, 9). The following is an example.

$$2 R_2NH + CS_2 \rightarrow [R_2NH_2]^+[R_2NCSS]^-$$

By using an alkali-metal hydroxide *in situ* as a proton acceptor, the alkali-metal dithiocarbamate salts, having various degrees of hydration, may be obtained (17).

$$R_2NH + CS_2 + MOH \xrightarrow{H_2O} R_2NCS_2^-M^+ + H_2O$$

Due to instability in air, very few free dithiocarbamic acids have been isolated (18). Dithiocarbamates derived from primary amines are

TABLE I
Major Types of 1,1-Dithiolato Ligands

Ligand	Name	Abbreviation
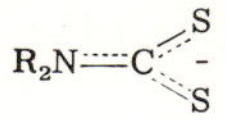 $R_2N-C(S)(S)^-$	Dithiocarbamate	R_2dtc
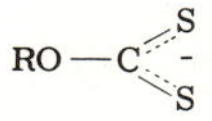 $RO-C(S)(S)^-$	Xanthates	RXant
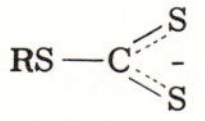 $RS-C(S)(S)^-$	Thioxanthate	RSxant
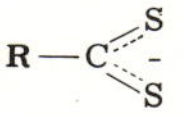 $R-C(S)(S)^-$	Dithiocarboxylate	
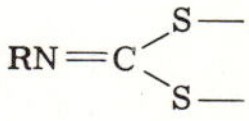 $RN=C(S^-)(S^-)$	Dithiocarbimate	
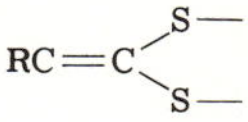 $RC=C(S^-)(S^-)$	1,1-Ethenedithiolate	
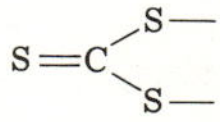 $S=C(S^-)(S^-)$	Trithiocarbonate	
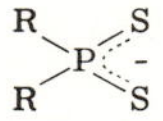 $R_2P(S)(S)^-$	Dithiophosphinate	dtp
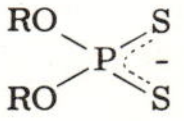 $(RO)_2P(S)(S)^-$	Dithiophosphate	
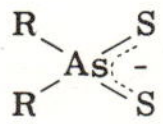 $R_2As(S)(S)^-$	Dithioarsinate	
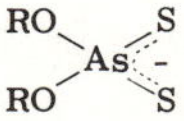 $(RO)_2As(S)(S)^-$	Dithioarsenate	

unstable, and, in the presence of base, are converted into the isothiocyanates (19). Disubstituted dithiocarbamates are more stable, but they decompose under acidic conditions (20). Thioformamides and quaternary ammonium dithiocarbamates have been obtained by heating a mixture of sulfur and formaldehyde in ethanol–water at 80° (21).

(b) O-Alkyldithiocarbonates (Xanthates)

The name xanthate, derived from the Greek "xanthos" (meaning blond), was coined by Zeiss in 1815, because the copper complexes that he isolated had a characteristic yellow color (22). Xanthates are formed by nucleophilic addition of an alkoxide ion to carbon disulfide.

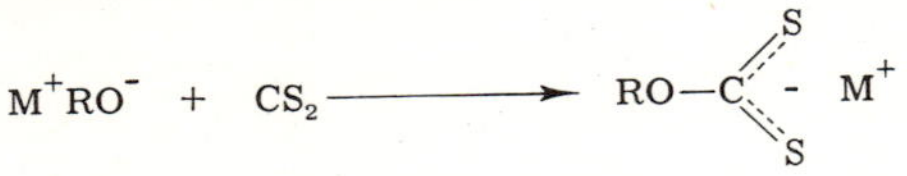

$$M^+RO^- + CS_2 \longrightarrow RO-C(\!\!\begin{array}{c}S\\S\end{array}\!\!)^- \; M^+$$

(M = an alkali metal)

Many alkali-metal xanthates are formed by direct xanthation of alcohols (23), but their chemistry is still vague. Acidification of the alkali-metal salts produces the unstable xanthic acids (24).

(c) Alkyltrithiocarbonates (Thioxanthates)

The alkali-metal salts of the thioxanthates are formed by a method analogous to that used for xanthates. Alkali-metal mercaptides react with carbon disulfide to form thioxanthate salts (25)

$$M^+RS^- + CS_2 \xrightarrow{THF} M^+RSCS_2^-$$

where THF = tetrahydrofuran. Air oxidizes these salts to disulfides (26). Unstable thioxanthic acids are formed upon acidification.

(d) Dithiocarboxylic Acids

These are strong, unstable acids. Oxidation to disulfides takes place readily. Several methods have been used to prepare the dithio acids (1), the most useful of which is the reaction of CS_2 with a Grignard reagent (27).

(e) 1,1-Ethenedithiolates

In the presence of base, bifunctional CH acids, CH_2XY, react with CS_2 to form a dithio acid or a 1,1-ethenedithiolate. The formation of the latter depends on the electron-withdrawing nature of X and Y, and the base used. The proposed mechanism is as follows.

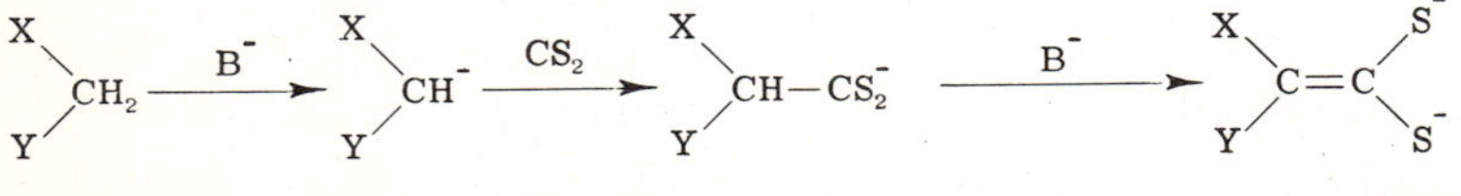

$$\underset{Y}{\overset{X}{\diagdown}}CH_2 \xrightarrow{B^-} \underset{Y}{\overset{X}{\diagdown}}CH^- \xrightarrow{CS_2} \underset{Y}{\overset{X}{\diagdown}}CH-CS_2^- \xrightarrow{B^-} \underset{Y}{\overset{X}{\diagdown}}C=C\underset{S^-}{\overset{S^-}{\diagup}}$$

Examples of X and Y include CO_2Me, CO_2Et, CN, H, NO_2, Ph, COMe, and ROC (1). A novel 1,1-dithiolate has been synthesized.

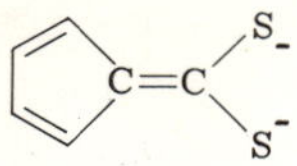

$$\text{(cyclopentadienyl)}C=C\begin{smallmatrix}S^-\\S^-\end{smallmatrix}$$

Such weakly basic nucleophiles as $(CN)_3C^-$ or $(NO_2)_3C^-$ will not react with CS_2 (28). Jensen and Hendriksen reported syntheses, reactions, and IR, NMR, and electronic spectra of 1,1-ethendithiols and their derivatives (29).

(f) Trithiocarbonates

Trithiocarbonates were reviewed by Reid (18) and Drager and Gattow (30). Reaction of S^{2-} and S_2^{2-} with CS_2 produces CS_3^{2-} and CS_4^{2-}, respectively. Air-sensitive Na_2CS_3 has been obtained by reaction of CS_2 with aqueous sodium hydroxide.

$$CS_2 + 2\ NaOH \rightarrow Na_2CS_2O + H_2O$$

$$3\ Na_2CS_2O \rightarrow 2\ Na_2CS_3 + Na_2CO_3$$

(g) Dithiocarbimates

Little is known about the chemistry of dithiocarbimates. The cyanodithiocarbimate salts $(NCNCS_2)^{2-}$ are the most extensively studied to date (31–34). They are obtained by the reaction of xanthine hydrase with a base (35).

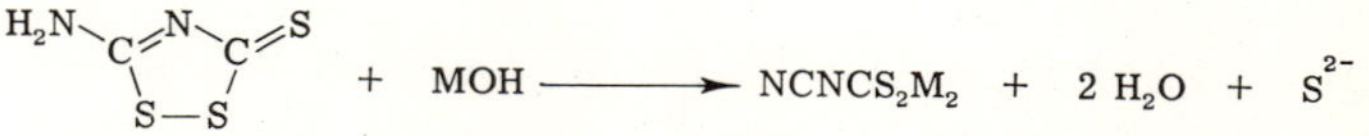

$$\begin{smallmatrix}H_2N\\ \\S\end{smallmatrix}\!\!C\!\!=\!\!N\!\!-\!\!C\!\!\begin{smallmatrix}\\=\!S\\ \\ \end{smallmatrix} + MOH \longrightarrow NCNCS_2M_2 + 2\ H_2O + S^{2-}$$

where M = a Group I metal. The calcium salt, $NCNCS_2Ca$, has been prepared by the action of calcium cyanamide on CS_2 in water (36).

III. Transition-Metal Complexes

The literature concerning the chemistry of transition-metal complexes containing 1,1-dithiolato ligands was extensively reviewed, up to 1968, by Coucouvanis (1). We attempt here to update that excellent account.

A. Titanium, Zirconium, and Hafnium

Relatively few complexes of the early transition metals with 1,1-dithiolato ligands have been prepared and characterized. This is consistent with their classification as "hard" or "class a" acceptors. Thus, there is little tendency to form complexes with "soft" sulfur-donor ligands. There has been a limited number of complexes synthesized either by modifying the nature of the metal ion or by using novel, preparative methods (37).

To date, the only bidentate 1,1-dithiolato ligands present in complexes of titanium, zirconium, and hafnium have been the N,N'-dialkyldithiocarbamates and the 2-alkylxanthates. As early as 1934, Dermer and Fernelius (38) reacted dibenzylamine and dibutylamine with carbon disulfide and $TiCl_3$ *in situ*, to produce yellow solids, $[Ti(R_2CNS_2)_4]$ (R = C_4H_9 and Bz). The formulas of the products suggested that the dithiocarbamate ion forms in solution. However, as successful reactions of dithiocarbamate salts with simple titanium salts do not appear to have been reported, an alternative explanation may be hypothesized, namely, the formation of $Ti(R_2N)_4$ *in situ*, followed by CS_2 insertion. This may be a more feasible explanation, as Bradley and co-workers (37, 39, 40) produced a series of tetrakis(dithiocarbamates) of Ti(IV), Zr(IV), and Hf(IV) by an insertion reaction of CS_2 with the $M(NR_2)_4$ complexes. Crystal-structure (41) and spectroscopic studies (42) indicated an 8-coordinate, dodecahedral structure for $[Ti(Et_2dtc)_4]$. Kirnickev (43) reported the formation of brown $[Ti(Me_2NCS_2)_2(Me_2N)]$ and $[Ti(Me_2NCS_2)(Me_3N)Cl_2]$ by a similar reaction. The carbon disulfide insertion-reaction has also been used to prepare a whole series of first-row transition-metal complexes of morpholine-4-carbodithioate (44).

Straightforward addition of anhydrous sodium dithiocarbamate to titanium(IV) chloride in refluxing dichloromethane produces a series of complexes $[Ti(S_2CNR_2)_nCl_{4-n}]$ (n = 2, 3, or 4; R = Me, i-Pr, i-Bu, or, when n = 3, Et) (45, 46). Molecular weight, conductance, and infrared-spectral data indicated that the complexes are monomeric nonelectrolytes containing bidentate dithiocarbamate ligands. The titanium atom has been assigned the coordination numbers of 6, 7, and 8 when n = 2, 3, and 4, respectively. Previously, the existence of 7-coordinate titanium(IV) had been postulated for only a few complexes (47). An X-ray diffraction study of $[Ti(S_2CNMe_2)_3Cl]$ revealed that the titanium has, indeed, a 7-coordinate, pentagonal, bipyramidal structure (48) (III). Dipole-moment measurements indicated that the 6-coordinate $[Ti(S_2CNR_2)_2Cl_2]$ complexes have cis configurations. The small "bites" and low charge of the dithiocarbamate ligands are especially suited for stabilization of higher coordination numbers in complexes with first-row transition-metals.

Modification of the acceptor properties of the metal atom may be achieved by using complexes containing π-cyclopentadienyl ligands.

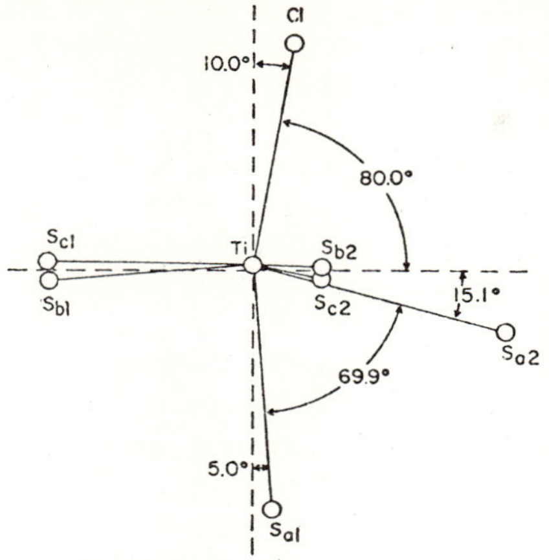

Thus, a series of bis-cyclopentadienetitanium(III) dithiocarbamate and xanthate complexes have been prepared by Coutts et al. (49–51) by reaction of the sodium salts of the ligands with [Cp$_2$Ti]Cl (Cp =π-cyclopentadiene) in air-free water under an inert atmosphere; the dithiocarbamate complexes are bright-green, and the xanthates are blue.

Both types of complex are extremely air-sensitive and are paramagnetic, with one unpaired electron per titanium atom. Their formulation as monomeric, symmetrical, bidentate, chelate complexes, (**IV**) and (**V**), has been established from spectral, magnetic, and molecular-weight data.

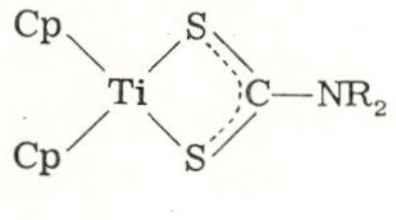

(IV)

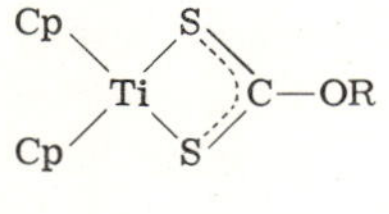

(V)

Coutts and Wailes (52, 53) synthesized a series of 5-coordinate, monocyclopentadiene dithiocarbamate complexes of titanium(III) and titanium(IV), [CpTi(S$_2$CNR$_2$)$_2$] (R =Me, Et, or Pr), prepared by ligand-exchange reactions between CpTiCl$_2$ and Na(S$_2$CNR$_2$). Infrared spectra and molecular-weight data suggested (52) structure (**VI**). CpTiX$_2$

(X =Cl or Br) are oxidized by thiuram disulfides, $R_2NC(S)S-SC(S)NR_2$

Cp
|
Ti
S / | \ S
 \\ S   S //
C   C
// \\   // \\
R—N   N—R
   \\   /
    R R
(VI)

(R =Me, Et, n-Bu), to monomeric, orange-red dithiocarbamates, $[CpTi(S_2CNR_2)X_2]$, which contain 5-coordinate titanium. It should be noted that several isomeric structures are possible for these compounds.

B. VANADIUM, NIOBIUM, AND TANTALUM

Until recently, there was little published material dealing with the chemistry of the Group V triad. This agrees with their classification as "class a" metal ions. Using new synthetic routes similar to those for the preparation of the titanium complexes, however, a number of V(III), V(IV), V(VI), Nb(IV), Nb(V), Ta(IV), and Ta(V) complexes have been prepared (54).

The first report of a group V metal complex containing 1,1-dithiolato ligands was (55) of the unstable $[V(Et_2dtc)_2(NO)]$, and following this, the majority of work was centered on the $VO_2^{2+}-N,N'$-dialkyldithiocarbamate systems. The monomeric $[VO(R^1R^2dtc)_2]$ ($R^1 = R^2$ =Me, Et, Pr^i, or pyrrolidine) complexes were prepared by McCormick (56, 57). Similar complexes (R =piperidine, R^1 =Me, R^2 =cyclohexyl, $R^1 = R^2$ =cyclohexyl) were prepared by Vigee and Selbin (58). Following an epr study on the oxovanadium(IV) dithiocarbamate complexes, however, McCormick et al. (59), after rigorously excluding oxygen during spectral measurement, found poor agreement with the previous spectral results, but reasonable agreement with those of Garif'yanov and Kozyrev (60), who reported epr spectra of the diethyl complex in solution without isolation of the solid. Thus, the results of Vigee and Selbin (58) probably reflect air oxidation of the VO^{2+} species.

Of the 24 lines expected for coupling between ^{51}V (I =7/2) and two equivalent ^{13}C (I = ½) nuclei, 18 were resolved in the epr spectrum of $VO(S_2^{13}CNEt_2)_2$, showing that the C(2s) orbital can also participate in transannular interactions (61). $[Cp_2V(S_2CNEt_2)BF_4]$ and dithiophosphate have been used in an extension of studies on the C_{4v} oxovanadium(IV) chelates to yield C_{2v} bis(cyclopentadienyl) complexes.

Transannular interaction via the electron-delocalization mechanism was found, but lessened by ~10–15% for the ligand superhyperfine splitting and 30–35% for the ^{51}V hyperfine splitting (62) in the epr spectrum. The crystal structure of [VOS$_2$CNEt$_2$)$_2$] shows that the molecular core has the expected C_{2v} symmetry [V–O = 159.1(4), V–S = 138.7(2)–241.0(2) pm] (63). Magnetic and spectral data provided evidence for a tetragonal, pyramidal structure (**VII**) for these complexes. Like many other coordinatively unsaturated, metal

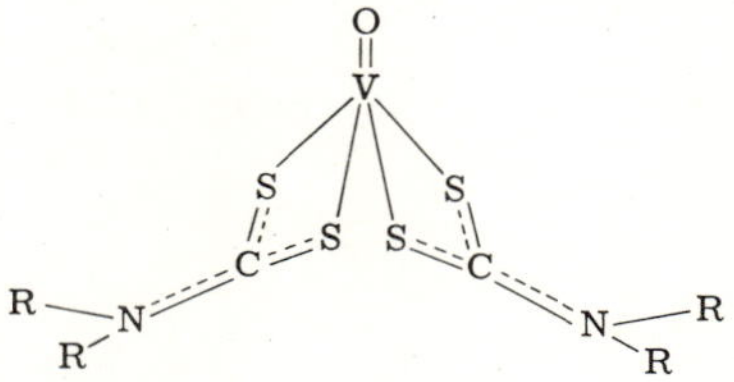

(VII)

dithio complexes, [VO(R$_2$dtc)$_2$] can undergo reaction with a Lewis base, and thus expand its coordination number. There is usually a correlation between the electronic properties of the ligands, and the ability to undergo Lewis base addition. Pyridine and 4-methylpyridine adducts of [VO(R$_2$dtc)$_2$] have been isolated and characterized as solids (64). Attempts to isolate dimethylsulfoxide adducts failed, although spectral evidence suggests adduct formation in solution. Hoyer et al. (65) have synthesized (Bu$_4$N)[VO{S$_2$C=C(CN)$_2$}$_2$] by reaction of VOSO·4H$_2$O and the ligand in methanol/water. This was the first report of a 1,1-ethendithiolato complex of vanadium. Vanadium oxocomplexes containing vanadium(III), (IV), (V) have been reported (42, 66, 67); [V(III)OL·H$_2$O], [V(IV)OL$_2$], and [V(V)O$_2$L] (HL = 3- and 5-phenyl and 3,5-diphenyl-pyrazolinedithiocarbamic acid were separated chromatographically (66), and [VC(R$_2$dtc)$_3$] and [NbO(R$_2$dtc)$_3$] have been prepared by reaction of sodium N,N'-dialkyldithiocarbamates with a metal oxo-salt (57, 67). X-Ray diffraction studies on the last two compounds showed (68) the environments of the metals to be 7-coordinate, pentagonal bipyramids (**VIII**). The

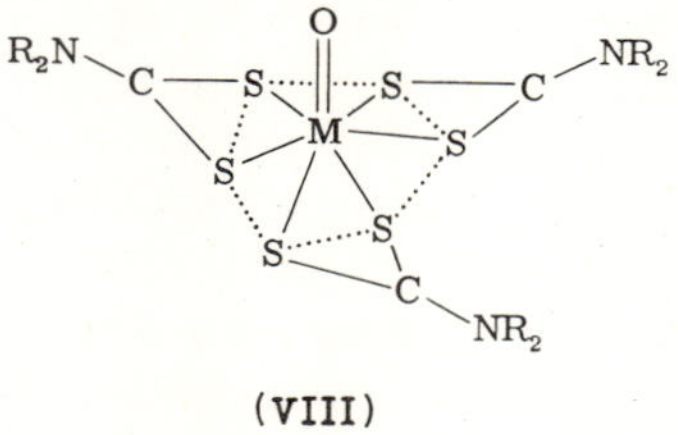

(VIII)

ability of VO^{2+} to form complexes with the sulfur ligands suggests that the double-bonded oxygen atom lessens the effective "hardness" of the V(IV) species. This modification has also been achieved by using π-bonded cyclopentadiene ligands in a manner analogous to that used for titanium. Thus, a series of green N,N'-dialkyldithiocarbamate complexes were prepared (69) by using the following routes.

(1) $Na(R_2dtc) + CpVCl_2 \longrightarrow [Cp_2V(R_2dtc)]Cl$ (soluble)
$\downarrow X^-$
$[Cp_2V(R_2dtc)]X$
$X = Ph_4B, BF_4, Pf_6, ClO_4$

(2) $Cp_2VCl_2 + 2 AgClO_4 \longrightarrow AgCl + [Cp_2V(ClO_4)_2]$
$\downarrow Na(R_2dtc)$
$[Cp_2V(R_2dtc)](ClO_4)$

The first xanthates of V(IV), $[Cp_2V(Rxanth)]$ (R = Me, Et, Pr^i, or Bu), isolated as purple solids were prepared by the same method (71). Kwoka et al. (70) were able to obtain an alkylenebis(dithiocarbamate) vanadyl complex, $VO[S_2CNH(CH_2)_nNHCS_2]$ for $n = 2$, but not for $n = 6$.

In the Me_2dtc complex, a unique, 15-line, epr spectrum was reported (69) that was peculiar only to the tetraphenylborate salt. This suggests a V–V interaction in the lattice. Electrochemical studies on these $[Cp_2VL]^+$ complexes (L = dithiocarboxylato ligand) shows two well defined, polarographic, reduction waves, and, for the process at most positive potential, the reversible formation of a V(III) species was postulated (72–74).

The pure vanadium(IV) dialkyldithiocarbamates, $[V(R_2dtc)_4]$, have been prepared (75) by using a method analogous to that used for the titanium analogs, i.e., CS_2 insertion into the V–N bonds in $[V(NR_2)_4]$. Physical studies indicated a dodecahedral, 8-coordinate, vanadium system (76–78). Infrared spectra of freshly prepared, and aged, compounds suggested a tetrachelated species with bidentate dithiocarbamates, and a species having unidentate ligands, respectively (79). Isomerism of this type had not previously been reported.

Recently, several papers have appeared that report the isolation of stable complexes of vanadium(IV) with dithiocarboxylato ligands (77, 80). Dithiocarboxylato salts ($PhCS_2^-$, p-$CH_3C_6H_4CS_2^-$, $CH_3CS_2^-$, $PhCH_2CS_2^-$) react with both VO^{2+} and V(III) species, to give stable VL_4 species. Magnetic, molecular-weight, and electronic, infrared- and epr-spectral data are consistent with 8-coordinate vanadium(IV) and a VS_8 chromophore. X-Ray crystallographic studies (78,

81, 82) confirmed this, and the structures approximate to dodecahedra with D_{2d} geometry. In the case of $V(MeCS_2)_4$, however, two different stereoisomers belonging to the dodecahedral subclasses 1d (symmetry D_{2d} and Vd (symmetry C_2) have been recognized (*82*).

Few complexes of niobium and tantalum with bidentate sulfur ligands have been reported, even though dithiocarbamates have been used for the determination of these metals (*83, 84*), [Nb(pyrroldtc)$_3$] was reported by Malissa and Kolbe-Rhode (*85*), and CS_2-insertion reactions with [Nb(NR$_2$)$_5$] and [Ta(NR$_2$)$_5$] result (*86*) in the reduction of Nb(V) to [Nb(R$_2$dtc)$_4$], and formation of [Ta(R$_2$dtc)$_5$]. Reaction of niobium tetrahalides with dithiocarbamato salts in a suitable solvent, such as acetonitrile, usually result in the formation of the 8-coordinate [Nb(R$_2$dtc)$_4$] complexes (*54, 87–89*). Using $TaCl_5$, the same reaction produced [Ta(R$_2$dtc)$_5$]; infrared spectra indicated the presence of one or more unidentate dithiocarbamate ligands in the complex (*54, 89*). From the reaction of niobium(V) and tantalum(V) halides and thiocyanates with sodium N,N'-dialkyldithiocarbamates in methanol, air-stable yellow, or white, crystals of [MX(OMe)$_2$(R$_2$dtc)$_2$] (X = Cl, Br, or NCS; R = Me, Et, or Bz) were isolated (*90*). X-Ray crystallographic studies on the niobium compounds showed that they are pentagonal bipyramids (*91*). Uvatova *et al.* (*92*) isolated two cationic complexes of niobium, [Nb(CH$_2$)$_4$NCS$_2$]$_4$ClO$_4$, and the corresponding pyrrolidine dithiocarbamate derivative, by reaction in the presence of ClO_4^- in acidic media. Holah *et al.* (*93, 94*) found that reaction of the metal pentahalides with Na(R$_2$dtc) in benzene and dichloromethane produced a new series of air-sensitive, substituted complexes of the type M(R$_2$dtc)$_3$X$_2$, M(R$_2$dtc)$_3$S, M(R$_2$dtc)$_2$X$_3$, and M(R$_2$dtc)$_2$X$_3 \cdot nC_6H_6$ (n = 0.7–1.0). The reaction scheme is depicted in Scheme 1. It was proposed that the formation of the sulfur-rich species M(R$_2$dtc)$_3$S takes place via the activated complex

$$[X(R_2dtc)_3M-S-\overset{\overset{\displaystyle S}{\|}}{C}-NR_2]^*$$

and it was suggested that the structure is [(R$_2$dtc)$_3$M=S]. It may, however, be possible that the product is similar to the sulfur-rich, perthio complexes previously isolated in complexes of other metals (*356–362, 439*).

C. Chromium, Molybdenum, and Tungsten

Many complexes containing the octahedral CrS_6 chromophore are known (*95*). The early preparative work was performed by Delepine (*96*) and Malatesta (*97*), who synthesized various chromium(III) dial-

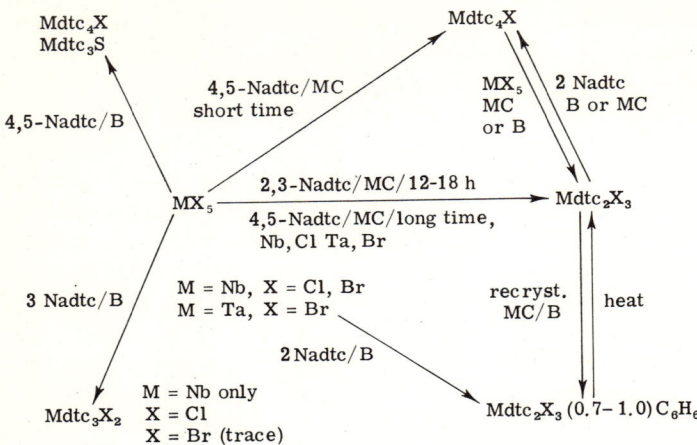

SCHEME 1. Summary of reactions between $NbCl_5$, $NbBr_5$, and $TaBr_5$ and Nadtc, B = benzene, MC = dichloromethane; all NbX_5/Nadtc B reactions also produce $Nbdtc_4$ (not shown), all MX_5/Nadtc/MC reactions also produce dtc_2CH_2 (not shown).

kyldithiocarbamates by using the straightforward reaction of an anhydrous chromium(III) salt with the ligand in a dry, organic solvent under acidic conditions (optimum pH, 5–6). Only [$Cr(Et_2dtc)_3$] (1, 85, 98, 99) and [$Cr(piperydyldtc)_3$] (100) have been systematically characterized. The [$Cr(R_2dtc)_3$] complexes are violet in color, and contain three unpaired electrons; the general properties are consistent with an octahedral CrS_6 chromphore (100). The infrared characteristics of uni- and bidentate dtc have been discussed (101) on the basis of a normal coordinate analysis of [$Cr(Et_2dtc)_3$].

From the reaction of the tetraalkylthiuram disulfides with $CrCl_3$ in absolute ethanol [$Cr(R_4tds)_2Cl_2$]Cl were (102, 103) isolated (R = Me, or Et; tds = tetraalkylthiuram disulfide). Bradley et al. (104) showed that CS_2 insertion into metal dialkyl amides produces [$Cr(Et_2dtc)_3$] in solution, but the complex was not isolated as a solid. The stereochemical rigidity of N,N-disubstituted dithiocarbamate complexes [$Cr(R^1R^2dtc)_3$] has been established by ^{1}H-NMR in $CHCl_3$ solution at temperatures up to 357 K. The relative rates of rearrangement by a trigonal, twist mechanism (resulting in optical inversion) have been established for a series of tervalent metal ions, and the order is (105) Ga, In, V > Mn > Cr.

Blue [$Cr(Rxant)_3$] (R = Me, Et, or L-menthyl) complexes have been prepared and characterized (103, 106, 107), and these complexes, together with some of the dialkyldithiocarbamate complexes, show spin-forbidden transitions, $^2E \leftarrow {^4A_2}$, $^2T_1 \leftarrow {^4A_2}$, and $^4T_2 \leftarrow {^4A_2}$, the last two respectively giving rise to phosphorescence and fluorescence (*108*). The [Cr(RSxanth)$_3$] (R = Et, Pr, or Bu) thioxanthate complexes, Cr(III) complexes of dithiocarboxylic acids [Cr(PhCS$_2$)$_3$], and [Cr(PhCH$_2$CS$_2$)$_3$], have been synthesized and characterized (*109–111*).

Only a few 1,1-dithiolate complexes of chromium(III) are known: [Pr$_4$N]$_3$[Cr(S$_2$C=C(CN)$_2$)$_3$] is formed as a green-brown solid by reaction of the ligand with a Cr(III) salt (*112*). The [Bu$_4$N]$_3$[Cr(S$_2$C=CHNO$_2$)$_3$] complex is similarly prepared, and both complexes have been shown to be effective insecticides (*113*). Recently, in a general study of trithiocarbonato complexes, [Cr(CS$_3$)$_3$]$^{3-}$ was isolated as its Ph$_4$P$^+$ and Ph$_4$As$^+$ salts (*114*).

Reaction of CrCl$_2$·2H$_2$O and anhydrous sodium diethyldithiocarbamate in water produces a bright-yellow-green solid, [Cr(Et$_2$dtc)$_2$], that is pyrophoric in air (*115*). X-Ray data show the complex to be isomorphous with Mn(II), Fe(II), and Zn(II) diethyldithiocarbamates. Solutions of [Cr(Et$_2$dtc)$_2$] can be oxidized to Cr(III) in air. These results, together with magnetic (the complex shows antiferromagnetic behavior) (*116*) and reflectance spectral data, indicated a CrS$_5$ donor grouping, i.e., square-pyramidal coordination of 5 sulfur atoms about each chromium atom in a dimeric unit (**IX**).

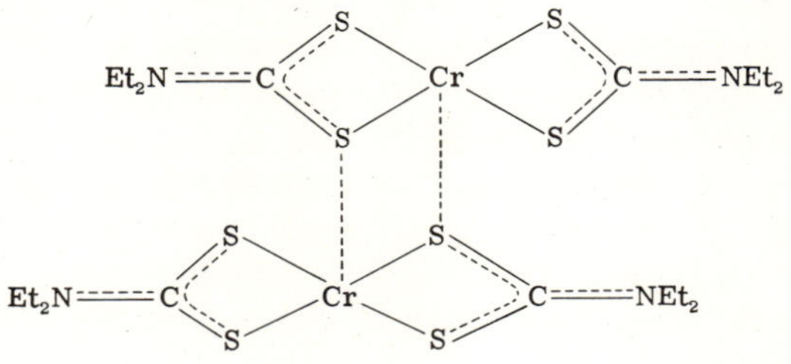

(IX)

A number of cationic Cr(I) species of the type depicted have been

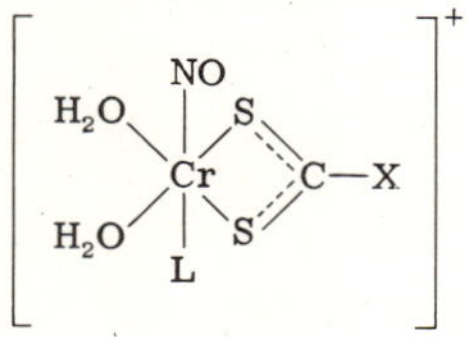

[X = OEt, or NEt$_2$;
L = AsR$_2$Ph, or PR$_3$ (R = Me, Et, Pr, or Bu), (EtO)$_2$PPh, EtOPPh$_2$, Et$_2$POEt, EtP(SEt)$_2$, P(SEt)$_3$, Et$_2$P(SPh), EtP(SPh)Ph, py, NH$_3$, hydroxylamine or ethylenediamine]

formed in solution by reaction of [Cr(NO)(H$_2$O)$_3$(S$_2$CX)] with the particular Lewis base, and studied by epr spectroscopy (117–121).

The dinitrosyl [Cr(NO)$_2$(Et$_2$dtc)$_2$] was isolated as an air-stable solid by Malatesta (15, 16), and by Connelly and Dahl (122), from the following reaction.

$$[Cr(NO)_2(MeCN)_4](PF_6)_2 \xrightarrow{As(Et_2dtc)_3} cis\text{-}[Cr(NO)_2(Et_2dtc)_2]$$

IR-spectral data indicated that the dinitrosyl has a cis conformation (123).

The complexes [Cr$_2$(S$_2$CNH(C$_2$H$_4$)CNHCS$_2$)] and [Cr$_2$(S$_2$CNH(C$_6$H$_{12}$)CNHCS$_2$)] have been synthesized by the action of solutions of the appropriate ligand in aqueous ethanol on aqueous ethanol solutions of Cr(III). The magnetic susceptibilities of these purple, amorphous compounds fit the Curie–Weiss law with very small, negative values of θ, implying the presence of weak, antiferromagnetic interactions between the metal centers (70). Chromium(III) complexes [Cr(R$_2$dtc)$_3$] have been isolated that contain the appropriate ligand and chromium acetate, and their temperature-dependent, ^{1}H-NMR spectra have been examined (124).

Complexes of molybdenum and tungsten with bidentate sulfur ligands have been investigated extensively. In recent years, the work in this field has been escalated by the impetus of designing models of such bioinorganic enzymes as nitrogenase and xanthine oxidase (125). The early work reviewed by Coucouvanis (1) dealt exclusively with the isolation of oxomolybdenum(V) and -(VI) species.

The dithiocarbamates form a series of oxomolybdenum complexes [MoO$_2$(R$_2$dtc)$_2$] and [Mo$_2$O$_3$(R$_2$dtc)$_4$]. The xanthates, however, only form [Mo$_2$O$_3$(Rxanth)$_4$]. The structure of the latter complexes (126) is shown in (X).

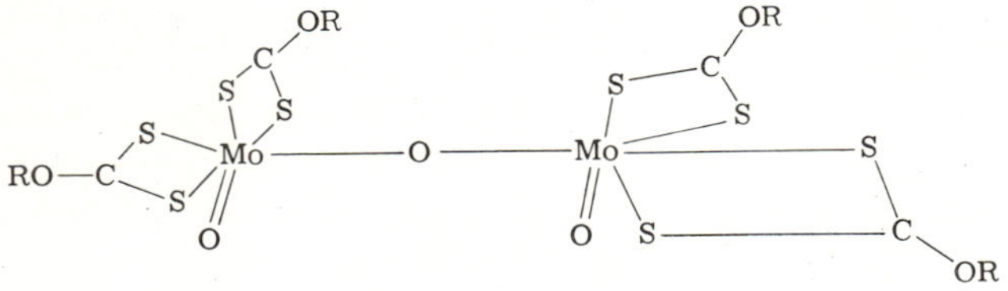

(X)

The two Mo=O bonds are disposed in a mutually cis orientation, and it is now considered that this is the only stable orientation for this type of species.

Synthesis of [Mo$_2$O$_3$(Rxanth)$_4$] and [Mo$_2$O$_3$(R$_2$dtc)$_4$] was achieved by reduction of MoO$_4^{2-}$ with SO$_2$ or sodium dithionite in the presence of

NaR$_2$dtc (97), and the synthesis, and electronic, IR-, and proton-NMR spectra of the compounds [MoO(RXant)$_2$](R=Pr^i,Bu^t) have been reported. The crystal structures of [MoO(Pr^{i}Xant)$_2$] shows the molecule to be a pseudo-square-pyramidal monomer, with one thioxanthate group exhibiting the usual bidentate geometry, whereas the second thioxanthate group displays unusual, nonclassic coordination, with the Mo significantly displaced from the S$_2$CS plane (127). An interesting reaction takes place between Mo$_2$O$_3$(Rxant)$_4$ and H$_2$S, or thiols, in benzene. The products are [MoS$_2$(Rxant)] and [MoS(R'S)(Rxant)], respectively, and it was suggested that they consist of molecular associations; [MoOS(Bu$_2$dtc)]$_2$ was prepared by treating [MoO$_2$(Bu$_2^n$dtc)$_2$] in the same way (128). Compounds of this type, and other oxomolybdenum dithiocarbamates, have been used as effective, extreme-pressure agents, antioxidants, and wear-inhibitors for lubricants (129, 130).

None of the earlier methods for the syntheses of [Mo$_2$O$_3$(R$_2$dtc)$_4$] were wholly satisfactory, as they tended to produce mixtures. An epr study of [Mo$_2$O$_3$(R$_2$dtc)$_4$] indicated the existence of two (presumably, sterically different) complexes in CHCl$_3$ solution, and this has been interpreted in terms of complicated redox mechanisms in solution (131). A recent innovation (132) produces pure [Mo$_2$O$_3$(R$_2$dtc)$_4$] by reaction of an aqueous solution of molybdenum(V) with NaR$_2$dtc (greater than 4 molar excess) at 0°. If the reaction is conducted under reflux, however, a new complex, [Mo$_2$O$_4$(R$_2$dtc)$_2$], can be isolated; it may be converted into the former complex by further reaction with Na(R$_2$dtc). The interrelationship and chemistry of these two complexes, and of the reactions of [Mo$_2$O$_3$(R$_2$dtc)$_4$] with carboxylic acids and pyridine (1, 132), suggest a possible model for the interaction of two molybdenum atoms at an active enzymic site. A similar, revised route was to produce pure μ-oxo-bis[bis(alkylxanthate)oxomolybdenum(V)], [Mo$_2$O$_3$(ROCS$_2$)$_4$], complexes (133) that react with alcohols, or H$_2$S, with loss of a xanthate ligand to produce a new series of di-μ-sulfido-bis(alkylxanthate)oxomolybdenum(V), [Mo$_2$O$_2$S$_2$(ROCS$_2$)$_2$]. Casey et al. (67) reported the formation of [Mo$_2$O$_3$(R$_2$dtc)$_4$] and [MoO$_2$(R$_2$dtc)$_2$] from NaR$_2$dtc and Mo(V) salts. Both hydrogen chloride and hydrogen bromide react (134) with [MoO(R$_2$dtc)$_2$] to produce 6-coordinate Mo^{IV} species [MoX$_2$(R$_2$dtc)$_2$]. Reactions of [MoO(Et$_2$dtc)$_2$], including abstraction of an oxygen atom from various substrates to form [MoO$_2$(Et$_2$dtc)$_2$] and [Mo$_2$O$_3$(Et$_2$dtc)$_4$], have been studied in detail by Mitchell and Searle (135), and some of these compounds, [MoO$_2$(R$_2^1$dtc)$_2$] (R^1 = Me, Et, or Ph), react (136) with phenyl-, methyl-, and benzoylhydrazine, to form [R^{2}N=N—Mo(R$_2^1$dtc)$_3$]. However, the analogous reactions with Me$_2$NNH$_2$ do not result in reduction giving the Mo^{VI} compounds, [MoO(N=NMe$_2$)-(R$_2$dtc)$_2$]. The ubiquitous [MoO$_2$(R$_3$dtc)$_2$] reacts with concentrated hydrohalogenic acids, to yield novel compounds [MoOX$_2$(R$_2$dtc)] (X = F, Cl, Br; R = Me, Et, Pr^n), and the crystal structure of [MoOCl$_2$(Et$_2$dtc)$_2$] confirmed 7-coordination (137).

A series of oxomolybdenumdipropyldithiocarbamates has been structurally characterized by X-ray studies (138); [MoO$_2$(Pr$_2$dtc)$_2$] has a distorted, octahedral structure with cis-oxo groups, whereas [MoO(Pr$_2$dtc)$_2$]O has a similar structure, with a single, bridging oxygen atom linking the two units; and [MoO(Pr$_2$dtc)$_2$] is square-pyramidal, with oxygen at the apex. The crystal structure of [MoO(Et$_2$dtc)$_3$]Mo$_2$O$_4$F$_6$ shows that the [MoO(Et$_2$dtc)$_3$]$^+$ cation has a distorted, pentagonal, bipyramidal geometry about the molybdenum (436).

The similarly structured 5-phenyl-1-pyrazoline dithiocarbamate complexes were prepared, separated chromatographically, and characterized by epr and IR spectroscopy (139). Workers in the USSR found that reaction of MoO^{3+} with NaR$_2$dtc (NaL) in concentrated hydrohalic acids results in the formation of the halo-substituted complexes cis-MoOXL$_2$, MoOL$_3$, MoOBr$_2$L(LNa), and MoOCl$_2$L, the structures of which were established from epr data (140). The oxomolybdenum(IV) complexes [MoO(Et$_2$dtc)$_2$] have been prepared by reduction of [Mo$_2$O$_3$(Et$_2$dtc)$_4$] with benzenethiol, sodium dithionite, or Zn dust (131). Physical measurements indicated that the complex has square-pyramidal geometry, similar to that of the analogous, vanadyl complex (**VII**). It appears that oxomolybdenum(IV) is preferentially stabilized by sulfur donors.

The crystal structure of di-μ-oxodioxobis(diethyldithiocarbamate)dimolybdenum(V) has been determined, and the Mo–Mo distance is (141) 258.0(1) pm. The related sulfur compound, di-μ-sulfidodithiobis(dibutyldithiocarbamato)dimolybdenum(V), [Mo$_2$S$_4$(Bu$_2^n$dtc)], has a similar structure (142) with the Mo atoms separated by 280.1(2) pm. This is one of the few examples where Mo is bonded to terminal sulfur. The compounds tetrabutylammonium di-μ-sulfidobis(oxo-1,1-dicyanoethylene-2,2-dithiolato)molybdenum(V) (143) (**Xa**) and μ-oxo-μ-sulfidobis[oxodipropylthiocarbamatomolybdenum(V)] (144) represent examples of the ability of sulfur to occupy bridging positions having the "cis-bent" geometry, the dihedral angle between the two basal planes of the pseudo-square-pyramidal molybdenum atoms being ~150°. Treatment of [MoO$_2$(μ-O$_2$(Et$_2$dtc)$_2$] with benzenethiol leads to the formation of the salt [Mo$_2$O$_2$(μ-SPh)$_2$—(μ-Cl)—(Et$_2$dtc)$_2$] [MoOCl$_4$-(H$_2$O)], which has been shown by X-ray diffraction to have a triply bridged, dinuclear cation (**Xb**) and a mononuclear anion (145).

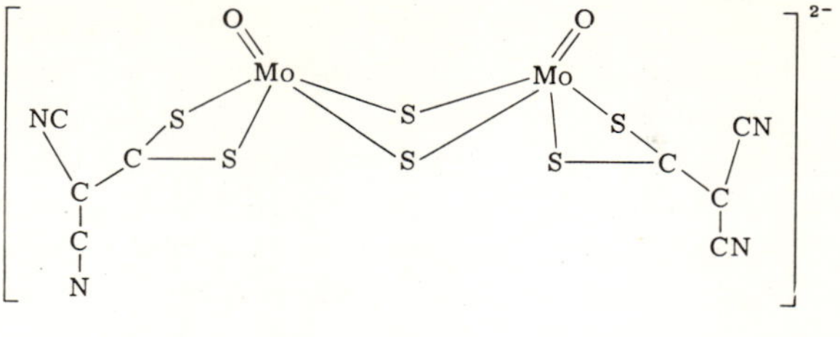

1,1-DITHIOLATO COMPLEXES OF THE TRANSITION ELEMENTS

(Xa)

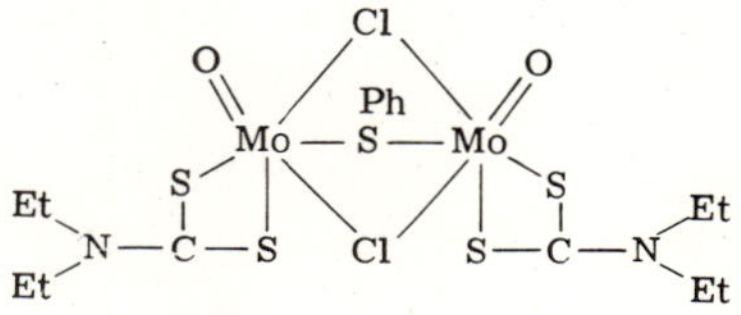

(Xb)

Sulfur ligands appear to be particularly adept at replacing oxygen ligands. Treatment of $[MoO_2(Pr^n_2dtc)_2]$ with H_2S produces the disulfur complex $[MoO(S_2)(Pr^n_2dtc)_2]$ in low yields, with other products of the reactions being compounds of the type shown in **Xc**, where X = S (*146*). In seeking an alternative route to

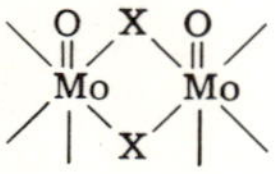

(Xc)

this disulfur complex, $[(Mo(CO)_2(Et_2dtc)_2]$ was treated with S_8, followed by oxidation with air, which produced small yields of the desired product, the main product being $[Mo(S_2O)(Et_2dtc)_2]_2$, in which S_2O ligands bridge the two metals (*146*). Treatment of $[Mo_2O_2(\mu\text{-O})_2(Pr^n_2dtc)_2]$ with H_2S in chloroform also gives $[Mo_2O_2(\mu\text{-O})(\mu\text{-S})(Pr^n_2dtc)_2]$, whereas in 1,2-dichloroethane at 80°C or by using P_4S_{10} in boiling xylene, all the oxide groups of the starting complex are replaced by sulfur, yielding $[Mo_2S_2(\mu\text{-S})_2(Pr^n_2dtc)_2]$. Three-atom-bridged complexes similar to **Xb** result if 2-mercaptoethanol, *o*-mercaptobenzoic acid, *o*-mercaptophenol (1 mol), or *N*-methylaminobenzenethiol are used as reactants, and when *o*-mercaptophenol (2 mol) is used, complete cleavage of the di-μ-oxo-bridge occurs, giving $[MoO(o\text{-}C_6H_4OS)(Pr^n_2dtc)_2]$. An excess of *o*-mercaptobenzene removes the oxide ligand completely to give $[Mo(o\text{-}C_6H_4XS)_2(Pr^n_2dtc)_2]$ (X = S, NH, or NMe) (*147*). Tertiary phosphines react with *cis*-$[MoO_2L_2]$ (L = R_2dtc or R_2dtp) to give the complexes $[MoOL_2]$ (*148*). Thiophenol in the presence of [MoO-

(R_2dtc)$_2$] reacts with diethyl azodicarboxylate, azobenzene, or dimethyl acetylenedicarboxylate to give the appropriate hydrazine and disulfide (*149*).

An exciting discovery of recent years has been the liquid-phase activation of molecular oxygen by MoOCl$_4$ and [MoO$_2$(R_2dtc)$_2$] (R = Et, Pr^n, Bu^i) (*150–152*). These complexes catalyze selective oxidation of PBu$_3$ to P(OBu)$_3$. McDonald and Shulman (*153*) reported that the reduction of a large excess of [MoO$_2$(Et$_2$dtc)$_2$] by PPh$_3$ in benzene solution provided a sensitive and accurate method for the determination of small quantities of PPh$_3$ by spectrophotometric assessment of [Mo$_2$O$_3$-(Et$_2$dtc)$_4$] produced in the reaction sequence as follows.

$$[\text{MoO}_2(R_2\text{dtc}) + \text{PR}_3^1 \rightarrow [\text{MoO}(R_2\text{dtc})_2] + \text{PR}_3^1\text{O} \quad (1)$$

$$[\text{MoO}(R_2\text{dtc})_2] + [\text{MoO}_2(R_2\text{dtc})] \rightarrow [\text{Mo}_2\text{O}_3(R_2\text{dtc})_4] \quad (2)$$

The kinetics of the oxygen-transfer reaction from [MoO$_2$(Et$_2$dtc)$_2$] to PPh$_3$ in MeCN solution have been monitored (*154*) by using stop-flow techniques at temperatures lying between 15 and 45°. The reactivity of coordinatively unsaturated oxo-Mo(IV) species towards a variety of multiple bonds has recently been recognized (*155*), and 1:1 adducts with several organic molecules and [MoO(R_2dtc)$_2$] have been isolated. Thus, tetracyanoethylene (TCNE) oxidatively adds to [MoO(Pr$_2^n$dtc)$_2$] to form a deformed, pentagonal, bipyramidal complex (*156*), [MoO(Pr$_2^n$dtc)$_2$(TCNE)], in which TCNE is considered to act as a bidentate ligand with considerable π-back donation from the metal, as reflected by the C=C and Mo—C distance of 228.5 and 147.3 pm, respectively (*156*).

The unsubstituted complexes [Mo(R_2dtc)$_4$] and [W(R_2dtc)$_4$] have been prepared by methods described earlier for synthesis of the vanadium-group complexes. They are 8-coordinate, with four bidentate dithiocarbamate ligands (*157*).

The 8-coordinate species [Mo(Et$_2$dtc)$_4$] can be obtained by reaction of Mo(CO)$_6$ with tetraethylthiuram disulfide (1:2) in acetone under N$_2$. The X-ray structure revealed square-antiprismatic coordination, with a crystallographic, twofold axis coinciding with the molecular pseudo $\bar{4}$ axis (*158*). The magnetism and spectra of [M(dtc)$_4$]$^{n+}$ (M = Mo or W; n = 0 or 1) have been interpreted in terms of dodecahedral symmetry (*159*).

The complex [Mo(Et$_2$dtc)$_3$] has been prepared (*160*) by reacting Na(Et$_2$dtc) with [Mo$_2$O$_2$Cl$_2$(H$_2$O)$_6$], having been previously prepared by reacting Mo(CO)$_4$ with tetraethylthiuram disulfide (*161*). The ionic halide complexes [Mo(Me$_2$dtc)$_4$]X (X = Br, Br$_3$, I, I$_3$, or I$_7$) are produced by oxidation of [Mo(R_2dtc)$_4$] with the appropriate halogen (*162*). The analogous series of Et$_2$dtc complexes, [Mo(Et$_2$dtc)$_4$]X (X = Cl, Br, I, or I$_3$) and [Mo(Et$_2$dtc)$_4$]XCHCl$_3$ (X = B, or I) have also been prepared (*163*). One of the syntheses provides the 7-coordinate complex [MoO(Et$_2$dtc)$_2$Br$_2$][Et$_2$NH$_2$]. [Et$_2$dtc] in MeCN cleaves the Mo=O bond in MoOCl$_3$, to form the 8-coordinate complex [Mo(Et$_2$dtc)$_4$]Cl, which was converted into the BPh$_4^-$ salt by treatment with NaBPh$_4$ (*164*). The electrochemistry of several of these Mo^V, as well as Mo^{IV} and Mo^{VI}, complexes of Et$_2$dtc has been studied in DMSO, DMF, and MeCN (*165*). A tungsten(V) analog [W(Et$_2$dtc)$_4$]Br has been shown by X-ray crystallography to involve an 8-coordinate, trigonal, dodecahedral WS$_8$ moiety (*166*). Very recently, several 8-coordinate, dodecahedrally structured complexes of molybdenum(IV) were synthesized (*167*); this was the first report of dithioaromatic or dithioaliphatic complexes of molybdenum. Electronic spectra revealed a higher degree of covalency in the metal-ligand bond than is found in any other 8-coordinate molybdenum(IV) species. The 6-coordinate [Et$_4$N]$_2$[Mo(CPDT)$_2$] (CPDT = cyclopentadienedithiocarboxylate) was obtained by reaction of the sodium dithiocarboxylate with MoCl$_4$ and Et$_4$NBr in MeCN (*168*).

Complexes of molybdenum in the lower valence-states of +2 and +3 have been produced only in the past two years. For the Mo(II) species, the usual starting-material is Mo$_2$(acetate)$_4$. Reaction of this with KS$_2$COEt in THF gives two products, a green complex tentatively assigned as [Mo$_2$(Etxant)$_4$], which solvates to form the red complex [Mo$_2$(Etxant)$_4$(THF)$_2$]. The structure of the latter complex was elucidated by X-ray analysis (*169*). Steele and Stephenson (*170*) were also able to synthesize a red, crystalline solid (methanol solution), which they formulated as [Mo(Etxant)$_2$]$_2$ (**XI**), and reacted this with Lewis bases, e.g., pyridine, to form [Mo(Etxant)$_2$L]$_2$. Thus, there appears to be a difference between the two compounds formulated as [Mo$_2$(Etxant)$_2$]$_2$ that

(**XI**)

may be due to two structurally different species. Molybdenum(II) dithiocarbamates have also been isolated (171, 176). The [Mo(R$_2$dtc$_2$)$_2$]$_n$ complex is formed as a green precipitate by using the reaction described previously with an excess of NaR$_2$dtc. In ethanol solution, or on standing in very dilute methanol, green solids, also of formula [Mo(R$_2$dtc)$_2$]$_n$, are obtained that have more-complicated NMR spectra than the previous complex; [Mo(Pr$_2^n$dtc)$_2$]$_2$ has been shown by X-ray structure determinations (171) to have the structure shown in **XII**.

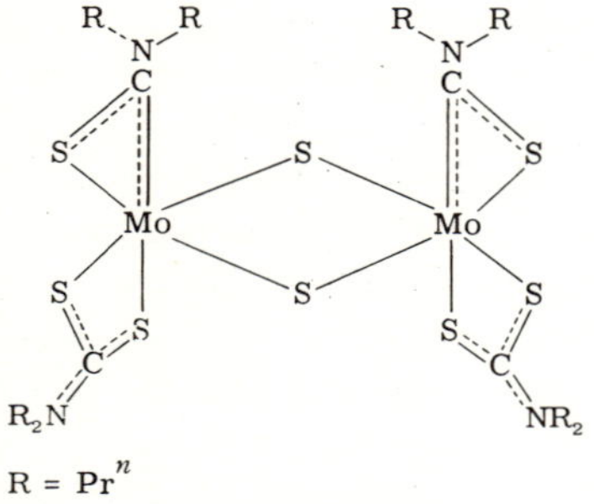

R = Pr^n

(XII)

This structure contains a very short Mo–C bond (206.9 pm) that has been attributed to a carbene type of bond. A report of a Mo(III) species with dithiocarbamate ligands used [Mo$_2$O$_2$Cl$_2$(H$_2$O)$_6$] as the starting material, and the product was [Mo(Et$_2$dtc)$_3$]$_2$ (172). This complex had previously been reported by Brown et al. (173). The structure was formulated as in **XIII**, with two bridging dithiocarbamate ligands (172). Butcher et al. (174) have shown that the electrochemical reduction of [MoO(R$_2$dtc)] occurs in two steps via a Mo^{IV} intermediate, while treatment of this complex with Cl$_2$ or Br$_2$ affords the 7-coordinate species [MoOX$_2$(R$_2$dtc)]. Willemse et al. (175) recently tried to prepare these compounds by using the procedure reported by Brown et al. (173), but were only able to isolate [Mo(R$_2$dtc)$_4$] despite exertion of considerable care and effort.

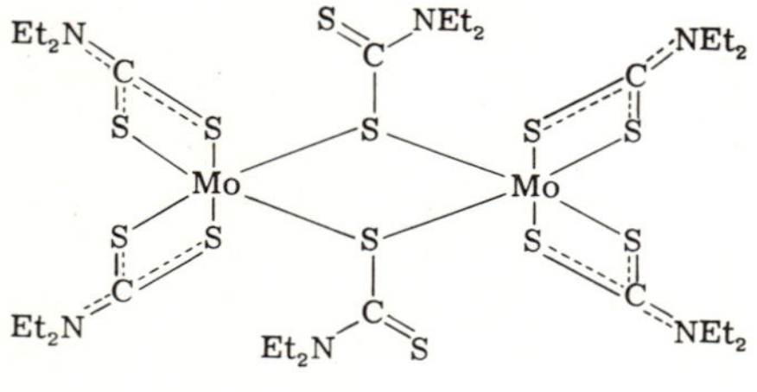

(XIII)

There has been great interest in molybdenum and tungsten nitrosyl and carbonyl dithiolate complexes; this has resulted mainly from interest in bioinorganic analogs and in organometallic derivatives. Thus, $Mo(CO)_3(Et_2dtc)_2$ and $Mo(CO)_2(Et_2dtc)_2$ constitute facile CO-carrying systems (176). When $[Mo(CO)_3)PPh_3)Cl_2]$ reacted with NaR_2dtc, it produced $[Mo(CO)_2(PPh_3)(R_2dtc)_2]$, oxidation of which gives molybdenum(VI) derivatives of empirical formula $[MoO_2(R_2dtc)_2]$, the properties of which differ from those reported earlier for compounds of the same formula (177).

Very few tungsten complexes with dithiolato ligands are known, although $[\pi\text{-}CpW(CO)_2(R_2dtc)]$ (R = Me, Et, or pip) have been prepared from $[\pi\text{-}CpW(CO)_3Cl]$ with NaR_2dtc (178) and $[Sn(Me)_3(R_2dtc)]$ (179). The 7-coordinate complex $[W(CO)_3(PPh_3)Et_2dtc)_2]$ reacts with $[Mo_2O_3\{S_2P(OEt)_2\}_4]$ to yield the 5-coordinate species $[WO(Et_2dtc)_2]$ (180). The complexes $[M(CO)_2(PPh_3)L_2]$ [M = Mo, W; L = R_2dtc, MeXant, $(EtO)_2PS_2$] have been prepared by a halide replacement-reaction of $[M(CO)_3(PPh_3)_2Cl_2]$ with the appropriate ligand. The routine synthesis of the tungsten complex was made possible by the development of an improved method of preparation of $[W(CO)_3(PPh_3)_2Cl_2]$ (181). Tungsten and molybdenum mono- and dinitrosyl complexes with dithiocarbamate (182–188) and 1,1-dicyano-2,2-dithiolate (186, 188) have been studied extensively in recent years. The cyclopentadiene rings in $[Mo(C_5H_5)_2(NO)(Me_2dtc)]$ and $[Ph_4P][Mo(C_5H_5)_2(NO)(i\text{-mnt})]$ show fluxional behavior at various temperatures as detected by NMR (186).

Several fluxional allylnitrosyl complexes of Mo, including $[(\eta^5\text{-}C_5H_5)Mo(NO)(CO)(C_3H_5)(S_2CNMe_2)]$ and $[(\eta^5\text{-}C_5H_5)Mo(NO)(C_3H_5)(S_2CNMe_2)]$ have been studied, and crystal structures obtained (188). $[Mo(Bu_2^n dtc)_3(NO)]$ contains 7-coordinate molybdenum and has a pentagonal, bipyramidal structure (189).

The cationic nitrosyl complexes $[M(NO)(CO)_3 diphos]PF_6$ (M = Mo, or W) react with NaX salts (X = halide, S_2CNMe_2, or S_2CNEt_2) to form the corresponding $[M(NO)(CO)_2(diphos)X]$ derivative (183). With phosphine ligands (L) in $CHCl_3$, $[M(NO)(CO)L(diphos)Cl]$ is obtained, but, in acetone, $[M(NO)(CO)_2L_2(diphos)](PF_6)(acetone)$ and $[M(NO)(CO)(diphos)_2]PF_6$ are formed (183). The synthesis and ^1H- and ^{13}C-NMR-spectral behavior, over a temperature range, of $[Mo(\eta^5\text{-}C_5H_5)(6\text{-}C_5H_5)(NO)(R_2dtc)]$(R = Me or Bu^n) have very recently been reported (190). An unusual μ-nitrido complex has been reported, $[Mo_3(N)_2(Etdtc)_9]^+$, which may be formulated as $[(dtc)_2Mo^{VI}\equiv NMo^{VI}(dtc)_3 \leftarrow N\equiv Mo^{VI}(dtc)_3]^{3+}$ because of the asymmetric Mo–N distances (191). An intermediate complex containing a nitrene is formed when dry HCl is reacted with $[Mo(Etdtc)_2(N_3)(Me_2SO)(NO)]$ to yield ammonia (160).

D. Manganese, Technetium, and Rhenium

The ease of oxidation of manganese(II) to manganese(III) in the dithiocarbamate complexes prevented early workers from isolating a pure [Mn(R$_2$dtc)$_2$] complex. Using rigorous deoxygenation techniques and an inert atmosphere, Fackler and Holah (*115*) isolated the yellow [Mn(Et$_2$dtc)$_2$] by reaction of NaEt$_2$dtc with MnCl$_2$ in water. The complex is isomorphous with the divalent Cu, Zn, Fe, and Cr analogs, and probably has the dimeric 5-coordinate, square-pyramidal structure (**VII**) described earlier. X-Ray powder photography of [Mn(Et$_2$dtc)$_2$], prepared in an alcoholic medium, shows the compound to be isostructural with the square-planar nickel analog (*192*). Epr spectroscopy and low-temperature magnetic susceptibility indicate that the compound has a pure spin quartet ground-state. The synthesis and properties of some ethylxanthate complexes, [Mn(Etxant)$_2$] and [Mn(Etxant)$_3$]$^-$, have been reported (*193*), and reaction of these with such Lewis bases as phenanthroline and 2,2'-bipyridine results in the formation of Mn(base)$_3$(Etxant)$_3$, which is similar to the nickel system (*194*). The inability to obtain epr spectra for the complexes [Mn(Et$_2$dtc)$_2$] and [Mn(Buxant)$_2$] is puzzling, and the stereochemistry is thus put in doubt (*195*). It is possible that oxidation had taken place in solution before, or during, the recording of the spectrum.

Manganese(III) complexes with dithiocarbamate ligands have been known, and studied frequently, since the early part of this century (*1*), and there is still interest in their physical properties. Deviations from octahedral symmetry in [Mn(Et$_2$dtc)$_3$] are reflected in the electronic spectrum (*196*). This was confirmed by an X-ray structural study that showed that the [MnS$_6$] chromophore exhibits a large distortion from D$_3$ point symmetry, and this was attributed to the ^{5}E electronic ground-state of the molecule, which would be expected to show a large Jahn–Teller distortion. Golding and Lehonen (*197*) showed that there is a simple relationship between the $-$NCH$_2-$ proton hyperfine interaction observed in the NMR spectra of ferric and manganic tris(dithiocarbamates) and the half-wave polarographic potential (E$_{1/2}$) of these complexes. From this, it was inferred that the redox processes for these compounds take place through the nitrogen atom. Electrochemical studies (*198, 199*) of the reaction of Mn^{n+} and R$_2$dtc in aprotic solvents established that the dithiocarbamate ligand forms tris-chelated complexes of manganese in three oxidation states, namely, II, III, and IV. Thus, [Mn(R$_2$dtc)$_3$] undergoes single, one-electron, oxidation and reduction steps at a platinum electrode.

$$[\text{Mn}^{\text{IV}}(\text{R}_2\text{dtc})_3]^+ + e^- \rightleftharpoons [\text{Mn}^{\text{III}}(\text{R}_2\text{dtc})_3] \rightleftharpoons [\text{Mn}^{\text{II}}(\text{R}_2\text{dtc})_3]^- - e^-$$

Titration of Mn^{2+} with pyrolidzyldithiocarbamate (R_2dtc) has been monitored by electrochemical means, and this showed (200) the existence of [Mn(R_2dtc)] and [Mn(R_2dtc)]$^-$.

Only a few cationic complexes of Mn(IV) are known (201). Saleh and Straub (202) bubbled BF_3 through a solution of [Mn(R_2dtc)$_3$] in dichloromethane for five minutes, to produce [Mn(R_2dtc)$_3$]BF_4 complexes. Golding et al. (203) were able to isolate four explosive, air-sensitive [Mn(R_2dtc)$_3$]ClO_4 complexes (R_2 = Me_2, Et_2, pyridyl, or piperidyl) by reacting [Mn(H_2O)$_6$](ClO_4)$_2$ in acetone with Mn(R_2dtc)$_3$ in benzene. They also prepared the tetrafluoroborates in an analogous manner. Although the complexes were found to be difficult to study, the crystal and molecular structure of [Mn(pipdtc)$_3$]ClO_4 was determined. The manganese atom is surrounded (203) by three dithiocarbamate ligands in approximately D_3 symmetry.

Complexes of manganese with the 1,1-dithiolates have been restricted to the (Pr_4^nN)$_3$[Mn(i-mnt)$_3$] complex (112) and nitrosyl or carbonyl derivatives, e.g., [π-CpMn(NO)(S$_2$C=X)]$^-$ (X = C(CN)CO$_2$Et, C(CN)CONH$_2$, NCN, C(CN)$_2$, or (HNO$_2$) (110, 204, 205). These complexes undergo electrochemical, one-electron oxidations and reductions to afford the neutral or dianionic species.

Manganese carbonyl complexes containing dithiocarbamate ligands have been synthesized by the method that Abel and Dunster (179) described earlier for tungsten complexes. A general method of preparing dithiocarboxylate complexes of the Group VII elements utilizes CS_2 insertion into M—R or M—H σ-bonds. Thus, the following complexes have been prepared by this method: [M(CO)$_4$(RCS$_2$)] (M = Mn; R = Me, Ph, p-MeC$_6$H$_4$; M = Re; R = Me, Ph, p-MeC$_6$H$_4$, p-ClC$_6$H$_4$, Bz, or Ph$_3$C) (206–208) and [Re(HCS$_2$)(CO)$_2$(phosphine)$_{1,2}$] (phosphine = PPh$_3$, diphos, dpm) (209–211). In the latter complexes, with bidentate phosphines, CS_2 insertion causes the phosphine to become monodentate. Treatment of [Mn(CO)$_5$Br] with Me$_2$dtc$^-$ produces [Mn(CO)$_4$(Me$_2$dtc)], which can undergo (200) thermally induced dimerization to afford [Mn(CO)$_3$Me$_2$dtc)$_2$]$_2$.

Interest in the chemistry of rhenium with sulfur-containing ligands has increased in the past few years. In 1969, Hieber and Rohm (213) reacted [Re(CO)$_5$Cl] with thiols, and thiocarboxylic and mercaptocarboxylic acids, and, among the products, they identified [Re(CO)$_3$RS$_2$]$_2$ complexes. Using the same starting material, Rowbottom and Wilkinson (214) isolated five dithiocarbamato derivatives from the reaction with tetraethylthiuram disulfide. The main product was the monomeric, diamagnetic [Re(CO)$_2$(R$_2$dtc)$_3$], and its IR spectrum contained a single, carbonyl stretching-frequency at 1870 cm^{-1}, as well as the characteristic bands of chelated dithiocarbamate groups. The complex is very inert, and it was suggested that the molecular structure is that of rhenium(III) with a bicapped, trigonal, prismatic geometry (**XIV**).

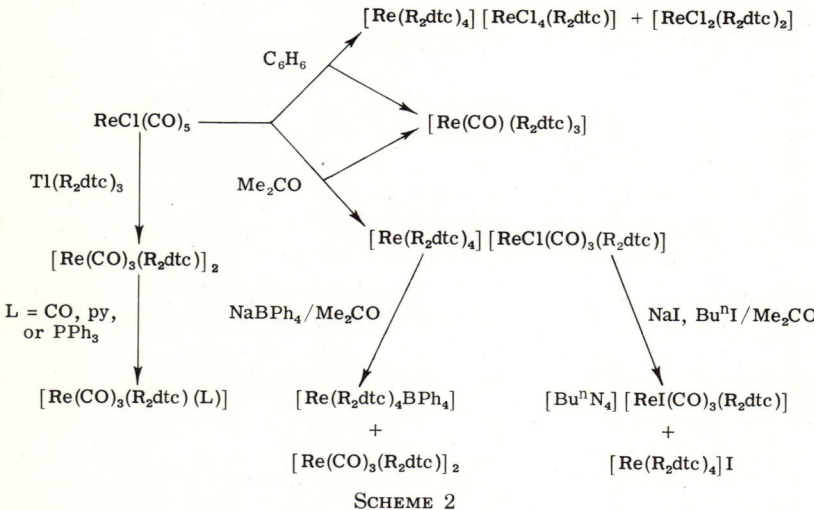

(XIV)

[Re(Et$_2$dtc)$_4$]$^+$[Re(CO)$_3$Cl(R$_2$dtc)]$^-$ was characterized, from the same reaction, by polarography. Subsequent work led to the isolation of numerous, interrelated rhenium dithiocarbamates (215) (see Scheme 2).

SCHEME 2

The structure of [Re(Et$_2$)dtc$_3$(CO)] consists of monomeric Re with a distorted, pentagonal, bipyramidal geometry, having an axial CO group. This is the first, clear-cut example of monomeric 7-coordinate rhenium(III), as other examples usually involve Re–Re bonds (216).

Oxorhenium complexes have appeared in a number of reports. Tisley et al. (217) produced orange-brown [Re$_2$O$_3$(Et$_2$dtc)$_4$] from [ReCl(PPh$_3$)$_2$], [Re$_2$O$_3$py$_4$Cl$_4$], or [Re$_2$O$_2$(CH$_3$)$_4$Cl$_2$] with NaEt$_2$dtc in acetone. The almost linear O=Re–O–Re=O system has been identified by X-ray crystallography *(217, 218)* (**XV**).

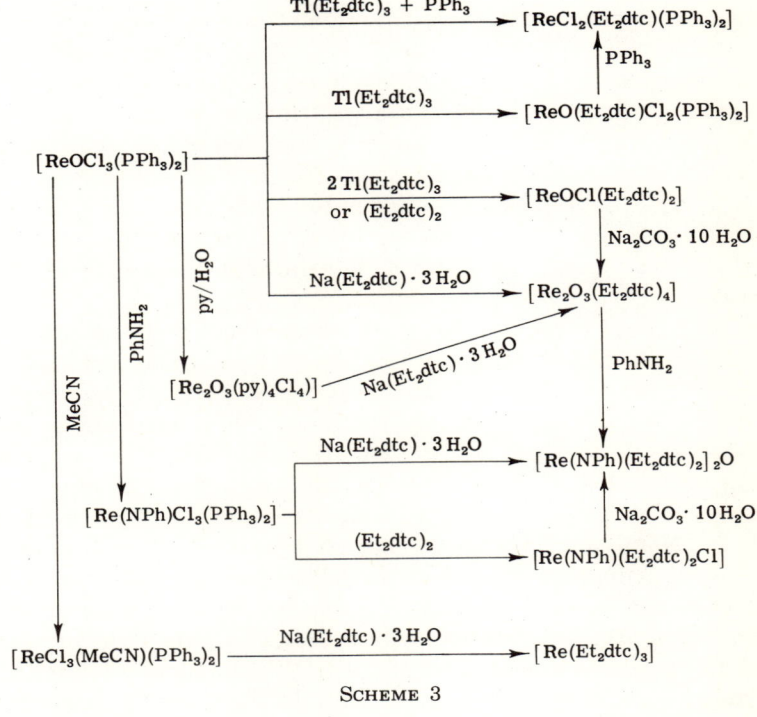

(XV)

Rowbottom and Wilkinson *(219)* synthesized a series of new, and known, oxo-, chloro-, unsubstituted dithiocarbamates of Re(V) and Re(III) (see Scheme 3).

The only nitrido complex known is the square pyramidal [ReN(Et$_2$dtc)$_2$], synthesized *(220)* from [ReNCl$_2$(PPh$_3$)$_2$]. The same structural type is suggested by epr for [Mn(S$_2$CX)$_2$(NO)] (X = OEt, or NEt$_2$) and [ReO(Et$_2$dtc)$_2$] *(221, 222)*. An interesting series of novel dinitrogen complexes, [Me$_2$PhP)$_3$(N$_2$)-(R$_2$dtc)Re], and their reactions with

SCHEME 3

hydrogen halides, have recently been studied as part of the design of models for nitrogenase (223).

The reaction of [Re(CO)$_5$]$^-$ with CS$_2$ and methyl iodide proceeds to give the trithiocarbonate derivatives [Re(CO)$_4$S$_2$C(SMe)]. When [ReBr(CO)$_5$] is used instead of MeI, the dinuclear complex [Re(CO)$_4$S$_2$CSRe(CO)$_5$] is isolated (224). The product of the reaction of [CF$_3$Re(CO)$_5$] with CS$_2$ has been identified by X-ray crystallography (225) as the trithiocarbamato complex [(OC)$_4$ReS$_2$CSRe(CO)$_{4 2}$SCS$_2$Re (CO)$_4$]. The crystal structure of [Re(PhCS$_2$)(CO)$_4$] confirmed the bidentate nature of the dithiocarboxylate group (212).

The only technetium complex has been reported (226) is [Tc(CO)$_4$(R$_2$dtc)]. There appear to be few reports of rhenium or technetium complexes with 1,1-dithiolato ligands.

E. Iron, Ruthenium, and Osmium

The earlier literature containing work on 1,1-dithio complexes of iron has been covered in depth by Coucouvanis (1), and, since 1968, interest in these iron complexes has increased.

Ferrous complexes with the dithio ligands are difficult to isolate, as they are unstable to air oxidation (227). Magnetic susceptibility and Mössbauer spectral data (227, 228) show [Fe(R$_2$NCS$_2$)] to be of two structural types: (a) intramolecular, antiferromagnetic dimers, e.g., [Fe(Et$_2$dtc)$_2$]$_2$ having each iron atom surrounded by five sulfur atoms, with square pyramidal geometry as in (IX); and (b) antiferromagnetic polymers, e.g., [Fe(Me$_2$dtc)$_2$]$_n$, believed to involve octahedral coordination about the metal (227). The crystal structures of a number of ferrous dithio complexes have been determined. These include (229) [Fe(Et$_2$dtc)$_2$].

Although the bis(dithiocarbamate) complexes of Fe(II) are relatively unstable to air oxidation, early studies (12, 15) produced stable adducts of NO and CO. Both 5-coordinate [Fe(NO)(R$_2$dtc)$_2$] and 6-coordinate [Fe(NO)$_2$(R$_2$dtc)$_2$] complexes are known. There has been considerable interest in the mode of attachment of the NO molecule, as there are six possibilities (see Scheme 4). (A) and (B) represent valence-bond structures of the linear Fe–NO bond. Structure (C) involves a symmetric Fe–NO π-bond. Structure D illustrates the bent mode of attachment, in which nitrosyl is coordinated to the metal through the nitrogen atom, but the Fe–NO bond-angle differs greatly from 180°. Structures (E) and (F) are valence-bond formalisms of an unsymmetrical, metal–NO π-bond. The structure of [Fe(NO)(R$_2$dtc)$_2$] (R = Me or Et) has been shown (230, 231) to be square pyramidal, with four sulfur atoms in

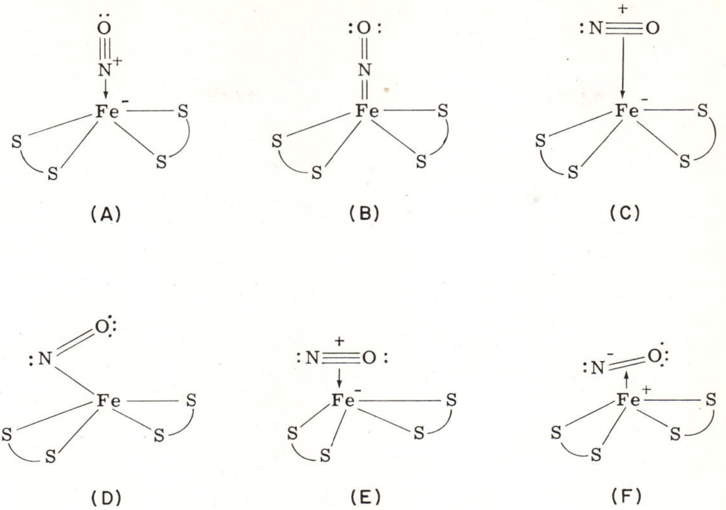

Scheme 4

(A), (B), (C), (D), (E), (F) — structures of Fe nitrosyl/dithiolate complexes with various Fe–N–O bonding modes.

the basal plane. The nitrosyl is bonded through the nitrogen atom, and the length of the Fe–N bond in both compounds is ~170 pm, which indicates some multiple bonding. There is some controversy as to whether the Fe–NO group is linear, or bent. Infrared and ESCA studies have been reported (232, 233) for [Fe(NO)(Me$_2$dtc)$_2$].

Buttner and Feltham (234) synthesized several new nitrosyl and carbonyl complexes of iron. These include cis-[Fe(Et$_2$dtc)$_2$(I)(NO)], cis-[Fe(Et$_2$dtc)$_2$(CO)$_2$], and the dimeric S-bridged species [Fe(Et$_2$dtc)(CO)$_2$SCH$_3$]$_2$. They also found that the compound previously identified as a dinitrosyl iron complex contains one NO group and one nitro group, [Fe(Et$_2$dtc)$_2$(NO$_2$)(NO)]. Cotton and McCleverty (235) prepared [Fe(CO)$_2$(R$_2$dtc)$_2$] (R = Me, or Et) and [(π-Cp)Fe(CO)(Me$_2$dtc)] by reacting NaR$_2$dtc with metal carbonyl halides, or by oxidation of metal carbonyls with tetraalkylthiocarbamoyl disulfides. The latter method was also used by Brennan and Bernal (189) and Bernal and co-workers (236) for the preparation of bis(cyclopentamethylenedithiocarbamate) Fe^{II} dicarbonyl. New dimeric carbonyls of iron dithiocarbamates were produced by Abel and Dunster (179).

Trivalent iron dithiocarbamate complexes have been extensively studied, because of the existence of "spin equilibria" in these complexes. Table II outlines the tris(1,1-dithiocarbamate) iron(III) complexes and, some of their physical properties.

TABLE II
Tris(N,N-Disubstituted dialkyldithiocarbamato)iron(III) Complexes

R¹	R²	MP (°C)	Solid μ_B	Solution^a μ_B
H	H			
Me	H			
Et	H			
Pr^n	H			
Bz	H			
Me	Me	300	4.17	4.20c
Et	Et	252–255	4.24	4.41c
Pr^n	Pr^n	167–168	4.48	4.24b
Pr^i	Pr^i	285–288	2.62	2.34b
Bu^n	Bu^n	146–151	5.32	4.34b
Bu^i	Bu^i	167–168	3.02	2.88cb
sec-Bu	sec-Bu	132		2.20cb
Amyl^n	Amyl^n	146–148		4.32cb
Amyl^i	Amyl^i	199–200	4.30	4.37b
Hexyl^n	Hexyl^n	164–166	3.52	4.42b
Heptyl^n	Heptyl^n			4.32b
Octyl^n	Octyl^n			
n-C$_{12}$H$_{25}$	n-C$_{12}$H$_{25}$			
n-C$_{16}$H$_{33}$	n-C$_{16}$H$_{33}$			
Bz	Bz	214–222	4.02	3.60b
Me	Ph	237–247	2.99	3.33c
Et	Ph	238–244	4.70	3.63c
Pr^n	Ph	212–214	4.68	3.55b
Pr^i	Ph			
Amyl^i	Ph	208–210	3.36	3.43b
Me	Bz			4.06cb
Et	Bz			4.22cb
Me	Bu^n			4.26cb
Et	Bu^n			4.40cb
Cyclopentyl	Cyclopentyl			
Cyclohexyl	Cyclohexyl	285–287	2.75	4.16c
Allyl	Allyl	106–107	4.40	4.34b
Piperidyl		284–293	4.01	4.16c
2-Me-Piperidyl				3.73cb
Piperazyl				
Phenylpiperazyl				4.06cb
Pyrrole		280	5.83	5.82b
Morpholinyl				4.02cb
Thiomorpholinyl		350	4.03	

^{a} Solvent: b, benzene; c, chloroform.

The majority of octahedral ferric complexes exhibit simple Curie or Curie–Weiss magnetic behavior (i.e., magnetic susceptibility $\propto 1/T$). They can be classified as either "high spin" or "low spin." In high-spin complexes, the lowest term (ground state) is $^6A_{1g}$, which corresponds to the $t_{2g}^3 e_g^2$ configuration. The low-spin complexes have the $^2T_{2g}$ term as

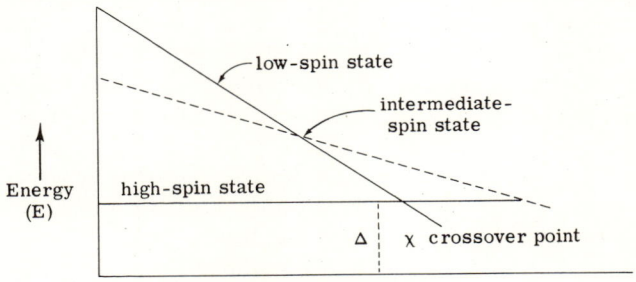

the lowest, and this corresponds to the t_{2g}^5 configuration. Whether the $^6A_2$ or the $^2T_2$ term is the ground state depends on the magnitude of the ligand field, Δ. There is a balance of the energy terms Δ and π, the "mean pairing energy," which determines whether the complex is high-spin or low-spin. For $\Delta > \pi$, a high-spin ground-state results, and, for $\Delta < \pi$, a low-spin ground-state results. The transition from high to low spin on increasing Δ can be shown diagrammatically.

There will exist an equilibrium between the two states. If the energy between the two states, E, is of the order of kT, then the relative populations of the two states will vary with the temperature of the sample. In the Fe(III) dithiocarbamate series of complexes, [Fe(R₁R₂dtc)₃], ΔE can be varied by suitable choice of substituents R_1 and R_2. Although these are well removed from the FeS_6 molecular core, they can appreciably affect the electronic parameters of the central iron atom and of the surrounding crystal field of the sulfur atom by way of the conjugated system of the ligand (237). The results of the spin crossover are reflected in magnetic susceptibility data.

Martin and co-workers (238) were the first to recognize the possibility of polar resonance forms of the type

$$R_2\overset{+}{N}C=C\begin{matrix}\diagup S^-\\ \diagdown S^-\end{matrix}$$

and they suggested that they were predominant in high-spin complexes; this agrees with the fact that IR and X-ray studies have shown that

$$R\overset{+}{O}=C\begin{matrix}\diagup S^-\\ \diagdown S^-\end{matrix}$$

is unimportant in describing the structure of the low-spin xanthate complexes. Reaction of py with [FeL$_3$] (L = MeXant, EtXant, S$_2$PPh$_2$, or S$_2$Pcyclohexyl$_2$) gives bright-yellow compounds with magnetic moments (μ_{eff} = 4.9–5.0 μ_B) and Mössbauer spectra that strongly indicate iron(II). It is most likely (239) that they are of the type [Fe(L)$_2$(py)$_2$]. Ewald et al. (240) proposed that the principal trend influencing the magnetic behavior of the iron(III) dithiocarbamates is an increasing R–N–R bond-angle resulting from steric interactions between the nitrogen substituents. Eley et al. (241), however, conducted a study the results of which indicated that the primary function of the substituent, R, is as an electron-releasing group. They found that steric interactions appear to be relatively unimportant, except when a secondary carbon atom is the substituent. These workers and others (242) prepared a series of iron(III) dialkyldithiocarbamates in which the amino nitrogen atom is a member of a ring group, e.g., morpholine, 2-methylpyridine, or 3-methylpiperidine. On the basis of the evidence obtained from magnetic measurements, they concluded that a $^6A_1 - {}^2T_2$ crossover situation exists, similar to that for the N,N'-dialkyl substituted dithiocarbamates. More recently, however, correlations between molecular geometry and temperature-dependent magnetic behavior of [Fe(R$_2$dtc)$_3$] complexes have been found (243).

When the Fe–S distances in all of the known Fe(III) dithiocarbamates are plotted against their magnetic moments, a smooth curve is obtained (see Fig. 1). Mössbauer spectral studies (244, 245) on [Fe(R$_2$dtc)$_3$] (R$_2$ = Bu$_2^i$, Pr$_2^i$, Me$_2$, piperidyl, or Bz$_2$) showed that the isomer shift is nearly identical in all of the complexes, and the temperature-dependence was explained through the existence of the almost equienergetic $^6A_1$ and $^2T_2$ ground-states in thermal equilibrium. The Fe–S bond contraction, on going from high to low spin, was originally attributed to a distortion from D$_3$ to C$_3$ symmetry, and spin–orbit

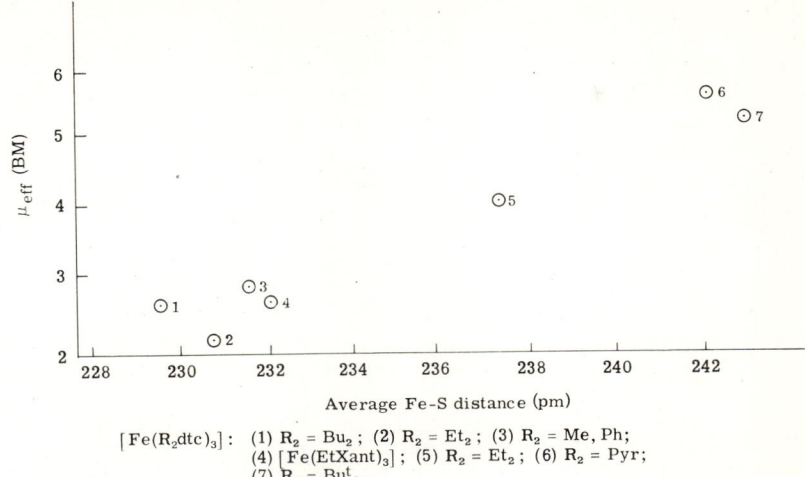

[Fe(R₂dtc)₃]: (1) $R_2 = Bu_2$; (2) $R_2 = Et_2$; (3) $R_2 = Me, Ph$;
(4) [Fe(EtXant)₃]; (5) $R_2 = Et_2$; (6) $R_2 = Pyr$;
(7) $R_2 = Bu^t{}_2$.

FIG. 1. Plot of effective magnetic moment vs. iron–sulfur bond length in [Fe(R₂dtc)₃] complexes.

interactions inducing Jahn–Teller splitting in the ground state (237, 246), but this has since been discounted (243). In the epr spectra of some iron(III) dithiocarbamate complexes exhibiting spin equilibria, separate signals were observed from high- and low-spin forms (247, 248). The tris(dtc)iron(III) spin equilibrium has been reinvestigated (249) by variable-temperature determination of magnetic susceptibility (4.2–296 K), infrared spectra (30–300 K) and epr (4.2–85 K).

It is likely that these compounds will receive increasing attention as inorganic chemists, biochemists, and crystallographers (437) wrestle with the problems associated with spin changes in biological systems containing iron. Additional examples of $\overline{CH_2CH_2XCH_2CH_2CS_2^-}$ (X = S, or NMe) have (250) magnetic moments of 4.01 and 4.03μ_B. Paramagnetic anisotropy measurements on the single crystals of a series of halobis(diethyldithiocarbamate)iron(III) in which the ferric ion is an intermediate-spin state have been made (251).

A number of iron(III) xanthates and thioxanthates have been prepared (1). The structure of the predominantly low-spin [Fe(Rxant)₃] (R = Et, or SBu) complexes show features (246, 252) of coordination about the iron atom that are significantly different from those of [Fe(Bu₂dtc)₃], and it is estimated that there is 10–30% of double-bond character in coordinated S₂CX (X = OR, or SR), which is appreciably less than that of the dithiocarbamates (40–50%). Reactions of NaR_2dtc and KS_2OEt with Fe(II) halides in the presence of 2,2'-bipyridyl (bipyr) and 1,10-phenanthroline (phen) produce products similar to those found (194) for Mn(II). The magnetic properties of the thioxanthate complexes of Fe(III) indicate that a spin-equilibrium exists, with the low-spin form preponderating, as for the xanthates (253). The alkylthioxanthate complexes are thermally unstable, and, at room temperature, readily eliminate CS_2 to afford dimeric, diamagnetic compounds.

$$2\ Fe(RSXant)_3 \xrightarrow{THF} [Fe(SR)(S_2CSR)_2]_2 + 2\ CS_2$$

The crystal structure of one of the foregoing complexes (R = Et) was determined by Coucouvanis et al. (253) in a study of inorganic analogs of nonheme proteins; it is dimeric with bridging mercaptide and thioxanthate groups (**XVI**). The short Fe–Fe distance (261 pm) and the diamagnetism suggest a Fe–Fe single bond in the complex. The same workers (254) prepared a series of S-bridged, dimeric complexes of Fe(III), $[Fe(RSXant)_2(SR)]_2$ (R = Et, Pr^n, Bu^n, Bu^t, or Bz), and found that the *tert*-butyl thioxanthate is thermally more stable, due to steric factors, and a dimer of this complex was not isolated. The crystal

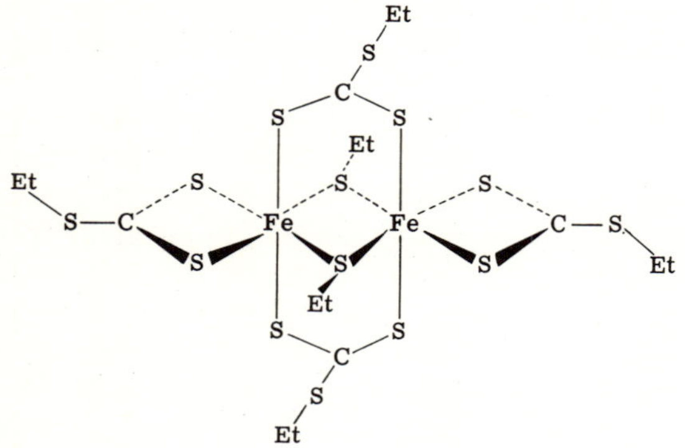

(**XVI**)

structure of the *tert*-butyl thioxanthate (255) shows approximately octahedral configuration around Fe, but with distortion due to the small "bite" of the dithio ligands. $[Fe(S_2CSR)_3]$ compounds (R = Et, Pr^n, or Bu^n) have been synthesized as 6-coordinate, octahedral complexes (256). The ethyl derivative is diamagnetic, which contradicts earlier work (252), and this is ascribed to spin-pairing via Fe–Fe of Fe–S–S–Fe interactions between molecules, or to a change in symmetry due to packing. Lippard et al. (255), however, suggested that the material that Pelezzi (256) had used had probably dimerized. A number of tris(dtc)iron(III) and -cobalt(III) complexes have been examined, and it was found that the inclusion of polar solvents appears to be a common phenomenon in crystals of these compounds; however, the iron complexes tend to lose solvent on standing. The structures of the tris(4-morphdtc) complexes of both metals were determined, and, although the [CoS$_6$] geometry is identical with that of previously reported complexes, the dimensions of the Fe complex (Fe–S = 44 pm, S–Fe–S = 72.5°) are atypical (257).

There have been but few reports of iron complexes with dithiocarboxylate ligands. Ferrous complexes are not very stable, and have only been reported qualitatively (258). The most stable dithiocarboxylate complexes of iron(III) are [Fe(C$_6$H$_5$CS$_2$)$_3$], the dithiobenzoate, and [Fe(p-MeC$_6$H$_4$CS$_2$)$_3$], the dithiotoluate; these were prepared by Coucouvanis and Lippard (259, 260), who, in the course of their study, also prepared "sulfur-rich" [Fe(PhCS$_3$)$_2$(PhCS$_2$)] and [Fe(p-tolCS$_3$)(p-tolCS$_2$)] complexes. The crystal structure of the latter shows iron in a distorted, octahedral environment of six sulfur atoms. It also appeared that one of the chelate rings had undergone expansion, to form a five-membered, chelate ring containing three sulfur atoms, a perthiotoluate complex (**XVII**).

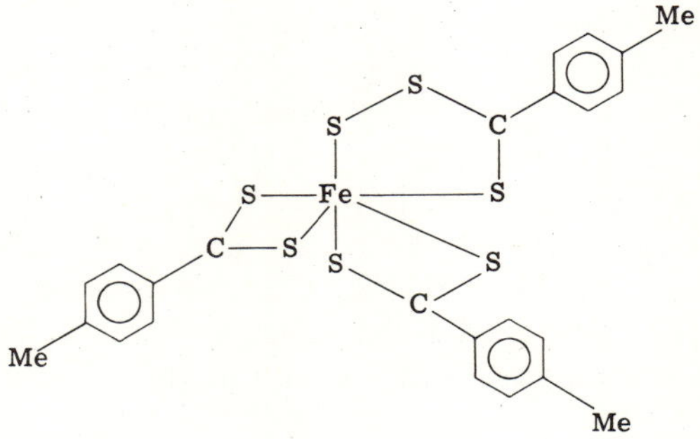

(XVII)

The preparation and characterization, and Mössbauer and epr spectra of Fe(II) and Fe(III) complexes of cyclopentadienyldithiocarboxylate have been reported recently (261).

Several complexes of iron with the 1,1-ethenedithiolates have been isolated. These are mainly tris-Fe(III) complexes with the 1,1-dicyano-2,2-ethylenedithiolate ligand (1). Recently, however, Coucouvanis et al. (262) synthesized a new 1,1-dithiolate ligand (**XVIII**) from the reaction of CS_2 with malonic ester in the presence of a base. The new lig-

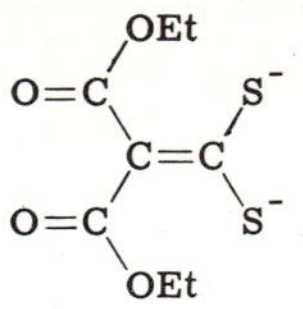

$$\begin{array}{c} O=C(OEt)\\ C=C(S^-)_2\\ O=C(OEt) \end{array}$$

(XVIII)

and, 1,1-dicarboethoxyethylene-2,2-dithiolate dianion (DED^{2-}) can be represented by extensively delocalized structures. By reaction of the

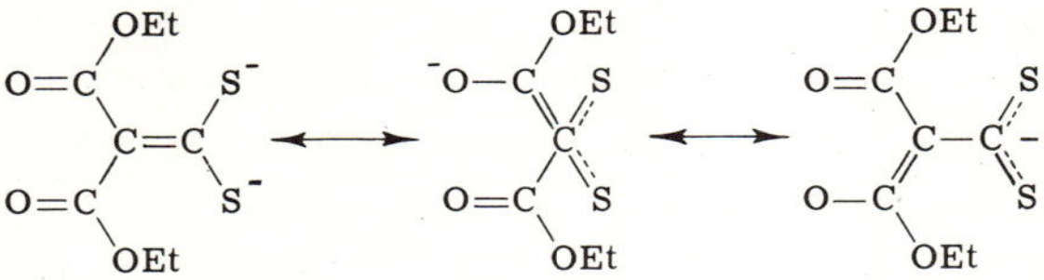

[three resonance structures shown]

potassium salt of DED with $Fe(ClO_3)_3 \cdot H_2O$ in the presence of $BzPh_3PCl$ in dichloromethane, a brown solution forms, from which an oil, and then a solid complex (262), $[BzPh_3P]_2[Fe(DED)_3]$, may be pentane-extracted. The complex has been characterized by Mössbauer and IR-spectral, polarographic and X-ray methods, and is octahedral, iron(IV) having a $^3T_1$ ground-state. The geometry of the MS_6 moiety in the structure can be described as originating from a trigonal prism that suffers individual rotations of the chelating ligands around the C_2 axes. One of the rotations is more severe than the other two, causing (263) a lowering of the overall symmetry to C_2.

Compounds of the type $[FeX(R_2dtc)_2]$ have been obtained by treating $[Fe(R_2dtc)_3]$ complexes with concentrated hydrohalic acids. $[FeCl(Et_2dtc)_3]$ has been studied by Hoskins and White (264); it has a square pyramidal structure, with the chlorine atom at the apex, and with the Fe atom situated 62 pm above the basal plane of the four sulfur atoms. A similar structure is found (265) for the monoiodo derivative $[FeI(Et_2dtc)_2]$. The chloro complex has been synthesized (266) by the following reaction.

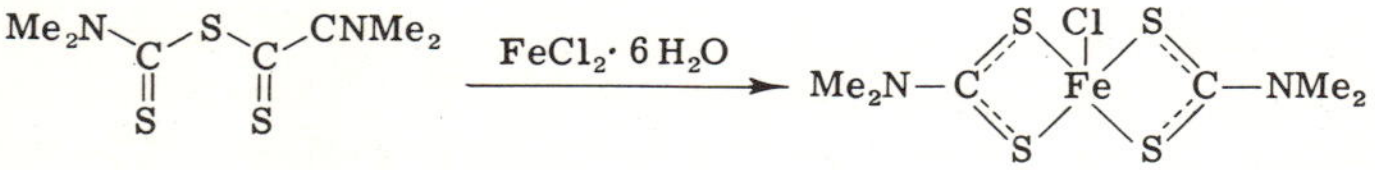

$$Me_2N-C(=S)-S-C(=S)-CNMe_2 \xrightarrow{FeCl_2 \cdot 6H_2O} Me_2N-C\begin{pmatrix}S\\S\end{pmatrix}Fe(Cl)\begin{pmatrix}S\\S\end{pmatrix}C-NMe_2$$

Reactions of $[Fe(Et_2dtc)_3]^n$ ($n = +1$, or 0), $[FeI(Et_2dtc)_2]$, and $[Fe(Et_2dtc)_2]$ with dppe, CNR, and PPh_3 have recently been described (267). Treatment of $[Fe(Et_2dtc)_3]^+$ with PPh_3 afforded (267)

[Fe(Et₂dtc)₃], PPh₃S, and [PPh₃{C(NEt₂)S}]⁺. The He(I) photoelectron spectra of [Fe(Et₂dtc)₃] has been reported. *(268)*.

Iron(IV) complexes are very rare. During the course of investigations into the synthesis and properties of iron dithiocarbamate complexes as model compounds for such naturally occurring, iron–sulfur proteins as ferredoxin, Pasek and Straub *(269)* isolated several cationic tris(dithiocarbamate) complexes of Fe(IV). The species [Fe(R₂dtc)₃]BF₄ (R = Et, Pr^i, pyrollidine, or cyclohexyl) and [Fe(Et₂dtc)₃]PF₆ are readily prepared by bubbling gaseous BF₃ for 30 sec through a benzene solution of [Fe(R₂dtc)₃] in the presence of air, or by reaction of KPF₆ with an acidic solution of [Fe(R₂dtc)₃] in acetone, followed by prolonged bubbling (5 h) of dry air through the mixture. Golding and co-workers *(270)* also produced cationic tris(dithiocarbamate) complexes of iron(IV), namely, [Fe(R₂dtc)₃]X (X = ClO₄ and FeCl₄), by reaction of Fe(ClO₄)₃ or FeCl₃ with [Fe(R₂dtc)₃]. The structure of the iron(IV) complex [Fe(pyrroldtc)₃](ClO₄) has been reported, and the [FeS₆] core was shown to be intermediate between octahedral and trigonal prismatic *(271)*.

Previous reactions of FeCl₃ and tetramethylthiuram disulfide and [Fe(R₂dtc)₃] or [Fe(R₂dtc)₂Cl] with Cl₂ or Br₂ resulted in the formation of complexes of the following type. Iron(IV) and iron(II) complexes con-

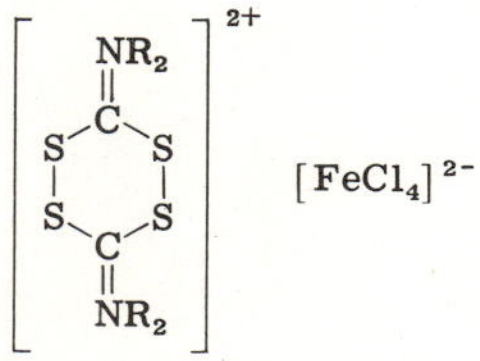

$$\left[\begin{array}{c} NR_2 \\ \| \\ S^{-C}{\diagdown}S \\ | \quad\quad | \\ S_{\diagdown C}{-}S \\ \| \\ NR_2 \end{array} \right]^{2+} [FeCl_4]^{2-}$$

taining dithiocarbamate ligands have also been produced electrochemically *(270, 272–274)*. [Fe(Et₂dtc)₃] exhibits reversible, one-electron oxidation and reduction steps. This method of synthesizing

$$[Fe(Et_2dtc)_3]^+ \underset{+e^-}{\overset{-e^-}{\rightleftarrows}} [Fe(Et_2dtc)_3] \underset{-e^-}{\overset{+e^-}{\rightleftarrows}} [Fe(Et_2dtc)_3]^-$$

[Fe(Et₂dtc)₃]BF₄ has the advantage that the product is completely free from by-products and is isolable in yields of 90%. The relationship between $E_{1/2}$ values and electronic structures has been described in terms of $^6A_1 - {}^2T_2$ crossover equilibrium *(273)*. The rate of e⁻ transfer between [Fe(Me₂dtc)₃] and [Fe(Me₂dtc)₃]BF₄ has been measured by ²H-NMR, line-broadening experiments *(275)*. Mössbauer-spectral data have been published *(276)* for a series of (R₂dtc) complexes of iron(IV).

A report of cationic, mixed-ligand complexes of iron(III) containing dithiocarbamate ligands appeared recently (277). The complexes, [FeL(R$_2$dtc)$_2$]$^+$ (L = CNR, diphos, or PPh$_3$) were obtained by reaction of L with [Fe(R$_2$dtc)$_3$]$^+$, involving a ligand-displacement reaction (277).

A number of mixed 1,1-dithiolato and 1,2-dithiolene complexes of iron have been prepared and characterized (278–280), namely, [Fe(R^{1}R^{2}dtc)$_2$(S$_2$C$_2$X$_2$)] (R^1 = Et, or Ph; R^2 = Et, or Me; X = CN, or CF$_3$). They are formed on treating the bis(dithiocarbamato)iron complex with the Na$_2$mnt, followed by successive oxidation with copper(II) in air in MeCN (278). Alternatively, the 1,2-dithiolene complex can be treated with the dithiocarbamate ligand, followed by sulfite reduction (279, 280). The complexes have room temperature magnetic moments of ~2.5μ_B, indicating low-spin iron(III). The crystal structure of [Fe(Et$_2$dtc)$_2$(tfd)] shows FeS$_6$ coordination of distorted octahedral symmetry, and two limiting, nonequivalent valence-bond structures can be drawn for this complex (ignoring the relatively trivial resonance of the dithiocarbamate ligand).

dithioketonic form   dithiolene form

Relatively few complexes of ruthenium with 1,1-dithio ligands are known. The earlier, sparse reports of these complexes are included in Coucouvanis's review (1). Only two tris(dithiocarbamato) complexes of ruthenium have been reported, the [Ru(Et$_2$dtc)$_3$] complexes by reaction of K$_2$RuCl$_6$ with NaR$_2$dtc in water (281), and the [Ru(Et$_2$dtc)$_3$]BF$_4$ complex by reaction of the former complex with BF$_3$ in the presence of air (282). However, a recent crystal-structure determination (283, 284) of a complex has shown that the product from this reaction is [Ru$_2$(Et$_2$dtc)$_5$]BF$_4$, a Ru(III) species (**XIX**). The complex contains a Ru–Ru bond and two types of bridging dithiocarbamate ligand. The nature of one of the bridging dithiocarbamate ligands is novel, as no other structures studied to date contain this ligand bonding in this mode. Duffy and Pignolet, and Pignolet (285, 286) studied the ^{1}H-NMR spectrum and crystal structure of [Ru(Et$_2$dtc)$_3$], which is intermediate between trigonal antiprismatic and trigonal prismatic, approximating to D$_3$ symmetry, and is stereochemically nonrigid. The only other pure,

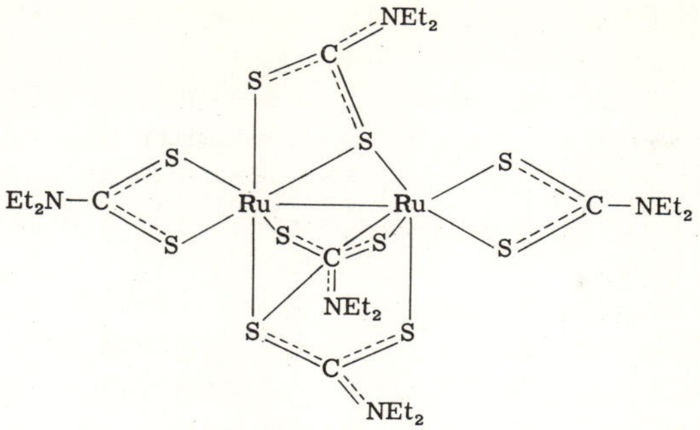

(XIX)

1,1-dithio complex of ruthenium prepared (287) is $(Ph_4As)_3[Ru(i\text{-}mnt)_3]$.

A number of ruthenium(II) complexes have been prepared. Cole-Hamilton and Stephenson isolated cis-$[Ru(Me_2dtc)_2L_2]$ (L = PPh_3, PMe_2, Ph, $PPhMe_2$, or $P(OPh)_3$) from Ru(II) and Ru(III) tertiary phosphine and phosphite complexes with $NaMe_2dtc$, and found that they undergo rearrangements (288).

New dithiocarbamato and O-alkyl dithiocarbonato derivatives $[M(S-S)_2(PPh_3)_2]$, $[M(S-S)_2(CO)(PPh_3)]$, $[MH(S-S)(CO)(PPh_3)_2]$ (M = Ru, or Os; S—S = R_2dtc, or RXant; R = Me, or Et), $[OsCl(R_2dtc)(CO)(PPh_3)_2]$ and $[IrH_2(S-S)(PPh_3)_2]$ have been prepared by treating the appropriate chloro or carboxylato complexes with sodium salts of the ligand (289). "Insertion" of CS_2 into the metal–hydrogen bonds has been employed (438) in synthesizing a range of new dithioformato complexes of Ru, Os, and Ir, including $[MX(S_2CH)(CO)(PPh_3)_2]$,(two)—$[MX(RN\text{---}CH\text{---}S)(CO)(PPh_3)_2]$ (M = Ru, or Os; X = Cl, Br, or $OCOCF_3$), $[M(S_2CH)_2(PPh_3)_2]$,$[IrCl_2(S_2CH)(PPh_3)_2]$, $[MH(RN\text{---}CH\text{---}S)(CO)(PPh_3)_2]$, and $[Ru(RN-CH-S)_2(PPh_3)_2]$.

An electrochemical study of several Et_2dtc complexes of Ru(III) and Ru(IV), including $[(CO)Ru(dtc)_2]$, $Ru(dtc)_3$, and $ClRu(dtc)_3$, using dc, ac, and cyclic, voltammetric techniques in solvents MeCN, Me_2Cl_2, and propylene carbonate has been very recently reported (290).

The dimeric $[Ru_2N(Et_2dtc)_4]Cl$ was prepared by Griffith and Pawson (441). An organometallic, Ru(II) adduct $[Ru(Me_2dtc)_2(1,5\text{-COD})]$ (1,5-COD ≡ bicyclo[2.2.1]hepta-2,5-diene) was produced by Powell (292). Other organometallic derivatives have been synthesized (293, 294) by CS_2 insertion into Ru bonds in such complexes as $[RuCl(CO)(PCy_3)_2(H)]$ and $[RuH_4(PPh_3)_3]$. The product of the latter reaction, [Ru(S$_2$CH)$_2$(PPh$_3$)$_2$], has been characterized as a dithioformato complex (295). A number of Ru(II) carbonyl complexes containing dithiocarbamate ligands were prepared by treating a solution of a ruthenium carbonyl with a solution of the particular dithiocarbamate, e.g., [Ru(Et$_2$dtc)$_2$(CO)] (R = Et or Me) and [Ru(Bz$_2$dtc)$_2$(CO)$_2$].

Most of the reports of osmium complexes containing 1,1-dithio ligands are concerned in the use of dithiocarbamate complexes for the analytical determination of the metal (296); a violet color forms when OsO$_4$ and NaR$_2$dtc are mixed in aqueous solution (84).

Addition of CS$_2$ to [OsHCl(CO)(PCy$_3$)$_2$] gives two different products, depending upon the conditions employed. In benzene, the product β-[OsCl(HCS$_2$)(CO)(PCy$_3$)$_2$] is indicated on the basis of infrared data [ν(C–S) bands at 917 and 790 cm^{-1}], whereas, if the reaction is conducted in the solid state, an additional product [OsHCl(CS$_2$)(CO)(PCy$_3$)$_2$], containing terminally S-bonded CS$_2$ [ν(C–S) = 1510 cm^{-1}] was isolated (294).

The reactions of mer-[OsCl$_3$(PMe$_2$Ph)$_3$] with Me$_2$dtc$^-$, MePS$_2^-$, Ph$_2$PS$_2^-$, and EtXant have been studied, and some of the products isolated included (297) cis-[Os(PMe$_2$PH)$_2$(Me$_2$dtc)$_2$], mer-[Os(PMe$_2$Ph)$_3$(Me$_2$dtc)], fac-[OsCl(PMe$_2$Ph)$_3$(Me$_2$dtc)], fac-[Os(OEt)(PMe$_2$Ph)$_3$(Me$_2$dtc)], and [Os(PMe$_2$Ph)$_2$(Me$_2$dtc)(EtXant)]. From measurements of the ionic products of the sparingly soluble R$_2$dtc (R = Et, Bu^n, or Bz)·Os^{IV} complexes [OsO$_2$(R$_2$dtc)$_2$], it has been concluded that these species are of high stability (298).

F. Cobalt, Rhodium, and Iridium

The 1,1-dithio complexes of cobalt have been studied extensively (1). Most of the literature is concerned with Co(III) compounds, as the complexes with divalent cobalt are extremely air-sensitive and have only been synthesized in acidic solution under rigorously deoxygenated conditions. The recent complexes of cobalt with 1,1-dithio ligands are listed in Table III, together with some of their physical properties.

Recently, Preti and co-workers (291) prepared Co(II) complexes with piperidine (pipdtc), thiomorpholine (Timdtc), and N-methylpiperazine-4-carbodithioate (Me-Pzdtc) (**XX**).

$$\begin{array}{c} H_2C-CH_2 \quad\quad S \\ X \quad\quad N-C \\ H_2C-CH_2 \quad\quad S \end{array}$$

X = CH$_2$-pipdtc;
X = S-Timdtc;
X = NCH$_3$-Me-pzdtc.

(**XX**)

TABLE III
1,1-DITHIOCHELATES OF $Co^{a,b}$

Compound	Melting point (°C) (d = decomp)	Color
[Co(pipdtc)$_2$]	350	Green
[Co(Me-Pzdtc)$_2$]	304	Green
[Co(Thiomorphdtc)$_2$]	350	Green
[Co(C$_6$H$_8$NS$_2$)$_2$]	350d	Brown
[Co(Morphdtc)$_3$]	—	—
[Co(EtSxant)$_3$]	—	—
[Co(PrSxant)$_3$]	—	—
[Co(BuSxant)$_3$]	—	—
[Co(Et$_2$NCH$_2$)$_2$HCS$_2$)]	Oil	Green
[Co(Et$_2\overset{+}{N}$HCH$_2$)$_2$NCS$_2$)]Cl$_6$	Decomp.	Green
(Et$_4$N)$_2$[Co(C$_3$H$_4$CS$_2$)$_2$]	—	Brown
[Co(Me$_2$dtc)$_3$]BF$_4$		
[Co(Et$_2$dtc)$_3$]BF$_4$		
[Co(Pr$_2^i$dtc)$_3$]BF$_4$		
[Co(cyclohexyl$_2$dtc)$_3$]BF$_4$		Green
[Co(MeBzdtc)$_3$]·2 H$_2$O	—	—
[Co$_2$(Me$_2$dtc)$_5$]BF$_4$	—	—
[Co$_2$(Et$_2$dtc)$_5$]BF$_4$	—	—
[Co$_2$(Bz$_2$dtc)$_5$]BF$_4$	—	—
[Co$_2$(pyrrolidyldtc)$_5$]BF$_4$	—	—
[Co$_2$(MeBu^{n}dtc)$_5$]BF$_4$	—	—
(Ph$_2$I)$_3$[Co(S$_2$CNCN)$_3$]	72d	Yellow-green
Co[Co(S$_2$C=CHNO$_2$)$_2$]	—	—
(Bu$_4$N)$_2$[Co(S$_2$C=C(CN)$_2$]	157–164	—

[a] D. H. Brown and D. Venkappaya, *J. Inorg. Nucl. Chem.* **35**, 2108 (1973).
[b] P. Thomas and O. Poveda, *Z. Chem.* **10**, 153 (1971).

These complexes are stable (presumably to aerial oxidation), which is, perhaps, surprising, in view of the fact that the analogous complexes of iron(II) could not be isolated, oxidation to iron(III) taking place in each case. A number of interesting poly(bisdithiocarbamate) complexes of cobalt(II) and nickel(II) (**XXI**) have been prepared and

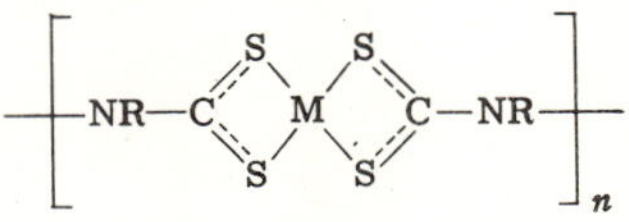

$$\left[-NR-C\begin{smallmatrix}S\\ \\S\end{smallmatrix}M\begin{smallmatrix}S\\ \\S\end{smallmatrix}C-NR- \right]_n$$

M = Co, or Ni; R = (CH$_2$)$_2$,
(CH$_2$)$_6$, or (C$_6$H$_4$–C$_6$H$_4$)

(**XXI**)

their IR spectra studied (*299*). In addition, some suggestions concerning the nature of the bonding in cobalt(II) and nickel(II) xanthates have been made (300).

The 1,1-ethylenedithiolate ligands appear to stabilize cobalt(II) species. Thus, Lewis and Miller (113) synthesized $Co[Co(S_2C=CHNO_2)_2]$ and $(Bu_4N)[Co(S_2C=C(CN)_2)]$, and found them to be active insecticides. The disodium salt of cyclopentadienedithiocarboxylic acid reacts with $CoBr_2$ in anhydrous acetonitrile, from which can be isolated (301) $[Et_4N]_2[Co(C_5H_4CS_2)_2]$, of D_{2h} symmetry. The dinegatively charged 1,1-dithiolate ligands are probably more effective in stabilizing a cobalt(II) complex than the uninegative dialkyldithiocarbamate ligands. Cobalt nitrosyl complexes have been known for some time (302). A three-dimensional structure determination of the complex $[Co(NO)(Me_2dtc)_2]$, and an improved synthesis have been reported (303). The coordination geometry about the cobalt atom is tetragonal pyramidal, with the NO group at the apex, and the Co atom situated 52 pm "above" the basal plane of the four sulfur atoms. The NO group is disordered in such a way that the oxygen atom alternatively lies above one or other of the Co–S bonds of the same dithiocarbamate ligand. The reaction of this complex with oxygen in the presence of pyridine or a phosphine (B) resulted (304) in the formation of complexes of the type $[Co(Me_2dtc)(NO_2)(B)]$. From kinetic data, the following mechanism was proposed.

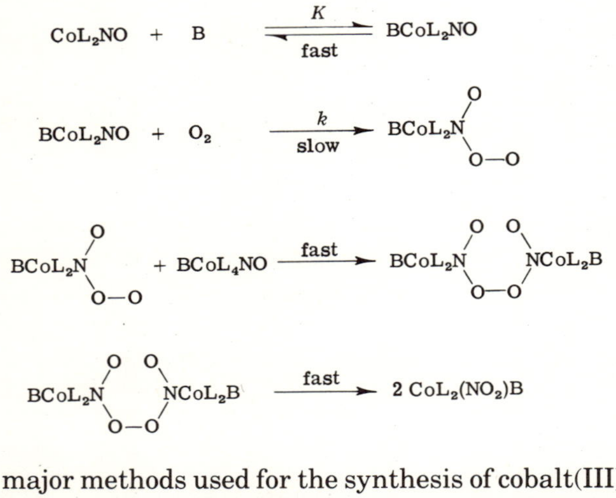

$$CoL_2NO + B \underset{\text{fast}}{\overset{K}{\rightleftharpoons}} BCoL_2NO$$

$$BCoL_2NO + O_2 \xrightarrow[\text{slow}]{k} BCoL_2N\begin{matrix}O\\ \diagdown\\ O\!-\!O\end{matrix}$$

$$BCoL_2N\begin{matrix}O\\ \diagdown\\ O\!-\!O\end{matrix} + BCoL_4NO \xrightarrow{\text{fast}} BCoL_2N\begin{matrix}O\quad O\\ \diagdown\;\diagup\\ O\!-\!O\end{matrix}NCoL_2B$$

$$BCoL_2N\begin{matrix}O\quad O\\ \diagdown\;\diagup\\ O\!-\!O\end{matrix}NCoL_2B \xrightarrow{\text{fast}} 2\, CoL_2(NO_2)B$$

The two major methods used for the synthesis of cobalt(III) dithiocarbamates are (a) treatment of a cobaltous salt with aqueous NaR_2dtc in the presence of air, or (b) oxidation of a cobalt(II) salt with tetraalkyldithiuram disulfides. In recent years, a complex with morpholine-4-carbodithioate (Mdtc), $[Co(Mdtc)_3]$, has been prepared and characterized (*44*); the reaction involved formation of an adduct with the parent base, morpholine, followed by CS_2 insertion. Complexes of Ti(IV), VO^{2+}, Cr(III), Mn(III), Ni(II), Cu(I), Zn(II), and Cd(II) were prepared by the same method.

The alleged preparation of the supposed cobalt(II) complex $Na[Co(Et_2dtc)_3]$ described by D'Ascenzo and Wendlandt (*305*) has been repeated by Holah and Murphy (*306*), who identified the product as $[Co(Et_2dtc)_3]$. Complexes of cobalt(III), nickel(II), and palladium(II) salts with cationic, dithiocarbamate ligands have been synthesized (*307*). Reaction of the secondary amine $(Et_2N(CH_2)_2)_2NH$ with CS_2 produces

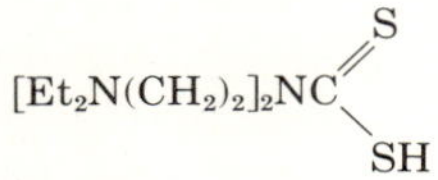

$$[Et_2N(CH_2)_2]_2NC\diagup^{S}_{SH}$$

(L), the IR spectrum of which indicated that it exists in a zwitterionic form. Thus, reaction of L with one mole, or two moles, or protonic acid produces L^+ and L^{2+}, respectively (**XXII**).

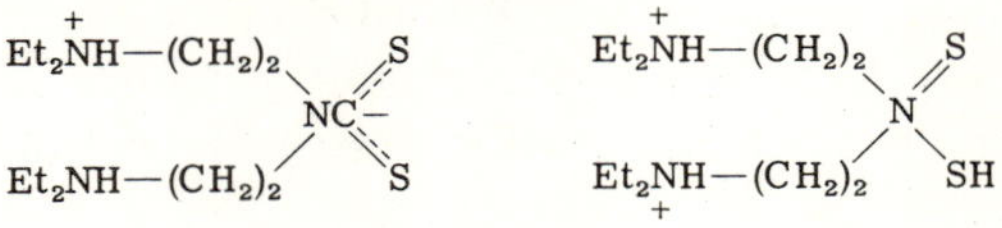

$$\begin{array}{cc}
Et_2\overset{+}{N}H-(CH_2)_2 \diagdown \quad \diagup S & Et_2\overset{+}{N}H-(CH_2)_2 \diagdown \quad \diagup S \\
\quad\quad NC^- & \quad\quad N \\
Et_2\overset{+}{N}H-(CH_2)_2 \diagup \quad \diagdown S & Et_2\overset{+}{N}H-(CH_2)_2 \diagup \quad \diagdown SH \\
(L^+) & (L^{2+})
\end{array}$$

(**XXII**)

Reaction with one mole of base produces the dithiocarbamato anion, L^- (**XXIII**).

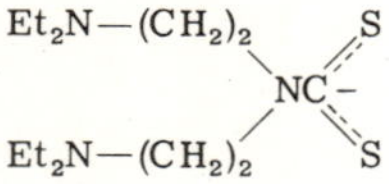

$$\begin{array}{c}
Et_2N-(CH_2)_2 \diagdown \quad \diagup S \\
\quad\quad NC^- \\
Et_2N-(CH_2)_2 \diagup \quad \diagdown S \\
(L^-)
\end{array}$$

(**XXIII**)

Complexes $[Co(L^-)_3]$ and $[Co(L^+)_3]Cl_6$ have been obtained. The presence of a positive charge on the dithiocarbamato ligand has little effect on the coordination properties of the ligands or complexes, except for an alteration of solubility characteristics. It is surprising that work on these zwitterionic, sulfur-containing ligands does not appear to have been extended, as they are potential, biological-model systems. The complexes $[Co(RSXant)_3]$ (R = Me, Et, Pr^n, Bu^t, and Bz) and $[Co(RSXant)_2(SR)]_2$ (R = Et, or Pr^n) have been reported (*308*), and the elimination of CS_2 from the tris(thioxanthato) complexes to give the dimers has been directly confirmed, and the reaction found to be first-order. The structure of the solid ethyl thioxanthato dimer was found to be centrosymmetric, with bridging SEt groups, and ^{1}H-NMR-spectral data indicated that this structure also exists in solution. The structure of tris(ethylthioxanthato)cobalt(III)·Co(S$_2$CSEt)$_3$ shows Co–S (mean) = 226.6 pm, and LS–Co–S (mean) = 76.2°. The discrepancy between these results (309) and those previously reported (310) led Li and Lippard (309) to suggest that the earlier determination had, in fact, been performed on the chromium(III) analog. Villa et al. (311) also redetermined the structure of the compound, and found it to be isostructural with the chromium(III) complex, and the apparent discrepancies between the previously reported Co–S distance and the spectroscopic data have been explained. The single-crystal structure of the arylxanthate complex of cobalt(III), [Co(S$_2$COC$_6$H$_2$-2,4,6-Me$_3$)$_3$], shows that the pseudooctahedral CoS$_6$ unit has (312) a mean Co–S distance of 227.0(6) pm. The structure is quite similar to that of the aliphatic analog [Co(S$_2$COEt)$_3$] reported by Merlino (442).

Cobalt(III) complexes containing mixed chelating ligands have been produced. Reaction of potassium bis[biuretocobaltate(III)], K$_2$[Co(bi)$_2$] with R$_2$dtc$^-$ or Rxant$^-$ at 0° produces (313) the blue–violet [Co(bi)$_2$(S—S)]$^{2-}$ ion (S—S = R$_2$dtc or RXant). If the reaction is performed above 0° in the presence of water, the products are [Co(bi)$_2$(S—S)$_2$]$^-$ and biuret.

A number of complexes of cobalt(III) with 1,1-dithiolato ligands are known. Since 1969, however, only the isolation and reactions of the first dithiocarbimate complex of cobalt(III), [Ph$_2$I]$_3$[Co S$_2$CN(CN)$_3$], the diphenyliodonium tris(N-cyanodithiocarbimato)cobaltate salt, has been reported (314). Pyrolysis of this compound showed that the reactive sites in the anion are the coordinated sulfur atoms of the same ligand. Partial reduction of the metal ion to cobalt(II), and liberation of PhS$_2$ and PhI, occur simultaneously. In pyridine, ligand phenylation occurs, but the arylated ligand decomposes. In pyridine, nonarylated ligands dissociate into thiocyanate, accompanied by reduction to cobalt(II).

Several complexes with cobalt in the unusually high oxidation state of (+4) were reported (198, 202, 280) in 1974. All of the complexes reported were prepared by reaction of [Co(R$_2$dtc)$_3$] with BF$_3$ or Et$_2$OBF$_3$ in the presence of air. The complexes were formulated (202, 280) as [Co(R$_2$dtc)$_3$]BF$_4$ (R = Me, Et, Pr^i, or cyclohexyl), but Hendrickson and Martin (198) suggested that dimeric [Co$_2$(R$_2$dtc)$_5$]BF$_4$ (R$_2$ = Me$_2$, Et$_2$ pyrrolidyl, MeBu^n, or Bz$_2$), Co(III) complexes, form that are analogous with the ruthenium(III) complex discussed earlier.

The dynamic stereochemistries of M(dtc)$_3$ and [M(dtc)$_3$] (M = Fe, Co, or Rh) complexes have been studied (315). The cobalt complex is non-rigid, but the mechanism of optical inversion could not be determined. The Rh complex is stereochemically rigid up to 200°. The optical inversion of (+)$_{546}$ [Co(pyr-dtc)$_3$] in chloroform has been studied, by loss of optical activity, by polarimetry (316).

The diene complex [(diene)Rh(dtc)] [diene = (C$_8$H$_{14}$)$_2$, or 1,5-cod] is oxidized in the presence of an excess of dtc to [Rh(dtc)]. One of these ligands can be replaced (317) by HCl, to give [Rh(dtc)$_2$Cl]$_2$. The preparation of other [Rh(dtc)$_3$] complexes derived from cyclic amines has also been reported (318). Treatment of [(cod)Rh(MeCN)$_2$]$^+$ with [NO]PF$_6$ yields green [Rh(NO)(MeCN)$_4$](PF$_6$)$_2$. The dtc complexes were found to exhibit low N–O stretching-frequencies, e.g., 1545 cm^{-1} for [Rh(NO)(Me$_2$dtc)$_3$]PF$_6$ (319).

The literature concerning the preparation of 1,1-dithio complexes of rhodium and iridium is relatively sparse. Wilkinson and co-workers (320, 321) reported that replacement of chloro groups in tertiary phosphite complexes of rhodium with various alkali-metal dithio-acid salts produces [Rh(Me$_2$dtc)$_3$], [Rh(R$_2$dtc)$_3$(PPh$_3$)] (R = Me, or Et), [Rh(Me$_2$dtc)(PPh$_3$)$_2$], [Rh(Me$_2$dtc)(CO)PPh$_3$], [Rh(Me$_2$dtc)$_3$(CO)PPh$_3$], and (Rh(Et$_2$dtc)$_2$(PPh$_3$)$_2$]BF$_4$. [Rh(CO)(PRh$_3$)(Me$_2$dtc)$_3$] contains two monodentate and one bidentate dithiocarbamate ligands, and [Rh(PPh$_3$)(Me$_2$dtc)$_3$] contains one monodentate ligand and two bidentate ligands. Recently Cole-Hamilton and Stephenson produced a series of new dithiocarbamate and xanthate complexes by the same route (322), and postulated a detailed mechanism for the overall reaction.

The crystal and molecular structure of K[RhCl$_2$(S$_2$CO)(PMe$_2$PH)$_2$]·3H$_2$O show that there are two independent anions per unit cell, with crystallographic C$_i$(I) symmetry. They are essentially identical, and the planar dtc ligands have (323) Rh–S = 237 pm, C–S = 172.5 pm, C–O = 125 pm; S–Rh–S = 73.5°, S–C–S = 111°, and S–C–O = 125°.

The new air-stable complex [Rh(PPh$_3$)(Et$_2$dtc)$_2$]BF$_4$ was prepared (324) by treating Rh$_2^{4+}$ with Et$_2$dtc$^-$ and PPh$_3$. The rhodium(IV) complexes [Rh(R$_2$dtc)$_3$]BF$_4$ were prepared during the work with the cobalt(IV) and ruthenium(IV) complexes (see earlier). Rhodium carbonyl derivatives containing dithiocarbamate ligands have been prepared by Abel and Dunster (179).

The first report (325) of iridium(III) dithiocarbamate complexes appeared in 1973. The procedure is straightforward, consisting in addition of an excess of NaR$_2$dtc to sodium chloroiridate in aqueous methanol. On being kept for several days, an orange–yellow precipitate forms. Eleven complexes (R = Me, Et, Pr^n, Pr^i, Bu^n, Bu^u, amyl, cyclohexyl, pyrrole, piperidine, and $EtOH_2$) were isolated in this way. Reaction of $[Ir(CO)(PR_3)Cl]$ with $\{(CF_3)C=CS_2\}_2$, and subsequent isolation of the product, $[Ir(CO)(PR_3)_2(Cl)\{S_2C=C(CF_3)_2\}]$, represent the first reported preparation of an iridium or rhodium complex with a 1,1-dithiolene ligand (326). The crystal structure of $[Ir(Et_2dtc)_3)]$ shows the complex to be isostructural (327) with the cobalt(III) analog having a mean Ir–S of 236 pm.

G. Nickel, Palladium, and Platinum

The chemistry of complexes of members of the nickel triad with 1,1-ligands has been studied extensively (1).

Nickel 1,1-dithio complexes with the metal in the unusually low oxidation state of +1 have been prepared by Garif'yanov and Luchkina (328), who studied the esr spectra of $[Ni(NO)(S_2CX)]$ (X = NEt_2, or OEt), and observed a signal, g = 2.103, in toluene solution. Using ^{61}Ni-enriched samples, hyperfine structure was observed, and the structure is probably square or tetragonal pyramidal. $[\pi\text{-}CpNi(PBu_3)_2]Cl$ reacts with RCS_2^- (R = Ph, Et, or Bz), to form $[\pi\text{-}CpNi(PBu_3)_2(SC(S)R)]$, which, on treatment with HCl, yields (329) $[\pi\text{-}CpNi(SCR)]$. The complex is a Ni(I) species, and was found to revert to the intermediate complex on treatment with PBu_3 in hexane.

The nickel(II) dithiocarbamate complexes are neutral, water-insoluble, usually square-planar, species, and they have been studied extensively by a range of physical techniques. The usual methods for the synthesis of dithiocarbamate complexes have been employed in the case of Ni(II), Pd(II), and Pt(II). In addition, McCormick and co-workers (330, 331) found that CS_2 inserted into the Ni–N bonds of $[Ni(aziridine)_4]^{2+}$, $[Ni(aziridine)_6]^{2+}$, and $[Ni(2\text{-methylaziridine})_4]^{2+}$ to afford dithiocarbamate complexes. The diamagnetic products are probably planar, but they have properties typical of dithiocarbamate complexes, and IR- and electronic-spectral measurements suggested that they may be examples of N,S-, rather than S,S-, bonded dithiocarbamates. The S,S-bonded complexes are however, obtained, by a slow rearrangement in methanol. The optically active N-alkyl-N(α-phenethyl)dithiocarbamates of Ni(II), Pd(II), and Cu(II) (**XXIV**) have been synthesized, and the optical activity was found to be related to the anisotropy of the charge-transfer transitions (332).

An interesting, halogen-exchange reaction takes place when $[Ni(Et_2dtc)_2]$ is refluxed in 1,2-dichloroethane with an excess of α,α'-dibromo-o-xylene and α-bromo-α'-chloro-o-xylene. The products, α,α'-

M = Cu, Ni, or Pd; R = H, Et, Pr^n, or Bz; and
* denotes optically active carbon atom.

(XXIV)

dichloro-o-xylene and some α-bromo-α'-chloro-o-xylene, were not found, however, in the absence of [Ni(Et$_2$dtc)$_2$], or when [Ni(Etxant)$_2$] was used. Red–violet [Ni(Et$_2$dtc)$_2$Br(Bu$_3^n$P)] and dark-violet [Ni(Et$_2$dtc)$_2$Cl(PPh$_3$)] have been prepared by reaction of the respective nickel halide, phosphine, and NaEt$_2$dtc·3H$_2$O in ethanol (333).

Resonance-Raman spectra of Cu(II) and Ni(II) diethyldithiocarbamates have been reported recently (334). Both spectra show the preferential intensity enhancement of four bands in the M–S stretching- and SCS, SCN, MSC, and SMS bending-regions at 367–157 cm^{-1}. A thermodynamic study of the addition of pyridine or 4-Me-pyridine to bis(penthiobenzato)nickel(II) and [Ni(Bz$_2$dtc)$_2$] in C$_6$H$_6$ solution showed that the variation of base-adduct stabilities of the NiS$_4$ complexes with four-membered chelate rings is due primarily to entropy effects (335). A correlation between isotropic contact shifts and Taft σ* values for some pyridine-base adducts of nickel(II) bis(o-alkyldithiocarbonates) and bis(β-diketonates) has been found (350). Several X-ray crystallographic, structural determinations of nickel(II) dithiocarbamate complexes have been performed in recent years (236, 244). Thus, the structures of [Ni(R^{1}R^{2}dtc)$_2$] (R^1 = R^2 = Me (344), Pr^i (343), Pr^n (135, 137), H (336); R^1 = Me, R^2 = Ph (341); R^1 = H, R$_2^2$, = Pr^i (343); R^{1}R^2 = (CH$_2$)$_4$ (342); and R^{1}R^2 = (CH$_2$)$_6$ (338) have been determined. Deviations from planarity in the coordination plane were explained by crystal packing and hydrogen-bonding interactions (344, 336). The structure of cis-bis(diisopropyldithiocyanate)nickel(II) is the first example of the cis configuration in complexes of this kind. The adoption of a cis stereochemistry is due to the strong hydrogen-bonding between the HN hydrogen atom and the sulfur atoms in the neighboring molecules (343). Fackler et al. (345) recently prepared three classes of unsymmetrical dithiocarbamate complex of the nickel triad elements: (i) M(R$_2$dtc) (PR$_3$)X (M = d, or Ni; R = Bu^i, or Et; R = alkyl, or aryl; X = Cl, Br, I, SCN, or SR); (ii) Pt(R^{1}R^{2}dtc)(PMe$_2$Ph)$_2$X (R^1 = Me; R^2 = Ph; X = Cl,BPh$_4$, or PF$_6$); and (iii) (R^{1}R^{2}dtc)$_3$. In the first two classes, magnetic nonequivalence can be observed in the ^{1}H-NMR spectra of the liquids, whereas, in the third class, both cis and trans isomers are observed. For the Ni(II) compounds, partial halide exchange appears to be responsible for the observation of magnetically equivalent R groups in structurally nonsymmetric dtc complexes in the presence of free phosphine, or upon increase in temperature.

The nickel(II) xanthate complexes are diamagnetic, square planar species that are slightly soluble in water, the solubility decreasing with increasing size of the hydrophobic alkyl groups. The position of the $^1A_g \rightarrow {}^1B_{1g}$ transition in these complexes indicates that the thioxanthate ligands have smaller ligand field-splitting potentials than the corresponding xanthates (1).

The only new complexes of Ni(II) containing only xanthate ligands to have been reported in the recent literature include (160) the aryl xanthate complex $[Ni(S_2COC_6H_4\text{-}4\text{-}tert\text{-}Bu)_2]$. The palladium(II) analogs $[Pd(S_2COC_6H_2\text{-}2,4,6\text{-}Me_3)_2]$ have also been studied by single-crystal, X-ray crystallography. The coordination about both the Ni(II) and Pd(II) complex is planar, with the phenyl ring approximately perpendicular to this plane (160). Only one report of the formation of thioxanthate complexes of Ni(II) has appeared (256). $[Ni(S_2CSR)_2]$ (R = Et, Pr, or Bu) were prepared by reaction of S_2CSR^- with a nickel(II) salt in ethanol. Attempts to crystallize the ethyl complex from $CHCl_3\text{-}Et_2O$ gave a dimeric complex, $[Ni(S_2CSEt)(SEt)]_2$, which was found to contain two equivalent, square planar NiS_4 species; the -SEt groups bridge the two nickel atoms (**XXV**). The short Ni–Ni distance (276 pm) may indicate the presence of the first example of metal–metal bond in a binuclear Ni(II) complex. A kinetic study of the spontaneous elimination of CS_2 to afford $[M(S_2CSR)(SR)]_2$ (M = Ni^{II}, Pd^{II}, or Pt^{II}) and $[M(S_2(CSR)_2(SR)]_2$ [M = Co(III), or Fe(III)] was undertaken by Fackler and co-workers (346, 347). Radioactive labeling of the mercaptide sulfur atom of the thioxanthate complexes showed that it forms the bridge in the dimeric complexes.

(XXV)

The majority of the work on xanthates of divalent nickel has, in recent years, been centered on the formation of base adducts with $[Ni(Rxant)_2]$. Carlin and Siegel (348) and Daktenieks and Graddon (349) reported the formation of paramagnetic $[Ni(Etxant)_2B_2]$ or $[Ni(Et_2xant)_2B]$, where B = pyridine, 4-methylpyridine, bipyridyl, or

2,9-dimethyl-1,10-phenanthroline. An additional study showed that [Ni(EtXant)$_2$] forms 1:2 adducts with pyridine and 4-picoline in nitrobenzene solution (*350*). Furthermore, Carlin and Siegel (*348*) observed that pyridine derivatives lacking an α-substituent form only the [Ni(Etxant)$_2$B$_2$] adduct, but α-substitution led to steric hindrance at the coordination site. Triphenylphosphine and other bulky bases formed only a monoadduct [Ni(Etxant)$_2$B]. Steric hindrance and the amount of electron density on the metal ion were found to have important effects on the thermal stability of the adducts (*351*). The effect of replacing pyridine by 4-methylpyridine is to increase the stability constant of the adduct about fivefold; this is due to the greater base-strength of 4-methylpyridine. 2,2′-Bipyridine forms a bidentate-bound adduct, and the large adduct-formation constants (K > 10^5) shows a typical, chelate effect (*352*). Base-adduct formation in nitrobenzene was also studied by ^{1}H-NMR spectroscopy (*353*), and the equilibrium constant for the reaction

$$[\text{Ni(Etxant)}_2] + 2\text{ B} \overset{K}{\rightleftharpoons} [\text{Ni(Etxant)}_2\text{B}_2]$$

may be related to the NMR-spectroscopic parameters $\Delta\nu_0$ (observed isotropic shift in the presence of Lewis base), $\Delta\nu_p$ (isotropic shift of 1:2 adducts), and α ($= \Delta\nu_0/\Delta\nu_p$, the degree of interaction)

$$K = [\{\text{Ni(Etxant)}_2\text{B}_2\}]/[\{\text{Ni(Etxant)}_2\}][\text{B}]^2 = \alpha/(1-\alpha)([\text{B}] - 2\ \alpha[\{\text{Ni(Etxant)}_2\}])^2$$

The equilibrium constants thus determined were found to be smaller than those previously reported, no doubt due to the use of a polar solvent. In three papers by Kruger and Winter (*354–356*), two polymorphs of the adduct of [Ni(Etxant$_2$] with 4,4′-bipyridine were reported, one of which is capable of forming clathrate compounds by trapping a molecule of cyclohexane, C$_6$H$_6$, Et$_2$O, Me$_2$CO, CHCl$_3$, CCl$_4$, or cyclopentadiene. The 1,10-phenanthroline adduct also forms 1:1 clathrates with C$_6$H$_6$, C$_6$H$_3$Me, CHCl$_3$, and C$_6$H$_5$Cl. The inclusion compounds are formed from solutions of the respective solvents, or from solvent vapors. The same workers also successfully isolated the first two adducts of [Ni(Etxant)$_2$] with oxygen as the donor atom. Bright-green, 2:1 adducts of [Ni(Et$_2$xant)$_2$] with DMSO and pyridine N-oxide were isolated. A most interesting reaction that takes place in the thermal decomposition of nickel xanthates is the formation of olefins (*357*).

The chemistry of the dithiocarboxylate complexes of nickel(II) has been investigated extensively. Interest in recent years has been mainly in the further investigation of the "sulfur-rich" species, the perthiocarboxylates, and the unusual structures discovered in the dithiocarboxylate complexes (*359*). The violet complex formed by reaction of [Ni(C$_6$H$_5$CS$_2$)$_2$] with sulfur or polysulfide, [Ni(C$_6$H$_5$CS$_2$)$_2$]$_2$, originally formulated as a disulfide-bridged, nickel(II) dimer, has been reformulated (358) as the monomeric perthiocarboxylate complex on the basis of physical and chemical similarities with the analogous bis(dithiocumato)nickel complex (**XXVI**).

(XXVI)

The large dihedral angle between the plane of the "sulfur-rich," chelate ring and that of its attached benzene ring is striking, and suggests a decreased, π-electron interaction in this ligand compared with the dithiocumate ligand bonded to the same nickel atom. Photochemical reaction of sulfur with [Ni(S$_2$CC$_6$H$_4$CHMe$_2$)$_2$] produces the same complex (**XXV**). Treatment of the "sulfur-rich" species with Ph$_3$P results in specific removal of the sulfur atom adjacent to the carbon atom in the Ni–S–S–C linkage (360). This had been observed in a mass-spectrometric study of [Ni(S$_2$CPh)(S$_3$^{34}SCPh)], which was formed (361) from the addition of ^{34}S-enriched sulfur to [Ni(S$_2$CPh)$_2$]. Spectrophotometric and NMR-spectroscopic, kinetic studies of "sulfur-rich" dithioarylates of Ni(II), Pd(II), Pt(II), and Zn(II) indicated S-atom mobility in these complexes (362). The kinetic data for the reaction of PPh$_3$ with [Ni(S$_2$CC$_6$H$_4$CMe$_2$)(S$_3$CC$_6$H$_4$CMe$_2$)] are consistent with a mechanism that involves kinetically controlled attack of PPh$_3$ on the complex at the sulfur atom adjacent to the carbon atom in the disulfide linkage.

A crystal-structure determination on [Ni(PhCH$_2$CS$_2$)$_2$] showed evidence of a Ni–Ni bond (Ni—Ni distance, 256 pm) in a bridging, acetate-cage, binuclear complex (363). Each nickel atom is 5-coordinate and is in a tetragonally distorted, square-pyramid; spectroscopic evidence for a Ni–Ni bond has been obtained (364). The polarized crystal spectra showed more bands than predicted for a mononuclear, diamagnetic, square-planar nickel(II), and the spectra are indicative of substantial overlap of the d-orbitals between the two nickel atoms. The bis(dithiobenzation)nickel(II) complex was found to exhibit unusual spectrochemical behavior (365).

In nickel and palladium dithiobenzoato complexes, four-membered chelate rings are formed (366), whereas, in the corresponding phenyldithio acetates [M$_2$(S$_2$CCH$_2$Ph)$_4$], the dithio ligands act as bridging groups between the two metal atoms, with the formation of binuclear units (367). The molecular structure of the latter compounds shows that each metal atom is coordinated to four sulfur atoms and to the other metal atoms in a square-pyramidal geometry. Other evidence for a metal–metal bond was obtained from a comparison of the metal–sulfur distances with those in nickel and palladium square-planar complexes having comparable bond-angles about the sulfur and the metal. The bis-pyridine adduct, [Ni(PhCH$_2$·CHS$_2$)$_2$(py)$_2$] has been prepared, and the X-ray structure showed monomeric units consisting of distorted octahedrons about the nickel atoms, with the pyridine molecules cis-coordinated (368). This is in contrast to the nickel–nickel-bonded bridging of binuclear structures found (363) in tetrakis(phenyldithioacetate)dinickel(II).

The infrared spectra have been reported for polycrystalline samples of [Ni(S$_3$CPh)$_2$] and [Zn(S$_3$CPh)$_2$]. The nickel compound has two molecules per unit cell on sites of C$_i$ symmetry, and assignment of a weak absorption at 655 cm^{-1} to ν(S–S) has been made (369).

An X-ray crystal-structure determination (370) revealed that mixed, nickel coordination exists in the trimeric complex, [Ni(S$_2$CPh)$_2$]$_3$. The trimeric structure is similar to that found (371) in [Pd(PhCS$_2$)]. The structure contains one molecule of type A linked centrosymmetrically through short Ni–S bridges to two molecules of type B, the three molecules being closely parallel (**XXVII**). Bonamico and co-workers (372)

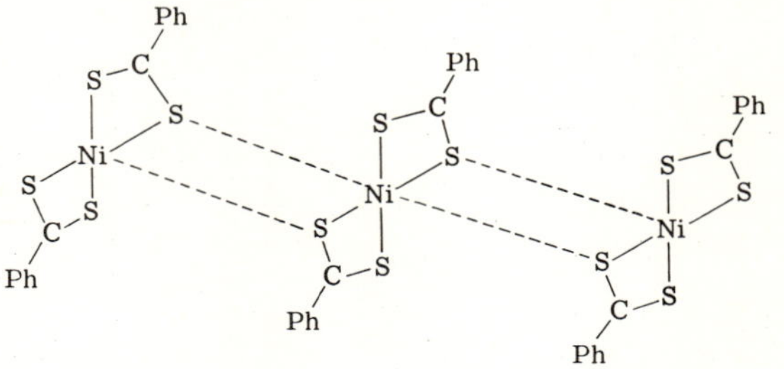

(XXVII)

also determined the crystal structure of bis(dithiopivalato)nickel(II), [Ni(C$_5$H$_9$S$_2$)$_2$].

Nickel(II) complexes with most of the 1,1-dithiolate ligands are known. Recently, the new complexes, K$_2$[Ni(S$_2$C=CHNO$_2$)$_2$] (113, 373), [Ag(PPh$_3$)$_2$]$_2$[Ni(i-mnt)$_2$] (374), and (Ph$_2$BzPh)$_2$[Ni(DED)$_2$] (375) have been prepared and studied. The complex [Ph$_2$I]$_2$[Ni(S$_2$CNCN)$_2$] was prepared as part of a study of iodonium salts of complex anions (376). Pyrolysis phenylated the two sulfur atoms of the same ligand, as verified by the isolation and identification of N-cyanophenyldithiocarbimate, (PhS)$_2$CNCN. As a result of phenylation, the configuration about Ni changed from planar to octahedral, resulting in a polymeric species.

Rate constants for the substitution reactions of square-planar dithiophosphates and dithiocarbonate complexes of Ni(II), Pd(II), and Pt(II), with ethylenediamine and cyanide ion as nucleophiles, have been measured in methanol. The results were compared with those obtained in previous investigations, and interpreted in terms of the stabilities of 5-coordinate species that are formed prior to substitution (377).

The preparation and chemistry of nickel trithiocarbonate complexes have been studied in detail, and both $[Ni(CS_3)_2]^{2-}$ and $[Ni(CS_4)_2]^{3-}$ have been isolated (379). Shul'man and co-workers (379) reprepared the known $[Ni(NH_3)_3(CS_3)]$ and $[Ni(en)_3]CS_3$, and the kinetics of the reaction of Ni(II) with the trithiocarbonate ion in methanol was studied; the results confirmed the previously proposed ion-pair mechanism (380).

$$Ni^{2+} + CS_3^{2-} \rightarrow NiCS_3$$

$$NiCS_3 + CS_3^{2-} \rightarrow Ni(CS_3)_2^{2-}$$

$$Ni(CS_3)_2^{2-} + CS_3^{2-} + 1/2\ O_2 \rightarrow Ni(CS_3)(CS_4)^{2-} + COS_2^{2-}$$

A re-examination of $[Ni(CS_3)(NH_3)_3]$ showed (381) it to be $[Ni(NH_3)_6][Ni(CS_3)_2]$. An interesting, dimeric nitrosyl complex containing a bridging CS_2^{2-} group (**XXVII**) was prepared in 100% yield by Brunner (382) by prolonged (30 h) reaction of $BaCS_3$ with $[(Ph_3P)_2Ni(NO)I]$ in ethanol.

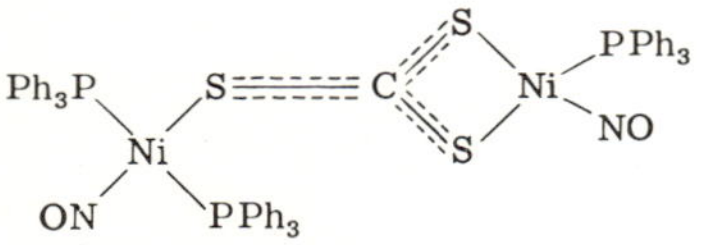

Ph$_3$P S══════ C ⟨S⟩ Ni ⟨PPh$_3$, NO⟩
     Ni
ON   PPh$_3$ S

(XXVIII)

The electronic spectra of a range of dithio- and perthiocarboxylato-nickel(II) complexes and their pyridine adducts show the presence of a variety of structures in solution, but complete interpretation of the spectra was prevented by lack of a complete MO treatment of these complexes (378).

Recent literature contains work in which complexes of nickel(IV) with 1,1-dithio ligands have been isolated. Brinkhoff (383) successfully oxidized $[Ni(Et_2dtc)_2]$ to $[Ni(Et_2dtc)_3]I_3$ with iodine, and detected a nickel(III) species in $[Ni(Etxant_2X)]$ (X = Cl, or Br) by epr spectroscopy during the oxidation of $[Ni(Etxant_2)]$ with chlorine and bromine. $[Ni(R_2dtc)_3]X$ (X = ClO_4^- and $FeCl_4^-$) is prepared by reaction of $[Ni(R_2dtc)_2]$ with $Fe(ClO_4)_3$ and $FeCl_3$, respectively (270) Solozhenkin and Kopitsaya (384) detected by epr a paramagnetic, nickel(III) dithiocarbamate species formed from oxidation of $[Ni(R_2dtc)_3]$ with tetramethylthiuram disulfide at the boiling point of benzene, no spectrum is obtained, owing to oxidation of Ni(III) to Ni(IV). Brinkhoff and coworkers (385) had previously reported a similar reaction, and the isolation of [Ni(Bu$_2$dtc)$_3$X], which they characterized as an octahedral Ni(IV) species. Structurally, the coordination in [Ni(Bu$_2$dtc)$_3$]$^+$ is similar to that in the [Ni(Etxant)$_3$]$^-$ anion, containing high-spin nickel(II). Nigo and co-workers (386) isolated [Ni^{IV}(Et$_2$dtc)$_2$]Br$_2$ by oxidation of [Ni(Et$_2$dtc)$_2$] with Br$_2$ in CS$_2$ solution. On the basis of spectroscopic and polarographic data, the complex was characterized as a very rare example of a low-spin, nickel(IV) species having a planar, chelate structure. [Ni(Bu$_2$dtc)$_2$] reacts with I$_2$ at $-30°$ in ether, to give Ni[Bu$_2$dtc)$_2$I], although the magnetic moment of the complex (1.33 B.M.) is somewhat low for low-spin nickel(III); epr evidence indicated square-pyramidal coordination (387). Very recently, Lachenal (388) and Hendrickson et al. (389) studied the oxidation and reduction processes for Ni(II) and Ni(IV) dithiocarbonate complexes. Lachenal (388) prepared [Ni(Et$_2$dtc)$_3$]BF$_4$ by an electrochemical method. The [Ni(R$_2$dtc)$_3$]$^+$ cations react with isocyanides or phosphines to form [NiL$_2$(R$_2$dtc)]$^+$ or [Ni(diphos)(R$_2$dtc)]$^+$. Reaction with PPh$_3$ produces (390) [Ni(R$_2$dtc)$_2$], Ph$_3$PS, [Ph$_3$PC(=S)NR$_2$]$^+$, and some [Ni(PPh$_3$)$_2$(R$_2$dtc)$_2$]$^+$.

McCleverty and Harrison recently studied (391) the reaction of [Ni^{IV}(dtc)$_2$]cations with various Lewis bases, and found that treatment of [Ni(R$_2$dtc)$_3$][PF$_6$] R = Et, or Bu^n) with Lewis bases [L = CNR^1 (R^1 = Pr^i, Bu^t, or p-C$_6$H$_4$Cl), p-MePH$_2$ or ½ dppe] gave [NiL$_2$(R$_2$dtc)]PF$_6$, and, when L = CNR^1, the thiuram disulfide was isolated. Reaction with PPh$_3$ yielded PPh$_3$S, [PPh$_3${C(NR$_2$)S}]PF$_6$ and Ni(dtc)$_2$, with the isolation of only very low yields of [Ni(PPh$_3$)$_2$(R$_2$dtc)][PF$_6$].

Considerably less is known about the chemistry of palladium and platinum 1,1-dithio complexes. Of late, there has been only one report that dealt with the synthesis of a large number of palladium dithiocarbamates (392). Twenty-five yellow palladium dithiocarbamate complexes were obtained by reaction of PdCl$_2$ with NaR$_2$dtc in methanol solution. Several other reports have appeared in which a few dithiocarbamate complexes of palladium were synthesized. Thus, the novel [Pd{(OH)$_2$dtc}$_2$], which is soluble in water, was isolated (393). The synthesis of optically active palladium(II) complexes of N-alkyl-α-phenethyldithiocarbamates, similar to (**XXIV**), via the reaction between the optically active amine, CS$_2$, and PdCl$_2$, has been described. From ORD and CD spectra, it has been established that the vicinal contribution of a remote, asymmetric carbon center could give rise to optical activity of the d—d transitions of palladium (394). Carbon disulfide has been shown to insert into the Pt–F bond of [PtF(PPh$_3$)$_3$]HF$_2$, and X-ray studies indicated the structure (**XXIX**).

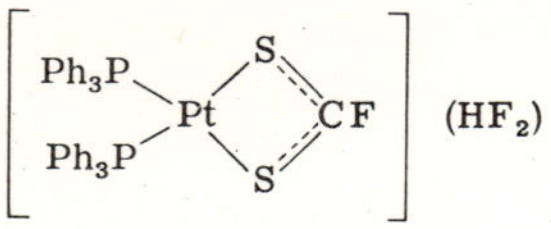

$$\left[\begin{array}{c} Ph_3P\diagdown\diagup S\diagdown \\ PtCF \\ Ph_3P\diagup\diagdown S\diagup \end{array}\right] (HF_2)$$

(XXIX)

The mixed complexes of the type [M(mnt)(R_2dtc)]⁻ (M = Pd, or Pt; R = Et, Bu, or Ph) have been prepared, and voltammetry in CH_2Cl_2 revealed that the complexes undergo one-electron oxidation; the values of the half-wave potentials are intermediate between those of the unmixed complexes (*395*). The halide-substituted complexes [MX$_2$(Bu$_2$dtc)$_2$] (M = Pd, or Pt; X = Br, or I) have been prepared by Willemse and co-workers (*396*) and the platinum complex was isolated as cis and trans isomers. ESCA measurements showed the trans isomer to be a Pt(IV) species. With tetrabutylthiuram disulfide, these complexes give [M(Bu$_2$dtc)$_3$]X. A novel dimeric complex (**XXX**) is formed (*397*) from the reaction of NaEt$_2$dtc with [Pt(CH$_3$)(π-1,5-cyclooctadiene)L]PF$_6$ (L = 4-Mepy, AsPh$_3$, PPh$_3$, *p*-NCC$_6$H$_4$OCH$_3$, or CNEt).

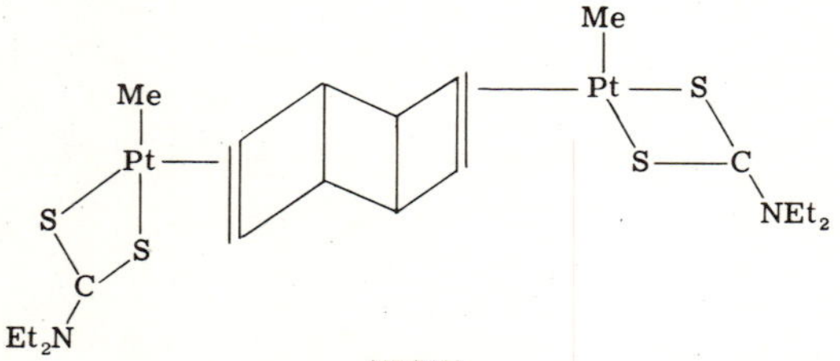

(XXX)

The reaction of [M(S—S)$_2$] complexes (S—S⁻ = R$_2$dtc, Rxant; M = Pd, Pt) with tertiary phosphines occurs by a stepwise cleavage of the M–S bonds to generate the 4-coordinate compounds [M(S—S)$_2$PR$_3$] and [M(S—S)(PR$_3$)$_2$](S—S), which respectively exhibit unidentate/bidentate and bidentate/ionic modes of bonding of the dithioacid groups (*398, 399*). Reactions of platinum(II) xanthates with the xanthate ion (*400*) produce a series of compounds, Ph$_4$As[Pt(Rxant)$_3$] (R = Et, or Pr^i), Ph$_4$As[M(S$_2$CO)(Rxant)] (R = Et, Pr^i, or Me; M = Pr, or Pd), and (Ph$_4$As)$_2$[Pt(S$_2$CO)$_2$], and the following scheme was proposed for the formation of the second type of complex.

Very few studies on complexes of palladium and platinum with 1,1-dithiolato ligands have been reported recently. The electronic absorption spectra of the Et$_2$dtc complexes of Pd and Pt, as well as Zn, Cu, Fe, Co, and Mn, have been measured in MeCN, EtOH, and *n*-heptane, with diffuse-reflectance spectra also being determined (*401*). In 1969 and

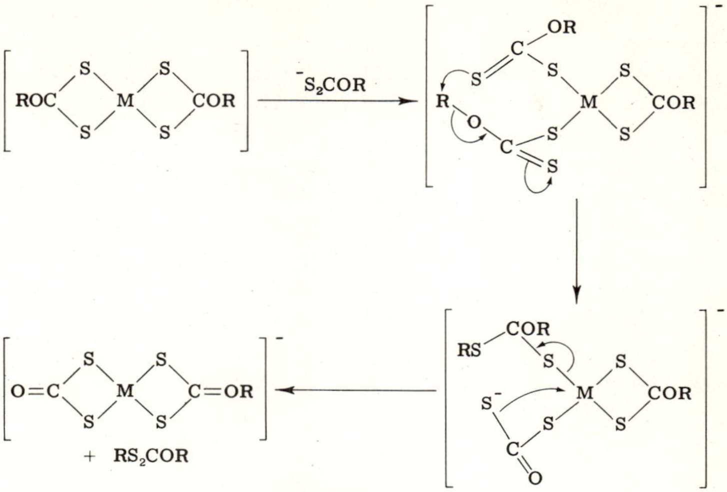

SCHEME 5

1970, the first, stable, transition-metal dithiocarbonate (CS_2O) complexes were reported (402, 403).

Reaction of the complexes [$PtS_2(COR)_2$] with K[S_2COR] (R = Et, or Pr^i), followed by the addition of [$AsPh_4$]Cl, generates [$AsPh_4$][$Pt(S_2COR)_3$]. Variable-temperature ^{1}H-NMR studies indicated rapid unidentate–bidentate exchange at ambient temperature. When the palladium(II) analog [$Pd(S_2COEt)_2$] reacts with K[S_2COEt] and [$AsPh_4$]Cl, the main product is [$AsPh_4$][$Pd(S_2COEt)$]. Reaction of [$Pt(S_2COR)_2$] with K[S_2COR] (R = Me, or CH_2Ph) and [$AsPh_4$]Cl generates [$AsPh_4$][$Pt(S_2CO)(S_2COMe)$] or [$AsPh_4$]$_2$[$Pt(S_2CO)_2$], both of which gave [$PtL_2(S_2Co)$] on addition of various Lewis bases L (L = PPh_3, PMe_2Ph, or ½ dppe) (404).

Facile replacement of the unidentate [S_2PPh_2]$^-$ group by X^-, to give [$PdX(PR_3)(S_2PPh_2)$], occurs when [$Pd(PR_3)(S_2PPh_2)_2$] (PR_3 = PPh_3, or PMe_2Ph) reacts with an excess of AgX in acetone (X = Cl, Br, I, or SCN). Similar reactions with [$Pt(PR_3)(S—S)_2$] (S—S$^-$ = [S_2PPh_2]$^-$ or [Etdtc]$^-$) are not very efficient, and a better, general method of synthesizing [$MX(PR_3^-)(S—S)$] complexes (M = Pd, or Pt; X = Cl, Br, or I; S—S$^-$ = [S_2PMe_2]$^-$, [Etdtc]$^-$, [Pr_2^i dtc], or [S_2COEt]$^-$) is be reaction of equimolar amounts of [{$MX_2(PR_3)$}$_2$] and alkali-metal dithioacid salts (405). Prolonged reaction of [$Pd(S_2PMe_2)_2$] with an excess of $PPh_2(OR)$(R = Me, or Et) in either $MeCl_2$ of benzene gives the 4-coordinated complex [$Pd(S_2PMe_2)_2(PPh_2O)\{PPh_2(OH)\}$], shown by spectroscopic evidence and X-ray structural analysis to contain the symmetrically hydrogen-bonded $Ph_2 \cdot POHOPPh_2$ ligand (406).

Bis complexes of pyrrole-N-carbodithiolate (L) with Pt(II), Pd(II), and Co(II), as well as Cu(II), have been reported. A mixed chlorine-ligand complex of Pt(IV), $[PtCl_2L_2]$, was also prepared. Infrared spectra of these complexes indicated an exocyclic single C–N bond, as opposed to similar complexes containing other dithiocarbamate ligands (407).

The reactions of platinum(II) complexes $[Pt(L)Cl_3]^-$, $[Pt(L)_4]Cl_2$ (L = NH_3, or py) with an excess of aqueous Na_2CS_3 yield $[Pt(L)CS_3]$ and $[Pt(L)_3CS_3]$; heating the latter results (408, 409) in the formation of $[PtCS_3]$ and $[Pt(py)_2CS_3]$. Trithiocarbonates, $(CH_2)_nCS_3$ (n = 2, or 3), oxidatively add to $[PtL_4]$ (L = PPh_3, or PPh_2Me) to yield (410) cis complexes as shown (**XXXI**).

$$[PtL_4] \ + \ S=C\diagup^S_S\diagdown(CH_2)_n \xrightarrow{C_6H_6} \begin{array}{c} L \diagdown \diagup^{\overset{S}{\|}}C-S \\ Pt \\ L \diagup \diagdown S-(CH_2)_n \end{array}$$

(**XXXI**)

The first example of the synthesis of a novel metallodithiocarbonylate ligand complex, $[Cl(Ph_3P)_2Pt(CS_2)Pt(Ph_3P)_2]BF_4$, has been reported (411). $[L_2Pt\{(CF_3)_2C:CS_2\}]$ (L = PPh_3, or PPh_2Me) complexes have been prepared by reaction of low-valent platinum–phosphine complexes with the ligand (326). A novel synthesis of platinum dithiocarbimidato complexes was recently evolved (412) by Haszeldine and co-workers: $[Pt(S_2C=NR^1)(PPh_3)_2]$ complexes (R^1 = Ph, or Me) were obtained in 30–60% yields by a sulfur abstraction-reaction between $[PtR(PPh_3)_2]$ (R = $CH_2:C:C:CH_2$, or $CF_3CH=C=CH_2$), $[Pt(PPh_3)_4]$, or $[Pt(PhNCS)(PPh_3)_2]$ and R'NCS, to yield crystalline products $[Pt(S_2C=NR')(PPh_3)_2]^+CF_3CO_2^-$. Kinetics and mechanisms of substitution reactions of platinum dithiolato complexes $[Pt(S-S)_2]^{2-}((S-S)^{2-} = S_2C:C(CN)_2^{2-}$, mnt^{2-}, $S_2C=CHNO_2$, $S(O)C=C(O)S^{2-}$, $NC-N=CS_2$, and $NC(Ph)C:CS_2^{2-}$) with both unidentate and bidentate nucleophiles have been studied in aqueous solution (413). Reactions were found to be second-order overall (first-order in substrate, and first-order in nucleophile). The results permit a comparison of the relative reactivities of dithiolato complexes of Ni, Pd, and Pt. Except for $[Pt S(O)C:C(O)S_2]^{2-}$, two steps were observed for the substitution reactions with CN_3^-, the first step being substitution of one dithiolato ligand by two CN^- ions, and the second, the substitution of the second dithiolato ligand.

NMR spectra have been used for comparing the solution structure of Pt[Bu$_2^i$dtc]$_2$ (PMe$_2$Ph$_2$) with its solid-state structure. The low-temperature, solution structure is that of the cation [Pt(Bu$_2^i$dtc)(PMePh)$_2$]$^+$ with a free dithiolate anion, compared to the structure of the crystalline solid, which is *trans*-PtS$_2$P$_2$. The structure of Pt(S$_2$CO)(Ph$_3$P)$_2$] and the diphos analog show a *cis*-PtS$_2$P$_2$ coordination (*414*).

The isolation, separation, and chemistry of dithio- and perthioarylcarboxylate complexes of Ni(II), Pd(II), and Pt(II) were reported in two complementary reports (*381, 415*). The perthiocarboxylate complexes have also been obtained by oxidative addition of sulfur to the dithiocarboxylic acid complexes. The abstraction of the sulfur atom adjacent to carbon by PPh$_3$ was again observed, and rationalized as follows.

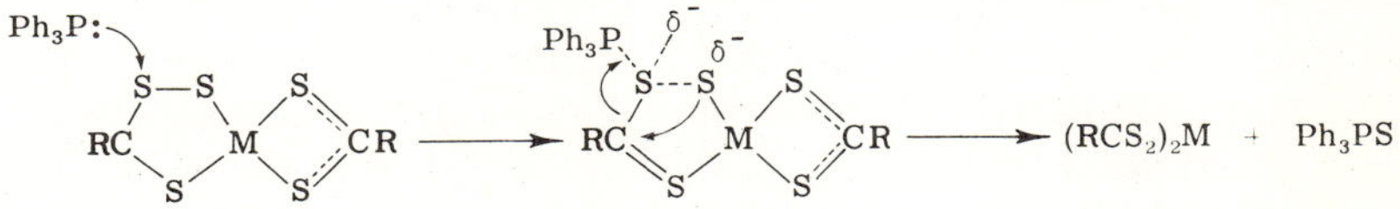

X-Ray structural analysis of [Pt(dithiocumato)$_2$] revealed a dimeric structure containing two bridging and two terminal dithiocarboxylate ligands (**XXXII**). The proximity of the two platinum atoms (278 pm) suggested that a Pt–Pt bond may be present (*415*).

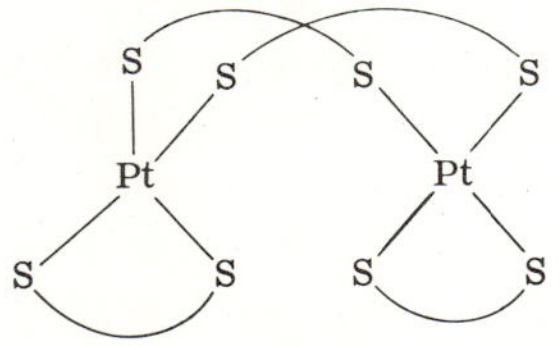

(**XXXII**)

H. Copper, Silver, and Gold

The complexes of the 1,1-dithio ligands with the Group I transition metals have been studied in considerable detail, and have been extensively reviewed by Coucouvanis (*1*). Interest in the synthetic chemistry of these complexes has been maintained over recent years, but a large proportion of the work on these complexes has been concerned with physical studies, particularly by esr spectroscopy.

Solid-state, esr spectra of [Cu(Et$_2$dtc)] and [Cu{S$_2$P(OPr^i)$_2$}$_2$] dissolved in coordinating and noncoordinating solvents have been compared with single-crystal and powdered samples diluted with the corresponding complexes of divalent nickel and zinc. With noncoordinating solvents, the ESR spectra are similar to those obtained from the solid samples, where weak self-association is present. On the basis of the results of this study, it was suggested that in the ground state the unpaired electron is in a hybrid orbital of the $dx^2 - y^2$, dz^2, and 4s atomic orbitals of copper (416). The effect of varying the solvent, and the temperature, on esr parameters and linewidths of $[Cu(Et_2dtc)_2]$ have been reported (417). An esr study of the Cu(II) complex of pyrrole-N-carbodithiolate, taken together with the optical transition for this complex, showed a very covalent σ and π metal–sulfur bond (407).

Dithiocarbamate complexes of copper have been sythesized at a high rate. Reports of new complexes include the morpholine-4- (44), thiomorpholine, N-methylpiperazine-4-, and piperidine- (291) dithiocarbamates. Novel, polymeric complexes of the type $\{Cu(pipdtc)_2\}$ $(CuBr)_n$ ($n = 4$, or 6) and $\{Cu(pipdtc)_2\}$ $(CuCl)_4$ have been prepared by reactions of $[Cu(pipdtc)_2]$ with the respective copper halide in $CHCl_3$–EtOH (418). The crystal structures of the polymers are known to consist of sheets of individual $[Cu(pipdtc)_2]$ molecules linked to polymeric CuBr chains via Cu–S bonds. A series of copper(I) dtc complexes have been the subject of a ^{63}Cu and ^{65}Cu NQR-spectral study (440).

Unlike 2-thiazolidinethione and its simpler derivatives, the 3 - alkyl - 5 - hydroxy - 5 - (1,2,3,4 - tetrahydroxy - n - butyl) - 2 - thiazolidine thiones [which are specific, photometric reagents for Cu(II)] undergo rearrangement in the presence of Cu(II) ions, to form complexes of dithiocarbamate. The analytical specificity is explained by the inability of most other metal ions to effect this rearrangement (419).

Oxidation of copper dithiocarbamate complexes of Cu(I) with chlorine or bromine yields paramagnetic compounds with the composition $CuX_3(R_2dtc)$, in which the copper has a formal oxidation-state of (IV), but the structure of these compounds is obscure (385). The synthesis and crystal structures of $[Cu_2(Et_2dtc)_2Cl_2]$ and $[Cu_3(Et_2dtc)_2Cl_3]$ have been reported. The latter is an example (420) of a complex containing the unusual mixed valence Cu^{II} and Cu^{I}. Both Cu(I) and Cu(III) are present in complexes of the type $[Cu_7Br_7(Bu_2dtc)_2]$, $[Cu_5Br_5(Pr_2dtc)_2]$, $[Cu_5Br_5(Et_2dtc)_2]$, and $[Cu_3Br_3(Me_2dtc)_2]$. The Cu(III) is present in the $[Cu(dtc)_2]^+$ cation, and the Cu(I) is in the haloanion (421). $[Cu(R_2dtc)_2]ClO_4$ and $[Cu(R_2dtc)_2]FeCl_4$ have been prepared (270) by reaction of $FeCl_3$ and $Cu(ClO_4)_2$ with copper dithiocarbamate, and the reaction of $Cu(Bu_2dtc)_2$, MBr_2 (M = Zn, Cd, or Hg), and bromine in stoichiometric quantities, or the reaction of $[M(Bu_2dtc)_2]$ and $[Bu_4bitt][Cu_2Br_6]$ (Bu_4bitt = 3,5-bis-(N,N-dialkyliminium)-1,2,4-trithiolane) $[Cu_3(Bu_2dtc)_6][MBr_3]_2$. The cadmium-complex cation contains one copper atom having distorted octahedral geometry, the other being square pyramidal; the anion is $Cd_2Br_6^{2-}$. ESR measurements (422) suggested that the unpaired electron is localized on one copper atom, and that the cation is best described as $[\{Cu^{III}(Bu_2dtc)_2\}_2\{Cu^{II}(Bu_2dtc)_2\}]^+$. The $[Cu(Pr_2^n dtc)_2]$ species is dimeric in the solid state, the copper atoms having tetragonal-pyramidal coordination (423); $[Cu(PhMedtc)_2]$ is monomeric, and isostructural with its nickel(II) analog.

The crystal structure of $[Cu(Me_2dtc)_2]$ shows that it possesses a center of symmetry, with the copper octahedrally co-ordinated to six S atoms, two Cu–S bonds being longer than the other four (424). Choi and Wasson (425) showed that there is only Cu–S bonding in $[Cu(acdc)_2]$ (acdc = 2-amino-1-cyclopentadienyl-1-dithiocarboxylate).

Of late, few complexes of copper with 1,1-dithiolate ligands have appeared. A red–brown, air-sensitive complex with the new, cyclopentadiene-dithiocarboxylate ligand was readily obtained by reaction with $CuBr_2$ in degassed THF or acetonitrile (426), and NMR data indicated a large amount of charge residing on the cyclopentadiene ring system, but, from esr and electronic-spectral data, it was concluded that, as expected, the out-of-plane π-bonding was much more covalent than that in the copper dithiocarbamate systems. The in-plane π- and σ-bondings were similar. As the out-of-plane bond is anti-bonding in character, and is very covalent, the bonding molecular orbital is similar, indicating that the dithiolate is a strong, π-bonding ligand. The delocalization of charge onto the ring system can be shown as follows.

2-Aminocyclopentene-1-dithiocarbamic acid (LH) forms a series of complexes, $MI_2(M = Ni^{II}, Co^{II},$ or $Cu^{II})$, in which the ligands are S,S-bonded (427). Diethanoldithiocarbamic acid forms $Cu[(HOC_2H_4)_2NCS_2]_2$, which is a monomeric, planar, CuS_4 chromophore in solution, but Cu $\cdots$ S interaction probably occurs between neighboring molecules in the solid state (428).

A number of complexes of copper with 1,1-dithiolenes are known; they are interesting, inasmuch as they form (1) polynuclear species, e.g., $[Cu_4(i\text{-mnt})_3]_2^{4-}$. Recently, a copper(III) complex of 1,1-dicarboethoxy-2-ethylenedithiolate (DED^{2-}) was prepared (375) by oxidation of aqueous solutions of $K_2[Cu(DED)_2]$ with a 10–15% excess of Cu(II) or H_2O_2, and of $(BzPh_3P)_2[Cu(DED)_2]$ with I_2. The possibility of this system as a model for the Cu^{III}/Cu^{I} system in D-galactose oxidase has been pointed out. Lewis and Miller (113) also prepared $M[Cu(S_2C:CHNO_2)_2]$ (M = Cu, or Zn) and $Cu[Cu\{S_2C:C(CN)_2\}_2]$, and found that they are effective insecticides.

A variety of 1,1-dithio complexes of silver are known, but, lately, there have been few reports concerning the chemistry of these species. However, the chemistry of the gold 1,1-dithio complexes has been probed in depth. The preparation of a series of dithiocarbamate complexes of Au(III) was reported by Van der Linden (429). Thus, $Au(R_2dtc)_2^+ X^-$ (R = H, Me, Et, Pr^n, Bu^n, or Ph; X = Br, ClO_4, PF_6, $AgBr_2$, or BPh_4), $[Au(tdt)_2]$, and $[Au(mnt)_2]$ have been prepared. It was found that, in the solid state, some of the anions have a distinct effect on the C–N stretching-frequency that is absent in $CHCl_3$ solution. X-Ray studies showed that, in some cases, the anion is near to the nitrogen atom of the cation and the effect on the $\nu(C-N)$ is due to polarization of the C–N bond. The infrared and ^{1}H-NMR spectra of $[(PPh_3)_2Cu(Et_2dtc)]$ have been studied (430).

The complexes of Ph_3P and Cu(I)- and Au(I)-dithiocarbamates have been reinvestigated (431). An equilibrium

$$(Ph_3P)_2M(R_2dtc) \rightleftharpoons (Ph_3P)M(R_2dtc) + Ph_3P$$

was indicated by the results, and the complexes originally formulated as $[(Ph_3P)_2M]^+[R_2dtc^-]$ may not be constituted as indicated. The Au(III) complexes $[Au(Bu_2dtc)_2]Br$ and $[Au(Bu_2dtc)_2]^+[AuBr_4]^-$ have been prepared and characterized (432). A spectrophotometric study (393) showed the formation of $[Au(R_2dtc)Cl_2]$, $[Au(R_2dtc)_2Cl]$, and $[Au(R_2dtc)_3]$ by reaction of Au(III) species with R_2dtc^- in acidic solution. The Bergendahls (433) studied the equilibrium of Au^{III}/Au^I and Ag^{II}/Ag^{III} systems with Bu_2dtc^-,

$$[(Bu_2dtc)_3M] \rightleftharpoons [M(Bu_2dtc)_2] + 0.5\,[Bu_2NCS_2]_2$$

and the esr spectra support the formation of the paramagnetic species $[Au(Bu_2dtc)_2]$ from mixtures of $LiBu_2dtc$ and $[AuBr_4]^-$. The structure of $[Au(Pr^ndtc)_2]_2$ is dimeric with tetragonal-pyramidal coordination at the Au atom (434); $[Au(Bu^idtc)_2]_2$ is known to be dimeric. The valence state, $5d^{10}$, of Au(I) is quite stable, and would not be expected to favor homonuclear Au–Au bonds. Many heteronuclear Au–metal bonds are known, but substantial electronegativity differences are involved. Nevertheless, the length of the Au–Au bond in $[Au(Bu_2dtc)_2]_2$ is 276 pm, compared with 288 pm in metallic gold, and is less than the span of the sulfur atoms in the ligands. The sulfur atoms are twisted out of the plane that the Au and carbon atoms form. From the Raman spectrum of this complex (435), an estimated bond-order of about one-fourth of its value in Au_2° was derived for Au–Au. The metal–metal interaction was understood to be due to charge transfer from the dithiocarbamate ligands to the Au(I) ions, with the partial formation of an Au_2° bond.

References

1. Coucouvanis, D., *Prog. Inorg. Chem.* **11**, 233 (1970).
2. Eisenberg, R., *Prog. Inorg. Chem.* **12**, 295 (1970).
3. Stokolosa, H. J., Wasson, J. R., and Woltermann, G. M.,*Fortschr. Chem. Forsch.* **35**, 65 (1973).
4. Thorn, G. D., and Ludwig, R. A., "The Dithiocarbamates and Related Compounds," Elsevier, New York, 1962.
5. Delepine, M., *Bull. Soc. Chim. Fr.* 5 (1958).
6. Glen, K., and Schwab, R., *Angew. Chem.* **62**, 320 (1950).
7. Busev, A. I., and Ivanyutin, M. L., *Zh. Anal. Khim.* **13**, 647 (1958). *Chem. Abstr.* **53**, 5954h (1959).
8. von Braun, J., *Ber.* **35**, 817 (1902).
9. Delepine, M., *Bull. Soc. Chim. Fr.* **3**, 643 (1908).
10. Cambi, L., and Szegoni, L., *Ber.* **64**, 2591 (1931).
11. Cambi, L., and Cortselli, C., *Gazz. Chim. Ital.* **66**, 779 (1936).
12. Cambi, L., *Z. Anorg. Chem.* **20**, 247 (1941).
13. Cambi, L., and Malatesta, *Chem. Ber.* **70**, 2067 (1977).
14. Malatesta, L., *Gazz. Chim. Ital.* **67**, 738 (1937).
15. Malatesta, L., *Gazz. Chim. Ital.* **70**, 541, 729 (1940).
16. Malatesta, L., *Gazz. Chim. Ital.* **70**, 734 (1940).
17. Takazi, S., and Tanaka, Y., *Yakugaku Zasshi* **69**, 298 (1949).
18. Reid, E. E., "Organic Chemistry of Bivalent Sulphur," Vol. 4, p. 140. Chemical Publ., New York, 1962.
19. Jensen, K. A., Athoni, M., Kagi, B., Larsen, C., and Pederson, C. T., *Acta Chem. Scand.* **22**, 1 (1968).
20. Lopatecki, L. E., and Newton, I., *Can. J. Bot.* **30**, 13 (1952).
21. Maier, L., *Helv. Chim. Acta* **53**, 1216 (1970).
22. Zeise, W. C., *Acad. R. Sci. Copenhagen* **1**, 1 (1815).
23. Vogel, A. I., "Practical Organic Chemistry," Longmans, New York, 1956.
24. Biilman, E., and Due, N. V., *Bull. Chem. Soc. Fr.* **35**, 384 (1924).
25. Hantzsch, A., and Bucerius, H., *Ber.* **59**, 793 (1926).
26. Twiss, D., *J. Am. Chem. Soc.* **49**, 491 (1927).
27. Kharasch, M. S., and Reimmuth, O., "Grignard Reactions of Nonmetallic Substances." Prentice-Hall, New York, 1956.
28. Gromper, R., and Topfl, W., *Chem. Ber.* **95**, 2861 (1962).
29. Jensen, K. A., and Hendriksen, L., *Acta Chem. Scand.* **22**, 1107 (1968).
30. Drager, M., and Gattow, G., *Angew Chem.* **80**, 954 (1968).
31. Amico, J. J. D., and Cambell, R. H., *J. Org. Chem.* **32**, 2567 (1967).
32. Frozin, E., and Van Gonez, D., *Ann. Chem.* **355**, 196 (1907).
33. Fleischer, A., *Ann. Chem.* **179**, 204 (1875).
34. Timmons, J. R., and Wittenbrook, L. S., *J. Org. Chem.* **32**, 1566 (1967).
35. Hatchard, W. R., *J. Org. Chem.* **28**, 2163 (1963).
36. Pera, J. D., U.S. Patent 2,816,136; *Chem. Abstr.* **52**, 5766 (1958).
37. Alyea, E. C., Bradley, D. C., Lappert, M. F., and Sanyer, A. R., *Chem. Commun.* 1064 (1969).
38. Dermer, O. C., and Fernelius, W. C., *Z. Anorg. Chem.* **221**, 83 (1934).
39. Bradley, D. C., and Gitlitz, M. H., *Chem. Commun.* p. 289 (1965).
40. Bradley, D. C., and Gitlitz, M. H., *J. Chem. Soc. A* p. 1152 (1969).
41. Colupietro, M., Vaciago, A., Bradley, D. C., Hursthouse, M. B., and Randell, I. F., *Chem. Commun.* p. 743 (1970).

42. Bradley, D. C., Randall, I. F., and Sales, K. D., *J. Chem. Soc., Dalton Trans.* p. 2228 (1973).
43. Kirnickev, W. A., U.S. Patent, 3,297,733 (1967).
44. Aravamudan, G., Brown, D. H., and Venkappayga, D., *J. Chem. Soc. A* p. 2744 (1971).
45. Alyea, E. C., Ramaswamy, B. S., Bhat, A. N., and Faẏ, R. C., *Inorg. Nucl. Chem. Lett.* **9**, 399 (1973).
46. Bhat, A. N., Fay, R. C., Lewis, D. F., Lindmark, A. F., and Strauss, S. H., *Inorg. Chem.* **13**, 886 (1974).
47. Clark, R. J. H., "The Chemistry of Titanium and Vanadium," Elsevier, Amsterdam, 1968.
48. Lewis, D. F., and Fay, R. C., *J. Am. Chem. Soc.* **96**, 3843 (1974).
49. Coutts, R. S. P., Wailes, P. C., and Kingston, J. V., *Chem. Commun.* 1170 (1968).
50. Coutts, R. S. P., Wailes, P. C., and Kingston, J. V., *Aust. J. Chem.* **23**, 463 (1970).
51. Coutts, R. S. P., Wailes, P. C., and Kingston, J. V., *Aust. J. Chem.* **23**, 469 (1970).
52. Coutts, R. S. P., and Wailes, P. C., *Aust. J. Chem.* **27**, 2483 (1974).
53. Coutts, R. S. P., and Wailes, P. C., *J. Organomet. Chem.* **84**, 47 (1975).
54. Smith, J. N., and Brown, T. M., *Inorg. Nucl. Chem. Lett.* **6**, 441 (1970).
55. Malatesta, L., *Gazz. Chim. Ital.* **71**, 615 (1941).
56. McCormick, B. J., *Inorg. Nucl. Chem. Lett.* **3**, 293 (1967).
57. McCormick, B. J., *Inorg. Chem.* **7**, 1965 (1968).
58. Vigee, G., and Selbin, J., *J. Inorg. Nucl. Chem.* **31**, 3187 (1969).
59. McCormick, B. J., and Bellott, E. M., Jr., *Inorg. Chem.* **9**, 1799 (1970).
60. Garif'yanov, N. S., and Kozyrev, B. M., *Teor. Eksp. Khim. Akad. Nauk SSSR* **1**, 525 (1965).
61. Stoklosa, M. J., and Wasson, J. R., *Inorg. Nucl. Chem. Lett.* **10**, 377 (1974).
62. Stocklosa, H. J., and Wasson, J. R., *Inorg. Nucl. Chem. Lett.* **10**, 401 (1974).
63. Henrick, K., Raston, C. L., and White, A. H., *J Chem. Soc., Dalton Trans.* p. 26 (1976).
64. McCormick, B. J., *Can. J. Chem.* **47**, 4283 (1969).
65. Kirmse, V. R., Dietzch, W., and Hoyer, E., *Z. Anorg. Allg. Chem.* **397**, 198 (1973).
66. Busev, A. I., Byr'ko, V. H., Polinskaya, M. B., and Kuz'mina, T. A., *Zh. Neorg. Khim.* **18**, 1557 (1973).
67. Casey, A. T., Mackey, D. J., Martin, R. L., and White, A. H., *Aust. J. Chem.* **25**, 477 (1972).
68. Dewan, J. C., Kepert, D. L., Raston, C. L., Taylor, D., White, A. H., and Maslen, E. N., *J. Chem. Soc., Dalton Trans.* p. 2082 (1973).
69. Casey, A. T., and Thackeray, J. R., *Aust. J. Chem.* **27**, 757 (1974).
70. Kwoka, W., Moyer, R. O., and Lindsay, R., *J. Inorg. Nucl. Chem.* **37**, 1889 (1975).
71. Casey, A. T., and Thackeray, J. R., *J. Inorg. Nucl. Chem.* **25**, 2085 (1972).
72. Bond, A. M., Casey, A. T., and Thackeray, J. R., *J. Chem. Soc., Dalton Trans.* p. 773 (1974).
73. Bond, A. M., Casey, A. T., and Thackeray, J. R., *Inorg. Chem.* **12**, 887 (1973).
74. Bond, A. M., Casey, A. T., and Thackeray, J. R., *Inorg. Chem.* **13**, 84 (1974).
75. Bradley, D. C., and Gitlitz, M., *Chem. Commun.* p. 289 (1965).
76. Alyea, E. C., and Bradley, D. C., *J. Chem. Soc. A* p. 2330 (1969).
77. Piovesana, O., and Furlani, C., *Chem. Commun.* p. 256 (1971).
78. Bonamico, M., Dessy, G., Fares, V., and Scaramuzza, L., *J. Chem. Soc., Dalton Trans.* p. 1258 (1974).
79. Bradley, D. C., Sales, K. D., and Moss, R. H., *Chem. Commun.* p. 1255 (1969).

80. Piovesana, O., and Cappuccilli, G., *Inorg. Chem.* **11,** 1543 (1972).
81. Bonamico, M., Dessy, G., Fares, V., Porta, P., and Scaramuzza, *Chem. Commun.* 365 (1971).
82. Fanfani, L., Nunzi, A., Zanazzi, P. F., and Zanzari, A. R., *Acta Crystallogr., Sect. B* **28,** 1298 (1972).
83. Sedivec, V., and Vasak, V., *Collect. Czech. Chem. Commun.* **15,** 260 (1950).
84. Gleu, K., and Schwab, R., *Angew. Chem.* **62,** 320 (1950).
85. Malissa, H., and Kolbe-Rhode, H., *Talanta* **8,** 841 (1961).
86. Bradley, D. C., and Gitlitz, M., *J. Chem. Soc. A* p. 1152 (1969).
87. Machin, D. J., and Sullivan, J. F., *J. Less Common Met.* **14,** 413 (1969).
88. Smith, J. N., and Brown, T. M., *J. Chem. Soc., Dalton Trans.* 1614 (1972).
89. Smith, J. N., Ph.D. Thesis, Arizona State University, 1972.
90. Panteleo, D. C., and Johnson, R. C., *Inorg. Chem.* **9,** 1248 (1970).
91. Moncrief, W. J., Panteleo, D. C., and Smith, E. N., *Inorg. Nucl. Chem. Lett.* **7,** 255 (1971).
92. Uvatova, K. A., Usatenko, Yu. I., Mel'nukova, N. V., and Klopova, Zh. G., *Zh. Neorg. Khim.* **16,** 2137 (1971).
93. Heckley, P. R., and Holah, D. G., *Inorg. Nucl. Chem. Lett.* **6,** 865 (1970).
94. Heckley, P. R., Holah, D. G., and Brown, D., *Can. J. Chem.* **49,** 1151 (1971).
95. Jorgensen, C. K., *Inorg. Chim. Acta* **2,** 65 (1968).
96. Delepine, M., *Compt. Rend.* **144,** 1125 (1907).
97. Malatesta, L., *Gazz. Chim. Ital.* **69,** 752 (1939).
98. Chatt, J., Duncanson, L. A., and Venanzi, L. M., *Suom. Kemistil. B* **29** (2), 75 (1956).
99. Jorgensen, C. K., *J. Inorg. Nucl. Chem.* **24,** 1571 (1962).
100. Price, E. R., and Wasson, J. R., *J. Inorg. Nucl. Chem.* **36,** 67 (1974).
101. Brown, D. A., Glass, W. K., and Burke, M. A., *Spectrochim. Acta, Part A* **32** 137 (1976).
102. Contreras, G., and Cortes, H., *Spectrochim. Acta,* **33,** 1337 (1971).
103. Galsbol, F., and Schaffer, C. E., *Inorg. Synth.* **10,** 42 (1967), and refs. cited therein.
104. Basi, J. S., Bradley, D. C., and Chiholm, M. H., *J. Chem. Soc. A* p. 1433 (1971).
105. Que, Jr., L., and Pignolet, L. H., *Inorg. Chem.* **13,** 351 (1974).
106. Merlino, S., and Sartori, E., *Acta Crystallogr. Sect. B* **28,** 972 (1972).
107. Watt, C. W., and McCormick, B. J., *J. Inorg. Nucl. Chem.* **27,** 898 (1965).
108. DeArmond, K., and Mitchell, W. J., *Inorg. Chem.* **11,** 181 (1972).
109. Furlani, C., and Luciani, M. L., *Inorg. Chem.* **7,** 1568 (1968).
110. Maltese, M., *J. Chem. Soc., Dalton Trans.* p. 2664 (1972).
111. Corraza, G., and Pellizi, C., *Inorg. Chim. Acta* **4,** 618 (1970).
112. Fackler, Jr., J. P., and Coucouvanis, D., *J. Am. Chem. Soc.* **88,** 3913 (1966).
113. Lewis, S. N., and Miller, G. A., U.S. Patent 3,499,388 (1969).
114. Mueller, A., Christophlieink, P., Tossidis, I., and Jorgensen, C. K., *Z. Anorg. Allg. Chem.* **401,** 274 (1973).
115. Fackler, Jr., J. P., and Holah, D. G., *Inorg. Nucl. Chem. Lett.* **2,** 251 (1966).
116. Larkworthy, L. F., and Patel, R. R., *Inorg. Nucl. Chem. Lett.* **8,** 139 (1972).
117. Garif'yanov, N. S., and Luchkina, S. A., *Dokl. Akad. Nauk SSSR* **189,** 543 (1969).
118. Garif'yanov, N. S., and Luchkina, S. A., *Teor. Eksp. Khim.* **5,** 571, (1969).
119. Garif'yanov, N. S., Troitskaya, A. D., Razurnov, A. I., Ovinnikov, I. V., Gurevich, P. A., and Kondrat'eva, O. I., *Dokl. Akad. Nauk. SSSR,* **196,** 1346 (1971).
120. Kondrateva, O. I., Troitskaya, A. D., Chadaeva, N. A., Chuikova, A. I., Usacheva, G. M., and Ivantsov, A. I., *Zh. Obshch. Khim.* **43,** 2087 (1973).

121. Garifyanov, N. S., Razumov, A. I., Gurevich, P. A., and Kondrat'eva, O. I., *Zh. Neorg. Khim.* **16,** 1059 (1971).
122. Connelly, N. G., and Dahl, L. F., *Chem. Commun.* 880 (1970).
123. Carlin, R. L., Canziani, F., and Batton, W. K., *J. Inorg. Nucl. Chem.* **26,** 898 (1964).
124. Golding, R. M., Healy, P. C., Colombera, P., and White, A. H., *Aust. J. Chem.* **27,** 2089 (1974).
125. Bowden, L., in "Bioinorganic Chemistry" (McAuliffe, C. A. ed.), Macmillan, New York, 1975.
126. Blake, A. B., Cotton, F. A., and Wood, J. S., *J. Am. Chem. Soc.* **86,** 3024 (1964).
127. Hyde, J., Venkatasubromanian, K., and Zubieta, J., *Inorg. Chem.* **17,** 414 (1978).
128. Moore, F. W., and Rice, F. R., *Inorg. Chem.* **7,** 2510, (1968).
129. Farmer, H. H., and Rowan, E. V., U.S. Patent 3,356,702 (1967).
130. Sakurai, T., and Kaguyama, H., *Jpn. Kokai* 73 56,202 (1973).
131. Kirmse, R., *Z. Chem.* **13,** 187 (1973).
132. Newton, W. E., Corbin, J. L., Bravard, D. C., Searles, J. E., and McDonald, J. W., *Inorg. Chem.* **13,** 1100 (1974).
133. Newton, W. E., Corbin, R. L., and McDonald, J. W., *J. Chem. Soc., Dalton Trans.* p. 1044 (1974).
134. Newton, W. E., Bravard, D. C., and McDonald, J. W., *Inorg. Nucl. Chem. Lett.* **11,** 553 (1975).
135. Mitchell, P. C. H., and Searle, R. D., *J. Chem. Soc., Dalton Trans.* p. 2552 (1975).
136. Bishop, M. W., Chatt, J., and Dilworth, J. R., *J. Organomet. Chem.* **73,** 159 (1974).
137. Dirand, J., Ricard, L., and Weiss, R., *J. Chem. Soc., Dalton Trans.* p. 278 (1976).
138. Ricard, L., Estienne, T., Karagiannidis, P., Toledano, P., Fischer, J., Mitschler, A., and Weiss, R., *J. Coord. Chem.* **3,** 277 (1974).
139. Byr'kro, V. M., Polinskaya, M. B., and Busev, A. I., *Zh. Neorg. Khim.* **18,** 2783 (1973).
140. Larin, M. G., Solozhenkin, P. M., and Semenov, E. V., *Dokl. Akad. Nauk SSSR* **214,** 1343 (1974).
141. Ricard, L., Martin, C., Wiest, R., and Weiss, R., *Inorg. Chem.* **14,** 2300 (1975).
142. Spivack, B., Doriand, Z., and Steifel, E. I., *Inorg. Nucl. Chem. Lett.* **11,** 501 (1975).
143. Gelder, J. I., and Enemark, J. H., *Inorg. Chem.* **15,** 1839 (1976).
144. Dirand-Colin, J., Ricard, L., and Weiss, R., *Inorg. Chim. Acta* **18,** L21 (1976).
145. Bunzey, G., Enemark, J. H., Walterman, G., Boston, D. A., and Haight, G. P., *J. Am. Chem. Soc.* **97,** 1616 (1975); Banzey, G., Enemark, J. H., Golder, J. I., Newton, W. E., and Yamanouchi, K., *Proc. Climax Molybdenum Int. Conf. Chem. Uses Molybdenum, 2nd, 1976,* p. 50.
146. Dirand-Colin, J., Schappacher, M., Ricard, L., and Weiss, R., *Proc. Climax Molybdenum Int. Conf. Chem. Uses Molybdenum, 2nd, 1976,* p. 46.
147. Newton, W. E., Chen, G. J.-J., and McDonald, J. W., *J. Am. Chem. Soc.* **98,** 5384 (1976).
148. Chen, G. J.-J., McDonald, J. W., and Newton, W. E., *Inorg. Chem.* **15,** 2612 (1976).
149. McDonald, J. W., Corbin, J. L., and Newton, W. E., *Inorg. Chem.* **15,** 2056 (1976).
150. Serree de Roch, I., Sajus, L., Barral, R., and Bacard, C., *Kinet. Katal.* **14,** 164 (1973).
151. Barral, R., Bocard, C., Serree de Roch, I., and Sajus, L., *Tetrahedron Lett.* p. 1693 (1972).
152. Barral, R., Bocard, C., Serree de Roch, I., and Sajas, L., *Kinet. Katal.* **14,** 164 (1973); *Chem. Alostr.* **79,** 10,281 (1973).
153. McDonald, D. B., and Shulman, J. I., *Anal. Chem.* **47,** 2023 (1975).
154. Durant, R., Garner, C. D., Hyde, M. R., and Mabbs, F. E., *J. Chem. Soc., Dalton Trans.* p. 955 (1977).

155. Schneider, P. W., Bravard, D. C., McDonald, J. W., and Newton, W. E., *J. Am. Chem. Soc.* **94,** 8640 (1972).
156. Ricard, L., and Weiss, R., *Inorg. Nucl. Chem. Lett.* **10,** 217 (1974).
157. Varadi, Z. B., and Nieuwpoort, A., *Inorg. Nucl. Chem. Lett.* **10,** 801 (1974).
158. Vanden Aalsvoort, J. G. M., and Beurskens, P. T., *Cryst. Struct. Commun.* **3,** 653 (1974); *Chem. Abstr.* **82,** 50,060 (1975).
159. Nieuwport, A., and Steggerda, J. J., *Recl. Trav. Chim. Pays-Bas* **95,** 289 294 (1976).
160. Mitchell, P. C. H., and Searle, R. D., *J. Chem. Soc., Dalton Trans.* p. 110 (1975).
161. Brown, D. A., Gordon, B. J., Glass, W. K., and O'Daly, C. J., *Proc. Int. Conf. Coord. Chem., 14th,* p. 646, 1972.
162. Nieuwpoort, A., Moonen, J. H. E., and Cras, J. A., *Recl. Trav. Chim. Pays-Bas* **92,** 1086 (1972).
163. Nieuwpoort, A., *J. Less Common Met.* **36,** 271 (1974).
164. McAuliffe, C. A., and Sayle, B. J., *Inorg. Chim. Acta* **12,** L7 (1975).
165. De Hayes, L. J., Faulkner, H. C., Doub, Jr., W., and Sawyer, D. T., *Inorg. Chem.* **14,** 2110 (1975).
166. Wijnhoven, J. G., *Cryst. Struct. Commun.* **2,** 637 (1973).
167. Piovasana, O., and Sestili, L., *Inorg. Chem.* **13,** 2745 (1974).
168. Kalbacher, B. J., and Bereman, R. D., *Inorg. Chem.* **14,** 1417 (1975).
169. Ricard, L., Karagiannidis, P., and Weiss, R., *Inorg. Chem.* **12,** 2179 (1973).
170. Steele, D. F., and Stephenson, T. A., *Inorg. Nucl. Chem. Lett.* **9,** 77 (1973).
171. Ricard, L., Estienne, J., and Weiss, R., *Chem. Commun.* 906, (1972); *Inorg. Chem.* **12,** 2182 (1973).
172. Mitchell, P. C. H., and Searle, R. D., *J. Chem. Soc., Dalton Trans.* p. 110 (1975).
173. Brown, D. A., Gordon, B. J., Glass, W. K., and O'Daly, C. J., *Proc. Int. Conf. Coord. Chem., 14th,* p. 646, 1972.
174. Butcher, R. J., Gunz, H. P., MacIagen, R. G. A. R., Powell, H. K. J., Wilkins, C. J., and Hian, Y. S., *J. Chem. Soc., Dalton Trans.* p. 1223 (1975).
175. Willemse, J., Cras, J. A., Steggerda, J. J., and Keijzers, C. P., *Struct. Bonding (Berlin)* **28,** 83 (1976).
176. Colton, R., Scollary, G. R., and Tomkins, I. B., *Aust. J. Chem.* **21,** 15 (1968).
177. Colton, R., and Rose, G. G., *Aust. J. Chem.* **23,** 1111 (1970).
178. Glass, W. K., and Shiels, A., *J. Organomet. Chem.* **67,** 401 (1974).
179. Abel, E. W., and Dunster, M. O., *J. Chem. Soc., Dalton Trans.* p. 98 (1973).
180. Chen, G. J.-J., McDonald, J. W., and Newton, W. E., *Inorg. Chim. Acta* **19,** 167 (1976).
181. Chen, G. J.-J., Yelton, R. D., and McDonald, J. W., *Inorg. Chim. Acta* **22,** 249 (1977).
182. Sartorelli, U., Zongales, F., and Canziani, F., *Chim. Ind. (Paris)* **49,** 751 (1967).
183. Connelly, N. G., *J. Chem. Soc., Dalton Trans.* p. 2183 (1973).
184. Davis, R., Johnson, B. F. G., and Al-Obaidi, K. H., *J. Chem. Soc., Dalton Trans.* p. 508 (1972).
185. Johnson, B. F. G., Al-Obaidi, K. H., and McCleverty, J. A., *J. Chem. Soc. A* p. 1668 (1969).
186. Kita, W. G., Lloyd, M. K., and McCleverty, J. A., *Chem. Commun.* p. 420 (1971).
187. James, T. A., and McCleverty, J. A., *J. Chem. Soc. A* p. 3308 (1970).
188. Bailey, N. A., Kita, W. G., McCleverty, J. A., Murray, A. J., Mann, B. E., and Walker, N. N. J., *Chem. Commun.* p. 592 (1974).
189. Brennan, T. F., and Bernal, I. *Inorg. Chim. Acta* **7,** 823 (1973).
190. Hunt, M. M., Kita, W. G., Mann, B. E., and McCleverty, J. A., *J. Chem. Soc., Dalton Trans.* p. 467 (1978).

191. Bishop, M. W., Chatt, J., Dilworth, J. R., Hursthouse, M. B., and Motevalli, M., *J. Chem. Soc., Chem. Commun.* p. 780 (1976).
192. Lahiry, S., and Anand, K. V., *Chem. Commun.* p. 1111 (1971).
193. Holah, D. G., and Murphy, C. N., *Can. J. Chem.* **49,** 2726 (1971).
194. Holah, D. G., and Murphy, C. N., *Inorg. Nucl. Chem. Lett.* **8,** 1069 (1972).
195. Garif'yanor, N. S., Kamonev, S. E., Kozyrev, B. M., and Ovichinnikov, I. V., *Dokl. Akad. Nauk. SSSR* **177,** 180 (1967).
196. Healy, P. C., and White, A. H., *J. Chem. Soc., Dalton Trans.* p. 1883 (1972).
197. Golding, R. M., and Lehonen, K., *Aust. J. Chem.* **27,** 2083 (1974).
198. Hendrickson, A. R., and Martin, R. L., *Chem. Commun.* p. 873 (1974).
199. Hendrickson, A. R., Martin, R. L., and Rhode, N. M., *Inorg. Chem.* **13,** 1933 (1974).
200. Tanaka, K., Miya-Uchi, Y., and Tanaka, T., *Inorg. Chem.* **14,** 1545 (1975).
201. Levason, W., and McAuliffe, C. A., *Coord. Chem. Rev.* **7,** 343 (1972).
202. Saleh, R. Y., Straub, D. K., *Inorg. Chem.* **13,** 3107 (1974).
203. Brown, K. L., Golding, R. M., Healy, P. C., Jessop, K. J., and Tennant, W. C., *Aust. J. Chem.* **27,** 2075 (1974).
204. McCleverty, J. A., James, T. A., Wharton, E. J., and Winscom, C. J., *Chem. Commun.* p. 933 (1968).
205. McCleverty, J. A., and Orchard, D. G., *J. Chem. Soc. A* p. 3315 (1970).
206. Linder, E., Grimmer, R., and Weber, H., *J. Organomet. Chem.* **23,** 209 (1970).
207. Lindner, E., Grimmer, R., and Weber, H., *J. Organomet. Chem.* **25,** 493 (1970).
208. Lindner, E., Grimmer, R., and Weber, H., *Angew. Chem. Int. Ed. Engl.* **9,** 639 (1970).
209. Albano, G. V., Bellon, P. L., and Ciani, G., *J. Organomet. Chem.* **31,** 75 (1971).
210. Freni, M., Guista, D., and Romiti, P., *J. Inorg. Nucl. Chem.* **33,** 1095 (1971).
211. Einstein, F. W., Enwall, E., Flitcroft, N., and Leach, J. M., *J. Inorg. Nucl. Chem.* **34,** 385 (1972).
212. Thiele, G., and Lichr, G., *Chem. Ber.* **104,** 1877 (1971).
213. Hieber, W., and Rohm, W., *Chem. Ber.* **102,** 2787 (1969).
214. Rowbottom, J. F., and Wilkinson, G., *Inorg. Nucl. Chem. Lett.* **9,** 675 (1973).
215. Rowbottom, J. F., and Wilkinson, G., *J. Chem. Soc., Dalton Trans.* p. 684 (1974).
216. Fletcher, S. R., and Skapski, A. C., *J. Chem. Soc., Dalton Trans.* p. 486 (1974).
217. Tisley, D. G., Walton, R. A., and Wills, D. L., *Inorg. Nucl. Chem. Lett.* **7,** 523 (1971).
218. Fletcher, S. R., and Skapski, A. C., *J. Chem. Soc., Dalton Trans.* p. 1073 (1972).
219. Rowbottom, J. F., and Wilkinson, G., *J. Chem. Soc., Dalton Trans.* p. 826 (1972).
220. Fletcher, R., and Skapski, A. C., *J. Chem. Soc., Dalton Trans.* p. 1079 (1972).
221. Garif'yanov, N. S., and Luchkina, S. A., *Izv. Akad. Nauk SSSR, Ser. Khim.* p. 471 (1969).
222. Garif'yanov, N. S., *Izv. Akad. Nauk SSSR Ser. Khim.* p. 1902 (1968).
223. Chatt, J., Crabtree, R. H., Dilworth, J. R., and Richards, R. L., *J. Chem. Soc., Dalton Trans.* p. 2358 (1974).
224. Hunt, J., Knox, S. A. R., and Oliphant, V., *J. Organomet. Chem.* **80,** C50 (1974).
225. Thiel, G., Liehr, G., and Lindner, E., *J. Organomet. Chem.* **70,** 427 (1974).
226. Hieber, W., Lux, F., and Herget, C., *Z. Naturforsch* **206,** 1159 (1965).
227. Larkworthy, L. F., Fitzsimmons, B. N., and Patel, R. R., *Chem. Commun.* 902 (1973).
228. de Vries, J. L. K. F., Trooster, J. M., and de Boer, E., *Inorg. Chem.* **12,** 2730 (1973).
229. Ileperuma, O. A., and Feltham, R. D., *Inorg. Chem.* **14,** 3042 (1975).
230. Davis, G. R., Mais, R. B. H., and Owston, P. G., *Chem. Commun.* p. 81 (1968).
231. Davis, G. R., Mais, R. B. H., Owston, P. G., Jarvis, J. A. J., and Kilbourn, B. T., *J. Chem. Soc. A* p. 1275 (1970).

232. Su, C. C., and Faller, J. W., *J. Organomet. Chem.* **84**, 53 (1975).
233. Folkeson, B., *Acta Chem. Scand. Ser. A* **28**, 491 (1974).
234. Buttner, H., and Feltham, R. D., *Inorg. Chem.* **11**, 971 (1972).
235. Cotton, F. A., and McCleverty, J. A., *Inorg. Chem.* **6**, 229 (1967).
236. Ricci, Jr., J. S., Eggars, C. A., and Bernal, I., *Inorg. Chim. Acta* **6**, 97 (1972).
237. Healy, P. C., and White, A. H., *J. Chem. Soc., Dalton Trans.* p. 1163 (1972).
238. Ewald, A. H., Martin, R. L., Ross, I. G., and White, A. H., *Proc. R. Soc. London Ser. A* **280**, 235 (1964).
239. Saleh, R. Y., and Straub, D. K., *Inorg. Chem.* **13**, 1559 (1974).
240. Ewald, A. H., Martin, R. L., Sinn, E., and White, A. H., *Inorg. Chem.* 1837 (1969).
241. Eley, R. R., Myers, R. R., and Duffy, N. V., *Inorg. Chem.* **11**, 1129 (1972).
242. Kokot, E., and Ryder, G. A., *Aust. J. Chem.* **24**, 649 (1971).
243. Leipoldt, J. C., and Coppens, P., *Inorg. Chem.* **12**, 2269 (1973).
244. Ricard, L., Johnson, C. E., and Hill, H. A. O., *J. Chem. Phys.* **48**, 5231 (1968).
245. Rasmussen, P. G., and Merrithew, P. B., *Inorg. Chem.* **11**, 325 (1972).
246. Hoskins, B. F., and Kelly, B. P., *Chem. Commun.* p. 45 (1970).
247. Cotton, S. A., and Gibson, J. F., *J. Chem. Soc.* 803 (1971).
248. Flick, C., and Gelerinter, E., *Chem. Phys. Lett.* **23**, 422 (1973).
249. Hall, G. R., and Hendrickson, D. N., *Inorg. Chem.* **13**, 607 (1976).
250. Po, H. N., Wang, W. K., and Chen, K. D., *J. Inorg. Nucl. Chem.* **36**, 3872 (1974).
251. Ganguli, P., Marathe, V. R., and Mitra, S., *Inorg. Chem.* **14**, 970 (1975).
252. Ewald, A. H., and Sinn, E., *Aust. J. Chem.* **21**, 927 (1968).
253. Coucouvanis, D., Lippard, S. J., and Zubieta, J. A., *J. Am. Chem. Soc.* **91**, 761 (1969).
254. Coucouvanis, D., Lippard, S. J., and Zubieta, J. A., *J. Am. Chem. Soc.* **92**, 3342 (1970).
255. Lewis, D. F., Lippard, S. J., and Zubieta, J. A., *Inorg. Chem.* **11**, 823 (1972).
256. Pelezzi, G. C., and Pelezzi, C., *Inorg. Chim. Acta* **4**, 618 (1970).
257. Healy, P. C., and Sinn, T., *Inorg. Chem.* **14**, 109 (1975).
258. Houben, J., and Pohl, H., *Ber.* **40**, 1303 (1907).
259. Coucouvanis, D., and Lippard, S. J., *J. Am. Chem. Soc.* **90**, 3281 (1968).
260. Coucouvanis, D., and Lippard, S. J., *J. Am. Chem. Soc.* **91**, 307 (1969).
261. Bereman, R. D., Good, M. L., Kalbacher, B. J., and Bultone, J., *Inorg. Chem.* **15**, 618 (1976).
262. Hollander, F. J., Pedelty, R., and Coucouvanis, D., *J. Am. Chem. Soc.* **96**, 4032 (1974).
263. Coucouvanis, D., Hollander F. J., and Pedelty, R., *Inorg. Chem.* **16**, 2691 (1977).
264. Hoskins, B. F., and White, A. H., *J. Chem. Soc. A* p. 1668 (1970).
265. Healy, P. C., White, A. H., and Hoskins, B. F., *J. Chem. Soc., Dalton Trans.* p. 1369 (1972).
266. Ojima, I., Onishi, T., Iwamoto, T., Inamoto, N., and Tamaru, K., *Inorg. Nucl. Chem. Lett.* **6**, 65 (1970).
267. McCleverty, J. A., McLuckie, S., Morrison, N. J., Bailey, N. A., and Walker, N. W., *J. Chem. Soc., Dalton Trans.* p. 359 (1977).
268. Cauletti, C., and Furlani, C., *J. Chem. Soc., Dalton Trans.* p. 1068 (1977).
269. Pasek, E. A., and Straub, D. K., *Inorg. Chem.* **11**, 259 (1972).
270. Golding, R. M., Harris, C. M., Jessop, K. J., and Tennant, W. C., *Aust. J. Chem.* **25**, 2567 (1972).
271. Martin, R. L., Rhodes, N. M., Robertson, G. B., and Taylor, D., *J. Am. Chem. Soc.* **96**, 3647 (1974).
272. Cauquis, G., and Lachenal, D., *Inorg. Nucl. Chem. Lett.* **9**, 1095 (1973).

273. Golding, R. M., Lethonen, K., and Ralph, B. J., *J. Inorg. Nucl. Chem.* **36,** 2047 (1974).
274. Chant, R., Hendrickson, A. R., Martin, R. L., and Rhodes, N. M., *Inorg. Chem.* **14,** 1894 (1975).
275. Palazzoto, M. C., and Pignolet, L. H., *Inorg. Chem.* **13,** 1781 (1974).
276. Golding, R. M., and Jessop, K. J., *Aust. J. Chem.* **28,** 179 (1975).
277. Groves, A. K. M., Morrison, N. J., and McCleverty, J. A., *J. Organomet. Chem.* **84,** C5 (1975).
278. Pignolet, L. H., Lewis, R. A., and Holm, R. H., *Inorg. Chem.* **11,** 99 (1972).
279. McCleverty, J. A., Orchard, D. G., and Smith, K., *J. Chem. Soc. A* p. 707 (1971).
280. Holm, R. H., Pignolet, L. H., and Lewis, R. A., *J. Am. Chem. Soc.* **93,** 360 (1971).
281. Cambi, L., and Malatesta, L., *Rend. Ist. Lomb. Sci. Lett. E. Sci. Mat. Nat.* 181 (1938); *Chem. Abstr.* **34,** 3201 (1940).
282. Graham, L. R., and O'Connor, M. J., *Chem. Commun.* p. 68 (1974).
283. Pignolet, L. H., and Mattson, B. M., *Chem. Commun.* p. 49 (1975).
284. Mattson, B. M., Heiman, J. R., and Pignolet, L. H., *Inorg. Chem.* **15,** 564 (1976).
285. Duffy, D. J., and Pignolet, L. H., *Inorg. Chem.* **13,** 2045 (1974).
286. Pignolet, L. H., *Inorg. Chem.* **13,** 2051 (1974).
287. Kirmse, R., Dietzsh, W., and Hayer, E., *Inorg. Nucl. Chem. Lett.* **10,** 819 (1974).
288. Cole-Hamilton, D. J., and Stephenson, T. A., *J. Chem. Soc., Dalton Trans.* 739 and 794 (1974).
289. Robinson, S. D., and Sahajpal, A., *Inorg. Chem.* **16,** 2718 (1977).
290. Wheeler, S. H., Mattson, B. M., Meissler, G. L., and Pignolet, L. H., *Inorg. Chem.* **17,** 340 (1978).
291. Marcotrigano, G., Pellacani, G. C., and Preti, C., *J. Inorg. Nucl. Chem.* **36,** 3709 (1974).
292. Powell, P., *J. Organomet. Chem.* **65,** 89 (1974).
293. Harris, R. O., Hota, N. K., Sadavoy, L., and Yuen, J. M. C., *J. Organomet. Chem.* **54,** 259 (1973).
294. Moers, F. G., ten Hoedt, R. W. M., and Langhout, J. P., *Inorg. Chem.* **12,** 2196 (1973).
295. Harris, O. R., Sadaway, L. S., Nyburg, S. C., and Pickard, F. H., *J. Chem. Soc., Dalton Trans.* p. 2646 (1973).
296. Bartodej, Z., *Chem. Listy* **48,** 1870 (1954).
297. Cole-Hamilton, D. J., and Stephenson, T. A., *J. Chem. Soc., Dalton Trans.* p. 2396 (1976).
298. Shuydka, L. F., Usatsenko, Y. D., and Tulyupa, F. M., *Russ. J. Inorg. Chem.* **18,** 396 (1973).
299. Dindoin, V. I., Shchekochikhin, Yu. M., and Keier, N. P., *Metody Issled. Katal. Katal. Reakts., Akad. Nauk SSSR, Sib. Otd., Inst. Katal.* **1,** 86 (1965). *Chem. Abstr.* **67,** 68,921 (1967).
300. Harcourt, R. D., and Winter G., *J. Inorg. Nucl. Chem.* **37,** 1039 (1975).
301. Bereman, R. D., and Kalbacher, B. J., *Inorg. Chem.* **12,** 2917 (1973).
302. Johnson, B. F. G., and McCleverty, J. A., *Prog. Inorg. Chem.* **7,** 277 (1966).
303. Enemark, J., and Feltham, R. D., *J. Chem. Soc. Dalton Trans.* 718 (1972).
304. Clarkson, S. G., and Basolo, F., *Inorg. Chem.* **12,** 1528 (1973).
305. D'Ascenzo, G., and Wendlandt, W. W., *J. Therm. Anal.* **1,** 423 (1969).
306. Holah, D. G., and Murphy, C. N., *J. Therm. Anal.* **3,** 311 (1971).
307. McCormick, B. J., Stormer, B. P., and Kaplan, R. I., *Inorg. Chem.* **8,** 2522 (1969).
308. Lewis, D. F., Lippard, S. J., and Zubieta, J. A., *J. Am. Chem. Soc.* **94,** 1563 (1972).
309. Li, T., and Lippard, S. J., *Inorg. Chem.* **73,** 1791 (1974).

310. Villa, A. C., Mafredotti, A. G., Guastini, G., and Nardelli, M., *Acta Crystallogr., Sect. B* **28,** 2231 (1972).
311. Villa, A. C., Manfredotti, A. G., Gustini, C., and Nardelli, M., *Acta Crystallogr., Sect. B* **30,** 2788 (1974).
312. Chen, H. W., and Fackler, Jr., J. P., *Inorg. Chem.* **17,** 22 (1978).
313. Brinkhoff, H. C., *Inorg. Nucl. Chem. Lett.* **7,** 413 (1971).
314. Ramaswamy, K. K., and Krause, R. A., *Inorg. Chem.* **9,** 2649 (1970).
315. Palazotto, M. C., Duffy, D. J., Edgar, B. L., Que, L., and Pignolet, L. H., *J. Am. Chem. Soc.* **95,** 4537 (1973).
316. Graham, L. R., Hughes, J. G., and O'Connor, M. J., *J. Am. Chem. Soc.* **96,** 2271 (1974).
317. de Croon, M. H. J. M., van Gaal, H. L., and van der Ent, A., *Inorg. Nucl. Chem. Lett.* **10,** 1081 (1974).
318. Marcotrigiano, G., Pallacani, G. C., Preti, C., and Tosi, G., *Bull. Chem. Soc. Jpn.* **48,** 1018 (1975).
319. Connelly, N. G., Green, M., and Kuc, T. A., *J. Chem. Soc., Chem. Commun.* 542 (1974).
320. O'Connor, C., Gilbert, J. D., and Wilkinson, G., *J. Chem. Soc. A* p. 84 (1969).
321. Mitchell, R. W., Ruddick, J. D., and Wilkinson, G., *J. Chem. Soc. A* p. 3224 (1969).
322. Cole-Hamilton, D. J., and Stephenson, T. A., *J. Chem. Soc., Dalton Trans.* p. 1818 (1974).
323. Gould, R. O., Gunn, A. M., and van den Hark, T. E. M., *J. Chem. Soc., Dalton Trans.* p. 1713 (1976).
324. Csenton, G., Heil, B., and Markro, L., *J. Organomet. Chem.* **39,** 217 (1972).
325. Sceney, G. C., and Magee, R. J., *Inorg. Nucl. Chem. Lett.* **9,** 595 (1973).
326. Green, M., Osborne, R. B. L., and Stone, F. G. A., *J. Chem. Soc. A* p. 944 (1970).
327. Raston, C. L., and White, A. H., *J. Chem. Soc., Dalton Trans.* p. 32 (1976).
328. Garif'yanov, N. S., and Luchkina, S. A., *Dokl. Akad. Nauk SSSR* **184,** 642 (1969).
329. Sato, F., and Sato, M., *J. Organomet. Chem.* **2,** 277 (1972).
330. McCormick, B. J., and Kaplan, R. I., *Can. J. Chem.* **48,** 1876 (1970).
331. McCormick, B. J., Kaplan, R. I., and Stormer, B. P., *Can. J. Chem.* **49,** 699 (1971).
332. Terent'ev, A. P., Rukhadze, E. G., Dunina, V. V., and Drobyshevskaya, E. V., *Dokl. Akad. Nauk SSSR* **195,** 380 (1970).
333. Maxfield, P. L., *Inorg. Nucl. Chem. Lett.* **6,** 693 (1970).
334. Tosi, L., and Garnier, A., *J. Chem. Soc., Dalton Trans.* p. 53 (1978).
335. Ang, L.-T., Graddon, D. P., Lindroy, L. F., and Prakash, S., *Austr. J. Chem.* **28,** 1005 (1975).
336. Villa, A. F., Gasparii, G. F., and Nardelli, M., *Acta Crystallogr.* **28,** 348 (1967).
337. Peyronel, G., and Pignedoli, A., *Acta Crystallogr. Sect. B* **24,** 433 (1968).
338. Agre, V. M., Shugam, E. A., and Rukhadze, E. G., *Chem. Abstr.* **69,** 71,311 (1968).
339. Newman, P. W. G., and White, A. H., *J. Chem. Soc., Dalton Trans.* p. 1460 (1972).
340. Newman, P. W. G., and White, A. H., *J. Chem. Soc., Dalton Trans.* p. 2239 (1972).
341. Martin, J. M., Newman, P. W. G., Robinson, B. W., and White, A. H., *J. Chem. Soc., Dalton Trans.* p. 2233 (1972).
342. Rasten, C. L., and White, A. H., *J. Chem. Soc., Dalton Trans.* p. 1790 (1974).
343. Previdi, J. C., and Krause, R. A., *Inorg. Nucl. Chem. Lett.* **7,** 647 (1971).
344. Peyronel, G., and Pignedoli, A., *Acta Crystallogr.* **23,** 398 (1967).
345. Fackler, Jr., J. P., Lin, I. J. B., and Andrews, J., *Inorg. Chem.* **16,** 450 (1977).
346. Andrews, J. M., Coucouvanis, D., and Fackler, Jr. J. P., *Inorg. Chem.* **11,** 493 (1972).
347. Fackler, Jr., J. P., and Zegarski, W. J., *J. Am. Chem. Soc.* **95,** 8566 (1973).

348. Carlin, R. L., and Seigel, A. E., *Inorg. Chem.* **9,** 1587 (1970).
349. Daktenieks, D. R., and Graddon, D. P., *Aust. J. Chem.* **24,** 2509 (1971).
350. Lim, Y. Y., and Chua, K. L., *J. Chem. Soc., Dalton Trans.* p. 1917 (1975).
351. Nanji, M., and Yamasaki, T., *J. Inorg. Nucl. Chem.* **32,** 2411 (1970).
352. Graddon, D. P., and Prakash, S., *Aust. J. Chem.* **27,** 2099 (1974).
353. Yau Yan, L., *Aust. J. Chem.* **27,** 213 (1974).
354. Kruger, A. G., and Winter, G., *Aust. J. Chem.* **24,** 161 (1971).
355. Kruger, A. G., and Winter, G., *Aust. J. Chem.* **24,** 1353 (1971).
356. Kruger, A. G., and Winter, G., *Aust. J. Chem.* **25,** 2497 (1972).
357. Fackler, Jr., J. P., Seidel, W. C., and Myron, M., *Chem Commun.* p. 1133 (1969).
358. Fries, D. C., and Fackler, Jr., J. P., *Chem. Commun.* p. 276 (1971).
359. Nay, K., and Joarder, D. S., *Inorg. Chim. Acta* **14,** 133 (1975).
360. Fackler, Jr., J. P., Fetchin, J. A., and Fries, D. C., *J. Am. Chem. Soc.* **11,** 1598 (1972).
361. Fackler, Jr., J. P., Fetchin, J. A., and Smith, J. A., *J. Am. Chem. Soc.* **92,** 2910 (1970).
362. Fackler, Jr., J. P., and Fetchin, J. A., *J. Am. Chem. Soc.* **92,** 2912 (1970).
363. Bonamico, M., Dessy, G., and Fares, V., *Chem. Commun.* p. 1106 (1969).
364. Furlani, C., Piovesana, P., and Tomlinson, A. A. G., *J. Chem. Soc., Dalton Trans.* p. 212 (1972).
365. Furlani, C., and Luciani, M. L., *Inorg. Chem.* **7,** 1586 (1963).
366. Bonamico, M., Dessy, G., Fares, V., and Scaramuzza, L., *J. Chem. Soc., Dalton Trans.* p. 2250 (1975).
367. Bonamico, M., Dessy, G., and Fares, V., *J. Chem. Soc., Dalton Trans.* p. 2315 (1977).
368. Bonamico, M., Dessy, G., Fares, V., Flamini, A., and Scaramuzza, L., *J. Chem. Soc., Dalton Trans.* p. 1743 (1976).
369. Maltese, M., and Giancarlo, G., *J. Chem. Soc., Dalton Trans.* p. 1601 (1977).
370. Bonamico, M., Dessy, G., and Fares, V., *Chem. Commun.* p. 324 (1969).
371. Bonamico, M., and Dessy, G., *Chem. Commun.* p. 483 (1969).
372. Bonamico, M., Dessy, G., Fares, V., and Scaramuzza, L., *Cryst. Struct. Commun.* **2,** 201 (1973).
373. Flamini, A., *J. Coord. Chem.* **4,** 205 (1975).
374. Coucouvanis, D., Baenziger, N. C., and Johnson, S. M., *Inorg. Chem.* **13,** 1191 (1974).
375. Hollander, F. J., Caffery, M. L., and Coucouvanis, D., *J. Am. Chem. Soc.* **96,** 4882 (1974).
376. Ramaswamy, K. K., and Krause, R. A., *Inorg. Chem.* **9,** 1136 (1970).
377. Hynes, M. J., and Brannick, P. F., *J. Chem. Soc., Dalton Trans.* p. 2281 (1977).
378. Furlani, C., Flamini, A., Scamellotti, A., Bellitto, C., and Pioresana, O., *J. Chem. Soc., Dalton Trans.* p. 2404 (1973).
379. Shul'man, V. M., Larionov, V., Podol'skaya, L. A., and Dederov, V. E., *Russ. J. Inorg. Chem.* **16,** 1024 (1971).
380. Mathews, R. J., and Moore, J. W., *Inorg. Chim. Acta* **6,** 359 (1972).
381. Burke, J. M., and Fackler, Jr., J. P. *Inorg. Chem.* **110,** 2744 (1972).
382. Brunner, H., *Z. Naturforsch. Teil B* **24,** 275 (1969).
383. Brinkhoff, H. C., *Recl. Chim. Trav. Pays-Bas* **90,** 377 (1971).
384. Solozhenkin, P., and Kopitsaya, N. I., *Dokl. Akad. Nauk. Tadzh. SSR* **12,** 30 (1969).
385. Brinkhoff, H. C., Cras, J. A., Steggerda, J. J., and Willemse, J., *Recl. Trav. Chim. Pays-Bas* **88,** 633 (1969).

386. Nigo, Y., Masuda, I., and Shinra, K., *Chem. Commun.* p. 476 (1970).
387. Willemse, J., Reuwette, P. H. F., and Cras, J. A., *Inorg. Nucl. Chem. Lett.* **8,** 389 (1972).
388. Lachenal, D., *Inorg. Nucl. Chem. Lett.* **11,** 101 (1975).
389. Hendrickson, A. T., Martin, R. L., and Rhode, N. M., *Inorg. Chem.* **12,** 2980 (1975).
390. Groves, A. K. M., Morrison, N. J., and McCleverty, J. A., *J. Organomet. Chem.* **84,** C5 (1975).
391. McCleverty, J. A., and Harrison, N. J., *J. Chem. Soc., Dalton Trans.* p. 541 (1976).
392. Sceney, C. G., and Magee, R. J., *Inorg. Nucl. Chem. Lett.* **10,** 323 (1974).
393. Tulyupa, F., Usatenko, Yu. I., Garos, Z. F., and Tkacheva, L. M., *Izv. Sib. Otd. Akad. Nauk SSSR Ser. Khim. Nauk* p. 110 (1970).
394. Terent'ev, A. P., Ruhdze, E. G., Dunina, V. V., and Drobysherskaya, E. V., *Dokl. Akad. Nauk. SSSR* **195,** 610 (1970).
395. Van der Linden, J. G. M., *J. Inorg. Nucl. Chem.* **34,** 1937 (1972).
396. Cras, J. A., Wignhoven, J. G., Beurskens, P. T., and Willemse, J., *Recl. Trav. Chim. Pays-Bas* **72,** 1199 (1973).
397. Manzer, L. E., *J. Chem. Soc., Dalton Trans.* p. 1535 (1974).
398. Steele, D. F., and Stephenson, T. A., *J. Chem. Soc., Dalton Trans.* p. 2124 (1973).
399. Allison, J. M. C., and Stephenson, T. A., *J. Chem. Soc., Dalton Trans.* p. 254 (1973).
400. Cornock, M. C., Steele, D. F., and Stephenson, T. A., *Inorg. Nucl. Chem. Lett.* **10,** 785 (1974).
401. Kurashvili, L. M., and Zavorakhina, N. A., *Zh. Prikl. Spektrosk.* **21,** 696 (1974); *Chem. Abstr.* **82,** 36,893 (1975).
402. Fackler, Jr., J. P., and Seidel, W. C., *Inorg. Chem.* **8,** 1631 (1969).
403. Hayward, P. J., Blake, D. M., Wilkinson, G., and Nyman, C. J., *J. Am. Chem. Soc.* **92,** 5873 (1970).
404. Cornock, M. C., Gould, R. O., Jones, C. L., Owens, J. D., Steele, D. F., and Stephenson, T. A., *J. Chem. Soc., Dalton Trans.* p. 496 (1977).
405. Cornock, M. C., and Stephenson, T. A., *J. Chem. Soc., Dalton Trans.* p. 501 (1977).
406. Barstow, T. J., and Whitefield, H. J., *J. Inorg. Nucl. Chem.* **36,** 97 (1974).
407. Bereman, R. D., and Nalewajek, D., *Inorg. Chem.* **16,** 2687 (1977).
408. Bapzargashieva, S. D., Arkhangel'skaya, O. T., Kukushkin, Yu. N., and Sibirskaya, V. V., *Zh. Neorg. Khim.* **18,** 1715 (1973).
409. Kukushkin, Y. N., Sibirskaya, V. V., Bapzargashieva, S. D., and Arkhangel'skaya, O. T., *Russ. J. Inorg. Chem.* **18,** 847 (1973).
410. Dobozynski, E. D., and Angelici, R. J., *J. Organomet. Chem.* **76,** 653 (1974).
411. Lisg, J. M., Dobozynski, E. D., Angelici, R. J., and Clardy, J., *J. Am. Chem. Soc.* **97,** 656 (1975).
412. Bowden, F. L., Giles, R., and Haszeldine, R. N., *Chem. Commun.* 578 (1974).
413. Hynes, M. J., and Moran, A. J., *J. Chem. Soc., Dalton Trans.* p. 2280 (1973).
414. Lin, I. J. B., Chen, H. W., and Fackler, Jr., J. P., *Inorg. Chem.* **17,** 394 (1978).
415. Burke, J. M., and Fackler, Jr., J. P., *Inorg. Chem.* **11,** 3000 (1972).
416. Yordanov, N. D., and Shopov, D., *J. Chem. Soc., Dalton Trans.* p. 883 (1976).
417. Herring, F. G., and Tapping, R. L., *Can. J. Chem.* **52,** 4017 (1974).
418. Golding, R. M., Rae, A. D., Ralph, B. J., and Sullogoi, L., *Inorg. Chem.* **13,** 2499 (1974).
419. Besson, W. D., and Du Proy, A. L., *J. Chem. Soc., Dalton Trans.* p. 1708 (1974).
420. Hendrickson, A. R., Martin, K. L., and Taylor, D., *J. Chem. Soc., Chem. Commun.* p. 843 (1975).
421. Cras, J. A., *Proc. Int. Conf. Coord. Chem., 16th, 1974,* p. 2.13b.

422. Cras, J. A., Willemse, J., Gal, A. W., and Hummelink, B. M. G., *Recl. Trav. Chim. Pays-Bas* **92,** 641 (1973).
423. Peyronel, G., Pignedoli, A., and Antoline, L., *Acta Crystallogr., Sect. B* **28,** 3596 (1972).
424. Einstein, F. W. B., and Field, J. S., *Acta Crystallogr., Sect. B* **30,** 2928 (1974).
425. Choi, S. M., and Wasson, J. R., *Inorg. Chem.* **14,** 1964 (1975).
426. Savino, P. C., and Bereman, R. D., *Inorg. Chem.* **12,** 173 (1973).
427. Thomas, P., and Poveda, A., *Z. Chem.* **10,** 153 (1971).
428. Annuar, B., Hill, J. O., and Magee, R. J., *J. Inorg. Nucl. Chem.* **36,** 1253 (1974).
429. van der Linden, J. G. M., *Recl. Trav. Chim. Pays-Bas* **90,** 1027 (1971).
430. Avdeef, A., and Fackler, Jr., J. P., *J. Coord. Chem.* **4,** 211 (1975).
431. Brinkhoff, H. C., Matthyssen, A. G., and Oomens, C. G., *Inorg. Nucl. Chem. Lett.* **7,** 86 (1971).
432. Beurskens, P. T., Cras, J. A., and van der Linden, J. G. M., *Inorg. Chem.* **9,** 475 (1970).
433. Bergendahl, T. J., and Bergendahl, E. M., *Inorg. Chem.* **11,** 638 (1972).
434. Hesse, R., and Jennische, P., *Acta Chem. Scand.* **26,** 3855 (1972).
435. Farrell, F. J., and Spiro, T. G., *Inorg. Chem.* **10,** 1606 (1971).
436. Dirand, J., Ricard, L., and Weiss, R., *Transition Met. Chem.* **1,** 2 (1975).
437. Perutz, M., Personal communication.
438. Critchlow, P. B., and Robinson, S. D., *J. Chem. Soc., Dalton Trans.* p. 1367 (1975).
439. Fackler, Jr., J. P., and Coucouvanis, D., *J. Am. Chem. Soc.* **89,** 1745 (1967).
440. Barstow, T. J., and Whitefield, H. J., *J. Inorg. Nucl. Chem.* **36,** 97 (1974).
441. Griffith, W. P., and Pawson, D., *J. Chem. Soc., Dalton Trans.* p. 1315 (1973).
442. Merlino, S., *Acta Cryst.* **B-25,** 2270 (1969).

GRAPHITE INTERCALATION COMPOUNDS

HENRY SELIG

Department of Inorganic and Analytical Chemistry,
Hebrew University of Jerusalem, Jerusalem, Israel

and

LAWRENCE B. EBERT

Exxon Research and Engineering Company, Linden, New Jersey

I. Introduction	281
II. Covalent Compounds of Graphite	283
A. Graphite Oxide	283
B. Carbon Monofluoride	284
III. Lamellar Compounds	285
A. Graphite–Alkali Metals	285
B. Graphite Acid Salts	289
C. Halogen Graphite Compounds	291
D. Noble-Gas Fluorides–Graphite	296
E. Graphite–Metal Halides	300
F. Graphite–Metal Oxides	314
IV. Residue Compounds	314
V. Applications of Intercalation Compounds	315
A. Chemical Reagents	315
B. Electrochemical Applications	316
C. Intercalation Compounds as Highly Conducting Materials	317
D. Catalysis by Graphite Intercalation Compounds	318
References	319

I. Introduction

Twenty years ago, Walter Rüdorff wrote a review for this series entitled "Graphite Intercalation Compounds" (*R1*). It was one of four definitive articles to come out in 1959 and 1960 (*H1, C1, U1*), a period of intense activity in graphite research. We have now again reached the "fever pitch," with not only the appearance of several new articles (*E1, H2, W1*) but also the convening of the first international conference dedicated exclusively to graphite compounds (*H3*). In the following, we shall concentrate on work performed between 1974 and the present, with particular emphasis on advances in the chemistry of graphite–acceptor compounds.

A glance at the structure of graphite, illustrated in Fig. 1, reveals the presence of voids between the planar, sp²-hybridized, carbon sheets. Intercalation is the insertion of ions, atoms, or molecules into this space without the destruction of the host's layered, bonding network. Stacking order, bond distances, and, possibly, bond direction may be altered, but the characteristic, lamellar identity of the host must in some sense be preserved.

The interlayer voids are frequently attacked, to yield a periodic sequence of filled and empty spaces. The stage of a compound is defined as the ratio of host layers to guest layers, so that a first-stage compound, in which every interlayer void is filled, is the most concentrated. There is evidence that the staging concept may be, to some extent, an idealization, as is illustrated in Fig. 2.

Historically, compounds of graphite have been placed in three categories, depending on the strength of interaction between reacting species and graphite:

(1) Covalent compounds, arising from the attack of strong oxidizing systems, such as fluorine or Mn(VII), on graphite. The aromatic planarity of the graphite sheet is destroyed, and a buckled, sp³-hybridized sheet is created.

(2) Lamellar compounds, arising from the attack of moderately strong reductants (such as potassium metal) or oxidants (such as AsF_5)

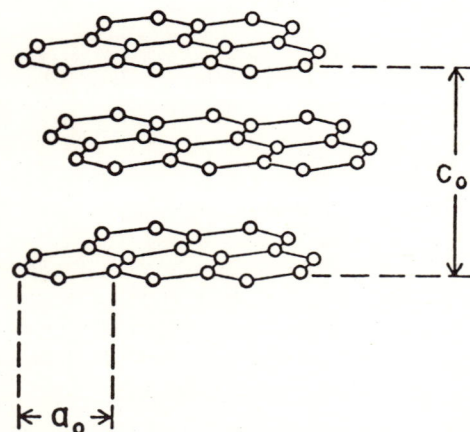

Fig. 1. The hexagonal modification of graphite. Typically, $a_0 = 2.46$ and $c_0 = 6.7$ Å, so that the spacing between adjacent, carbon planes is 3.35 Å.

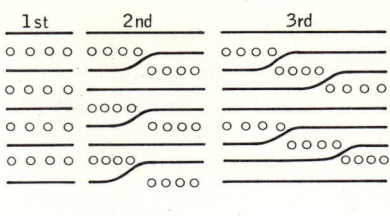

Fig. 2. A nonclassical view of staging proposed by Herold and co-workers (H3). From left to right are first-, second-, and third-stage compounds. (—, Carbon layer; oooo, alkali-metal layer.)

on graphite. The aromaticity of the graphite sheet is largely preserved, and the conductivity of the sheets (i.e., σ_a) increases dramatically.

(3) Residue compounds, arising from the decomposition of lamellar compounds by thermal or *in vacuo* treatment.

II. Covalent Compounds of Graphite

A. Graphite Oxide

Although it is the oldest of graphite compounds (B1), graphite oxide is still among the least understood. Made by the action of such strongly oxidizing systems as nitric acid–alkali chlorate, or sulfuric acid–sodium nitrate–permanganate, on graphite (H4), its formula is represented as $C_8O_2(OH)_2$ (C2) or $C_7O_4H_2$ (V1). Boehm (B2) suggested that graphite oxide formation is preceded by oxidative intercalation of the inorganic acid to yield a graphite "acid salt" of formula $C_n^+A^-(HA)_2$.

There are two schools of thought as to the structure of graphite oxide. Ortho or meta ether linkages have been postulated to enforce a puckering of planes (A1), whereas a keto–enol tautomerism was suggested to keep the carbon layers planar (C3).

With its oxygen functionality, graphite oxide has chemical properties more akin to those of layered disulfides or sheet silicates than to those of graphite (G1, T1, A2). Many studies have been of an extremely applied nature: the possibility of fluorination (L1, N1), redox potentials in the presence of hydrogen peroxide (V2), the apparent density (L2), the adsorption isotherms with nitrogen (L3), and the diffusion of Cs^+ in graphite oxide (R2).

B. Carbon Monofluoride

As a review on poly(carbon monofluoride) has recently appeared (*K1*), our treatment here will be cursory.

Made by direct combination of graphite and fluorine in the temperature range 420–630°C, carbon monofluoride is a white, nonconducting powder, chemically more similar to Teflon than to graphite.

As with graphite oxide, there are currently two views as to the structure of carbon monofluoride. Although detailed X-ray diffraction work suggested a chair arrangement of the sp^3-hybridized, carbon sheets (*M1*), second-moment calculations of the adsorption mode of the fluorine nuclear magnetic resonance suggested that a boat arrangement is more plausible (*E2*). The structures are illustrated in Fig. 3.

It is curious that the chair–boat problem, which is most associated with small, liquid-state molecules, arises in the context of solid-state research (*B3, I1*). Although the paucity of useful experiments militates against a definitive solution here (*E3*), the frequency independence of the NMR second moment (*E2*), the absence of an observable free-induction decay ($T_2 < 25$ μs) in the pulsed NMR spectrum (*E1*), and the smoothness of the absorption mode itself (*S1*), all argue against the

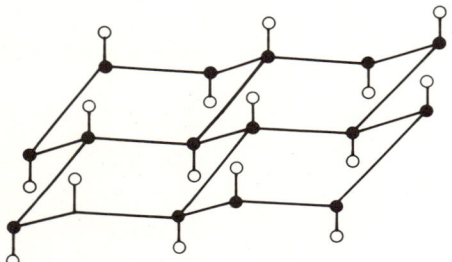

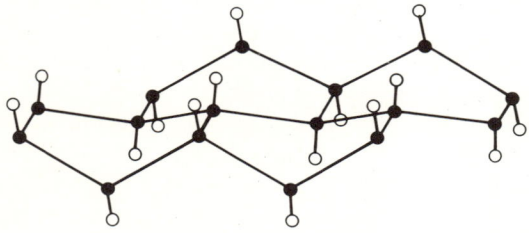

FIG. 3. The chair (top) and boat (bottom) models for the structure of poly(carbon monofluoride). (●, C; ○, F.)

presence of a F_2^- species in carbon monofluoride. Additionally, there are no peculiar features in the F_{1s} and F_{2s} ESCA profiles, and the C_{1s} profile indeed suggests that the boat structure more closely fits experimental data than the chair (*C4*).

Carbon monofluoride has found use as a lubricant. Studies by Fusaro and Sliney (*F1, F2*) and Gisser (*G2*) showed that carbon monofluoride is superior to molybdenum disulfide under many conditions. Grafting of monomers can further improve lubricity (*B4*). Electrochemical applications are discussed later.

Although the covalent compounds of graphite are thus important in their own right, they represent the extreme form of oxidative intercalation. The use of fluoride compounds to achieve highly conductive materials may ultimately lead to new forms of graphite fluoride (*S1*).

III. Lamellar Compounds

A. Graphite–Alkali Metals

First reported by Fredenhagen in 1926 (*F3, F4*), the graphite–alkali-metal compounds possess a relative simplicity with respect to other intercalation compounds. To the physicist, their uncomplicated structure and well defined stoichiometry permit reasonable band-structure calculations to be made (*S2, I2*); to the chemist, their identity as solid, "infinite radical-anions" frequently allows their useful chemical substitution for such homogeneous, molecular-basis reductants as alkali metal–amines and aromatic radical anions (*N2, B5*).

A number of synthetic procedures are available (*N2*). (*1*) For precisely defined stoichiometries, the isobaric, two-bulb method of Herold is preferred (*H5, H6, H2*). (*2*) To generate compounds suitable for organic synthesis work, graphite and alkali metal may be directly combined, and heated under inert gas (*P1, L4*). (*3*) Electrolysis of fused melts has been reported to be effective (*N2*). (*4*) Although alkali metal–amine solutions will react with graphite, solvent molecules co-intercalate with the alkali metal. Utilization of alkali metal–aromatic radical anion solutions suffers the same problem.

Whereas technique (*4*) works for all alkali metals, lithium and sodium behave differently from potassium, rubidium, and cesium with respect to graphite on direct combination. The last three react facilely with graphite, to form compounds C_8M (first stage) and $C_{12n}M$ (stage $n > 1$), but lithium reacts only under more extreme conditions of temperature or pressure, or both, to form compounds of formula $C_{6n}Li$ (*G3*,

$G4$, $Z1$). Sodium reacts to form $C_{64}Na$, although ternary systems with other alkali-metal compounds are known ($A3$, $B6$).

There are a number of graphite–electron-donor systems that are formally analogous to the graphite–alkali metals. Intercalation compounds of graphite with barium ($G5$), with strontium, calcium, and samarium ($G6$), and with europium and ytterbium ($G7$) may be formed by direct combination of the elements. Craven ($C5$) had prepared intercalation compounds of graphite with cerium, samarium, gadolinium, terbium, dysprosium, holmium, erbium, thulium, and ytterbium by utilization of metal–ammonia solutions. In a different vein, both Besenhard ($B7$) and Simonet ($S3$) reported the electrochemical intercalation of tetraalkylammonium cations into graphite, thereby creating donor compounds without the use of metals.

Historically, the C_8M graphite–alkali-metal compounds have been considered to have a hexagonal, two-dimensional network, as illustrated in Fig. 4, leading to an orthorhombic, Bravais lattice through an $A\alpha A\beta A\gamma A\delta$ stacking of respective graphite (A) and alkali metal (α) planes. The C_6M graphite–metal compounds also have sixfold, in-plane, metal coordination, but at a distance of 4.3 Å [graphite a_o × sqr (3)], instead of the C_8M distance of 4.9 Å (graphite a_o × 2). Through a detailed study of multiplicity factors and relative intensities of ($hk0$) and $hk1$) reflections, Lagrange ($L5$) suggested the existence of three, distinct, C_8M, orthorhombic phases, arising from a-axis displacements in the all-metal layers, to yield $A\alpha A\beta A\gamma A\delta$, $A\alpha A\gamma A\delta A\beta$, and $A\alpha A\delta A$-

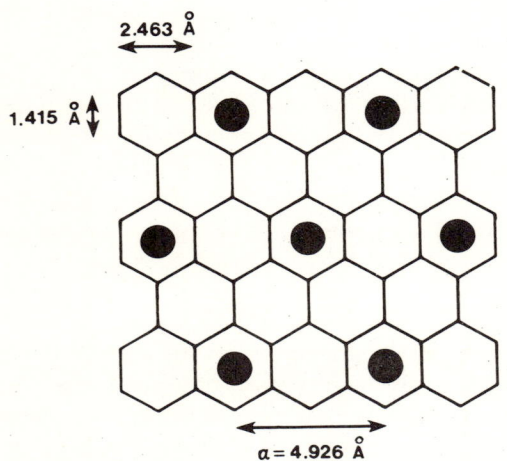

Fig. 4. The in-plane arrangement of a C_8X network typical of many donor and acceptor compounds of graphite.

βAγ stackings along the c axis. Combining this behavior, which arises from "defects" within the ab plane, with the work of Thomas (E4, T2), which suggested stacking disorder along the c axis, it may be concluded that intercalation compounds of graphite may not, in the general case, be entirely well defined structurally. Consistent with this concept is the uncertainty that surrounds the in-plane structure of the second-stage, $C_{12}M$ compounds. Whereas Rüdorff and Shulze (R3) proposed a straightforward, in-plane, metal network, of spacing 4.9 Å (R3), Parry and co-workers concluded that this structure is incorrect (P2, P3).

Raman spectroscopy has been used as a structural probe of graphite–alkali-metal compounds (N3, E5). Pure graphite has a sharp reflection of E_{2g} symmetry at 1582 cm^{-1} (T3), but alkali-metal-intercalated graphite can reveal modes at different energies, arising from the distinction of environment of a graphite plane directly bounded by two graphite planes, by one graphite plane and one alkali-metal plane, or by two alkali-metal planes. As a given stage would have a well-defined number of each of the symmetry environments, the Raman experiment offers an alternative approach to X-ray diffraction for the determination of stage information. For molecular compounds, such as intercalated, metal amines, spectral position and polarization of the Raman modes could be used as a subtle probe of *molecular* structure, although preliminary work on the acceptor compound C_xAsF_5 (D1, S4) did not, in fact, reveal molecular vibrations.

With the increased structural knowledge of graphite–alkali-metal compounds have come electronic, band-structure calculations on both C_8K (I2) and C_6Li (H7). The study of the potassium compound revealed the coexistence of isotropic, three-dimensional carriers with extremely two-dimensional carriers, and suggested the possibility of charge-density waves, in analogy to established behavior in layered dichalcogenides (T4). The investigation of the lithium compound revealed a possible covalent contribution to interlayer bonds, consistent with the frequent observation of lithium carbide in C_6Li synthesis (N2, G3).

The electronic properties of graphite–alkali-metal compounds have been the subject of a review by Fischer (F5), so we shall not discuss them in depth. One of the most promising techniques is optical spectroscopy, first used by Hennig (H8), to determine the number of carriers from the position of the plasma edge (K2). The optical properties of $C_{24}M$ (blue in reflection) and C_8M (bronze in reflection) are those expected from the "alloying" of graphite (with an inferred plasma edge in the infrared) and alkali metal (with an edge in the ultraviolet). This "sweep" of the plasma edge across the visible spectrum [from 1.4 to

2.6 eV for $C_{36}Rb$ to C_8Rb (*G8*)] is similar to that found for the alkali-metal–tungsten bronze system M_xWO_3 (*D2*).

The color of the intercalation compound may be a useful guideline in monitoring chemistry. The bronze, first-stage C_8K reacts with hydrogen to form a blue, second-stage $C_8KH_{2/3}$, which may then be further intercalated by alkali metal to form a first-stage compound $C_{16}MK_2H_{4/3}$ (*G9*). Conversely, when such aromatic molecules as phenanthrene or perylene (*B8*), or benzene in tetrahydrofuran (*B9*), react with C_8K, radical ions are created, and the resultant solid is again blue.

Comparison of physical and chemical measurements suggested something of the true identity of compounds customarily written as C_8M. Measuring the Knight shift of ^{133}Cs in C_xCs compounds, Carver (*C6*) inferred a 55% ionization of cesium in C_8Cs, but a 100% ionization of cesium in $C_{24}Cs$ and more-dilute compounds; this was directly consistent with Hennig's previous optical data (*H8*), which had suggested 67% ionization in stage 1 graphite–alkali-metal compounds, and 100% ionization for higher stages. Later, Mössbauer measurements (*C7*) on the C_xCs system suggested a 50% ionization for the first stage and a 100% ionization for higher stages, a result qualitatively confirmed by specific-heat measurements (*M2*) and electron-spin resonance observations (*P4*).

For chemical comparison, the reaction of C_8M with water can distinguish between M^+ and M^0 (*B10, B11*). If, as an example, C_8K were entirely atomic, we should expect

$$C_8K + H_2O \rightarrow 8\,C + KOH + 0.5\,H_2. \tag{1}$$

Were the potassium entirely ionic, we should expect

$$C_8^-K^+ + H_2O \rightarrow C_8^-H^+ + KOH. \tag{2}$$

Using these two reactions as extreme cases, hydrogen evolution can be monitored to infer a value of potassium ionicity. Although initial reports (*B10*) suggested a maximum yield of hydrogen of 16% of the theoretical value of reaction (1), corresponding to 84% ionization of potassium, later reports (*E6, E7*) suggested yields as high as 40% of the theoretical (60% ionized), a value consistent with physical measurements. Thus, in going from $C_{24}^+M^-$ to "$C_8^{-0.6}M_{0.6}^+M_{0.4}^0$", a great deal of *atomic* alkali-metal character is added to the compound, and this should result in enhanced chemical activity.

The reductive capability of C_8K has been a subject of interest (*L4*). Uses for C_8K include the reductive cleavage of carbon–sulfur bonds (*S5*), the reductive alkylation of nitriles and esters (*S6*), and the reductive alkylation of aldehydes and ketones (*S7*). The activity of C_8K has been compared to that of high-surface-area, alkali-metal dispersions in the reductive alkylation of ketones (*H9*). Bergbreiter (*B11*) evaluated reaction mechanisms for the interaction of C_8K with protic acids, alkyl and aryl halides, and alkyl sulfonic esters. The potential of C_8K in inorganic reductions has also been evaluated (*U2, U3*).

B. Graphite Acid Salts

The acid salts, known since 1841 (*S8*), are among the best characterized graphite compounds. Their properties have been discussed in recent reviews (*H10, E1*). The acid salts can be prepared either by oxidation with a chemical reagent, or electrochemically. The resulting compounds are ionic in nature, with electrical charges balanced between acid anions and carbon macrocations. They usually also include neutral acid molecules. Intermediate stages are formed, and critical threshold-concentrations are needed for intercalation to occur (*U4*). The general stoichiometry is $C_m^+A^-\cdot nHA$. Acid salts of the oxygen acids sulfuric, chlorosulfonic, fluorosulfonic, selenic, perchloric, and nitric acids have been known for some time (*B12, R4*), as has the salt $C_{24}^+HF_2^-\cdot 4\,HF$ (*R5*). On the other hand, layer compounds of type C_nHF ($n = 4-8$), prepared by reaction of HF with graphite without oxidation (*O1*), could not be confirmed in the authors' laboratory. Conversions of one type of salt to another can sometimes be accomplished by double decomposition (*B12*). Values close to $m = 24$ are repeatedly encountered and seem to represent a structure with relatively close packing. Because of the instability of these compounds in air or water, they are difficult to analyze; this, as well as the presence of defect sites, may account for different values reported for the number (n) of neutral molecules. For bisulfate, it has been reported as 2 (*R4*) or ~2.5 (*A4, A5, H11*), whereas, for nitric (*N4*), perchloric (*A4, F6*), and trifluoromethanesulfonic (*H11*) acids, $n = 3, 2$, and 1.64, respectively. The neutral acid molecules are presumed to participate in hydrogen bonding between themselves and the anions (*A5*), in contrast to Hennig's theory, which suggested a role as spacers to lessen electrostatic repulsion (*H12, D3*). There are, in fact, indications that the incorporation of neutral acid molecules may be altogether unnecessary (*B13*). Simple lamellar salts of the type C_nX were obtained by anodic oxidation of solutions of $LiClO_4$, $NaBF_4$, and KPF_6 in nonaqueous, aprotic solvents (*B13*).

Thermodynamic information has been obtained in different stages of graphite bisulfate (*A5*). The results have been interpreted in terms of a model previously applied to alkali-metal-graphite compounds. Part of the bonding energy results from electrostatic attractions of bisulfate ions to positive carbon layers. It is unlikely, however, that such effects can account for long-range stacking-orders found between different intercalate layers in the low-temperature phase of graphite nitrate (*N4*). Rather, ordering has been attributed to Brillouin-zone effects and redistribution of energy surfaces associated with the latter (*N4*). Order–disorder transitions occur around $-20°C$ (*N4*); these have been confirmed by differential thermal analysis, by a discontinuity of T_1 in proton magnetic resonance spectra (*A6, A7*), and by changes of line width in esr spectra (*K3*). A further transition at $-140°C$ may be due to freezing of rotational moments of nitrate ions (*A6, A7*).

The salts most studied are the graphite nitrates. These occur in several stages that have been formulated as $C_{6n}HNO_3$ ($n = 1, 2, 3$) with a repeat distance $I_c = 7.8 + 3.35 (n - 1)$ Å (*F7*). A general formulation of $C_{(8+2x)n}$ $N_2O_5 \cdot x$ H_2O ($n = 1, 2, 3$; $0 \leq x \leq 3$) has been proposed, leading to two series $C_{5n}HNO_3$ with $I_c = 7.80$ Å and $C_{6n}HNO_3$ with $I_c = 6.55$ Å, which are interconvertible (*F8*). The final composition appears to depend on the nitric acid concentration, but, probably, all contain at least some nitrate ions. Graphite nitrates and other acid salts display metallic behavior, as shown by measurements of various electrical properties, such as resistivity and the thermoelectric power (*U5*), magnetoresistance, and the Hall effect (*U6, U7*). The thermal expansion, specific heat, and magnetic properties of second-stage graphite nitrate confirm metallic properties (*I3*). Similar results were obtained from optical-reflectance studies (*F7*), which were compared with dc transport properties, although not being consistent with a simple Drude model.

Changes in electrical resistivity (*V3*) and mechanical properties (*V3, V4*) of graphite fibers upon nitration have been studied. Increases in elastic modulus, and decreases in tensile strengths, have been related to removal of boundary dislocations by the intercalation process proposed elsewhere (*N4*).

NMR studies on graphite–phosphoric acid showed simultaneous, motional narrowing of both ^1H and ^{31}P resonances above 225 K, indicating high mobility of phosphoric acid in the compound (*E8*). Chlorosulfonic acid is inserted alone into graphite in the presence of many inorganic chlorides. The reaction temperature and stage seem to be related to the redox potential of the $M^{n+}-M$ couple (*M3*).

Graphite acid salts have been considered in terms of possible practical applications. Their potential as useful electrodes in high-energy-density batteries has been reviewed (*A8*), as well as their potential usefulness as chemical reagents (*B14*).

C. Halogen Graphite Compounds

Reactions of graphite with the halogens are of such a variegated nature that they are best discussed element by element. The reaction of fluorine is rather different in nature from those with the other halogens, and, in the limit, leads to compounds in which the aromatic graphite layer-structure is totally destroyed, leading to a perfluorocyclohexane net. Regarding the other halogens, a comparison of their interatomic distances [Cl_2 (1.99), Br_2 (2.28), I_2 (2.66), ICl (2.33), and IBr (2.47 Å)] with the distance separating centers of adjacent hexagons in the graphite layers (2.46 Å) suggested that epitaxy may play an important role in compound formation (*H13, H14*). Thus, iodine does not intercalate with graphite at all. The existence of graphite–chlorine is moot; and if such compounds exist at all, they are relatively unstable, and exist only at low temperatures.

On the other hand, bromine compounds are well known, and their formation and properties have been most thoroughly studied. Epitaxy is not the only factor affecting intercalation, however (*H14*); it does not explain why some compounds (ICl) form first stages, whereas others (IBr, Br_2) form only second stages. Other factors suggesting a role in compound formation are the relative electronegativities of the halogens and the graphite (*D4*), as well as polarizabilities of the former (*R6*). The amount of charge transfer to or from an atom directly over the carbon undoubtedly plays a role as well (*H13*).

Chlorine–Graphite.—According to Hennig (*H15*), graphite reacts slowly with chlorine at low temperatures accompanied by a drastic diminution in the resistance of the graphite. Similar conclusions were reached by Juza *et al.* (*J1*), who reported that the reaction of chlorine at $-78°C$ removes the anomalous diamagnetism of the graphite. The reaction rate is strongly dependent on particle size and temperature, reaching a maximum at $-12°C$. Above $0°C$, no chlorine uptake was noted (*J1, J2*). A limiting stoichiometry C_8Cl is supposedly attained (*H15, J1*). X-Ray evidence for chlorine–graphite was adduced by Juza and Seidel (*J3*). They suggested a second-stage compound, having $a = 2.45$ Å and $c = 10.09$ Å, indicating a 50.5% expansion of the lattice constant c over graphite. Their starting material was, however, $C_{104}Br$ rather than virgin graphite. The composition determined at $-57.7°C$ by isothermal tensimetry was $C_{8.5}Cl$, the deviation from ideal C_8Cl being attributed to lattice imperfections. Others have been unable to duplicate these results (*T5*), although they may have encountered difficulties in compound transfer. According to Herold (*H6*), the chlorine uptake by graphite at low temperature is due to adsorption only, no intercalation taking place. Hooley (*H16*) has come to similar conclusions based on adsorption-isotherm measurements over a wide temperature-range. He also pointed out that changes in susceptibility do not necessarily prove intercalation. Although the existence of chlorine–graphite is debatable, there seems to be agreement that chlorine–graphite residues are formed. Moreover, it appears that the reaction can be catalyzed by the presence of HCl or by using residue compounds as starting materials (*J4*).

Iodine–graphite.—In contrast to the situation with chlorine, there is general agreement that iodine is not intercalated into graphite (*H6, H15*). However, reaction of iodine with a bromine residue led to a marked decrease in its diamagnetic susceptibility (*J4*). *Bromine–graphite.*—As early as 1933, the uptake of bromine by graphite was shown to lead to a compound of stoichiometry C_8Br (*F9*). X-Ray investigations by Rüdorff showed (*R7*) that C_8Br has one bromine layer for every two graphite layers, the expanded layer-spacing becoming 7.05 Å. The existence of this second-stage compound, as well as stages poorer in bromine, namely, $C_{12}Br$ and $C_{16}Br$, was confirmed by Herold (*H6, B15*). Single crystals of natural graphite treated with bromine vapor were examined by electron and X-ray diffraction (*E9, E10*). The saturated compound is, indeed, second-stage, with $I_c = 10.3$ Å (*E9*). Upon desorption, a relatively stable fourth stage, $C_{28}Br$, is formed, having a repeat distance $I_c = 17.0$ Å. The bromine layers are well-defined, two-dimensional structures with $a_0 = 8.5$ Å (along [$10\bar{1}0$]) and $b_0 = 17.25$ Å (along [$1\bar{2}10$]). The interatomic distances in the bromine layers are virtually identical with those of molecular bromine, indicating that van der Waals forces play no less a role than charge transfer in determining the configurations between layers (*E10*). Although the bromine molecules have precise orientations with respect to adjacent, graphite layers, relative positions could not be uniquely determined. The formation of a $C_{14}Br$ phase was also reported in which bromine layers are intercalated into every second layer-spacing of graphite. As the compound C_8Br is formed from the latter without alteration of layer sequences, additional bromine appears to enter already existing bromine layers.

In a subsequent study (*S9*), isotherms of bromine on pyrolytic graphite showed the presence of several phases $C_{4n}Br$ ($n = 2$ to 5). X-ray studies confirmed these to be stages 2 to 5, respectively. At intermediate concentrations, X-ray patterns showed mixtures of higher and lower stages. The density and configuration of intercalated bromine molecules were believed to be the same in all stages. Other structural types of lower density reported by Eeles and Turnball (*E10*) were believed to be metastable states.

An electron-microscope study (*H17*) showed the existence in bromine–graphite of isolated dislocations bounding intercalated reactant layers. Such migrating loops of occluded reactant are believed to account for residues retained upon decomposition. The existence of crystallites having grain boundaries within larger graphite flakes has also been postulated to account for isotopic exchange-rate of bromine with bromine–graphite (*S10*).

An electron-diffraction study showed reversible changes in pattern as a function of temperature (*C8, C10*). These were interpreted as order–disorder transitions for the intralayer structure, occurring for bromine–graphite at 108°C. The intralayer intercalate structure was found to be independent of intercalate concentration. Therefore, a concentration increase changes the number of intercalate layers without changing the arrangement within an intercalate layer. Evidence for ordering persists up to the fifth stage, which contradicts the well-known concept (*H18*) that, in residues, the intercalant resides at the structural defects (*C9*). Another reversible transition at 194°C, having a large hysteresis effect, was identified with charge-density waves (*C8*).

Simultaneous weight-change and diamagnetic-anisotropy measurements upon adsorption and desorption of bromine in PG have been carried out (*M4*). They showed that final susceptibility values are reached well before full bromine saturations, and remain until 80% desorption is attained.

Many attempts have been made to elucidate the intercalation mechanism. No clear picture has yet emerged, but it is obvious that the nature of the graphite, as well as the manner in which the intercalation process is carried out, strongly influence the experimental results. Saunders, Ubbelohde, and Young (*S11*) postulated that the absorption process involves the withdrawal of electron charge from all layers by bromine adsorbed by carbon edge atoms. This "unpins" the layers, allowing the penetration of more bromine molecules. They used conductivity measurements to show that the threshold pressure is lower for more perfect graphite. Confirmation of threshold lowering was obtained on even more perfect graphite specimens (*U8*). The rate of intercalation of natural graphite by bromine was measured over a large range of crystal areas and thicknesses (*H19*). Results indicated that diffusion is independent of thickness over a factor of two, and all layers are attacked randomly. Further experiments (*H20*) on marked-off

HOPG cylinders showed that intercalation starts close to the basal planes and proceeds thence to the central layers. Cylinders capped at both ends with glass caps and grease are not intercalated. The layer system is split into a number of discs at a rate that increases with the degree of alignment of layer planes. For non-heat-treated PG, the threshold pressure increases with the number of layer planes in the sample. Similar results confirming this mechanism have been obtained with metal halides (*H13*). A mechanism of intercalation based on these results has been discussed by Hooley (*H22*).

One of the cardinal problems regarding bromine–graphite (as well as other intercalates) has been the extent to which intercalate molecules are ionized. Hennig (*H15*) found that the presence of bromine in graphite drastically lowers the resistance and magnetoresistance, the Hall effect being similar for graphite bisulfate of the same resistivity. He concluded that the compound should be formulated $C_n^+Br^-\cdot 3\ Br_2$, i.e., only one out of seven bromine atoms is ionized. The fact that only part of the molecules is ionized was explained by mutual electrostatic repulsion of negative ions requiring uncharged "spacers" between them. The electrical conductivity has been reinvestigated many times, and it depends on the manner of intercalation and the type of graphite (*U8, U9, B16, T6, S12*). Although the results differ in detail, there appears to be general agreement that some intercalate ionization occurs. Blackman *et al.* (*B16*) concluded that, at high dilutions, the bromine is fully ionized, but, as the concentration increases, the total charge-transfer continues, but fractional charge transfer per bromine falls below unity. Others (*D5*) reinterpreted Hennig's Hall-effect data (*H15*) according to the rigid-band model, and concluded that the fraction of ionized, molecular intercalate is $\nu = 0.025$, consistent with infrared magnetoreflection results (*P5*). Raman-scattering experiments (*S13, D1*) showed that the molecular vibration attributed to intercalated bromine molecules is shifted to 242 from 323 cm^{-1} in the free molecule. The low frequency is attributed to coupling with the E_{2g_2} mode for carbon atoms in the adjacent plane. Weak fine-structure observed at 77 K was attributed to ionized, intercalate species. A summary of all experimental evidence related to halogen ionization showed it to be consistent with low intercalate ionization ($\nu = 0.02$) and preservation of molecular identity of the halogen intercalates (*D6*). Hennig's considerably higher value ($\nu = 0.29$) can be reconciled with this if charged, molecular, bisulfate ions are screened in the graphite layers bounding the intercalate layers, i.e., only a fraction of the charge associated with intercalate layers will contribute to delocalized charge-density in the graphite layers.

Graphite Intercalation Compounds with Interhalogens

Given the proper conditions, all known halogen fluorides that can be isolated in stable form at room temperature will form layer compounds with graphite. Iodine pentafluoride was first reported to form an intercalate of composition $C_{18}IF_5$ (*O2*). Later work appeared to confirm this, suggesting a first-stage compound, $C_{8.7}IF_5$, as well (*S14*). It has now been shown that IF_5 intercalation is catalyzed by HF (*S15*) and such Lewis acids as BF_3 (*S16*). Both thermogravimetric and mass-spectrometric studies showed that IF_5 is evolved as a function of temperature, in stages (*S15*). Broad-line, ^{19}F-NMR spectra showed a resonance that could not be identified with any known, pentavalent iodine fluoride species.

The reaction of iodine heptafluoride is accompanied by reduction to the pentavalent state (*S15*). The intercalation process can thus be described in terms of the reaction

$$x\,C + y\,IF_7 \rightarrow C_x(IF_5)F_{2y} + (y - 1)\,IF_5.$$

Infrared spectra and ^{19}F-NMR spectroscopy showed the presence of IF_5 and covalently bonded fluorine. Grafoil turns white upon intercalation with IF_7; this is reminiscent of graphite fluoride, $CF_{1.12}$ (*L6*). The IF_7 intercalate also evolves IF_5 upon heating, but at much higher temperatures than C/IF_5; this has been attributed to the lowered mobility of IF_5 in the fluorinated matrix, which may no longer be planar. At 450°C, considerable amounts of fluorocarbons are evolved.

A layer compound, $C_{8.9}BrF_3$, stable up to 179°C, has been reported (*O2*). Fluorine NMR and infrared spectra have been interpreted as indicating that the BrF_3 is present in molecular form. In another study (*S15*), a stoichiometry $C_{8.9}BrF_3$ was observed on the basis of weight increase of the original graphite, but considerable bromine evolution was observed, and chemical analyses confirmed that the graphite was partially fluorinated. Thermal stabilities and ^{19}F-NMR shifts also differed sharply from those reported earlier. It is evident that experimental conditions, including types of graphite used, play a decisive role in the intercalation process.

A study of the intercalation of BrF_5 led to inconclusive results (*S15*). The difficulty may result from reduction of BrF_5 to BrF_3, which itself intercalates.

Of the chlorine fluorides, only the monofluoride does not intercalate in the neat state. It does, however, intercalate in the presence of HF (*S15*), BF_3, or PF_5 (*S16*). Chlorine trifluoride intercalates into graphite with simultaneous fluorination of the graphite lattice (*S15*), releasing

ClF in the process. Upon heating, only chlorine and fluorocarbons are emitted. In the presence of HF, a different product is obtained (*O3*). The green-colored compound formed, with liberation of ClF, has the composition $C_{14}F \cdot ClF_3 \cdot 3\,HF$. Infrared spectra have been interpreted in terms of the structure $C_{14}^+ H_2 F_3^- ClF_2^+ HF_2^-$. However, some of the infrared bands ascribed to $H_2F_3^-$ and HF_2^- ions could be assigned to covalent C–F bonds in the intercalate. Reactions of $C_{14}F \cdot ClF_3 \cdot 3HF$ with the acids CF_3COOH, CH_3COOH, $HClO_4$, and HNO_3 gave products in which the three HF molecules were replaced by one molecule of the respective acid (*N5*). Reaction of graphite with solutions of cesium fluoride in ClF_3 gives a compound of stoichiometry $C_4F \cdot ClF_3$ (*N6*). Interaction of graphite with solutions of antimony pentafluoride in ClF_3 leads to a product of composition $C_8F \cdot SbF_5 \cdot ClF_3$ (*O4*). On the basis of infrared spectroscopy, the structure $C_8^+ SbF_6^- \cdot ClF_3$ has been proposed.

Although iodine does not intercalate, and the situation regarding chlorine is controversial, ICl does intercalate, to form C_8ICl and higher-stage compounds (*R6, H15, T5*); this has been attributed to the importance of polarizability of the molecule, combined with its high electron-affinity (*R6*). The threshold for ICl intercalation is even considerably lower than that for Br_2 (*U8*). Upon intercalation, the graphite resistance decreases markedly (*H15*). A theory relating the a-axis conductivity, σ_a, to intercalate concentrations, X, has been developed by Fischer (*F10*) using some simplifying assumptions. The predicted behavior that σ_a is proportional to $X^{1/2}$ seems to hold approximately only for ICl, however; this has been attributed to the unusually large layer-spacings and correspondingly weak interactions between layers (*F10*). The identity period, $I_c = 21.15$ Å, emcompasses two carbon planes (*R6*). The ICl layers become disordered above the transition temperature of 43°C (*C10*).

Iodine monobromide forms an intercalate with limiting composition $C_8I_xBr_{1-x}$ (*C9*) when natural graphite is used. Higher stages $C_9I_xBr_{1-x}$ and $C_{12}I_xBr_{1-x}$ have also been shown to exist (*C9*). An order–disorder transition within IBr layers occurs at ~60°C (*C10*).

Ternary compounds $C_8Br_xCl_{1-x}$ have been synthesized. The configuration having $x = 0.55$ seems to be especially stable, and the limiting value in the bromine-poor regime is $x = 0.33$ (*F11*).

D. NOBLE-GAS FLUORIDES–GRAPHITE

It has been pointed out on a number of occasions that the chemistry of the noble gases resembles in many respects that of its neighboring elements, namely, the halogens. Reasoning based on this analogy led to the synthesis of the first intercalate with xenon oxide tetrafluoride (*S14*). Since then, intercalates have been prepared with all known binary fluorides of the noble gases, xenon and krypton (*S1, S18, S19, N7, R8*). These are listed in Table I.

The identification of the valence state of the intercalated xenon species has been facilitated by ^{19}F-NMR spectroscopy (*E11*). It has been shown (*F12*) that both the fluorine chemical-shift and the ^{129}Xe–^{19}F spin–spin coupling-constant are strongly dependent on the oxidation state of the central xenon atom (see Fig. 5). By using this technique, it has been determined that, upon intercalation, some of the noble-gas fluorides are reduced to the next lower oxidation state (XeF_4 and XeF_6), whereas others (XeF_2 and $XeOF_4$) are intercalated without attendant reduction. However, the proviso must be added that, in the case of XeF_2, reports from different laboratories conflict as to the nature of the intercalation process. Nikolaev *et al.* (*N7*) reported that graphite intercalates with solutions of XeF_2 in anhydrous hydrogen fluoride (AHF) leading to a product of composition $C_{17}XeF_2 \cdot 1.3$ HF, with no oxidation of graphite taking place. This material was reported to undergo oxidation at 155°, with formation of a mixture of graphite and a new product, $C_9F \cdot XeF_2 \cdot 3$ HF. The latter decomposes at 470°, liberating, *inter alia*, xenon and fluorocarbons. The presence of XeF_2 and HF in the product was shown by infrared spectroscopy.

In another study (*R8*), it was found that graphite does not intercalate with neat XeF_2 or with solutions of XeF_2 in acetonitrile. However, reaction with solutions of XeF_2 in AHF led to copious xenon evolution, indicating that oxidation does take place, even at room temperature. Broad-line, ^{19}F- and ^{1}H-NMR spectra (*E11*) showed the presence of both XeF_2 and HF in the product, but no definite stoichiometry could be as-

TABLE I

INTERCALATES OF GRAPHITE WITH NOBLE-GAS FLUORIDES

Formula	Stage	Color[a]	Ref.
$C_{8.7}XeOF_4$	1		*S14*
$C_{19}XeF_6$			*S18*
$C_{28}XeF_4$	2		*S19*
$C_{40}XeF_4$	3		*S19*
$C_9F \cdot XeF_2 \cdot 3$ HF		Brown	*N7*
$C_{17}XeF_2 \cdot 1.3$ HF			*N7*
C_mKrF_n ($n > 2$)			*S1*
$C_xF(HF)_y(XeF_2)_z$		Blue in the presence of HF	*R8*

[a] Where not otherwise indicated, no color change occurred upon intercalation.

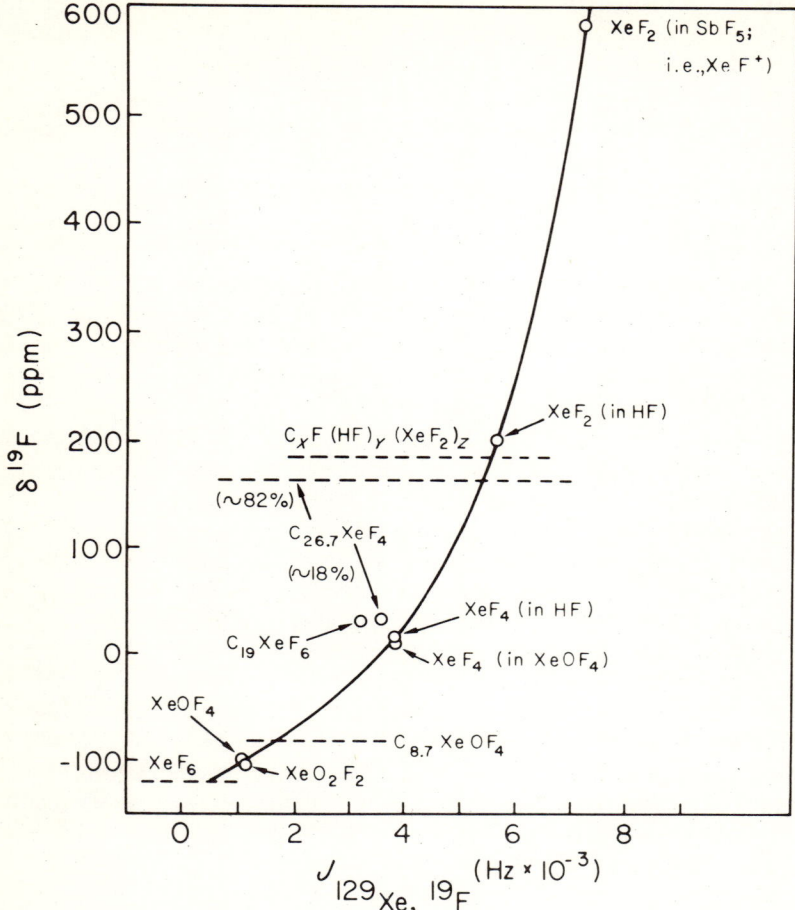

FIG. 5. Dependence of the spin–spin coupling-constant and the ^{19}F chemical shift on the oxidation state of the central xenon atom.

signed, because of the variability of the products. It is interesting that, in the presence of liquid HF, the solid product has a blue color reminiscent of that of such graphite acid salts as $C_{24}^{+}HF_{2}^{-}\cdot(HF)_{4}$. The origin of the discrepancy is unknown, although it may be related to the different types of graphite used.

Xenon tetrafluoride forms intercalates with graphite at room temperature (S19). Powder and Grafoil have both been used, and products have been isolated having stoichiometries $C_{28}XeF_{4}$ and $C_{40}XeF_{4}$ that apparently correspond to the second- and third-stage compounds, respectively. Analysis of the oxidation power of these intercalates showed only two equivalents of iodine liberated per mole of intercalate (*S19*); this is consistent with ^{19}F-NMR measurements that showed that the major species in the intercalate is XeF_2, although some XeF_4 is present as well (*S19*). Both thermogravimetric measurements (*S19*) and mass spectra (*S20*) showed that, at room temperature, the second-stage compound liberates slight amounts of xenon difluoride as well as traces of XeF_4. Xenon difluoride evolution ceases at ~100°. The rest of the xenon is held very tenaciously, and xenon evolution becomes appreciable only above 350°. At this temperature, fluorocarbon emission also becomes appreciable, reaching a maximum at ~380°. This feature is common to many of the inorganic fluoride intercalates, and apparently corresponds to the decomposition of covalent, carbon–fluoride bonds formed either during the initial intercalation or subsequent fluorination of the graphite matrix (*S20*). Preliminary investigations showed that the intercalate may be useful as a moderate fluorinating agent in organic chemistry.

Xenon hexafluoride forms with graphite an intercalate of nominal stoichiometry, $C_{19}XeF_6$ (*S18*). Both the ^{19}F chemical-shift and the $^{129}Xe-^{19}F$ coupling-constant showed that the intercalated species is XeF_4. This conclusion was confirmed by ^{129}Xe Mössbauer measurements (*P6*), which showed little or no XeF_6 present, but, instead, a signal corresponding to XeF_4 or XeF_2, based on its quadrupole splitting and absence of isomer shifts. XeF_2, can however, be excluded, on the basis of NMR spectra. Because the compound exhibits an oxidation power corresponding to hexavalent xenon, it has been tentatively proposed that it has the formula $(C^+_{9.5}F^-)_2XeF_4$ (*R8*). In analogy to the behavior of graphite–antimony pentafluoride intercalation-compounds (*E12*), the fluorine line-widths of "$C_{19}XeF_6$" are narrow (0.1–0.2 G) to relatively low temperatures (*E11*, *R8*), suggesting the possibility of translational motion of the inserted species. The intercalate appears to be relatively stable thermally. Thermogravimetric analysis indicated that substantial weight-loss occurs only above 400°C (*R8*). Mass spectrometry showed that the minor weight-loss at lower temperatures is accompanied by evolution of XeF_4 up to ~150° (*S20*). At higher temperatures, only xenon and fluorocarbons are evolved, probably as a result of fluorination of the graphite lattice by xenon fluoride, and subsequent decomposition of the graphite fluoride formed (*S20*). The XeF_6 intercalate has been used as a conveniently handled, moderate, fluorinating agent in organic chemistry (*S18*, *R8*). It has been shown to be particularly useful as a mild, selective, fluorinating agent of carcinogenic, polycyclic, aromatic hydrocarbons (*A9*).

The intercalate of graphite with xenon oxide tetrafluoride was the first one isolated for this series (S14), based on the analogy of the latter with iodine pentafluoride. Xenon oxide tetrafluoride forms a compound with stoichiometry $C_{8.7}XeOF_4$. The compound is stable at 0°, but decomposes slowly above room temperature, presumably to a higher-stage compound (R8). Because variable amounts of compounds other than $XeOF_4$ are released at higher temperatures, the second-stage stoichiometry could not be clearly defined. Appreciable amounts of $XeOF_4$ can be recovered from the graphite up to 80° (S20). Above this temperature, substantial amounts of molecular oxygen are liberated in addition to minor amounts of XeO_2F_2 and XeF_2. Above 200°, mass spectra showed that decomposition proceeds with the emission of xenon, fluorocarbons, and carbonyl fluorides (S20), as well as some CO_2. As COF_2 emission proceeds well beyond that of xenon, the formation of graphite oxyfluoride is indicated, a product of this type having been described previously (L6). Broad-line, fluorine-NMR measurements showed that the chemical shift of the intercalated species corresponds to that of neat $XeOF_4$ (E11), this being in harmony with the mass-spectral measurements. C_9XeOF_4 has also been shown to be a potentially useful fluorinating agent of organic compounds (R8). In addition, it reacts with anthracene to form substantial yields of anthraquinone (R8).

In contrast to xenon difluoride, krypton difluoride reacts directly with graphite to form intercalates of variable composition, having the general formula C_mKrF_n ($n > 2$) (S1). The krypton difluoride both intercalates into, and partially fluorinates, the graphite lattice. X-Ray powder photographs showed the absence of the graphite line at 3.35 Å and the presence of new lines having a c-axis, repeat-distance of $4.22n$ Å (n = an integer) distinct from that of either CF or C_4F. Also, wide-line ^{19}F-NMR measurements showed a spectrum different from those of these species, although it was not possible to distinguish between the presence of either KrF^+ or KrF_2 as the intercalated species. It is unlikely, however, that the trapped, krypton species is present as atomic krypton, because of the relatively high temperature at which decomposition occurs.

E. Graphite–Metal Halides

The graphite–metal halides constitute the most populous group of intercalation compounds. Most of the investigative efforts have been directed towards the metal chlorides, particularly $FeCl_3$, whereas considerably less is known about the metal bromides (S21). Compounds with metal iodides are unknown. Metal fluoride compounds, unknown prior to 1972, have since sprung into prominence, because some of them (AsF_5, SbF_5) display theoretical and practical interest owing to their unusually high metal-like conductivities (F13). Nevertheless, the group as a whole appears to be among the least understood in terms of chemical bonding.

A number of general, synthetic techniques have been developed for preparing graphite compounds.

(1) Direct combination of graphite with the metal halide in a sealed tube in the presence of chlorine or bromine at temperatures giving reasonable reaction rates. As a corollary to this method, compounds may be prepared by heating graphite and metal, or graphite and metal oxide, together in the presence of halogen.

(2) In order to define more precisely the conditions for compound formation, the two-temperature method has been used (*H6*). Here, the graphite is maintained at a higher temperature than the intercalant, in order to avoid condensation of an excess on the graphite. Holding either the graphite or the intercalant at constant temperature while varying the temperature differential allows the determination of intercalation or deintercalation isotherms or isobars, giving information on stage formation and hysteresis effects (*H13, H23*).

(3) Intercalation from solutions in nonaqueous solvents (*S21*). This method may suffer from the drawback that final stoichiometries may not correspond to equilibrium conditions, because of partial leaching out of metal halide. For this reason, some chlorides can be intercalated only from solvents in which they have limited solubility (*L8*). It has often been the practice to wash intercalates with solvents to remove the excess of intercalant; this may lead to stoichiometries lower than the original ones. The two-ampoule method may, therefore, be preferable (*H24*).

(4) Intercalation via complexation in the vapor phase (*S21*). Thus, for example, the nonvolatile $CoCl_2$ can be intercalated in the presence of $AlCl_3$ at 500°C, because of the enhanced volatility of the former. Graphite–$CoCl_2$ containing only traces of $AlCl_3$ is obtained.

At this time, no all-inclusive rule can be given that will predict whether a given compound will intercalate or not. Most of the information available seems to have been obtained empirically. Such analogies as similar chemical properties have been helpful. The many factors that influence the intercalation process have been surveyed by Herold (*H14*). In Tables II–VI are listed metal halides considered to intercalate into graphite, together with some structural information (*S21, R9*). Several general characteristics have been ascribed to intercalat-

TABLE II
Chlorides Claimed to Intercalate Graphite

Chloride	Composition	State	I_c (Å)	Ref.
Group Ib				
$CuCl_2$	$C_{4.9}CuCl_2$	1	9.40	*R9*
$AuCl_3$	$C_{12.6}AuCl_3$	1	6.80	*V10*
	$C_{2.52}AuCl_3$	2	10.15	*V10*
	$C_{37.8}AuCl_3$	3	13.50	*V10*
	$C_{50.4}AuCl_3$	4	16.85	*V10*
Group IIb				
$MgCl_2$	$C_{12.18}MgCl_2$	1(+2)	9.50	*S28*
	$C_{26-27}MgCl_2$	2(+3)	12.85	*S28*
	$C_{40}MgCl_2$	3	16.26	*S28*
$ZnCl_2$	$C_{16.6}ZnCl_2$	3	16.29	*S22*
$CdCl_2$	$C_{11.1}CdCl_2$	1	9.51	*R9, S28*
	$C_{6.6}CdCl_2$	1	9.63	*C13*
$HgCl_2$	$C_{21-26}HgCl_{2.1}$	3	16.45	*R9*
Group IIIa				
$AlCl_3$	$C_9AlCl_{3.3}$	1	9.54	*R10*
	$C_{18}AlCl_{3.3}$	2	12.83	*R10*
	$C_{31-50}AlCl_3$	4	19.67	*R10*
$GaCl_3$	$C_9GaCl_{3.2}$	1	9.56	*R11*
	$C_{18}GaCl_{3.5}$	4	19.69	*R11*
$InCl_3$	$C_{16-20}InCl_3$	2	12.83	*R11*
	$C_{25-31}InCl_3$	3	16.23	*R11*
	$C_{32-45}InCl_3$	4	19.69	*R11*
$TlCl_3$	$C_{8.5}TlCl_{3.3}$	1	9.77	*S21*
	$C_{18.5}TlCl_{3.2}$	2	13.12	*S21*
Group IVb				
$ZrCl_4$	$C_{23.8}ZrCl_{4.15}$	3	16.32	*S21*
$HfCl_4$	$C_{45.7}HfCl_{4.77}$	3	15.87	*S21*
Group Va&b				
$SbCl_5$	$C_{12}SbCl_5$	1	9.42	*M6, M9, F14*
	$C_{24}SbCl_5$	2	12.72	*M6, M9, F14*
	$C_{36}SbCl_5$	3	16.07	*M6, M9, F14*
	$C_{48}SbCl_5$	4	19.42	*M6, M9, F14*
	$C_{12}SbCl_{3.9}$	1	9.50	*S21*
$NbCl_5$	$C_{40}NbCl_{5.2}$	3	16.21	*S21*
$TaCl_5$	$C_{27}TaCl_5$	3	16.19	*S21*
$CrCl_3$	$C_{21}CrCl_3$	2	12.80	*V6*
	$C_{22-29}CrCl_3$	3	16.15	*V11*
$MoCl_5$	$C_{18.5}MoCl_5$	2	12.54	*N8*
	$C_{27}MoCl_5$	3	16.02	*N8*
WCl_6	$C_{70}WCl_6$	5	23.02	*R9*
UCl_5	$C_{37}UCl_5$	2	12.87	*R9, B17*
		1	9.62	*B17*

TABLE II (Continued)

Chloride	Composition	Stage	I_c (Å)	Ref.
GROUP VIIb				
$MnCl_2$	$C_{6.4}MnCl_{2.06}$	1	9.51	S22
	$C_{12}MnCl_{2.07}$	2	12.88	S22
$ReCl_4$	$C_{13}ReCl_{4.3}$	1	11.78	S21
GROUP VIII				
$FeCl_2$	$C_{4.7}FeCl_2$	1	9.56	P7
	C_9FeCl_2	1	9.51	N9
	$C_{15.8}FeCl_2$	2	12.85	N9
$FeCl_3$	$C_{5.9}FeCl_3$	1	9.37	R12
	$C_{20}FeCl_3$	2	12.80	R12
		2	12.73	H31
	$C_{23-29}FeCl_3$	3	16.21	R12
		3	16.08	H31
	C_7FeCl_3	1		H23
	$C_{12}FeCl_3$	2		H23
	$C_{31}FeCl_3$	4	19.45	F13b
$CoCl_2$	$C_{5.5}CoCl_{2.07}$	1	9.50	R9
	$C_{15}CoCl_{2.07}$	2	12.85	R9
$NiCl_2$	$C_{13}NiCl_{2.04}$	2	12.71	S22
$RuCl_3$		2	12.60	S21
$OsCl_3$	$C_{12.2}OsCl_3$	2	12.70	S21
$PdCl_3$	$C_{14.8}PdCl_2$	3	16.70	S21
$PtCl_4$	$C_{42-51}PtCl_{4.5}$	3	16.06	B20

ing materials: (1) The metal is generally in a high oxidation state. (2) For a metal in a given oxidation state, intercalation is promoted by increasing electronegativity of the ligand. (3) The crystal structure of the intercalant does not possess oriented bonds in all three dimensions (C1, E1).

The presumed nonintercalation of a given compound may, however, be due to the fact that appropriate experimental conditions for intercalation have not yet been found. Thus, some of the rare-earth chlorides originally considered not to intercalate (C11, V5), do, in fact, do so in the presence of a complexing agent (S21). In addition, the role of chlorine in compound formation has been the subject of controversy. Whereas Croft (C1) considered the presence of an excess of chlorine to be nonessential, it has since been shown to be a *sine qua non* for compound formation (D3, R10, H13, R11, S22, B17, H25). Moreover, contrary to earlier assumptions (R10), chlorine does not act as a catalyst, but is incorporated into the graphite to a greater or lesser extent (R11, D3). In cases where the presence of chlorine is *apparently* not required

TABLE III
Bromides

Bromide	Composition	Stage	I_c (Å)	Ref.
$AlBr_3$	$C_9AlBr_3 \cdot Br_2$	1	10.24	S12
	$C_{18}AlBr_3$	2	13.40	B21
	$C_{24}AlBr_{3.3}$	2	13.35	S12
	$C_{33}AlBr_3$	4	20.10	B21
$GaBr_3$	$C_{13-16.5}GaBr_3 \cdot Br_{2.5}$	2	13.38	B21
$AuBr_3$		1	6.90	B21
$TlBr_3$	$C_{18.6}TlBr_{3.4}$	2	13.40	S21
$CdBr_2$	$C_{15}CdBr_{2.06}$	2	13.30	S21
	$C_{28.6}CdBr_{2.1}$	3	16.62	S21
$HgBr_2$	$C_{23.8}HgBr_{2.1}$	3	16.62	S21
UBr_5	$C_{38}UBr_{5.1}$	2(+3)	13.28	S21, S29
UBr_5	$C_{58}UBr_{4.7}$			S29
$FeBr_2$	$C_{14.2}FeBr_{2.1}$	1	9.90	S26
		2	13.25	
		5	22.8	
$FeBr_3$	$C_{23}FeBr_3$	2	12.90	S26

(WCl_6, $SbCl_5$, $FeCl_3$, $CoCl_2$), the intercalants themselves supply chlorine, due to decomposition under the reaction conditions (H13). Where intercalation is carried out from a solvent, chlorine may be supplied by decomposition of the latter, e.g., its presence in $SOCl_2$ has been detected spectroscopically (B17). However, certain solvents, such as nitromethane, appear to take the place of chlorine in promoting intercalation (H25).

TABLE IV
Rare Earth Chlorides (S21)

Chloride	Composition	Stage	I_c (Å)
$SmCl_3$	$C_{56.6}SmCl_{3.2}$		
$EuCl_3$	$C_{37.1}EuCl_3$	3	16.30
$GdCl_3$	$C_{23.3}GdCl_{3.1}$	3	16.42
$TbCl_3$	$C_{18.7}TbCl_{3.2}$	2	16.01
$DyCl_3$	$C_{19}DyCl_3$	2	12.90
$HoCl_3$	$C_{20.3}HoCl_{3.1}$	2	12.91
$ErCl_3$	$C_{23.3}ErCl_{3.1}$	3	16.35
$TmCl_3$	$C_{23.7}TmCl_3$	3 + 2	16.47 (13.10)
$YbCl_3$	$C_{25.4}YbCl_{3.1}$	3	16.29
$LuCl_3$	$C_{34.8}LuCl_3$	4	19.48
YCl_3	$C_{28.6}YCl_3$	3 + 2	16.54 (12.98)

TABLE V

FLUORIDES

Fluoride	Composition	Stage	I_c (Å)	Ref.
TiF_4	$C_{19-24}TiF_4$	2	15.10	B18
NbF_5	$C_{16.6-17.2}NbF_5$	2	11.76	M5
TaF_5	$C_{17.6}TaF_5$	2	11.76	M5
	$C_{22.4}TaF_5$	3	15.14	M5
AsF_5	$C_{10}AsF_5$	(1 + 2)		H32
	C_8AsF_5	1	8.10	F15
	$C_{16}AsF_5$	2	11.45	F15
	$C_{24}AsF_5$	3	14.80	F15
	$C_{32}AsF_5$	4	18.15	F15
	$C_8^+ AsF_6^-$	1	8.06	B22
SbF_5	$C_{6-6.5}SbF_5$	1	8.46	M5, T10
	$C_{12}SbF_5$	2	11.76	M5, T10
	$C_{18}SbF_5$	3	15.11	M5, T10
	$C_{24}SbF_5$	4	18.45	M5, T10
WF_6	$C_{14\pm1}WF_6(HF)_x$	2	11.70	H33
MoF_6	$C_{11\pm1}MoF_6$	1	8.42	H33
UF_6	$C_{9.1}UF_6$			B23
	$C_{13\pm3}UF_6$	2	11.90	E14
OsF_6	$C_8^+ OsF_6^-$	1	8.06	B22

The nonintercalation of $AlCl_3$, except in the presence of Cl_2, as well as its low intercalation threshold have been used as a basis for determining the extent of intercalation and distinguishing it from amounts added by other mechanisms, such as adsorption (H26).

Whereas most fluorides intercalate directly, some of the less reactive ones (TiF_4, TaF_5, NbF_5) also require the presence of free chlorine (B18, M5).

Structural information on graphite-metal halides is mostly limited to the determination of stage and repeat distances along the c-axis. Detailed electron and X-ray diffraction studies have been carried out only on graphite compounds with $FeCl_3$ (C12, E4, T2), $CrCl_3$ (V6), $MoCl_5$ (J5), and $SbCl_5$ (M6). Several generalizations can be made. The orientation of the graphite layers remains that of the hexagonal modification ABAB, in contrast to alkali-metal-graphites and graphite salts, where lateral displacements are observed. The structures of the free chlorides are largely preserved upon intercalation. In fact, for many metal chlorides, the c-axis expansion remains relatively constant between 9.4 and 9.78 Å (R9). This follows from the close packing of the two hexagonal arrays of chlorides facing the graphite layers that dictate the volume requirements. As in the free metal chlorides, the metal

TABLE VI
Intercalation of Metal Chlorides from Nonaqueous Solvents

Solvent	Chloride	Reaction time	Metal chloride (%)	State	Ref.
CH_3NO_2	$FeCl_3$	6 h (50°C)	36.8	2	G11
	$FeCl_3$	24 h (25°C)	64.6		H25
	$FeCl_3$	3 h (80°C)	36.3	2 + 3	S21
$C_2H_5NO_2$	$FeCl_3$	24 h (50°C)	20.4		G11
$SOCl_2$	UCl_5	3 d	61.7	1	B17
	From UO_3	3 d	48.3	2	B17
	$AlCl_3$	8 h	39.6	2	B17
	$NbCl_5$	12 h	34.1	3	B17
	$TaCl_5$	12 h	43.5	3	B17
	$MoOCl_4$	6 h	23.6	3	B17
	$ZrCl_4$	17 h	31.0	3	S21
	$HfCl_4$	4 d	38.8	3	S21
	$PtCl_4$	2 d	38.0	3	S21
	$AuCl_3$	6 d	46.4	2	S21
SO_2Cl_2	$FeCl_3$	2 d	35.2	3	S21
	$AlCl_3$	12 h	45.8	1	S21
	$AuCl_3$	3 d	47.6	2	S21
CCl_4	$SbCl_5$	64 h	64.4	1	S21
$CCl_4 + Cl_2$	$SbCl_3$	48 h	50.8	2	S21
	$FeCl_3$	2 d	35.8	3	S21
	$AuCl_3$	3 d	38.2	2	S21
	$AlCl_3$	3 h	40	2	L8
	$FeCl_3$	24 h	20		L8
	$NiCl_2$	72 h	24.5		L8
	$PtCl_4$	24 h	15		L8
	$CuCl_2$	48 h	22		L8
	UCl_4	72 h	10		L8
HSO_3Cl	$AuCl_3$	15 d		2	M3
	BCl_3		$BCl_3 + HSO_3Cl$	3	M3
	$SbCl_5$		$SbCl_5 + HSO_3Cl$	2	M3

fits into the octahedral sites created by the chlorides. There exists considerable stacking disorder, strict ordering proving to be more the exception than the rule (*S21, M7, M8*). An apparent exception is graphite–$SbCl_5$ (*M6*), in which intercalated layers are in epitaxy with the carbon layers, leading to a crystallographic, ideal formula $C_{98} \cdot 8\ SbCl_5$. The presence of certain impurities can lead to almost completely ordered, higher stages without affecting the $FeCl_3$ content (*M14*). Metz and co-workers (*S23*) developed a domain theory according to which a number of ordered structures transform into one another. The whole gamut of single stages is traversed, but with preference for certain stages. The mechanism has been detected up to stage nine, and accounts for hysteresis effects in the isotherms.

Attempts to elucidate the bonding have concentrated mainly on graphite–$FeCl_3$. This intercalate is especially suitable as a model compound, because the magnetic and Mössbauer properties of the iron nucleus constitute excellent probes for electronic structure and environment of the latter.

The role of chlorine in intercalation has been discussed. Earlier theories that the chlorine intercalates, and thus opens the layer spaces for penetration of metal halides, are untenable, because chlorine itself does not intercalate. Dzurus and Hennig (*D3*) found that chlorine (or Br_2 or I_2) is, nevertheless, indispensable for intercalation. Hall coefficient data showed that $FeCl_3$ and $AlCl_3$ give acceptor types of compounds. Electrical properties being similar to those of graphite bisulfate or graphite–bromine, these authors adopted such formulas as $C_n^+Cl^-FeCl_2 \cdot 3\ FeCl_3$ or $C_n^+Cl^- \cdot x\ AlCl_3$ ($30 < n < \infty$). Rüdorff and Landel (*R11*) preferred the formulations $C_n^+[FeCl_3]^- \cdot x\ FeCl_3$ or $C_n^+[AlCl_4^-] \cdot x\ AlCl_3$.[1] On the basis of the electrostatic nature of these compounds, Hennig (*H1*) proposed a model to predict free energies of formation of intercalates, using a modified Born–Haber cycle. The method is not easily applicable to the complex halides, too many parameters being unknown (*E1*). According to Metz and co-workers, (*S23*), a hitherto-neglected, thermodynamic parameter that should be included is the mechanical-stress energy required for deforming domains of different stages into one another.

One of the problems in applying thermodynamic cycles is knowing the species involved. The original formulation by Hennig, which presupposes the presence of 25% of $FeCl_2$ in graphite–$FeCl_3$, is unacceptable in light of subsequent Mössbauer measurements. Most of these exhibited only a single resonance at room temperature, characteristic of trivalent iron in an octahedral environment, as in $FeCl_3$ (*L9, H27, F13a, G10, T7, T8, O5, H28, J6*). A slightly higher isomer shift indicated partial transfer of π-electrons from the graphite conduction-band to the d-shell of iron. The resulting structure has been formulated as $C_n^{\delta+}(FeCl_3)^{\delta-}$, and definitely excludes such others as $C_n^+(FeCl_4)^-$ or these proposed earlier (*D3, R11*). The slight charge-transfer to iron is considered to account for the decreased thermal stability of $FeCl_3$ intercalated in graphite, compared to that of bulk $FeCl_3$ (*H28*), shifting the equilibrium $Fe_2Cl_6 \rightleftharpoons 2\ FeCl_2 + Cl_2$ to the right. Nevertheless, $FeCl_3$ intercalated in graphite is stabilized, in that it does not react with anhydrous hydrogen fluoride up to 250°C, whereas $FeCl_3$ reacts readily at room temperature (*B34*). The ternary compound

---

[1] However, magnetic-susceptibility measurements on other graphite–metal chlorides indicated that electron transfer is insufficient to account for reduction of the cation (*R9*).

$C_{31}FeCl_3(N_2O_5)_{1.7}$ has the same Mössbauer spectrum as graphite–$FeCl_3$ (*F13b*).

The reduction products of graphite–metal halides have also attracted widespread interest as potential catalysts, or battery electrodes. Samples of graphite–$FeCl_3$ reduced in H_2 or N_2 at 350° show the presence of two Mössbauer, quadrupole doublets of unequal intensities (*L9, H27, G10*). The product may not be a true intercalate, as the isomer shift is identical with that of anhydrous $FeCl_2$, indicating the absence of π-electron transfer (*L9*). The presence of two doublets seems to indicate two Fe^{2+} sites. One of these doublets has been related to $FeCl_2$ in proximity to Cl_2 trapped in certain sites, changing the electric-field gradient (*H30*). Similar results were obtained by Jadhav et al. (*J6*), who showed that, at still higher temperatures, further reduction, to α-Fe, occurs. However, even up to 1000°C, reduction by this method is incomplete (*V7, S24*). A graphite intercalation-compound of composition $C_{4.7}FeCl_2$, close to the theoretical limit, $C_{4.22}FeCl_2$, has been obtained by reduction of $C_{7.1}FeCl_3$ by $Fe(CO)_5$ under CO pressure (*P7*). Other reducing agents, such as lithium biphenyl (*S24*) or solvated electrons (Na–liq NH_3) (*K4, K5*) lead to iron metal. In some cases, π-complexes of graphite with Fe have been reported (*S24, V8*). The completely reduced product is probably not a true intercalate. Such compounds with transition metals are considered unlikely (*B19*), because the latter have high ionization-potentials and lattice energies. The formation of two-dimensional, intercalated layers would thus be energetically unfavorable (*B19*). The structure of Fe magnetic-layers between graphite networks is inhomogeneous, and characterized by the presence of supermagnetic clusters. More-drastic treatment produced layered systems consisting of ferromagnetic α-Fe clusters separated by graphite layers (*S25*). The iron in these layers is considerably more resistant to oxidation than free iron (*K5*). Heating graphite–$FeCl_3$ with potassium at 350°C also produces mixtures of graphite–$FeCl_2$ and α-Fe (*T8*). Similar products were obtained by electrochemical reduction of first-stage graphite–$FeCl_3$ (*T9*). On the other hand, other workers found evidence only for graphite, α-Fe, and KCl in the product (*B19*). The resulting products showed little catalytic activity for ammonia production (*B19*). In fact, the purported catalytic activity of graphite–$FeCl_3$ derives from the $FeCl_2$ produced by thermal decomposition of $FeCl_3$, which diffuses to the surface, rather than from its existence as an intercalate (*P8*); this is reminiscent of the purported catalytic activity of graphite–CrO_3 (*L10*), which is really due to Cr_3O_8 (*E13*).

The graphite–bromine–iron system has also been studied. No stoichiometric compounds are formed, and the ratios Fe:Br:C can be varied independently (S26). The ^{57}Fe Mössbauer spectra of these systems have been studied (S27). In contrast to graphite–metal halides, graphite–AuCl$_3$ is readily reduced to metallic gold (V7, V9). No intercalation compounds with AuCl or Au are formed. Lamellar compounds of a number of other transition metals with graphite have been reported (V8). The reduction product of graphite–MoCl$_5$ is said to be a sandwich type of π-complex similar to dibenzene–chromium. Catalytic properties were ascribed to the nickel compound (V8).

1. Graphite Compounds with Group V Pentafluorides

Lalancette and LaFontaine (L11) were the first to report the intercalation of SbF$_5$ into a graphite lattice to give a first-stage compound, C$_6$SbF$_5$, as well as higher-stage compounds. They reported a new X-ray line at 11.10 Å, evidently for the second-stage compound. The use of these compounds as mild reagents for exchange of halogen in organic chlorides was proposed. This lessened activity of intercalated SbF$_5$ may be ascribed to the role of graphite as an inert solvent, as suggested by Ebert et al. (E12), who reported wide-line, ^{19}F-NMR spectra showing line-narrowing well below the freezing point of pure SbF$_5$. This behavior was ascribed to translational motion of the SbF$_5$ species, rotational motion being restricted by the presumed polymeric nature of the intercalated SbF$_5$. Other workers have also reported compounds of the type C$_{6n}$SbF$_5$ to C$_{6.5n}$SbF$_5$ (n = stage), with periodicities I$_c$ = 8.41 + 3.35 (n − 1) (M5, T10); this differs from earlier periodicity values (L11), as well as from others reported in the literature (O4, S30, E12). The discrepancies in periodicity values may arise from the methods of sample preparation, the types of graphite, and the purity of the SbF$_5$ used. However, the presence of HF, the most common impurity in SbF$_5$, seems to have little effect on the nature of the intercalated product, other than affecting the rates of intercalation (T10). Whereas pure SbF$_5$ intercalates in the vapor state, no intercalation takes place in the presence of HF; this may however, be due, to suppression of SbF$_5$ concentration by the high vapor-pressure of HF. Contact with liquid SbF$_5$–HF accelerates the intercalation process (T10).

The HF–SbF$_5$ system is known to be a superacid (H34). The possible relevance of this to the intercalation process was pointed out by Vogel (V12), who first reported on the extremely high electrical conductivity of graphite–SbF$_5$ measured normal to the crystallographic c-axis. The measured conductivity was approximately 40 times that of pristine graphite, and 50% greater than that of pure copper. Other workers disputed this high conductivity value, claiming it to be only about one-third that of copper (*T10, F14*). The metallic behavior was also confirmed by measurements of the temperature coefficient of the resistivity of graphite–SbF_5 (*F14*). The discrepancies in reported conductivities are apparently attributable to deficiencies in the original experimental method (*V12, V13*). In fact, the original high value was later discounted by the same workers (*F16*), who pointed out that, for quasi-two-dimensional materials of high anisotropy, such as graphite and some of its compounds, such standard techniques as the classical, 4-point-bridge method lead to serious errors in measured electrical conductivities, and should be replaced by r.f. techniques (*F16, Z2*). Various methods for conductivity measurements have been critically compared (*Z2*). Nevertheless, graphite–SbF_5 does indeed exhibit, in the a-plane, very high electrical conductivity, which may have important technological implications. Composite wires containing graphite–SbF_5, when drawn down, show a strong tendency for the c-axis of the intercalation compound to orient itself normal to the wire axis (*S30*). Optical-reflectance measurements (*H35*) also confirmed the metallic character of these compounds, showing high reflectances below a plasma edge and definite minima above it.

Another point of contention has been the extent to which, if any, SbF_5 is reduced to SbF_3 upon intercalation. Although chemical analyses have shown an F:Sb ratio of 5:1 (*L11, M5*), ^{121}Sb Mössbauer measurements (*B24*) indicated partial reduction of Sb(V) to Sb(III). On the other hand, mass-spectral measurements as a function of temperature (*S15*) showed only SbF_5, evolved in stages, with no fluorocarbons emitted at any time. The latter are usually an indication of partial reduction of the intercalant and fluorination of the graphite host. Wide-line, ^{19}F-NMR chemical-shifts are consistent with either SbF_5 or SbF_6^-, but not with SbF_3, but the occurrence of fluorine exchange could produce minor amounts of trivalent species (*E11*); this point is thus still controversial, and will be alluded to again.

The formation of intercalation compounds of AsF_5 with Grafoil and graphite powder was first reported by Selig and co-workers (*H32*). The blue compounds obtained by prolonged pumping at room temperature had a stoichiometry $C_{10}AsF_5$. Later work (*F15*) with HOPG showed that AsF_5 intercalates spontaneously, to yield compounds of stoichiometry $C_{8n}AsF_5$ (n = stage) with repeat distances, $I_c = 4.75 + 3.35\ n$ (Å), the lattice expansion being consistent with the size of an AsF_5 molecule calculated from covalent radii. In fact, the c-axis repeat-distance is shorter than the comparable one for graphite–SbF_5 by the approximate difference between the covalent radii of Sb and As. On leaving the first-stage compound exposed to dry nitrogen, a composition $C_{10}AsF_5$ is attained, indicating that the earlier reported stoichiometry (*H32*) corresponds to mixtures of first- and second-stage compounds. AsF_5 intercalation is unique, in that it exhibits spontaneous staging under isothermal conditions. Studies of the physical properties of $C_{8n}AsF_5$ are considerably facilitated by this property, as well as by the fact that, in contrast to SbF_5, no edge fraying or exfoliation occurs. Higher-stage compounds are more stable, and more easily handled, and milder reaction-conditions (lower temperature, shorter reaction times) may be used. Unusually well defined correlations exist between increases in c-axis thickness-expansion, the c-axis repeat-distances, and the electrical conductivity up to stage 3. Higher stages are nonhomogeneous (*F15*).

As in the case of SbF_5 intercalation, there are differences of opinion as to whether AsF_5 is reduced upon intercalation. According to Falardeau *et al.* (*F15*), direct correlations between weight loss and tensimetric measurements of gases evolved upon deintercalation indicated that AsF_5 is unaltered in the intercalation process. It should be pointed out, however, that their criterion for AsF_5 purity (molecular weight determinations $\pm 1\%$ of the theoretical) is still consistent with the presence of up to 5% of AsF_3. More-direct impurity-determinations by infrared spectroscopy (*H32*) showed that small proportions ($<5\%$) of AsF_3 are formed upon intercalation. Bartlett and co-workers (*B22*) even claimed a high degree of ionization, and a formula of $C_8^+AsF_6^-$ for the intercalate. Although their product was obtained by oxidation of graphite single-crystals by $O_2^+AsF_6^-$, several experimental results appeared to indicate similarities to graphite intercalated directly with AsF_5. The unit-cell parameters are not significantly different, and K-shell, absorption-edge spectra showed the presence of identical peaks, characteristic of As(V), for $C_{10}AsF_5$, $C_8^+AsF_6^-$, and other AsF_6^- salts. In addition, $C_{10}AsF_5$ showed an absorption-edge peak consistent with As(III). They concluded that the following reaction takes place upon intercalation:

$$3\ AsF_5 + 2\ e^- \rightarrow 2\ AsF_6^- + AsF_3.$$

There are, however, differences of opinion regarding the justification of using these K-shell, absorption-edge spectra for assignments to specific oxidation-states (*H36*). Both symmetry and ligands can affect absorptions by as much as 10 eV, and the absence or presence of a center of symmetry can determine whether a given transition is forbidden or allowed. However, a direct comparison of $C_{10}AsF_5$ with pure AsF_5 has now been made, and it suggested that these problems are minimal (*B35*).

Perhaps the most convincing argument in favor of intercalant ionization derives from the analogous reaction of graphite with OsF_6 (*B22*). The product, formulated as $C_8^+ OsF_6^-$, obeys the Curie–Weiss law, and has a magnetic moment characteristic of OsF_6^-(V) salts.

Arguments against substantial AsF_5 reduction were based on the small proportion of AsF_3 released upon intercalation (*H32, F15*), and ^{19}F-NMR measurements that showed the absence of AsF_3, although not distinguishing between AsF_5 and AsF_6^-(*E11*). Moreover, optical-reflectance measurements (*H29*) indicated fractional ionization of only ~5% in the first-stage compound. These conflicting results have been discussed by Fischer (*F17*), who argued that the strong metallic character of these compounds may not be dominated by charge-transfer effects, but, rather, by structural variables.

Electrical-conductivity measurements (*F16, F18, V13*) confirmed the metallic character of graphite–AsF_5. The second stage has a peak, a-axis conductivity of $6.3 \times 10^5 \, \Omega^{-1} \, cm^{-1}$, i.e., marginally higher than that of copper ($5.8 \times 10^{-5} \, \Omega^{-1} \, cm^{-1}$); this value is considered to be conservative, as sample imperfections tend to decrease the conductivity. The c-axis conductivity decrease monotonically with progressive intercalation (*F16*), a feature shared with other acceptor compounds. The resulting anisotropy is $\sigma_a/\sigma_c > 10^6$, compared to 2×10^3 for pristine graphite.

Because, on balance, the experimental evidence appears to favor relatively low carrier-generation (*H29*), the high a-axis conductivity seems to result from a proportionally smaller decrease in mobility over that of the parent graphite. Infrared optical-reflectance measurements (*H29, H35*) confirmed the metallic character of the compounds. Interpretation of the spectra in terms of a one-carrier model yielded high optical conductivities, in good agreement with dc measurements. Similar conclusions were reached on the basis of resistivity and magnetoresistance data (*Z3*). Further confirmation of the metallic character of graphite–AsF_5 was obtained from conduction electron-spin resonance (*K3*). The results implied a very small density of states at the Fermi energy. The temperature dependence of the line widths show order–disorder transitions of the intercalant layers, implying delocalization of conduction electrons.

Of the other Group V pentafluorides, PF_5 does not intercalate (*H32, H35*), and BiF_5 has not been studied, because it is a nonvolatile solid that decomposes to BiF_3 at higher temperatures. In the presence of chlorine or bromine, NbF_5 and TaF_5 yield $C_{16.6}NbF_5$ and $C_{17.6}TaF_5$ (*M5*).

2. *Reactions of Graphite with Metal Hexafluorides*

Hexafluorides run the gamut of reactions with graphite, from no intercalation except in the presence of a catalyst (WF_6), through intercalation with partial reduction (UF_6), to intercalation with complete reduction (OsF_6).

Interest in UF_6 as a nuclear material first stimulated a study of intercalation of UF_6 into graphite (*M10*). Based on weight increases of graphite, the nominal stoichiometries $C_{16.8}UF_6$ and $C_{9.8}UF_6$ were obtained with natural and artificial graphites, respectively. Desorption of UF_6 was strongly dependent on initial conditions of formation, becoming appreciable only above 300°C. Later work (*B23*) confirmed these results, although a stoichiometry $C_{9.1}UF_6$ was reported. However, it was found that very little UF_6 could be recovered, the evolved gases being fluorocarbons, or, at higher temperatures, UF_4, indicating the reaction:

$$C_{9.1}UF_6 \rightarrow C_{8.6}UF_4 + \tfrac{1}{2} CF_4.$$

Magnetic-susceptibility measurements showed the presence of UF_4 in heated samples. Ebert *et al.* (*E14*) reported a nominal stoichiometry $C_{13}UF_6$, magnetic-susceptibility measurements indicating partial reduction at room temperature, with ~10% of the uranium species present as U(IV). Wide-line NMR demonstrated the presence of both U(VI) and U(IV).

In contrast, MoF_6 has been found to intercalate without reduction, giving first- and second-stage compounds, depending on reaction times (*O6*, *H33*). Stoichiometries of C_8MoF_6 (*O6*) and $C_{(11\pm1)m}MoF_6$(*H33*) have been reported, with a c-axis repeat-distance $I_c = 5.0 + 3.35n$ (n = stage). Most of the MoF_6 can be desorbed as a function of temperature.[2] Tungsten hexafluoride intercalates only in the presence of gaseous or liquid Cl_2, F_2, or HF, giving a second-stage compound formulated as $C_{14\pm1}WF_6 (HF)_x$ ($x \geq 0$), as it was not determined whether HF is inserted as well (*H33*).

At room temperature, osmium hexafluoride yields a blue material of approximate composition C_8OsF_6 (*B22*). Its magnetic susceptibility obeys the Curie–Weiss law, with a magnetic moment, $\mu_{\text{eff}} = 3.5$ BM,

---

[2] The ^{19}F-NMR spectrum of C_nMoF_6 is, however, quite different from that of pure MoF_6 (*E1*).

corresponding to OsF_6^-. The repeat distance is 8.06 Å, consistent with alignment of the threefold axis of the OsF_6^- species with the graphite c-axis.

The reactions of hexafluorides with graphite may thus involve both intercalation and oxidation, a phenomenon that has been encountered to a greater or lesser degree with other fluorides, depending on their reactivities.

F. Graphite–Metal Oxides

Attention has now been accorded the graphite–chromium trioxide intercalation compound, considered to be of use as a cathode material in high-energy-density batteries (*A2*, *A10*, *G12*), as a reagent for the oxidation of primary alcohols to aldehydes (*L10*, *N10*), and as a reforming catalyst (*H37*). Whereas the synthetic method of Platzer (*P9*) is generally considered to yield a true intercalation compound, the direct, "dry" combination technique of Croft has aroused controversy (*E13*). Although X-ray diffraction, differential thermal analysis (*E13*), electron-spin resonance (*E11*), and electrochemical techniques (*E15*, *B25*) suggested that the Croft method yields reduced chromium oxides unaccompanied by intercalation of graphite, there is some evidence for the success of the dry method (*H38*). Unfortunately, characterization was made after heating with 6 M hydrochloric acid, which may have caused intercalation in the solution phase (*E16*). It so happens that all of the technical applications just discussed refer to the product arising from the dry synthesis, with the Platzer material being useless in each case; thus, it is possible that one of the more economically valuable, "intercalation" compounds is not an intercalation compound at all.

IV. Residue Compounds

True graphite intercalation compounds are frequently not stable in the absence of a given activity (pressure, concentration) of free intercalant, and thus decompose to lower stage and, ultimately, so-called "residue" compounds. This process can be accelerated by thermal or *in vacuo* treatment. Although intercalant binding in residue compounds has been considered to exist primarily at defects, electronic and magnetic properties of residue compounds often differ greatly from those of graphite (*B26*, *H39*).

Experimental results now suggest that the intercalant species in residue compounds may be ordered (*S31*, *I3*, *C10*). Detailed electron-diffraction and microscopy work by Chung on residue compounds arising either from highly oriented, pyrolytic graphite or natural crystals suggested that residue compounds have intralayer ordering similar to that of more concentrated, lamellar compounds. In particular, the previous concept that 99% of the residual intercalant resides on defect sites was rejected in favor of a model involving layers of intercalant. This new view is consistent with magnetoreflection (*C14*) and Raman (*S13*) results that indicated a similarity between lamellar compounds and residue compounds.

Although residue compounds are difficult to characterize experimentally, they should constitute only a minor perturbation on the band structure of pure graphite. Efforts to model the electronic properties in the dilute-concentration limit by perturbing the Slonczewski–Weiss–McClure model for graphite have been made (*D5*).

V. Applications of Intercalation Compounds

A. Chemical Reagents

Recent investigations have indicated that graphite compounds may become useful as chemical reagents; the results are summarized in several reviews (*W1, K6, K7, S32*). It is not always clear whether, in a given system, intercalates behave as catalysts or as chemical reagents. Moreover, it is often difficult to determine whether a given effect arises from the specific properties inherent in intercalation, or from the diluent effect of the graphite. The possible role of the latter as an inert solvent had been suggested earlier (*E12*), as applied to the system graphite–SbF_5, which was proposed as a mild reagent for halogen exchange in organic chlorides (*L11*). The advantages of intercalates could arise from slow release of the chemical reagent, thus moderating its activity; or from specific, stereochemical effects peculiar to graphite. From a technical standpoint, graphite compounds may be useful because they can be readily separated from the reaction medium, and can often be stored and handled more easily than the neat chemical reagent. The latter property has been used to advantage in fluorination reactions of aromatic compounds by graphite intercalated with XeF_6 (*S18, R8*), XeF_4 (*S19*), and $XeOF_4$ (*R8*). Graphite–XeF_6 seems to be a more selective fluorinating agent than XeF_2 (*A9*). Specificity has also been found in certain bromination reactions using bromine–graphite (*P10*). Other systems that have been studied include graphite–$SbCl_5$, which has particular advantages as a halogen-exchange reagent, with properties quite different from those of neat $SbCl_5$ (*B27*). Graphite bisulfate has been found to promote esterification of carboxylic acids with alcohols (*B14*). In some cases, yields of 80–90% were obtained without use of an excess of the reagents. It is assumed that the graphite compound acts both as an acid catalyst and a dehydrating agent. However, another study indicated that graphite bisulfate exerts no specific influence as a component, compared with that of free sulfuric acid (*B28*). Graphite bisulfate also promotes the formation of acetals or ethyl esters from ethyl orthoformate, as well as the nitration of aromatic compounds with nitric acid (*A11*).

The system studied most intensively is that of graphite–potassium. According to Lalancette *et al.* (*L10*), $C_{24}K$ causes condensation of benzene to biphenyl, as well as isomerization of 2-alkynes to 1-alkynes. It is not clear, however, whether reaction takes place within or outside of the graphite. Evidence that, for C_8K, reaction occurs within the matrix was obtained by Beguin and Setton (*B9*). The reaction is apparently driven by the high affinity of C_8K for hydrogen; this effect probably also plays a role in reactions of C_8K with such weak protic acids as water and alcohols. Only small proportions of hydrogen were found to be evolved, resulting in formation of partially reduced graphite (*B10*). For larger alcohols, however, larger proportions of hydrogen are released, suggesting that the reactivity of C_8K towards protonic acids is substrate-dependent (*B11*). The acid–base chemistry of rubidium–graphite and cesium–graphite is similar to that of C_8K (*B11*). Reduction of alkyl halides with C_8K was also studied (*B11*).

B. Electrochemical Applications

Intercalation compounds are logical candidates for application as reversible electrodes in cells. Possessing electroactive, but possibly nonconducting, species in electronic contact with the graphite matrix, the intercalation compound is expected to be insoluble in the electrolyte, conductive in the charged and discharged states, and mechanically stable with respect to cycling. Somewhat ironically, although this concept has been thoroughly explored (*A8, A10, A12, D7*), a nonconducting graphite compound has been the most economically successful.

Thus, although the use of graphite compounds as cathodes in electrochemical cells antedates the use of other layered materials (*B29, B30*), it was not until 1973, with the commercial introduction of the CF_x cathode in a lithium–nonaqueous electrolyte system (*F19, W2*), that graphite compounds achieved widespread publicity. Possessing a nominal voltage of 2.8 V, an energy weight density of 804 W-h/lb (CF) [or

1443 W-h/kg (LiCF)], and a volume density of 10 W-h/in.^{3}, the battery is currently marketed by Matsushita in six different capacities. The most popular of these, the BR-435, is a cylindrical cell 4 mm in diameter by 30 mm long; 4 million of these cells were produced in 1977 (*B31*).

Other covalent compounds of graphite have been investigated as cathode materials. Most thorough has been the work of the United States Army Electronics Command on graphite oxide, graphite oxyfluoride, and various graphite fluorides (*H40, H41, H42, L12, B32*). A new graphite fluoride compound, "C_2F", made from the action of ClF_3 and HF on graphite, has been suggested for yielding voltages and discharge characteristics superior to those of CF (*M11, M12, R13*).

Why do true intercalation-compounds not work better as reversible electrodes? In fact, the C_8K–K cell is reversible, but its room temperature emf of only 0.2 V (*A13*) is suggestive of the large "atomic" character of the intercalated potassium in C_8K. In contrast, in the Li–TiS_2 system (*W1*), in which the intercalated lithium is best considered ionic, there are observed an open-circuit emf of 2.5 V, complete reversibility, and an energy density of 455 W-h/kg ($LiTiS_2$). With intercalated oxidants, Lalancette found that bromine compounds can be used as electrodes in concentration cells, yielding voltages of the order of 50 mV, with thermal gradients of 60–75°C, in the range of 0–100°C (*L13*). If, however, a concentrated acceptor-compound, such as $C_{13}CrO_3$, is run against lithium, the oxidant, which is primarily *molecular* in graphite, leaves the graphite on reaction with lithium, or displacement by electrolyte-solvent molecules (*E16*). Thus, the concept of using graphite–acceptor compounds as cathodes fails, because of the large fraction of non-ionized species that is not in close, electronic contact with the graphite matrix. Existing as "spacers" to screen the coulombic repulsions among the intercalated ions, these molecules are reasonably mobile and may react to form a nonintercalated species, while solvent molecules from the electrolyte intercalate as new spacers.

For a more-detailed treatment of the use of intercalation compounds in electrochemistry, more-specialized reviews (*W1, E15, B33*) may be consulted.

C. Intercalation Compounds as Highly Conducting Materials

A great deal of excitement has been generated by the assertion that some intercalation compounds of graphite possess a conductivity greater than that of copper (*V10, F13, T11*). Much of this work was based upon earlier researches by Ubbelohde, who found that the *a*-axis conductivity of the semi-metal graphite increases, and develops a metallic temperature-dependence, on intercalation either by donors or acceptors; he thus referred to graphite intercalation compounds as "synthetic metals" (*U11*).

Currently, there is disagreement concerning the actual magnitude of the conductivity increase, but there is no doubt that an effect, most pronounced for such acceptors as AsF_5, does exist (*T10, V13*). Models evolved in order to account for the magnitude of the conductivity increase included intergraphite layer-separation (*F5*), the intercalant concentration (*F10*), and carrier-mobility enhancement (*F5*).

The actual utility of this discovery depends on the ability to go from hosts consisting of expensive, highly oriented, pyrolytic graphite to hosts composed of cheap graphite powders or fibers. Care must be taken on intercalation, because defects in such low-rank graphites may affect not only the intrinsic conductivity of the host (*Z4*) but may also serve as sites for oxidative reactions that may disrupt the host (*E11*).

D. Catalysis by Graphite Intercalation Compounds

Graphite compounds have been described as catalysts for ammonia synthesis from nitrogen and hydrogen (*I4, P11*), for Fischer–Tropsch chemistry (*M13, R14*), for paraffin isomerization (*R15*), and for Friedel–Crafts chemistry (*O7*).

Unfortunately, it is difficult to ascertain the identity of the actual catalytic species, and it is not clear whether catalysis by a true intercalation compound has been established. For instance, a frequent method for ammonia and Fischer–Tropsch catalyst generation is the following:

$$C_nMX_y + y\, M' \rightarrow \text{"}C_nM\text{"} + y\,(M'X),$$

where M = transition metal, M' = alkali metal, and X = halogen. This technique is more apt to lead to a finely dispersed metal than to an intercalated one, although the characterization of such materials is difficult (*B19, E1, E11, V1, V2, V14, V15*). A thorough discussion of the catalytic behavior of graphite compounds has appeared (*W1*).

Acknowledgments

Extreme appreciation is extended to K. Landsberg for literature searches, and to E. Frey and M. Hetrick for the preparation of the manuscript.

References

A1. Aragon de la Cruz, F., and Cowley, J. M., *Acta Crystallog.* **16,** 531 (1963).
A2. Adams, J. M., Thomas, J. M., and Walter, M. J., *J. Chem. Soc., Dalton Trans.* p. 1459 (1975).
A3. Asher, R. C., *J. Inorg. Nucl. Chem.* **10,** 238 (1958).
A4. Aronson, S., Lemont, S., and Weiner, J., *Inorg. Chem.* **10,** 1296 (1971).
A5. Aronson, S., Frishberg, C., and Frankel, G., *Carbon* **9,** 715 (1971).
A6. Avogadro, A., Bellvari, G., Borghesi, G., Sammogia, G., and Villa, M., *Nuovo Cimento Soc. Ital. Fis. B* **38,** 403 (1977).
A7. Avogadro, A., and Villa, M., *J. Chem. Phys.* **66,** 2359 (1977).
A8. Armand, M., and Touzain, P., *Mater. Sci. Eng.* **31,** 319 (1977).
A9. Agranat, I., Rabinovitz, M., Selig, H., and Lin, C. H., *Synthesis* p. 267 (1977).
A10. Armand, M., *in* "Fast Ion Transport in Solids" (W. van Gool, ed.), p. 665. North-Holland Publ., Amsterdam, 1973.
A11. Alazard, J. P., Kagan, H. B., and Setton, R., *Bull. Soc. Chim. Fr.* p. 499 (1977).
A12. Armand, M., U.S. Patent 3,956,194 (1976).
A13. Aronson, S., Salzano, F. S., and Belliafiore, D., *J. Chem. Phys.* **49,** 434 (1968).
B1. Brodie, B. C., *Philos. Trans. R. Soc. London* **149,** 249 (1859).
B2. Boehm, H. P., Eckel, M., and Scholz, W., *Z. Anorg. Allg. Chem.* **353,** 236 (1967).
B3. Brookeman, J. R., and Rushworth, F. A., *J. Phys. C* **9,** 1043 (1976).
B4. Brendle, M. C., *Wear* **43,** 127 (1977).
B5. Boersma, M. A. M., *Catal. Rev. Sci. Eng.* **10,** 243 (1974).
B6. Billaud, D., and Herold, A., *Bull. Soc. Chim. Fr.* p. 2715 (1974).
B7. Besenhard, J. O., *Carbon* **14,** 111 (1976).
B8. Beguin, F., and Setton, R., *Carbon* **10,** 539 (1972).
B9. Beguin, F., and Setton, R., *J. Chem. Soc., Chem. Commun.* p. 611 (1976).
B10. Bergbreiter, D. E., and Killough, J. M., *J. Chem. Soc., Chem. Commun.* p. 913 (1976).
B11. Bergbreiter, D. E., and Killough, J. M., *J. Am. Chem. Soc.* **100,** 2126 (1978).
B12. Bottomley, M. J., Parry, G. S., Ubbelohde, A. R., and Young, D. A., *J. Chem. Soc.* p. 5674 (1963).
B13. Besenhard, J., and Fritz, H. P., *Z. Naturforsch. Teil B* **27,** 1294 (1972).
B14. Bertin, J., Kagan, H. B., Luche, J. L., and Setton, R., *J. Am. Chem. Soc.* **96,** 8113 (1974).
B15. Bach, B., Bagouin, M., Bloc, F., and Herold, A., *C. R. Acad. Sci., Ser. C* **257,** 681 (1963).
B16. Blackman, L. C. F., Mathews, J. F., and Ubbelohde, A. R., *Proc. R. Soc. London Ser. A* **256,** 15 (1960).
B17. Boeck, A., and Rudorff, W., *Z. Anorg. Allg. Chem.* **397,** 179 (1973).
B18. Buscarlet, E., Touzain, P., and Bonnetain, L., *Carbon* **14,** 75 (1976).
B19. Bewer, G., Wichman, N., and Boehm, H. P., *Mater. Sci. Eng.* **31,** 73 (1977).
B20. Boeck, A., and Rudorff, W., *Z. Anorg. Allg. Chem.* **392,** 236 (1972).
B21. Balestri, C., Vangelisti, R., Melin, J., and Herold, A., *C. R. Acad. Sci., Ser. C* **279,** 279 (1974).
B22. Bartlett, N., Biagoni, R. N., McQuillan, B. W., Robertson, A. S., and Thompson, A. C., *J. Chem. Soc., Chem. Commun.* p. 200 (1978).
B23. Binenboym, J., Selig, H., and Sarig, S., *J. Inorg. Nucl. Chem.* **38,** 2313 (1976).
B24. Ballard, J. C., and Birchall, T., *J. Chem. Soc., Dalton Trans.* p. 1859 (1976).
B25. Besenhard, J. O., and Schollhorn, R., J. Electrochem. Soc. **124,** 968 (1977).

B26. Bach, B., Evans, E. L., Thomas, J. M., and Barber, M., *Chem. Phys. Lett.* **10,** 547 (1971).
B27. Bertin, J., Luche, J. L., Kagan, H. B., and Setton, R., *Tetrahedron Lett.* **9,** 763 (1974).
B28. Besenhard, J. O., *Z. Naturforsch. Teil B* **32,** 1210 (1977).
B29. Brown, B. K., *Trans. Am. Electrochem. Soc.* **53,** 113 (1928).
B30. Brown, B. K., and Storey, O. W., *Trans. Am. Electrochem. Soc.* **53,** 129 (1928).
B31. Brodd, R. J., Kozawa, A., and Kordesch, K. V., *J. Electrochem. Soc.* **125,** 271c (1978).
B32. Brauer, K., and Moyes, K. R., U.S. Patent 3,514,337 (1970).
B33. Bronoel, G., *Ann. Chim. (Paris)* **1** (2–4), 209 (1976).
B34. Boeck, A., and Rüdorff, W., *Z. Anorg. Allg. Chem.* **384,** 169 (1971).
B35. Bartlett, N., McQuillan, B., and Robertson, A. S., *Mater. Res. Bull.* **13,** 1259 (1978).
C1. Croft, R. C., *Q. Rev. Chem. Soc.* **14,** 1 (1960).
C2. Clauss, A., Plass, R., Boehm, H. P., and Hoffmann, U., *Z. Anorg. Allg. Chem.* **291,** 205 (1957).
C3. Carr, K. E., *Carbon* **8,** 245 (1970).
C4. Clark, D. T., and Peeling, J., *J. Polym. Sci., Polym. Chem.* **14,** 2941 (1976).
C5. Craven, W. E., "Intercalation of the Rare Earth Elements into Graphite and Dichalcogenides," M. S. Thesis, U S Air Force Institute of Technology, Air University, 1965, 68 pp. Wright-Patterson Air Force Base, Ohio.
C6. Carver, G. P., *Phys. Rev. Sect. B* **2,** 2284 (1970).
C7. Campbell, L. E., Montet, G. L., and Perlow, G. J., *Phys. Rev. Sect. B* **15,** 3318 (1977).
C8. Chung, D. D. L., Dresselhaus, G., and Dresselhaus, M. S., *Mater. Sci. Eng.* **31,** 107 (1977).
C9. Colin, G., and Herold, A., *C. R. Acad. Sci., Ser. C* **245,** 2294 (1957).
C10. Chung, D. D. L., *J. Electron. Mater.* **7,** 189 (1978).
C11. Croft, R. C., *Aust. J. Chem.* **9,** 184 (1956).
C12. Cowley, J. M., and Ibers, J. A., *Acta Crystallogr.* **9,** 421 (1956).
C13. Colin, G., and Durizot, E., *J. Mater. Sci.* **9,** 1994 (1974).
C14. Chung, D. D. L., and Dresselhaus, M. S., *Physica* **89b,** 131 (1977).
D1. Dresselhaus, M. S., Dresselhaus, G., Eklund, P. C., and Chung, D. D. L., *Mater. Sci. Eng.* **31,** 141 (1977).
D2. Dickens, P. G., and Whittingham, M. S., *Q. Rev. Chem. Soc.* **22,** 30 (1968).
D3. Dzurus, M. L., and Hennig, G. R., *J. Am. Chem. Soc.* **79,** 1051 (1957).
D4. Delhaes, P., *Mater. Sci. Eng.* **31,** 225 (1977).
D5. Dresselhaus, M. S., Dresselhaus, G., and Fischer, J. E., *Phys. Rev. Sect. B* **15,** 3180 (1977).
D6. Dresselhaus, G., and Dresselhaus, M. S., *Mater. Sci. Eng.* **31,** 235 (1977).
D7. Dey, A. N., U.S. Patent 3,998,658 (1976).
E1. Ebert, L. B., *Annu. Rev. Mater. Sci.* **6,** 181 (1976).
E2. Ebert, L. B., Brauman, J. I., and Huggins, R. A., *J. Am. Chem. Soc.* **96,** 7841 (1974).
E3. Ebert, L. B., "Characterization of Graphite Intercalated by Electron Acceptors," Ph.D. Thesis, Stanford University, 1975, 324 pp.
E4. Evans, E. L., and Thomas, J. M., *J. Solid State Chem.* **14,** 99 (1975).
E5. Eklund, P. C., Dresselhaus, G., Dresselhaus, M. S., and Fischer, J. E., *Phys. Rev. Sect. B* **16,** 3330 (1977).
E6. Ebert, L. B., *Bull. Am. Phys. Soc.* **23,** 185 (1978).

E7. Ebert, L. B., and Matty, L., INOR 18, 178th ACS National Meeting, Miami Beach, September 10–15, 1978.
E8. Ebert, L. B., Huggins, R. A., and Brauman, J. I., *Bull. Am. Phys. Soc.* **18,** 1578 (1973).
E9. Eeles, W. T., and Turnbull, J. A., *Nature* **198,** 877 (1963).
E10. Eeles, W. T., and Turnbull, J. A., *Proc. R. Soc. London Ser. A* **283,** 179 (1965).
E11. Ebert, L. B., and Selig, H., *Mater. Sci. Eng.* **31,** 177 (1977).
E12. Ebert, L. B., Huggins, R. A., and Brauman, J. I., *J. Chem. Soc. Chem. Commun.* p. 924 (1974).
E13. Ebert, L. B., Huggins, R. A., and Brauman, J. I., *Carbon* **12,** 199 (1974).
E14. Ebert, L. B., DeLuca, J. P., Thompson, A. H., and Scanlon, J. C., *Mater. Res. Bull.* **12,** 1135 (1977).
E15. Eichinger, G., and Besenhard, J. O., *J. Electroanal. Chem.* **72,** 1 (1976).
E16. Ebert, L. B., Preprints, American Chemical Society Petroleum Division 22 (1), 69 (1977).
F1. Fusaro, R. L., and Sliney, H. E., *ASLE Trans.* **13,** 56 (1970).
F2. R. L. Fusaro, and Sliney, H. E., *ASLE Trans.* **20,** 15 (1975).
F3. Fredenhagen, K., and Cadenbach, G., *Z. Anorg. Allg. Chem.* **158,** 249 (1926).
F4. Fredenhagen, K., and Suck, H., *Z. Anorg. Allg. Chem.* **178,** 353 (1929).
F5. Fischer, J. E., *in* "Physics and Chemistry of Materials with Layered Structures" (F. Levy, ed.), Vol. 6. D. Reidel, Dordrecht, 1979.
F6. Fuzellier, H., and Herold, A., *C. R. Acad. Sci., Ser. C* **276,** 1287 (1973).
F7. Fischer, J. E., Thompson, T. E., Foley, G. M. T., Guerard, D., Hoke, M., and Lederman, F. L., *Phys. Rev. Lett.* **37,** 769 (1976).
F8. Fuzellier, H., Melin, J., and Herold, A., *Mater. Sci. Eng.* **31,** 91 (1977).
F9. Frenzel, A., Dissertation, Technische Hochschule, Berlin, 1933.
F10. Fischer, J. E., *Carbon* **15,** 161 (1977).
F11. Furdin, G., Bach, B., and Herold, A., *C. R. Acad. Sci., Ser. C* **271,** 683 (1970).
F12. Frame, H. D., *Chem. Phys. Lett.* **3,** 182 (1969).
F13. Fischer, J. E., and Thompson, T. E., *Phys. Today* **31,** 36 (1978).
F13a. Freeman, A. G., *J. Chem. Soc., Chem. Commun.* p. 193 (1968).
F13b. Freeman, A. G., *J. Chem. Soc., Chem. Commun.* p. 746 (1974).
F14. Fuzellier, H., Melin, J., and Herold, A., *Carbon* **15,** 45 (1977).
F15. Falardeau, E. R., Hanlon, L. R., and Thompson, T. E., *Inorg. Chem.* **17,** 301 (1978).
F16. Foley, G. M. T., Zeller, C., Falardeau, E. R., and Vogel, F. L., *Solid State Commun.* **24,** 371 (1974).
F17. Fischer, J. E., *J. Chem. Soc., Chem. Commun.* p. 544 (1978).
F18. Falardeau, E. R., Foley, G. M. T., Zeller, C., and Vogel, F. L., *J. Chem. Soc., Chem. Commun.* p. 389 (1977).
F19. Fukuda, M., and Iijima, T., *in* "Power Sources 5" (D. H. Collins, ed.) p. 713. Academic Press, New York, 1975.
G1. Gamble, F. R., Osiecki, J. H., Cais, M., Pisharody, R., DiSalvo, F. J., and Geballe, T. H., *Science* **174,** 493 (1971).
G2. Gisser, H., Petronio, M., and Shapiro, A., *Am. Soc. Lubr. Eng., Proc.* **28,** 161 (1972).
G3. Guerard, D., and Herold, A., *C. R. Acad. Sci., Ser. C* **275,** 571 (1972).
G4. Guerard, D., and Herold, A., *London Int. Conf. Carbon Graphite, 4th,* p. 325 (1976).
G5. Guerard, D., and Herold, A., *C. R. Acad. Sci., Ser. C* **279,** 455 (1975).
G6. Guerard, D., and Herold, A., *C. R. Acad. Sci., Ser. C* **280,** 729 (1975).
G7. Guerard, D., and Herold, A., *C. R. Acad. Sci., Ser. C* **281,** 929 (1975).

G8. Guerard, D., Foley, G. M. T., Zanini, M., and Fischer, J. E., *Nuovo Cimento B* **38**, 410 (1977).
G9. Guerard, D., Lagrange, P., and Herold, A., *Mater. Sci. Eng.* **31**, 29 (1977).
G10. Grigutch, F. D., Hohlwein, D., and Knappwost, A., *Z. Phys. Chem. (Frankfurt am Main)* **65**, 322 (1969).
G11. Ginderow, D., and Setton, R., *C. R. Acad. Sci., Ser. C* **257**, 687 (1963).
G12. Gunther, R. G., D. T. 2502500 (1975); U.S. Application 453252 (1974).
H1. Hennig, G. R., *Prog. Inorg. Chem.* **1**, 125 (1959).
H2. Hulliger, F., *Phys. Chem. Mater. Layered Struct.* **5**, 52 (1976).
H3. Herold, A., and Vogel, F. L., *Mater. Sci. Eng.* **31** (1977).
H4. Hummers, W. S., and Offeman, R. E., *J. Am. Chem. Soc.* **80**, 1339 (1958).
H5. Herold, A., *C. R. Acad. Sci., Ser. C* **232**, 1489 (1951).
H6. Herold, A., *Bull. Soc. Chim. Fr.* p. 999 (1955).
H7. Holzworth, N. A. W., and Rabii, S., *Mater. Sci. Eng.* **31**, 195 (1977).
H8. Hennig, G. R., *J. Chem. Phys.* **43**, 1201 (1965).
H9. Hart, H., Chem, B., and Peng, C., *Tetrahedron Lett.* p. 3121 (1977).
H10. Hooley, J. G., *in* "Preparation and Crystal Growth of Materials in Layered Structures" (R. M. A. Lieth, ed.). Reidel, Dordrecht, 1977.
H11. Horn, D., and Boehm, H. P., *Mater. Sci. Eng.* **31**, 87 (1977).
H12. Hennig, G., *J. Chem. Phys.* **19**, 922 (1951).
H13. Hooley, J. G., *Carbon* **11**, 225 (1973).
H14. Herold, A., *Mater. Sci. Eng.* **31**, 1 (1977).
H15. Hennig, G., *J. Chem. Phys.* **20**, 1443 (1952).
H16. Hooley, J. G., *Carbon*, **8**, 333 (1970).
H17. Heerschap, M., Delavignette, P., and Amelinckx, S., *Carbon* **1**, 235 (1964).
H18. Hennig, G. R., *J. Chem. Phys.* **20**, 1438 (1952).
H19. Hooley, J. G., and Snee, J., *Carbon* **2**, 135 (1964).
H20. Hooley, J. G., Garby, W. P., and Valentin, J., *Carbon* **3**, 7 (1965).
H21. Hooley, J. G., *Carbon* **11**, 225 (1973).
H22. Hooley, J. G., *Mater. Sci. Eng.* **31**, 17 (1977).
H23. Hooley, J. G., and Bartlett, M. W., *Carbon* **5**, 417 (1967).
H24. Hooley, J. G., and Soniassy, R. N., *Carbon* **8**, 191 (1970).
H25. Hooley, J. G., *Carbon* **10**, 155 (1972).
H26. Hooley, J. G., *Carbon* **13**, 469 (1975).
H27. Hooley, J. G., Bartlett, M. W., Liengme, B. V., and Sams, J. R., *Carbon* **6**, 681 (1968).
H28. Hohlwein, D., Readman, P. W., Chamberod, A., and Coey, J. M. D., *Phys. Status Solidi B* **64**, 305 (1974).
H29. Hanlon, L. R., Falardeau, E. R., and Fischer, J. E., *Solid State Commun.* **24**, 377 (1977).
H30. Hooley, J. G., Sams, J. R., and Liengme, B. V., *Carbon* **8**, 467 (1970).
H31. Hohlwein, D., Grigutch, F. D., and Knappwost, A., *Angew. Chem.* **81**, 333 (1969).
H32. Hsu, L. C., Selig, H., Rabinovitz, M., Agranat, I., and Sarig, S., *J. Inorg. Nucl. Chem. Lett.* **11**, 601 (1975).
H33. Hamwi, A., Touzain, P., and Bonnetain, L., *Mater. Sci. Eng.* **31**, 95 (1977).
H34. Hyman, H. H., Quarterman, L. A., Kilpatrick, M., and Katz, J. J., *J. Phys. Chem.* **65**, 123 (1961).
H35. Hanlon, L. R., Falardeau, E. R., Guerard, D., and Fischer, J. E., *Mater. Sci. Eng.* **31**, 161 (1977).
H36. Hodgson, K., personal communications to L. B. Ebert.
H37. Harris, J. R., U.S. Patent 4,066,712 (1978).

H38. Hooley, J. G., and Reimer, M., *Carbon* **13**, 401 (1975).
H39. Hooley, J. G., *Chem. Phys. Carbon* **5**, 321 (1969).
H40. Hunger, H. F., and Heymach, G. J., *J. Electrochem. Soc.* **120**, 1161 (1973).
H41. Hunger, H. F., and Ellison, J. E., *J. Electrochem. Soc.* **122**, 1288 (1975).
H42. Hunger, H. F., and Heymach, G. J., U.S. Army Electronics Commands, ECOM 4047 (1972).
I1. Iwemura, H., *Tetrahedron Lett.* p. 615 (1976).
I2. Inoshita, T., Nakao, K., and Kaminora, H., *J. Phys. Soc. Jpn.* **43**, 1237 (1977).
I3. Inagaki, M., Rouillon, J. C., Fug, G., and Delhaes, F., *Carbon* **15**, 181 (1977).
I4. Ichikawa, M., Kondo, T., Kawase, K., Sudo, M., Onishi, T., and Tamaru, K., *J. Chem. Soc. J. Chem. Commun.* p. 176 (1972).
J1. Juza, R., Jonck, P., and Schmeckenbecher, A., *Z. Anorg. Allg. Chem.* **292**, 34 (1957).
J2. Juza, R., Schmidt, P., Schmeckenbecher, A., and Jonck, P., *Naturwissenschaften* **42**, 124 (1955).
J3. Juza, R., and Seidel, H., *Z. Anorg. Allg. Chem.* **317**, 73 (1962).
J4. Juza, R., and Schmeckenbecher, A., *Z. Anorg. Allg. Chem.* **292**, 46 (1957).
J5. Johnson, A. W. S., *Acta Crystallogr.* **23**, 770 (1969).
J6. Jadhav, V. G., Singra, R. M., Joshi, G. M., Pisharody, K. P. R., and Rao, C. N. R., *Z. Phys. Chem. (Frankfurt am Main)* **92**, 139 (1974).
K1. Kamarchik, P., and Margrave, J. L., *Acc. Chem. Res.* **11**, 296 (1978).
K2. Kittel, C., "Introduction to Solid State Physics" 4th ed., Chapter 8. Wiley, New York, 1971.
K3. Khanna, S. K., Falardeau, E. R., Heeger, A. J., and Fischer, J. E., *Solid State Commun.* **25**, 1059 (1978).
K4. Knappwost, A., and Metz, W., *Naturwissenschaften* **56**, 85 (1969).
K5. Klotz, H., and Schneider, A., *Naturwissenschaften* **49**, 448 (1962).
K6. Kagan, H. B., *Pure Appl. Chem.* **46**, 177 (1976).
K7. Kagan, H. B., *Chem. Tech.* p. 510 (1976).
L1. Lagow, R. J., Badachhape, R. B., Wood, J. L., and Margrave, J. L., *J. Chem. Soc., Dalton Trans.* p. 1268 (1974).
L2. Lopez-Gonzalez, J., *et al.*, *An. Quim.* **71**, 765 (1975).
L3. Lopez-Gonzalez, J., *et al.*, *An. Quim.* **72**, 759 (1976).
L4. Lalancette, J. M., Rollin, G., and Giraitis, A. P., *Can J. Chem.* **50**, 3058 (1972).
L5. Lagrange, P., Guerard, D., Herold, A., *Ann. Chim. (Paris)* **3**, 143 (1978).
L6. Lagow, R. J., Badachhape, R. B., Wood, J. L., and Margrave, J. L., *J. Chem. Soc., Dalton Trans.* p. 1268 (1974).
L8. Lalancette, J. M., Roy, L., and LaFontaine, J., *Can. J. Chem.* **54**, 2505 (1976).
L9. Liengme, B. V., Bartlett, M. W., Hooley, J. G., and Sams, J. R., *Phys. Lett.* **25**, 127 (1967).
L10. Lalancette, J. M., Rollin, G., and Dumas, P., *Can. J. Chem.* **50**, 3058 (1972).
L11. Lalancette, J. M., and LaFontaine, J., *J. C. S., Chem. Commun.* p. 815 (1973).
L12. Lagow, R. J., and Adcock, J. L., U.S. Army Electronics Command, ECOM-0166-F (1974).
L13. Lalancette, J. M., and Roussel, R., *Can. J. Chem.* **54**, 3541 (1976).
M1. Mahajan, V. K., Badachhape, R. B., and Margrave, J. L., *Inorg. Nucl. Chem. Lett.* **10**, 1103 (1974).
M2. Mitzutani, B., Kondow, T., and Massalski, T. B., *Phys. Rev. B* **17**, 3165 (1978).
M3. Melin, J., Furdin, G., Fuzellier, H., Vasse, R., and Herold, A., *Mater. Sci. Eng.* **31**, 61 (1977).
M4. Marchand, A., Rouillon, J. C., and Courtois d'Arcolliers, F., *Carbon* **9**, 347 (1971).

M5. Melin, J., and Herold, A., *C. R. Acad. Sci., Ser. C* **280,** 641 (1975).
M6. Melin, J., and Herold, A., *Carbon* **13,** 357 (1975).
M7. Metz, W., and Hohlwein, D., *Carbon* **13,** 84 (1975).
M8. Metz, W., and Hohlwein, D., *Carbon* **13,** 87 (1975).
M9. Melin, J., and Herold, A., *C. R. Acad. Sci., Ser. C* **269,** 877 (1969).
M10. Maire, J., *Proc. Conf. Peaceful Uses Atomic Energy,* 2nd. **28,** 392 (1958).
M11. Malachesky, P. A., Newman, G. H., and Shropshire, J. A., *Electrochem. Soc. Extended Abstr.* **77-2,** 35 (1977).
M12. Malachesky, P. A., and Newman, G. H., U.S. Patent 4,074,019 (1977).
M13. Mashinskii, V. I., et al., *Izv. Akad. Nauk SSSR, Ser. Khim.* **25,** 2018 (1976).
M14. Metz, W., and Schoppen, G., *Carbon* **16,** 303 (1978).
N1. Nazarov, A. S., *Zh. Neorg. Khim.* **21,** 2273 (1976).
N2. Novikov, Y. N., and Volpin, M., *Russ. Chem. Rev.* **40,** 733 (1971).
N3. Nemanich, R. J., Solin, S. A., and Guerard, D., *Phys. Rev. B* **16,** 2965 (1977).
N4. Nixon, D. E., Parry, G. S., and Ubbelohde, A. R., *Proc. R. Soc. London, Ser. A* **291,** 324 (1966).
N5. Nazarov, A. S., Yudanov, N. F., and Chicagov, Yu. V., *Russ. J. Inorg. Chem. (Engl. Transl.)* **21,** 1248 (1976).
N6. Nazarov, A. S., Makotchenko, V. G., and Yakovlev, I. I., *Zh. Neorg. Khim.* **23,** 1680 (1978).
N7. Nikolaev, A. V., Nazarov, A. S., and Makotchenko, V. G., *Izv. Sib. Otd. Akad. Nauk SSSR, Ser. Khim. Nauk* **3,** 62 (1976).
N8. Novikov, Y. N., Semion, V. A., Struchkov, Y. T., and Vol'pin, M. E., *Zh. Strukt. Khim.* **11,** 880 (1970).
N9. Novikov, Y. N., Vol'pin, M. E., Prusakov, V. E., Stukan, R. A., Goldanskii, V. I., Semion, V. A., and Struckhov, Y. T., *Zh. Strukt. Khim.* **11,** 1039 (1970).
N10. Nilsson, A., Palmquist, U., and Ronlan, A., *J. Electrochem. Soc.* **123,** 191C (1976).
O1. Opalovskii, A. A., Nazarov, A. S., and Uminskii, A. A., *Russ. J. Inorg. Chem. (Engl. Transl.)* **17,** 632 (1972).
O2. Opalovskii, A. A., Nazarov, A. S., Uminskii, A. A., and Chicagov, Yu. V., *Russ. J. Inorg. Chem. (Engl. Transl.)* **17,** 1227 (1972).
O3. Opalovskii, A. A., Nazarov, A. S., and Uminskii, A. A., *Russ. J. Inorg. Chem. (Engl. Transl.)* **17,** 1366 (1972).
O4. Opalovskii, A. A., Nazarov, A. S., and Uminskii, A. A., *Russ. J. Inorg. Chem. (Engl. Transl.)* **19,** 827 (1974).
O5. Ohhashi, K., and Tsujikawa, L., *J. Phys. Soc. Jpn.* **36,** 422 (1974).
O6. Opalovskii, A. A., Kuznetsova, Z. M., Chicagov, Yu. A., Nazarov, A. S., and Uminskii, A. A., *Russ. J. Inorg. Chem. (Engl. Transl.)* **19,** 1134 (1974).
O7. Olah, G. A., and Kaspi, J., *J. Org. Chem.* **42,** 3046 (1977).
P1. Podall, H., Foster, W. E., and Giraitis, A. P., *J. Org. Chem.* **23,** 82 (1958).
P2. Parry, G. S., and Nixon, D. E., *Nature* **216,** 909 (1967).
P3. Parry, G. S., Nixon, D. E., Lester, K. M., and Levene, B. C., *J. Phys. C* **2,** 2156 (1969).
P4. Poitrenaud, J., *Rev. Phys. Appl.* **5,** 275 (1970).
P5. Platts, D. A., Chung, D. D. L., and Dresselhaus, M. S., *Phys. Rev. B* **15,** 1087 (1977).
P6. Perlow, G. L., personal communication to H. Selig.
P7. Pritzlaff, B., and Stahl, H., *Carbon* **15,** 399 (1977).
P8. Parkash, S., Chakravartty, S. K., and Hooley, J. G., *Carbon* **15,** 307 (1977).
P9. Platzer, N., and de la Martiniere, B., *Bull. Soc. Chim. Fr.* p. 177 (1961).

P10. Page-Lecuyer, A., Luche, J. L., Kagan, H. B., and Mazieres, C., *Bull. Soc. Chim. Fr.* p. 1690 (1973).
P11. Postnikov, V. A., et al., *Izv. Akad. Nauk SSSR, Ser. Khim.* **24,** 2529 (1975).
R1. Rüdorff, W., *Adv. Inorg. Chem. Radiochem.* **1,** 223 (1959).
R2. Rodriquez, A., et al., *An. Quim.* **73,** 657 (1977).
R3. Rüdorff, W., and Shulze, E., *Z. Anorg. Allg. Chem.* **277,** 156 (1954).
R4. Rüdorff, W., and Hoffman, U., *Z. Anorg. Allgem. Chem.* **238,** 1 (1938).
R5. Rüdorff, W., *Z. Anorg. Allgem. Chem.* **254,** 319 (1947).
R6. Rüdorff, W., Sils, V., and Zeller, R., *Z. Anorg. Allg. Chem.* **283,** 298 (1956).
R7. Rüdorff, W., *Z. Anorg. Allg. Chem.* **245,** 383 (1941).
R8. Rabinovitz, M., Agranat, I., Selig, H., Lin, C. H., and Ebert, L., *J. Chem. Res.* (S) 216; (M) 2353 (1977).
R9. Rüdorff, W., Stumpp, E., Spriessler, W., and Siecke, F. W., *Angew. Chem.* **75,** 130 (1963).
R10. Rüdorff, W., and Zeller, R., *Z. Anorg. Allg. Chem.* **279,** 182 (1955).
R11. Rüdorff, W., and Landel, A., *Z. Anorg. Allg. Chem.* **293,** 327 (1958).
R12. Rüdorff, W., and Schulz, H., *Z. Anorg. Allg. Chem.* **245,** 121 (1940).
R13. Rao, B. M. L., and Malachesky, P. A., U.S. Patent 4,057,676 (1977).
R14. Rosynek, M. P., ERDA Reports FE-2467-1, FE-2467-2 (Avail. NTIS).
R15. Rodewald, P. G., U.S. Patents 3,962,133; 3,976,714; 3,984,352 (1976).
S1. Selig, H., Gallagher, P. K., and Ebert, L. B., *Inorg. Nucl. Chem. Lett.* **13,** 427 (1977).
S2. Swanson, R. M. F., "The Band Structure of Potassium Graphite," Ph.D. Thesis, Stanford University, 1969, 107 pp.
S3. Simonet, J., and Lund, H., *J. Electroanal. Chem.* **75,** 719 (1977).
S4. Solin, S. A., and Ebert, L. B., unpublished results.
S5. Savoia, D., Trombini, C., and Umani-Ronchi, A., *J. Chem. Soc., Perkin Trans. 1* p. 123 (1977).
S6. Savoia, D., Trombini, C., and Umani-Rochi, A., *Tetrahedron Lett.* p. 653 (1977).
S7. Savoia, D., Trombini, C., and Umani-Rochi, A., *J. Org. Chem.* **43,** 2907 (1978).
S8. Schaufhautl, P., *J. Prakt. Chem.* **21,** 155 (1841).
S9. Sasa, T., Takahashi, Y., and Mukaibo, T., *Carbon* **9,** 407 (1971).
S10. Salzano, F. J., and Aronson, S., *J. Inorg. Nucl. Chem.* **28,** 1343 (1966).
S11. Saunders, G. A., Ubbelohde, A. R., and Young, D. A., *Proc. R. Soc. London Ser. A* **271,** 499 (1963).
S12. Sasa, T., Takahashi, Y., and Mukaibo, T., *Bull. Chem. Soc. Jpn.* **43,** 34 (1970).
S13. Song, J. J., Chung, D. D. L., Eklund, P. C., and Dresselhaus, M. S., *Solid State Commun.* **20,** 1111 (1976).
S14. Selig, H., and Gani, O., *Inorg. Nucl. Chem. Lett.* **11,** 75 (1975).
S15. Selig, H., Sunder, W. A., Vasile, M. J., Stevie, F. A., Gallagher, P. K., and Ebert, L. B., *J. Fluorine Chem.* **12,** 397 (1978).
S16. Selig, H., and Fort, D., unpublished observation.
S18. Selig, H., Rabinovitz, M., Agranat, I., Lin, C. H., and Ebert, L. B., *J. Am. Chem. Soc.* **98,** 1601 (1976).
S19. Selig, H., Rabinovitz, M., Agranat, I., Lin, C. H., and Ebert, L. B., *J. Am. Chem. Soc.* **99,** 953 (1977).
S20. Selig, H., Vasile, M. J., Stevie, F. A., and Sunder, W. A., *J. Fluorine Chem.* **10,** 299 (1977).
S21. Stumpp, E., *Mater. Sci. Eng.* **31,** 53 (1977).
S22. Stumpp, E., and Werner, F., *Carbon* **4,** 538 (1966).

S23. Schoppen, G., Meyer-Spasche, H., Siemgluss, L., and Metz, W., *Mater. Sci. Eng.* **31,** 115 (1977).
S24. Stukan, R. A., Prusakov, V. A., Novikov, Yu. N., Vol'pin, M. E., and Goldanskii, V. I., *J. Struct. Chem. (USSR)* **12,** 567 (1971).
S25. Stukan, R. A., Novikov, Yu. N., Povitski, V. A., and Saluzin, A. N., *Sov. Phys. Sol. State* **14,** 2914 (1973).
S26. Stahl, H., *Z. Anorg. Allg. Chem.* **428,** 269 (1977).
S27. Stahl, H., *Z. Anorg. Allg. Chem.* **434,** 201 (1977).
S28. Stumpp, E., and Terlan, A., *Carbon* **14,** 89 (1976).
S29. Stumpp, E., and Niess, R., *Inorg. Nucl. Chem. Lett.* **14,** 217 (1978).
S30. Singhal, S. C., Schreurs, J., and Kuznicki, R. C., *Mater. Sci. Eng.* **31,** 123 (1977).
S31. Saito, M., and Tsuzuku, T., *Carbon* **15,** 347 (1977).
S32. Setton, R., *Mater. Sci. Eng.* **31,** 303 (1977).
S33. Sasa, T., Takahashi, Y., and Mukaibo, T., *Bull. Chem. Soc. Jpn.* **45,** 2250 (1972).
T1. Tennakoon, D. T. B., Thomas, J. M., Tricker, M. J., and Graham, S. H., *J. Chem. Soc., Chem. Commun.* p. 124 (1974).
T2. Thomas, J. M., Millware, G. R., Davies, N. C., Evans, E. L., *J. Chem. Soc., Dalton Trans.* p. 2443 (1976).
T3. Tuinstra, F., and Koenig, J. L., *J. Chem. Phys.* **53,** 1126 (1970).
T4. Thompson, A. H., *Comments Solid State Phys.* **7,** 125 (1976).
T5. Turnbull, J. A., and Eeles, W. T., *Conf. Indust. Carbon Graphite 2nd.* p. 173 (1965).
T6. Takahashi, T., Sasa, T., and Mukaibo, T., *Tanso* **52,** 199 (1969).
T7. Tominaga, T., Sakai, T., and Kimura, T., *Chem. Lett.* p. 853 (1974).
T8. Tricker, M. J., Evans, E. L., Cadman, P., Davies, N. C., and Bach, B., *Carbon* **12,** 499 (1974).
T9. Touzain, P., Chamberod, A., and Briggs, A., *Mater. Sci. Eng.* **31,** 77 (1977).
T10. Thompson, T. E., Falardeau, E. R., and Hanlon, L. R., *Carbon* **15,** 39 (1977).
T11. Taylor, D., *New Sci.* **75,** 593 (1977).
U1. Ubbelohde, A. R., and Lewis, F. A., "Graphite and its Crystal Compounds" Oxford Univ. Press (Clarendon) London and New York, 1960.
U2. Ungurenasu, C., and Palie, M., *J. Chem. Soc., Chem. Commun.* p. 388 (1975).
U3. Ungurenasu, C., and Palie, M., *Synth. React. Inorg. Met. Org. Chem.* **7,** 581 (1977).
U4. Ubbelohde, A. R., *Carbon* **7,** 523 (1969).
U5. Ubbelohde, A. R., *Proc. R. Soc. London, Ser. A* **304,** 25 (1968).
U6. Ubbelohde, A. R., *Proc. R. Soc. London, Ser. A* **309,** 297 (1969).
U7. Ubbelohde, A. R., *Proc. R. Soc. London, Ser. A* **321,** 445 (1971).
U8. Ubbelohde, A. R., *Carbon* **10,** 201 (1972).
U9. Ubbelohde, A. R., *Proc. R. Soc. London Ser. A* **327,** 289 (1972).
U10. Ubbelohde, A. R., *Nature* **268,** 16 (1977).
U11. Ubbelohde, A. R., *Carbon* **14,** 1 (1976).
V1. von Doorn, A. B. C., Groenewege, M. P., and deBoer, J. H., *K. Ned. Akad. Wet. B* **66,** 165 (1963).
V2. Voloshin, A. G., and Koleschikov, I. P., *Elektrokhimiya* **11,** 1903 (1975).
V3. Vogel, F. L., and Popowich, R., *in* "Petroleum Derived Carbons," (M. L. Deviney and T. M. O'Grady, eds.) *ACS Symp. Ser.* **21,** 411 (1975).
V4. Vogel, F. L., *Carbon* **14,** 175 (1976).
V5. Vickers, R. C., and Campbell, N. L., *J. Am. Chem. Soc.* **79,** 5897 (1957).
V6. Vangelisti, R., and Herold, A., *Carbon* **14,** 333 (1976).

V7. Vangelisti, R., and Herold, A., *C. R. Acad. Sci., Ser. C* **280,** 571 (1975).
V8. Vol'pin, M. E., Novikov, Yu. N., Lopkina, N. D., Kasatochkin, V. I., Struchkov, Yu. T., Kazakov, M. E., Stukan, R. A., Povitskij, V. A., Karimov, Yu. S., and Zvarikina, A. V., *J. Am. Chem. Soc.* **97,** 3366 (1975).
V9. Vangelisti, R., and Herold, A., *Mater. Sci. Eng.* **31,** 67 (1977).
V10. Vangelisti, R., and Herold, A., *C. R. Acad. Sci., Ser. C* **276,** 1109 (1973).
V11. Vangelisti, R., Furdin, G., Carton, B., and Herold, A., *C. R. Acad. Sci., Ser. C* **278,** 869 (1974).
V12. Vogel, F. L., *J. Mater. Sci.* **12,** 982 (1977).
V13. Vogel, F. L., Foley, G. M. T., Zeller, C., Falardeau, E. R., and Gan, J., *Mater. Sci. Eng.* **31,** 261 (1977).
V14. Vangelisti, R., and Herold, A., *C. R. Acad. Sci., Ser. C* **286,** 289 (1978).
V15. Vol'pin, M. E., Novikov, Yu. N., Postnikov, V. A., Shur, V. B., Bayerl, B., Kaden, L., Wahreh, M., Dmitrienko, L. M., Stulcan, R. A., and Nefed'ev, A. V., *Z. Anorg. Allg. Chem.* **428,** 231 (1977).
V16. Veraa, M. J., and Bell, A. T., *Fuel* **57,** 194 (1978).
W1. Whittingham, M. S., and Ebert, L. B., in "Physics and Chemistry of Materials and Layered Structures" (F. Levy, ed.), Vol. 6. D. Reidel, Dordrecht, 1979.
W2. Watanabe, N., and Fukuda, M., U.S. Patent 3,700,502 (1972).
Y1. Young, D. A., *Carbon* **15,** 373 (1977).
Z1. Zanini, M., Baso, S., and Fischer, J. E., *Carbon* **16,** 211 (1978).
Z2. Zeller, C., Foley, G. M. T., Falardeau, E. R., and Vogel, F. L., *Mater. Sci. Eng.* **31,** 255 (1977).
Z3. Zeller, C., Pendrys, L. A., and Vogel, F. L., *Bull. Am. Phys. Soc.* **23,** 220 (1978).
Z4. Zeller, C., Foley, G. M. T., and Vogel, F. L., *J. Mater. Sci.* **13,** 1114 (1978).

SOLID-STATE CHEMISTRY OF THIO-, SELENO-, AND TELLUROHALIDES OF REPRESENTATIVE AND TRANSITION ELEMENTS

J. FENNER,[*] A. RABENAU, and G. TRAGESER

Max-Planck-Institut für Festkörperforschung, Stuttgart, West Germany

I. Introduction	330
II. Group IB	332
A. Copper	332
B. Silver	338
C. Gold	342
D. Miscellaneous	348
III. Group IIB	351
A. Zinc and Cadmium	351
B. Mercury	351
IV. Group IIIB and Lanthanides	357
V. Group IVB	364
VI. Group VB	364
VII. Group VIB	370
A. Chromium	370
B. Molybdenum	370
C. Tungsten	377
VIII. Group VIIB	379
A. Manganese	379
B. Rhenium	379
IX. Group VIIIB	381
X. Group IIIA	382
A. Boron	383
B. Aluminum	383
C. Gallium	384
D. Indium	386
E. Thallium	388
XI. Group IVA	389
A. Carbon, Silicon, and Germanium	389
B. Tin	390
C. Lead	396

[*] Present address: Th. Goldschmidt AG, Goldschmidtstrasse 100, D-4300 Essen 1, West Germany.

XII. Group VA	400
A. Nitrogen and Phosphorus	400
B. Arsenic	401
C. Antimony and Bismuth	402
Appendix	412
References	413
Appendix References	425

I. Introduction

The aim of this review is to discuss the modern, solid-state chemistry of members of a well-defined class of compounds with respect to their preparative methods, structural features, and physical properties. The thio-, seleno-, and tellurohalides of the metallic elements seem to be well suited for this purpose. In contrast to such binary compounds as oxides and halides, they fit into the frame of the restricted space available, and, with the exception of alkaline and alkaline-earth elements, examples are found in all groups of the periodic system. The large group of oxyhalides, as well as the chalcogenide fluorides, have been excluded. Oxyhalides have been extensively reviewed (*182, 405*), and the properties of the fluorides are closely related to this group; in addition, the behavior differs in some respects from that of the chalcogenide halides treated here. A typical feature of the compounds under consideration is that of nonstoichiometry (*54, 296*). Besides, they may have a complicated, overall composition, due, for instance, to metal–metal bonds, and they may exhibit a broad range of existence over many mole percent, and, within such a phase width, the chemical potential of an elementary component can vary over powers of ten (*54, 298*). On the other hand, the numerous addition compounds and adducts, such as K_2TeBr_6, $TeCl_4 \cdot AlCl_3$, $SnCl_4 \cdot SCl_4$, and most molecular structures will not be treated.

As regards organization, the indicated compounds of the elements are discussed separately, or in groups as subdivisions of sections devoted to the different groups of the Periodic Table. Usually, after introductory remarks covering, if necessary, the historical background, the preparative methods are described, including the growth of single crystals. This topic becomes more and more important because of the need not only for small crystals for structure determinations but also for large crystals of high purity for the evaluation of physical properties. A number of preparative methods typical for solid-state chemistry and crystal-growth techniques are mentioned in the text, such as vapor growth, including chemical transport reactions (Section VI,5), hydrothermal syntheses (Section II,D,2), and melt growth techniques such as Bridgman–Stockbarger, flux, and others.

In the succeeding discussion, the chemical behavior of the respective compounds is given. An important role is played by the phase relationships in the respective systems, as they are a prerequisite not only for efficient crystal growth but also for the discussion of the thermodynamic properties. Among the crystallographic data, the occurrence of metal–metal bonds within the transition-element compounds, especially those of groups VB and VIB, should be mentioned, as these lead to the formation of metal clusters as building units of the structures. Almost the whole spectrum of physical properties, not only playing a role in modern, solid-state physics, but also of technical importance, is found among chalcogenide halides. Ionic conductors, which can be used as solid electrolytes in primary and secondary batteries (Section II,D,1), exist among the copper compounds. Semiconductors, among which are the group VA compounds of the SbSI type, play an important part, as these photoconductors exhibit ferroelectric, piezoelectric, nonlinear optical, electromechanical, and other properties. In this group, especially among the arsenide halides, are also found the vitreous semiconductors. Among the metallic conducting phases, such cluster compounds as $Mo_6S_6Br_2$ (Chevrel phases) are superconductors having extremely high, critical fields (Section VII,B,5).

Although chalcogenide halides are mentioned in the literature as early as the end of the last century, this class of compounds has excited growing interest in the scientific world during the past three decades only, as documented by the fact that, during the preparation of this article, two reviews of thio-, seleno-, and tellurohalides appeared (*8, 320*) that were, however, restricted to transition-metal compounds and to somewhat different aspects. The current literature will be cited as comprehensively as possible, with some exceptions, however, as, for instance, of the group VA compounds, where a discussion of the growing literature on the physical properties would exceed the scope of this article; the same applies to compounds of mercury. The older literature will be only partially mentioned, as the existence of many compounds is not well established at this time, and their inclusion would lead to some confusion. In such cases, we have referred to the respective volumes of Gmelin's famous "Handbook of Inorganic Chemistry." The reader interested in more-detailed information may readily find his way via the many references to the topic of interest.

The authors trust that one or more readers may even be stimulated, and start personal research-work in the many unexplored fields of this interesting class of compounds.

II. Group IB

A. Copper

Despite the tremendous amount of work on the binary compounds, copper chalcogenide halides were first reported in 1969 (*304*). Nine compounds of selenium and tellurium have been found, and they are listed in Table I. Copper sulfide halides are still unknown.

1. Preparative Methods

In polycrystalline form, the compounds are obtained by annealing stoichiometric amounts of the respective copper(I) halides with Se ($T = 300°C$) and Te ($T = 350°C$) in closed ampoules for some days. After grinding, the reaction product has to be treated once more in the same way (*307*), in order to complete the reaction. In some cases, small single crystals for X-ray measurements may also be isolated (*125*). Small single crystals of all chalcogenide halides may be obtained by hydrothermal synthesis in the respective halogen hydracids (*300*) (see Section II,D,2). The synthesis of copper chalcogenide halides by chemical-transport reactions (*336*) has been reported (*63, 65*). Large crystals of CuTeBr (*5*) and $CuSe_3I$ (*122*) have been obtained by the Bridgman–Stockbarger technique (*55*) (see Fig. 34).

2. Chemical Properties

Crystals of all of these compounds are stable in air and in alkaline solutions, and dissolve quickly in hot, conc. HNO_3 and H_2SO_4, the selenium compounds also dissolving in hot, alkaline solutions (*307*).

3. Phase Diagrams and Thermodynamics

The system $CuCl-Cu_2S$ is of the simple, eutectic type (*394, 400*), whereas the systems CuX–Se and CuX–Te (X = Cl, Br, or I) are pseudobinary cuts in the ternary diagram Cu–X–Se (Te) and show one (Se) and two (Te) intermediate compounds (*308*) (see Fig. 1). Both congruent and incongruent ternary phases are found, as shown in Fig. 2 for the system CuBr–Te (*299*). CuTeBr exhibits a first-order, displacive phase-transition with ΔH_u and ΔS_u of only 76.8 cal/g-mol and 0.222 cal/g-mol·K, respectively (*5*). Single crystals therefore retain their crystallinity during transition. This phenomenon is of interest, in that it allows the investigation of physical properties during the transition in a single-crystal specimen (*70*). The vapor above the copper se-

TABLE I

COPPER CHALCOGENIDE HALIDES

Compound	Symmetry	a (Å)	b (Å)	c (Å)	β (degrees)	Z	Structure ref.	Decomposition or melting point (°C)
CuSe$_2$Cl	Monoclinic	7.724	4.655	14.573	134.96	4	433	319
CuSe$_3$Br	Orthorhombic	14.363	4.488	7.696		4	424	338
CuSe$_3$I	Rhombohedral	14.083[a]		14.187[a]		18		394
CuTeCl	Tetragonal	15.63		4.78		16	123	400
CuTeBr	Tetragonal	16.417		4.711		16	63,64	430
	Orthorhombic[b]	23.22	23.22	14.13		96		
CuTeI	Tetragonal	17.170		4.876		16	125	442
	Orthorhombic[b]	24.258	24.292	14.611		96	123,126	
CuTe$_2$Cl	Monoclinic	8.207	4.935	15.279	134.92	4	120	~420
CuTe$_2$Br	Monoclinic	8.358	4.951	15.704	135.1	4		416
CuTe$_2$I	Monoclinic	8.672	4.881	16.493	135.0	4		400

[a] For the hexagonal cell. [b] For the superstructure.

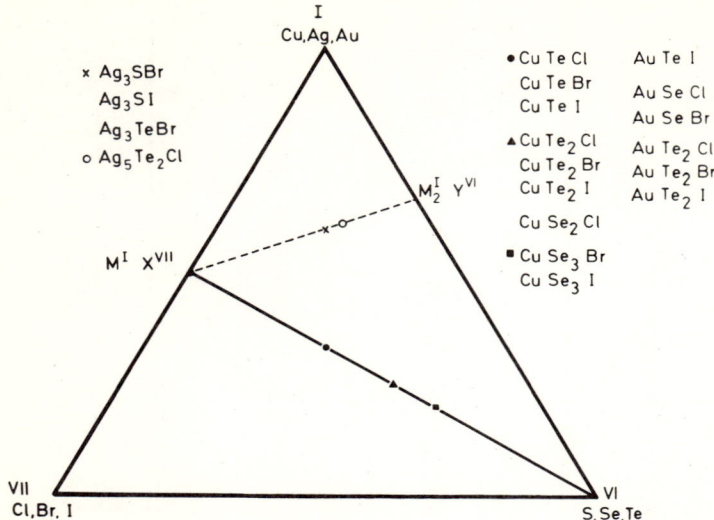

FIG. 1. Chalcogenide halides in ternary systems having the components: the Group IB elements Cu, Ag, and Au, the chalcogens S, Se, and Te, and the halogens Cl, Br, and I. They are indicated as M^I, Y^{VI}, and X^{VII}, respectively. (Redrawn from A. Rabenau, H. Rau, and G. Rosenstein, *J. Less-Common Metals* **21**, 395 (1970), Fig. 4, p. 401.)

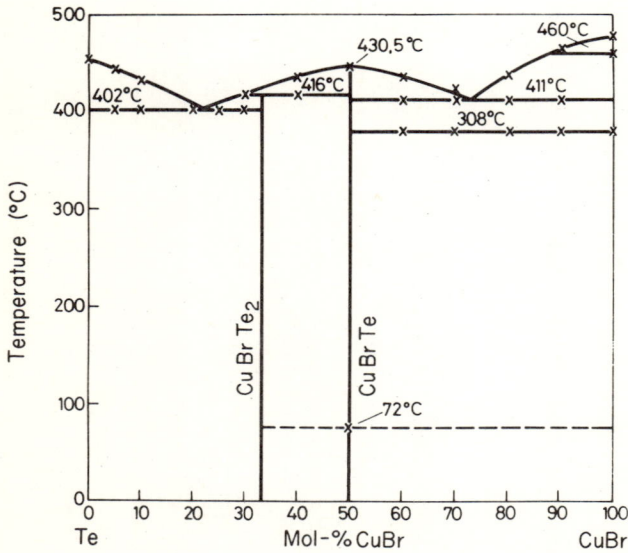

FIG. 2. Phase diagram for CuBr–Te.

lenium halides consists exclusively of selenium and p_{Se_2} can be measured by using the effusion method. From these data, the free energy of reaction can be calculated (307).

$$CuCl,c + 2\ Se,c = CuClSe_2,c;\ \Delta G° = -1800 + 2.3\ T\ cal/mol \quad (1)$$

$$CuBr,c + 3\ Se,c = CuBrSe_3,c;\ \Delta G° = -3100 + 4.0\ T\ cal/mol \quad (2)$$

$$CuI,c + 3\ Se,c = CuISe_3,c;\ \Delta G° = +1900 - 3.9\ T\ cal/mol \quad (3)$$

Due to the high heats of formation of the binary components, these small values seem to be representative for most of the chalcogenide halides discussed in this article.

4. Crystallographic Data

For crystallographic data, see Table I. The compounds of the type CuTeX and CuTe$_2$X (X = Cl, Br, or I), respectively, are isotypic, and their crystal structures have been determined. Copper has the oxida-

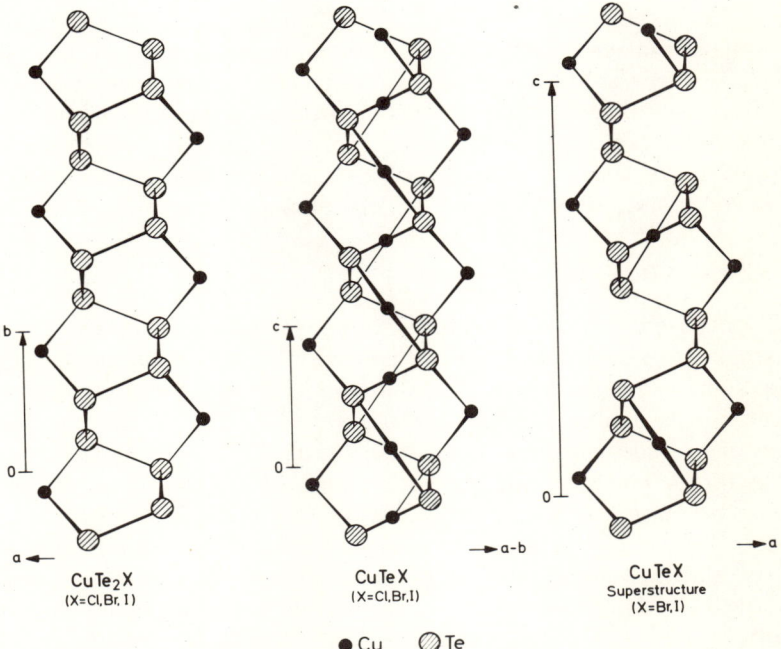

FIG. 3. Tellurium helices in copper telluride halides. (Redrawn from J. Fenner and H. Schulz, *Acta Cryst.* **B35**, 307 (1979), Fig. 2, p. 311.)

TABLE II

OCCUPATION PROBABILITIES OF THE COPPER ATOMIC SITES IN CuTeX COMPOUNDS

Atomic site	Number of equivalent positions	Occupation probabilities			
		CuTeI	CuTeBr (20 °C)	CuTeBr (100 °C)	CuTeCl
Cu(1)	8	0.34(1)	0.26(1)	0.35(2)	0.42(1)
Cu(2)	8	0.23(1)	0.19(1)	0.14(2)	0.07(2)
Cu(3)	16	0.52(1)	0.51(1)	0.60(1)	0.66(1)
Cu(4)	16	0.20(1)	0.26(1)	0.15(1)	0.08(1)

tion state +1, and is tetrahedrally surrounded by either four halogens, or two halogens and two telluriums. The main feature of these structures is infinite, pseudofourfold, tellurium helices, symmetry-related by crystallographic screw axes (see Fig. 3). These infinite, Te–Te bonds have only been found in cases where the formal oxidation state of tellurium lies between zero and unity, as in elemental tellurium (72) and the tellurium subhalides (188). In the CuTeX compounds, the copper atoms are distributed statistically over several crystallographic positions. This is the reason for the observed high copper ionic conductivity (4). The occupation probability of the copper sites is similar in CuTeBr and CuTe I (see Table II). These two compounds show a superstructure having a partial ordering of the occupation probabilities of the copper atomic sites (126). In the case of CuTeBr, this superstructure disappears with the phase transition at 72°C, and the occupation probabilities become similar to those of CuTeCl (123), as shown in Table II.

5. Physical Properties

All of these compounds exhibit a temperature-independent diamagnetism, as is to be expected for compounds of monovalent copper; this is of the same order of magnitude as for copper(I) halides (307). Spectral-reflectance measurements are presented in Fig. 4, a–c. Here, the logarithm of the reciprocal relative reflectance is plotted against photon energy (306). The steep increase above the level of the residual absorption is correlated with the band gap. In the case of the isotypic compounds CuTeX and $CuTe_2X$, the substitution of the halogen seems to have no significant influence on the band gap, as may be seen from Fig. 4, a and b, respectively. This result confirms a concept developed by Goodman (149) that, in the case of substitutional derivation (within a given structure), the energy gap should be determined by the weakest bond in the lattice, largely in terms of bond-electronegativity difference. This decreases in the sequence Cu–Cl, Cu–Br, Cu–I, and Cu–Te, the weakest bond being the Cu–Te. This principle no longer holds if the stoichiometry and the structure change (see Fig. 4c). The compounds of the type CuTeX show a high partial, Cu^+-ionic conductivity at rather low temperatures, with a negligible electronic contribution

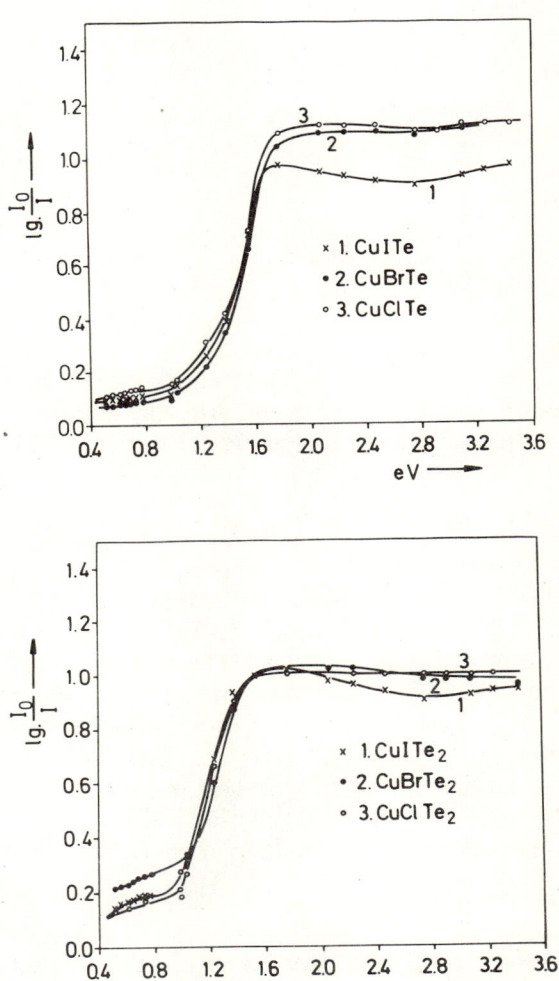

FIG. 4. Spectral reflectance measurements in copper telluride halides. (Redrawn from A. Rabenau, H. Rau, and G. Rosenstein, *Solid State Commun.* **7**, 1281 (1969), Fig. 1, p. 1281.)

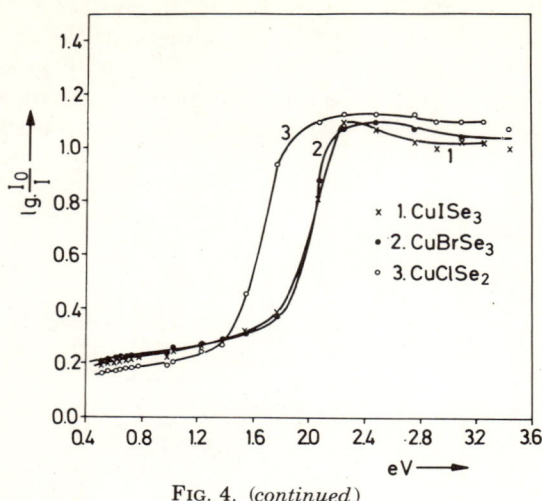

FIG. 4. (continued)

(4, 299, 430). The ionic conductivity is much higher than in the corresponding copper halides (see Fig. 5). At temperatures above 250°C, they enter the region of optimized ionic conductors (150).

B. Silver

The possible existence of a silver sulfide bromide was suggested from the results of conductivity measurements of aqueous Ag_2S–$AgBr$ mixtures, in 1955 (360). The compounds Ag_3SBr and Ag_3SI were first prepared by Reuter and Hardel in 1960 (315). The interest in these materials was stimulated by the discovery of their high ionic conductivity (316, 317, 319, 370). Little is known about the three other silver chalcogenide halides Ag_3TeBr (46, 49, 180), Ag_6TeBr_4 (180), and Ag_5Te_2Cl (46, 49, 53), which need further investigation. It is remarkable, however, that all of the silver chalcogenide halides lie on a cut formed by the respective silver(I) halide and silver(I) chalcogenide (see Fig. 1).

1. Preparative Methods

The sulfide bromide Ag_3SBr may be obtained by annealing of stoichiometric amounts of Ag_2S and $AgBr$ in closed, glass ampoules at 280°C. The reaction product is ground, and repeatedly treated in the same way (317); the end of the reaction is determined by powder patterns. In a similar way, the low-temperature modification of Ag_3SI, β-

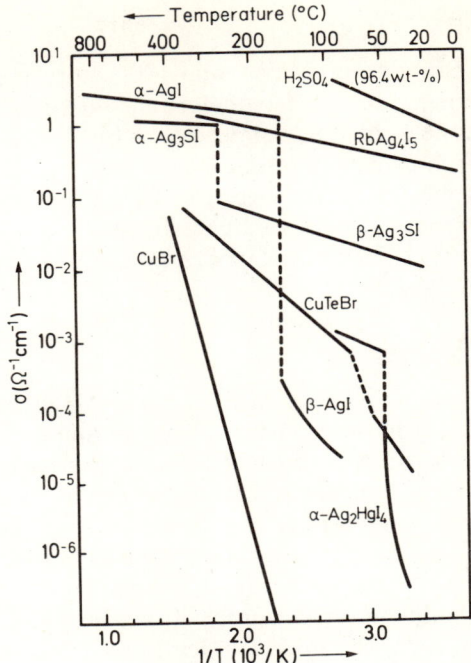

FIG. 5. Ionic conductivity of Group IB compounds.

Ag_3SI, is prepared at 210–215°C. In both cases, it takes several weeks to complete the reaction. The high-temperature modification α-Ag_3SI, which is stable above 235°C, can be obtained either by a treatment at 280°C instead of 210–215°C, as for β-Ag_3SI, or by annealing of the latter for some hundreds of hours at 280°C and quenching to room temperature (317). α-Ag_3SI forms in only 17 hours if the reaction takes place at 550°C under various sulfur pressures (370).

Ag_3TeBr and Ag_6TeBr_4 were obtained by heating the respective stoichiometric mixtures of Ag_2Te and $AgBr$ for 250 hours at 560°C, cooling, and annealing for 1400 hours at 350°C. They were identified by X-ray powder patterns (180). Their existence needs further confirmation. A compound having the probable composition Ag_5Te_2Cl was obtained during the investigation of the Ag_2Te–$AgCl$ system from Ag_2Te and $AgCl$ (53). The samples were tempered for 480 hours at 330°C in closed ampoules. Ag_3TeBr and Ag_5Te_2Cl have been prepared in a similar way, just by annealing at 350°C for 480 hours (49). All of these compounds were reported to be of the peritectic type, which may explain the uncertainty of the data acquired.

2. Chemical Properties

The black Ag_3SBr decomposes above 430°C into Ag_2S and $AgBr$. β-Ag_3SI has the same color, and transforms to α-Ag_3SI at 235°C (*317*). The latter shows no signs of decomposition in DTA measurements up to 800°C (*370*); if, however, a temperature gradient is applied, decomposition starts below this temperature, by sublimation of AgI. Both compounds decompose slowly in light. In such complex-forming solvents as thiosulfate, thiocyanate, or cyanide, the respective silver halide is dissolved (*317*).

3. Crystallographic Data

The structures of Ag_3SBr, β-Ag_3SI, and α-Ag_3SI have been determined on the basis of powder pattern, by trial and error (*318*). Ag_3SBr and β-Ag_3SI are isotypic, and crystallize in a structure that can be derived from the ideal, antiperovskite structure by displacement of silver ions from the face centers in the direction of the edges of the unit cell (see Fig. 6). α-Ag_3SI crystallizes bcc, and is homotypic with α-AgI and α-Ag_2S. These structures differ in two points. (*i*) The corners and the center of the body-centered, cubic cell are occupied by 2 I in α-AgI (see Fig. 7), 2 S in α-Ag_2S, and statistically by 1 S and 1 I in α-Ag_3SI, respectively. (*ii*) Within the 12 tetrahedral voids of the packing of ions,

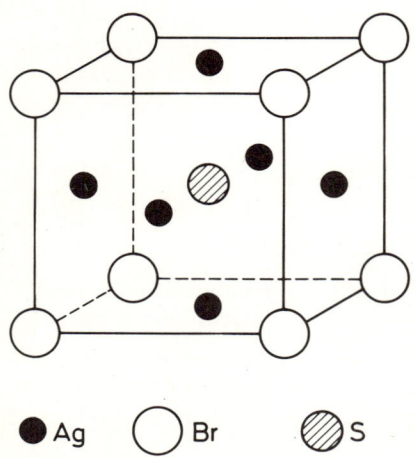

FIG. 6. Ideal antiperovskite structure of Ag_3SBr. (Redrawn from B. Reuter and K. Hardel, *Z. Anorg. Allg. Chem.* **340,** 168 (1965), Fig. 1, p. 171.)

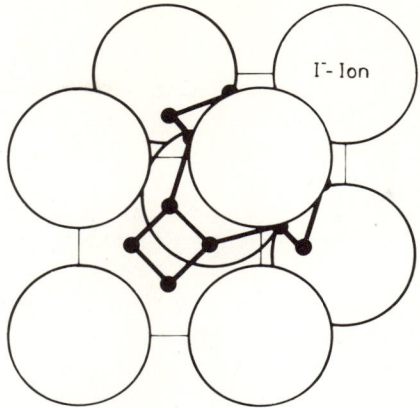

● Ag⁺ - lattice sites

FIG. 7. α-AgI structure, tetrahedral sites, and diffusion path of Ag⁺.

2 Ag in α-AgI (see Fig. 7), 3 Ag in α-Ag$_3$SI, and 4 Ag in α-Ag$_2$S are distributed statistically.[1] No crystallographic data are available for silver telluride halides.

4. Phase Diagrams

Stimulated by the search for optimized ionic conductors, phase-diagram investigations of silver(I) halide–silver(I) chalcogenide systems are numerous (see Table III). The phase diagrams for AgX–Ag$_2$Y (X = Cl, Br, and Y = S, Se Te) are summarized in Fig. 8. They are quasibinary cuts of the eutectic and peritectic type, respectively (46, 47, 49). As indicated by the dashed lines, not everything is yet clarified. The compound Ag$_3$SBr dissolves up to 20 mol-% of Ag$_2$S (317). According to Karbanov et al. (180), the system AgBr–Ag$_2$Te is not a quasibinary one, and the occurrence of a second ternary compound having the composition Ag$_6$TeBr$_4$ was claimed. These authors also found a high solubility of AgBr in Ag$_2$Te at ~500°C. The discrepancies have to be clarified (46). Whereas, in the system AgI–Ag$_2$S near room temperature, β-Ag$_3$SI is the only stable compound, which forms solid solutions containing up to 20 mol-% of Ag$_2$S, similar to Ag$_3$SBr, a very complicated phase-diagram was reported for temperatures above

---

[1] In contrast to the established concept, it has been found that, in α-AgI, silver is exclusively distributed on the tetrahedral voids (135, 69); it may be assumed that the same holds for α-Ag$_3$SI and α-Ag$_2$S.

TABLE III

Investigations of the Ternary Diagrams of $Ag-X^{VII}-Y^{VI}$
(X = Cl, Br, or I; Y = S, Se, or Te)

System	Ref.	System	Ref.	System	Ref.
$AgCl-Ag_2S$	46,47	$AgBr-Ag_2S$	46,47,317	$AgI-Ag_2S$	46,47,317,372,373
$AgCl-Ag_2Se$	46,49,52	$AgBr-Ag_2Se$	49,52	$AgI-Ag_2Se$	369
$AgCl-Ag_2Te$	46,49,53	$AgBr-Ag_2Te$	46,49,180	$AgI-Ag_2Te$	46,369

100°C (373), with additional ternary phases at 25 and 75 mol-% of Ag_2S, respectively, which decompose at room temperature to $AgI + Ag_3SI + Ag_2S$, respectively. The situation around the composition Ag_3SI is given in ref. (372). In the pseudobinary $AgI-Ag_2Se$ system, no compound was found at any temperature or composition (369). The diagram shows a peritectic reaction isotherm at 635°C, the peritectic composition being 45 mol-% of Ag_2Se. The same situation was proposed for the system $AgI-Ag_2Te$ (369), with values of 560°C and 30 mol-% of Ag_2Te, respectively. Complete solubility was found between Ag_3SBr and β-Ag_3SI by X-ray measurements (317).

5. Physical Properties

Ag_3SBr, β-Ag_3SI, and α-Ag_3SI are cationic conductors due to the structural disorder of the cation sublattices. Ag_3SI (see Fig. 5) has been discussed for use in solid-electrolyte cells (209, 371, 374, 414–416) because of its high silver ionic conductivity at rather low temperatures (see Section II,D,1). The practical use seems to be limited, however, by an electronic part of the conductivity that is not negligible (370), and by the instability of the material with respect to loss of iodine (415).

C. Gold

Gold telluride iodide, $AuTe_2I$, was the first example of a gold chalcogenide halide, and was found in 1969 (305). Systematic investigations confirmed the existence of at least six compounds: four telluride halides and two selenide halides (see Table IV). No sulfide halides have been reported.

1. Preparative Methods

The four tellurium compounds are obtained as polycrystalline samples by annealing stoichiometric amounts of the elements in sealed silica or Pyrex tubes at temperatures between 280 (AuTeI) and 350°C

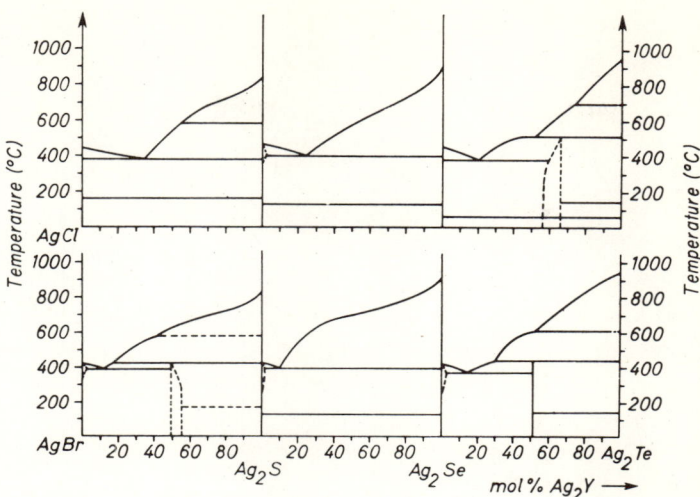

FIG. 8. Some phase diagrams for AgX–Ag$_2$Y (X = Cl or Br, and Y = S, Se, or Te). (Redrawn from R. Blachnik and G. Kudermann, *Z. Naturforsch.* **28B**, 1 (1973), Fig. 1, p. 2.)

(AuTe$_2$I) (*308*). In the case of the selenium compounds, the reaction of the elements remains incomplete, because of partial recrystallization of the highly active gold used for the syntheses (*309*). The best way of preparation proved to be hydrothermal synthesis in halogen hydracids in the arrangement given in Section II,D,2 (*308, 309*). Some compounds, for example, AuTe$_2$Cl, AuSeCl, and AuSeBr, are not formed in the presence of water.[2] For them, an anhydrous hydrogen halide, HCl or HBr, was used as the solvent, introduced into the ampoule via a vacuum system. The telluride halides are silvery white spears (see Fig. 9), or square crystals. Crystals grown in anhydrous media are of pronouncedly poorer quality, pointing out the role that water plays as a mineralizer.

2. *Chemical Properties*

The tellurium compounds are insoluble in dilute acids and alkali, and decompose in dilute nitric acid or conc. sulfuric acid. For melting or decomposition temperatures, see Table IV. No data are given for the selenium compounds, of which only a few, small crystals have as yet been obtained.

---

[2] In the case of selenium compounds in the presence of water, only gold selenide, AuSe, is formed (*309*).

TABLE IV
Gold Chalcogenide Halides

Compound	Symmetry	a (Å)	b (Å)	c (Å)	β (degrees)	Z	Space group	Ref.	Melting or decomposition point (°C)
AuITe	Monoclinic	7.313	7.624	7.255	106.26	4	$P2_1/c$	124,308	360
AuITe$_2$	Orthorhombic	4.056	12.579	4.741		2	Pmmb	155,308	440
AuBrTe$_2$	Orthorhombic	4.033	12.375	8.942		4	Cmcm	155,308	457
AuClTe$_2$	Orthorhombic	4.020	11.867	8.773		4	Cmcm	155,308	447
AuBrSe	Orthorhombic	6.77	12.22	7.38		8	Pnma	229	—
AuClSe	Orthorhombic	6.69	11.85	7.24		8	Pnma	229,309	—

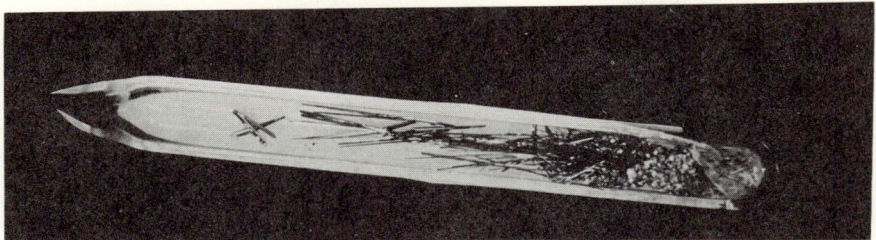

Fig. 9. Quartz glass ampoule with AuTe$_2$I and Au. [From A. Rabenau, *In* "Crystal Growth: an Introduction" (P. Hartmann, ed.), Fig. 7.1., p. 201. North-Holland Publ. Co., Amsterdam, 1973.]

3. Crystallographic Data

Crystallographic data are summarized in Table IV. All structures have been solved. AuTe$_2$I and AuTe$_2$Cl, the latter being isotypic to AuTe$_2$Br, have similar arrangements. They consist essentially of corrugated, two-dimensional nets of gold and tellurium atoms, with interleafing halogen atoms. The tellurium atoms form pairs, joined to successive gold atoms in a –Au–Te–Te–Au– sequence. Each gold atom is coordinated to four tellurium atoms, and each tellurium pair is likewise coordinated to four gold atoms (see Fig. 10). These considerations, and an unusually long gold–halogen distance, suggest that, structurally, the AuTe$_2$X compounds might conceivably be represented as [Au^{III}(Te$_2$)$_{4/4}$]$^+$X$^-$ (*155*).

Interesting relationships to other structures exist in AuTeI (*124*). The four bonds between Au and the nonmetal atoms (one terminal I atom and three bridging Te atoms) generate a two-dimensional net parallel to the b/c-plane (see Fig. 11). Adjacent nets are stacked on each other by lattice translations along the a axis. Geometrically, this net can be derived from the CdI$_2$ structure, with the metal atoms enclosed in octahedral voids (see Fig. 36, Section XII,C,4). This geometrical relationship may be of crystal-chemical meaning, inasmuch as BiTeI, a telluride iodide of analogous composition, has been reported, once, as a *true* isotype of CdI$_2$ (*108*), with disorder of Te and I atoms; and, once, with a hexagonal structure *similar* to that of CdI$_2$ (*390*), with an ordered distribution of all atoms, (see Section XII,C,4).

In AuSeBr and the isotypic AuSeCl, two gold atoms on mirror planes of the space group, and one nonmetal atom of each kind in general positions, form infinite ribbons parallel to the a axis. In these ribbons one Au atom is bonded to two Br atoms and two Se atoms, while the other Au makes four bonds to Se atoms only (see Fig. 12). In both cases, the

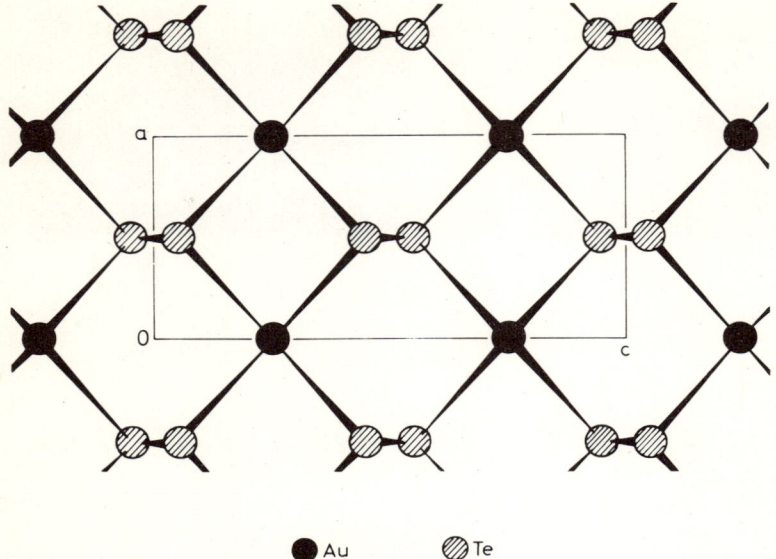

Fig. 10. AuTe$_2$Cl: Au–Te$_2$ net, shown perpendicular to the b axis. (Redrawn from H. M. Haendler, D. Mootz, A. Rabenau, and G. Rosenstein, *J. Solid State Chem.* **10**, 175 (1974), Fig. 1, p. 180.)

coordination is of a planar, distorted-square type. The ribbons are interlocked into sheets perpendicular to the b axis (*229*). The (sometimes distorted) square-planar coordination of gold in all gold chalcogenide halides shows that Au(III) is involved. In this oxidation state, gold forms four coplanar, covalent bonds (*407*). The gold compounds bear no relation to the copper(I) compounds of the same stoichiometry (see Fig. 1). The significant differences between these two systems may be summarized as follows (*308*). (*i*) The structure of corresponding compounds is different. (*ii*) In addition to (*i*), no detectable, mutual solubility has been observed. (*iii*) Whereas the cuts $M^I X^{VII} - Y^{VI}$ (see Fig. 1) in the copper systems are pseudobinary ones, this is not the case with the gold systems. (*iv*) The compounds of the type AuTe$_2$X (X = Cl, Br, I) exhibit a metallic type of conductivity, and the corresponding copper compounds are semiconductors.

4. Physical Properties

Resistivity measurements on single crystals of AuTe$_2$I and AuTe$_2$Br established the metallic character of the conductivity. Single crystals

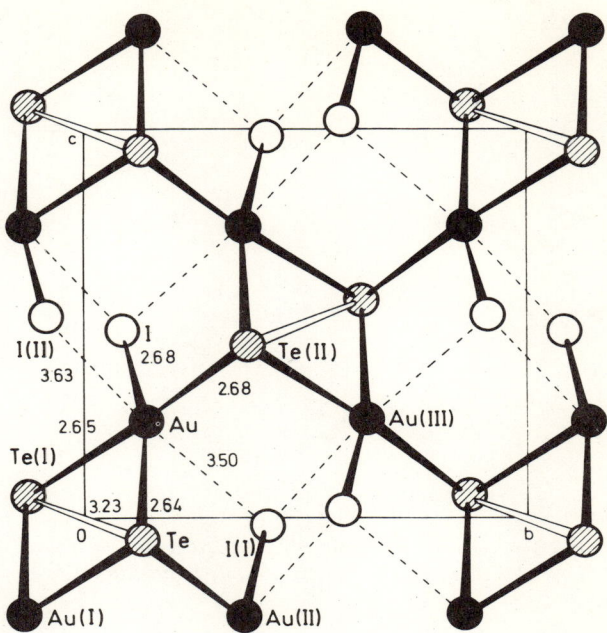

FIG. 11. Corrugated net of the AuTeI structure. (Redrawn from J. Fenner and D. Mootz, *J. Solid State Chem.* **24**, 367 (1978), Fig. 1, p. 368.)

of AuTe$_2$Cl were too small to be measured in the same way, but, qualitatively, they show the same behavior. For the (semiconducting) AuTeI, a band gap of ~0.9 eV results from measurements of the spectral reflectance (*308*). For the metallic conducting compounds AuTe$_2$I and AuTe$_2$Br, a small, temperature-independent diamagnetism was found: the feeble paramagnetism due to the charge carriers does not compensate for the diamagnetism of the compounds. AuTeI also shows a temperature-independent diamagnetism with $\chi_M = -0.14 \times 10^{-3}$, in good agreement with the value $\chi_M = -0.15 \times 10^{-3}$, calculated for the diamagnetic susceptibilities per gram-ion for Au^{3+}I$^-$Te^{2-} (*308*). AuTe$_2$I shows pronounced oscillations in the magnetoresistance at liquid-helium temperature (Shubnikov–de Haas effect). Measurements, on well developed, single crystals, of the angular dependence of the oscillations have been used to determine the Fermi surface. The layer structure of AuTe$_2$I is reflected by the high anisotropy, of at least 10:1, of one of the ellipsoidal, Fermi surfaces (*341, 342*).

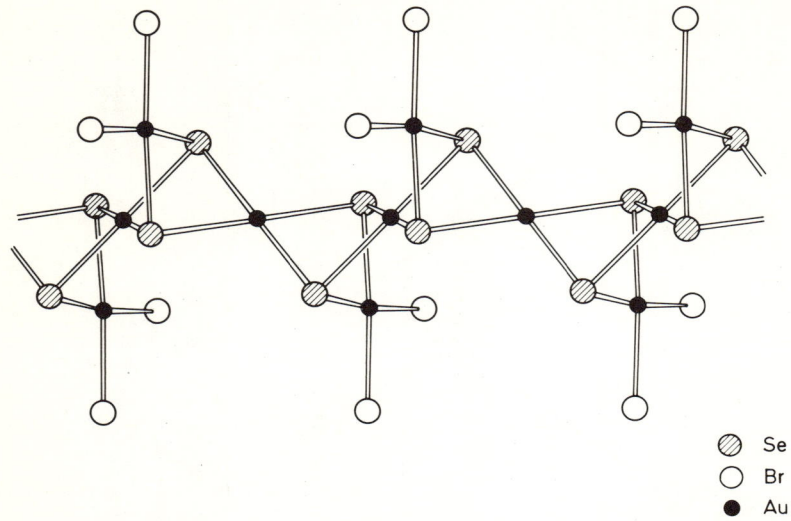

FIG. 12. Perspective view of an atomic ribbon of AuSeBr. (Redrawn from D. Mootz, A. Rabenau, M. Wunderlich, and G. Rosenstein, *J. Solid State Chem.* **6**, 583 (1973), Fig. 1, p. 584.)

D. MISCELLANEOUS

1. Solid State Electrolytes

The most important physical property exhibited by members of the Group IB chalcogenide halides (also, with respect to practical applications) is the enhanced, ionic conductivity found for Ag_3SI and the copper compounds of the type CuTeX. Solid-state electrolytes, where the electric current is due exclusively to the movement of ions, have been known for a long time. In such extreme cases as α-AgI, this conductivity is of the order of magnitude of that of concentrated, liquid electrolytes (see Fig. 5). Unfortunately, AgI loses this property because of a phase transition at 147°C. With the discovery of Ag_3SI by Reuter and Hardel (*316*), the first, optimized, ionic conductor at moderate temperatures was found (*150*). As may be seen from Fig. 5, this and other examples can be formally derived from the binary compound by a substitution in the respective, cation lattice (Ag_3SI, CuTeBr), or anion lattice ($RbAg_4I_5$).

Figure 13 shows the principle of a galvanic cell having solid-state electrolytes. By closing the outer circuit, e.g., by a load, silver from the

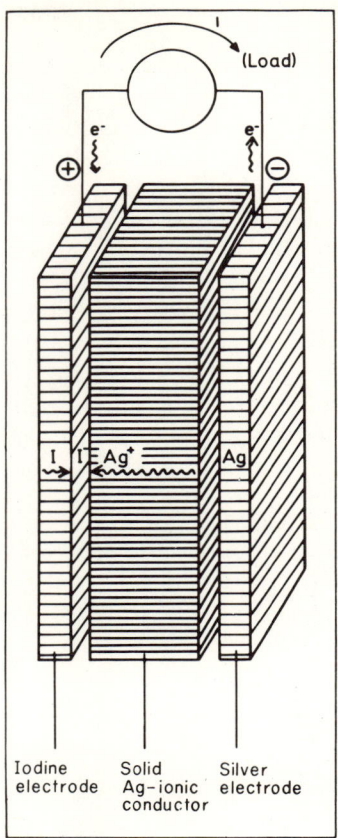

FIG. 13. Galvanic cell having a solid-state silver electrolyte.

cathode dissolves: $Ag \rightarrow Ag^+ + e^-$. The Ag^+ ions move through the solid electrolyte, e.g., Ag_3SI, causing an ionic current in the inner circuit. On the anode, an equivalent amount of I^- is formed by the reaction of iodine with the electrons from the outer circuit. On the interface anode–solid electrolyte, I^- combines with Ag^+. It is the chemical energy of the reaction $Ag + I \rightarrow AgI$ that is transformed directly into electrical energy. Besides many technical applications, such as solid-state batteries (209), the scientific significance of solid, ionic conductors for thermodynamic and kinetic investigations plays an increasing role. A number of review articles on this topic have been published (133, 135, 150, 167, 168, 408).

2. Hydrothermal Synthesis in Acid Solutions

Most of the chalcogenide halides can, in principle, be prepared by high-temperature reactions, e.g., by heating the respective elements or binary compounds (or both) together in sealed, quartz-glass ampoules. A certain amount of knowledge as to the thermal stabilities of the respective compounds is needed. Separation from other phases, however, often causes difficulties. The true composition has, therefore, to be known, making the method unsuitable for exploring new systems. Here, hydrothermal synthesis in acid solutions (*297, 300, 301*) has proved to be the most important tool, as it leads directly to isolated, single crystals of the ternary compounds, suitable for their identification by chemical analysis and X-ray investigation.

Many of the chalcogenide halides mentioned in the different Sections have been obtained in this way. As this method is not yet com-

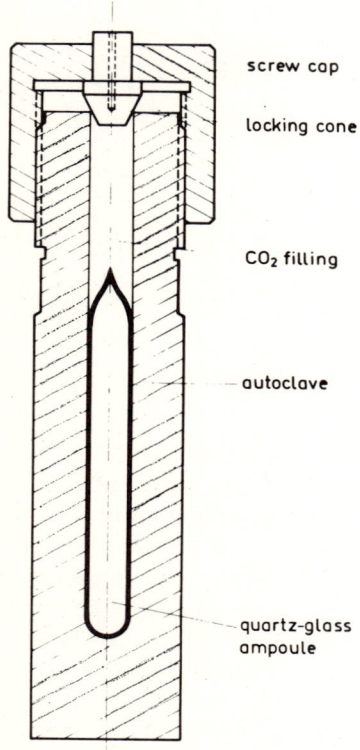

FIG. 14. Equipment for the hydrothermal method, with hydrohalic acids as solvents.

mon practice, the principle will be given. A quartz-glass ampoule having a diameter of ~15 mm is filled to between 50 and 70% of capacity with the respective hydrohalic acid. The acid is frozen with the aid of liquid nitrogen, and the components (elements or binary compounds, or both, not necessarily in stoichiometric proportions) are introduced into the ampoule. Then, the ampoule is evacuated, sealed, and placed in an autoclave (see Fig. 14). To prevent damage to the ampoule by the internal pressure developed during heating, the free volume of the autoclave is filled with solid carbon dioxide. The amount is calculated for an external pressure of ~2500 bar at the working temperature, a pressure that the ampoule can easily support. After about ten days, there have been formed crystals of sufficient size and quality that can be readily separated (see Fig. 9).

III. Group IIB

A. Zinc and Cadmium

No chalcogenide halides of zinc and cadmium are known. The phase diagrams of $CdS-CdCl_2$ (*7, 198, 210*), $CdSe-CdCl_2$ (*210, 314*), $CdTe-CdCl_2$ (*368*), $CdTe-CdBr_2$ (*368*), and $CdTe-CdI_2$ (*323*) are of a simple, eutectic type. The system $CdS-CdCl_2$ shows a range of solubility of CdS in solid $CdCl_2$ that extends to 5% of CdS at room temperature, and increases to a maximum of 12.5% of CdS at 500°C (*7, 210*).

B. Mercury

The most important mercury chalcogenide halides are of the type $Hg_3Y_2X_2$ (Y = S, Se, Te; X = Cl, Br, I). The corresponding sulfide halides have been known for over 150 years (*326*). Quite a lot of work has been performed concerning the preparation, structures, electronic and optical properties, and phototropic behavior of these compounds. Mercury chalcogenide halides of other compositions have been mentioned in the literature (*141*). As most of these compounds are not well established, they will not be treated in detail, with the exception of the latest contributions (see Table V).

1. *Preparative Methods*

The compounds $Hg_3Y_2X_2$ are formed as main products or by-products of many reactions. The most important, and most widely applicable, preparative routes are the following.

TABLE V

MERCURY CHALCOGENIDE HALIDES: BIBLIOGRAPHY[a]

Compound	Preparation, properties	Single crystals	Structure	Optical properties	Phototropy	Electric conductivity	Magnetism
$Hg_3S_2Cl_2$	15,20,290	66,189,312 381	10,66,110,132 291	20	377	378	377
$Hg_3S_2Br_2$	15,20,290	381	289	20,204	204,377	378	377
$Hg_3S_2I_2$	15,20,290	381	288	20,204,375,376	204,205,233,375–377,380	378	375
$Hg_3Se_2Cl_2$	15,20,290	189	290,292	20	310	—	—
$Hg_3Se_2Br_2$	15,20,290	—	290	20,204	204	—	—
$Hg_3Se_2I_2$	15,20,290	—	—	20,204	204,207	—	—
$Hg_3Te_2Cl_2$	15,20,290	189,242	290,292	20,242	—	—	—
$Hg_3Te_2Br_2$	20,290	189	290,292	20	—	—	—
$Hg_3Te_2I_2$	15,20,290	—	290	20	—	—	—
Hg_2YX (Y = S,Se; X = Cl,Br,I)	14	—	—	—	—	—	—
$Hg_5Te_2X_6$ (X = Cl,Br)	14.	—	—	—	—	—	—
Hg_2SCl_4	290	—	—	—	—	—	—
$Hg_3Te_2Br_3$	62	—	—	—	—	—	—
$Hg_3Te_2Br_8$	15	—	—	—	—	—	—
$Hg_3S_5Cl_2$	179	—	—	—	—	—	—
$Hg_2S_2I_2$	51	—	—	—	—	—	—
	313	—	—	—	—	—	—

[a] For a more detailed review of the literature up to 1965, see ref. *141*.

(i) Sintering of Hg_2X_2 and the chalcogen in a sealed tube. S, Se, and Te react at 250, 350, and 500°C, respectively. The initial chalcogen–HgX_2 ratios vary between 5:1 for S and Se, and 4:3 for Te. The reaction is complete after 10–15 h. The excess of chalcogen is removed by dissolution (S) in CS_2, or sublimation (Se, Te) (15, 20).

(ii) Annealing of a stoichiometric mixture of HgX_2 and HgY at 150°C in evacuated tubes for 12 h. A microcrystalline product is formed (290).

(iii) Reaction of HgX_2 with H_2Y by passing the gaseous H_2Y through an aqueous or organic solution of HgX_2. An excess of HgX_2 must be present throughout. The reaction between $HgCl_2$ and H_2Te has to be conducted in an atmosphere of an inert gas. $Hg_3Te_2Br_2$ and $Hg_3Te_2I_2$ cannot be prepared by this method (14, 59, 326).

Other methods of preparation, especially for $Hg_3S_2Cl_2$, have been mentioned. For instance, $HgCl_2$ reacts with S-donors as different as $Na_2S_2O_3$, CS_2, or CuS to afford $Hg_3S_2Cl_2$. $Hg_3Te_2Cl_2$ can also be prepared by the reduction of Te(IV) in the presence of Hg(II), with the aid of SO_2 in an aqueous HCl solution (50). A complete survey up to 1964 is to be found in ref. (141). Baroni (14) observed the intermediate formation of compounds of the composition Hg_2YX_2 (Y = S, Se; X = Cl, Br, I) when gaseous H_2Y, which was diluted by N_2, was passed over the surface of a 1% solution of HgX_2. According to Puff and Kohlschmidt, compounds of the composition $Hg_5Te_2X_6$ (X = Cl, Br) are formed when a mixture of HgTe and an excess of HgX_2 is tempered in evacuated tubes at 150°C for 12 h (290). On the other hand, the authors could not confirm the existence of the compounds $Hg_4S_3Cl_2$, $Hg_5S_4Cl_2$, and $Hg_6S_5Cl_2$, which had been described by Poleck and Goercki (276). Similarly, the existence of the compounds Hg_2SCl_4, $Hg_4S_3I_2$, $HgBr_2 \cdot xSe$, $Hg_3Te_2Br_3$, and $Hg_3Te_2Br_4$ does not seem to have been well established. For a résumé of the preparation and properties of these compounds, see (141). Leonova and Sviridov (206) have since reported the compounds $HgBr_2 \cdot nHgSe$ ($n = 0.1–2.1$), but their homogeneity was not definitely proved.

Growth of single crystals. Crystals of γ-$Hg_3S_2Cl_2$ (see *Crystallographic Data*) measuring $4 \times 4 \times 4$ mm³ can be grown over a period of 11 days by hydrothermal synthesis in the temperature range between 450 and 400°C, starting with HgS, and using 12 M HCl as the solvent (312). Single crystals of $Hg_3Y_2Cl_2$ (Y = S, Se, Te) and $Hg_3Te_2Br_2$ were obtained from polycrystallinic samples by hydrothermal synthesis in 25% aqueous HCl or HBr at 300–320°C (189). Crystals of α- and γ-$Hg_3S_2Cl_2$ (sizes: $1.4 \times 1.4 \times 1.4$ and $4.4 \times 1.9 \times 0.6$ mm³, respectively) were prepared in a chemical-transport reaction (see Section VI,5) by the action of HCl gas on HgS in a sealed ampoule exposed to a temperature gradient ranging from 450 to 345°C (66). The γ-modification was mainly formed. Its preponderance was further increased by the presence of an excess of HCl. Takei and Hagiwara (381) obtained single crystals of $Hg_3S_2Cl_2$ and $Hg_3S_2Br_2$ by a gas-phase reaction between HgS and $HgCl_2$–$HgBr_2$. The reaction mixture was heated in an evacuated tube to a temperature of 450°C, and, within 30 h, single crystals of a maximum length of 1 mm were deposited at the remote, colder part of the tube, whose temperature was kept between 160 and 200°C for the bromide, and at ~290°C for the chloride. In a similar way, crystals of $Hg_3S_2I_2$ can be grown. In an evacuated, T-shaped tube, HgS and HgI_2 are placed in opposite side-arms, and heated to 500 and 300°C, respectively. After 2 h, the temperature is uniformly lowered at a rate of 30–40°/h. Crystals of $Hg_3S_2I_2$ (size: $4 \times 2 \times 0.4$ mm^3) appear on the walls of the central tube, kept at a temperature of 150–175°C (381). Also, if H_2S gas, diluted with N_2, is passed over HgI_2 at 330°C, crystals of $Hg_3S_2I_2$ (size: $1.2 \times 0.5 \times 0.1$ mm^3) are deposited at the cool end (~150°C) of the reaction tube (381).

Nitsche (242) grew single crystals of $Hg_3Te_2I_2$ up to $4 \times 4 \times 4$ mm^3 in size by sublimation of the microcrystalline product, containing an excess of Te, in a sealed ampoule at 550°C. Within 10 days, crystals were formed at the colder (530°C) part of the ampoule.

2. Chemical Properties

The compounds $Hg_3Y_2X_2$ (Y = S, Se, Te; X = Cl, Br, I) are insoluble in water, dilute acids, and conc. HCl, and they are not attacked by these solvents. Conc. HNO_3 converts the sulfide chloride into the sulfide nitrate $Hg_3S_2(NO_3)_2$ (20, 290). With bases, rapid decomposition occurs, leading to the formation of oxide chacogenides (20), or a mixture of oxide and chalcogenide (111), a matter on which agreement has not yet been reached.

Solutions of HgSe and HgTe in $HgBr_2$ have been investigated by Jander and Brodersen (178, 179). From conductivity and cryoscopic measurements, they concluded that, at concentrations below 0.08 M, the chalcogenide is present in molecular form. With increasing concentrations, dissociation according to the equation $HgY + 2\ HgBr_2 = Hg_2YBr^+ + HgBr_3^-$ is followed by associative processes. The associates again dissociate into ionic species. HgTe, whose maximum solubility is 0.4 M, dissociates more strongly than HgSe. By extraction, from the solidified melts, of the excess of $HgBr_2$ with acetone, the compounds $Hg_3Se_2Br_2$ and $Hg_6Te_2Br_8$ were obtained.

3. Crystallographic Data

The cubic modifications of the mercury chalcogenide chlorides and $Hg_3Te_2Br_2$ are isotypic, and crystallize in the space-group $I2_13$ *(290, 292)*. The structure contains distorted $HgY_{2/3}X_{4/6}$ octahedra *(10, 132, 291, 292)* (see Fig. 15). However, from the observed bond lengths, it can more realistically be considered as being built up of a covalent, $(Hg_3S_2)^{2+}$ network containing isolated, ionic, interstitial chlorides. The net can be derived from pyramidal (SHg_3) groups. Each mercury atom is partitioned by two pyramids. The Hg–S distances of 2.42 Å and the S–Hg–S angles of 165.1° are close to the corresponding parameters for the spiral chains of cinnabar *(9)*. $Hg_3Se_2Br_2$ and $Hg_3S_2I_2$ both have orthorhombic symmetry, whereas $Hg_3Te_2I_2$ crystallizes in a monoclinic lattice *(288, 290)*. The $(Hg_3S_2)^{2+}$ polycation is the fundamental, structural element of these compounds, also *(288)*.

$Hg_3S_2Cl_2$ and $Hg_3S_2Br_2$ are polymorphic, affording a low-temperature α, a high-temperature β, and a metastable γ modification *(66, 289)*. β-$Hg_3S_2Cl_2$, like the α form, has a cubic symmetry, with an almost doubled lattice-constant. γ-$Hg_3S_2Br_2$ crystallizes in a tetragonal lattice. The other modifications are rhombic. The α and β modifications of $Hg_3S_2Cl_2$ undergo a reversible, mutual transformation at ~300°C.

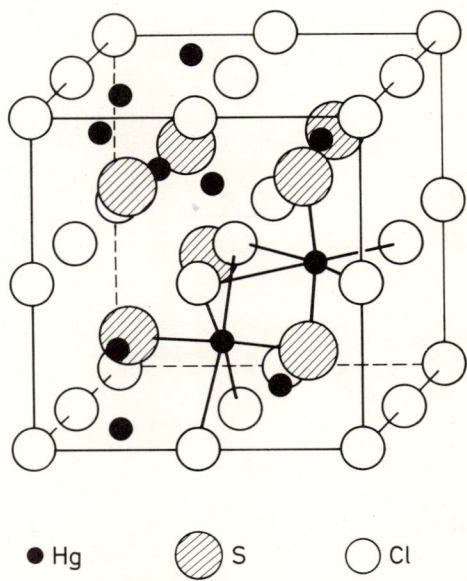

FIG. 15. The crystal structure of α-$Hg_3S_2Cl_2$. (Redrawn from A. J. Frueh and N. Gray, *Acta Cryst.* **B24,** 156 (1968), Fig. 1, p. 157.)

γ-$Hg_3S_2Cl_2$ has an OD-structure consisting of irregularly staggered layers (110). As in the α modification, a $(Hg_3S_2)^{2+}$ network, built up of (Hg_3S) pyramids that share Hg atoms, is formed. Four Cl^- again complete the coordination sphere of Hg to a distorted octahedron, but now, the (Hg_3S) net extends in only two dimensions. The layers are held together by weak Hg–Cl bonds.

Table VI summarizes the structural data on mercury chalcogenide halides.

4. Physical Properties

The phototropic behavior of the mercury chalcogenide halides has received particular attention. This interest was stimulated by the hope that these materials might allow the preparation of photolayers capable of repeated use for the production of images.

When exposed to daylight, the sulfide and selenide halides $Hg_3Y_2X_2$ are blackened within a few minutes. This black color reversibly disappears when the sample is heated to 90 to 120°C, or stored in the dark for several days (204, 375–377). The nature of this phototropic behavior has now been widely investigated by analytical, spectroscopic, structural, magnetic, EPR, and radiotracer investigations (205, 233, 375–377, 379, 380, 382). During irradiation of the compounds, electrons belonging to S^{2-} or I^- ions are excited to upper states. The result-

TABLE VI

CRYSTALLOGRAPHIC DATA FOR MERCURY CHALCOGENIDE HALIDES

Compound	Symmetry	a (Å)	b (Å)	c (Å)	β (deg.)	Z	Ref.
α-$Hg_3S_2Cl_2$	Cubic	8.94(1)				4	*10,290–292*
		8.949(2)				4	*66,132*
		8.937				4	*289*
β-$Hg_3S_2Cl_2$	Cubic	17.993				32	*289*
γ-$Hg_3S_2Cl_2$	Orthorhombic	9.094	16.843	9.349		8	*289*
		9.081(6)	16.82(1)	9.328(5)		8	*110*
α-$Hg_3S_2Br_2$	Orthorhombic	36.89	18.00	9.28		32	*289*
β-$Hg_3S_2Br_2$	Orthorhombic	18.22	9.19	9.24		8	*289*
γ-$Hg_3S_2Br_2$	Tetragonal	13.14		8.89		8	*289*
$Hg_3S_2I_2$	Orthorhombic	9.78	18.68	9.43		8	*288*
$Hg_3Se_2Cl_2$	Cubic	9.06				4	*290,292*
$Hg_3Se_2Br_2$	Orthorhombic	9.42	9.74	8.87		4	*290*
$Hg_3Te_2Cl_2$	Cubic	9.33				4	*290,292*
$Hg_3Te_2Br_2$	Cubic	9.54				4	*290,292*
$Hg_3Te_2I_2$	Monoclinic	9.78	14.09	14.14	96	8	*290*

ing, neutral S or I atoms diffuse toward the crystal surface, and leave as gaseous I_2 or SO_2. Some of the excited electrons neutralize Hg^{2+} ions. The Hg atoms thus formed also diffuse toward the crystal surface, and aggregate there as colloidal droplets that are responsible for the black color. Other excited electrons are captured by the positive holes, which were created by the leaving anions, to form F-centers. Formation of F-centers is irreversible, as is shown by the persisting, paramagnetic moment of the crystals (375, 382), but the printed-out mercury evaporates in the dark, and the crystals regain their original color. According to the results of the radiotracer experiments, considerably more sulfide than iodide is oxidized during irradiation (233, 380). The blackening during the irradiation of $Hg_3Se_2I_2$ with light from a Hg-arc lamp is accompanied by a weak photocurrent (90 pA) (310).

The absorption and diffuse-reflection spectra of the $Hg_3Y_2X_2$ – compounds are characterized by an absorption edge in the region lying between 420 and 500 nm (204, 376), corresponding to band gaps of 2.1–3.2 eV (20, 205, 242). The forbidden-zone width is mainly determined by the halogen, and decreases with increasing atomic weight of both the halogen and the chalcogen (20). The electric conductivities of the sulfide halides are of the order of 10^{-10} $\Omega^{-1}\text{cm}^{-1}$ at room temperature, and 10^{-6} at 150°C. The activation energies of 0.95–1.4 eV are smaller than the optically measured band-gaps, but exhibit the same relative orders (378).

The compounds $Hg_3Y_2Cl_2$ are diamagnetic, as expected (375, 377). The coulomb energies of their lattices have been calculated by Simon and Zeller (358).

IV. Group IIIB and Lanthanides

The first compound of this series, CeSI, was reported by Carter (68) in 1961, and later discussed by Dagron (93). It was obtained by the reaction of iodine with cerium sulfide at 430°C, or by direct synthesis from the elements at 500°C. This was the start of a detailed investigation of this group of compounds mainly by Dagron and co-workers. The present situation is presented in Table VII. No scandium compounds are known thus far, and the same is true for selenium and tellurium halides of these elements.

1. Preparative Methods

The best way of preparation seems to be the successive reaction of the respective elements (94), as described in detail for the sulfide bro-

TABLE VII

Chalcogenide Halides of Group IIIB and Lanthanides: Bibliography

Compound	Ref.	Compound	Ref.	Compound	Ref.
LaSI	92,94,96,128	LaSBr	92,94–96,128	LaSCl	92,94,96,128
α-CeSI	68,92–94,96,115,128	CeSBr	92,94–96,128	CeSCl	92,94,96,128
PrSI	94	α-PrSBr	94–96,128	PrSCl	94,96,128
		β-PrSBr	95,128,334		
NdSI	94,96,128	NdSBr	94–96,128,334	NdSCl	94,96
SmSI	94,96,128,335	SmSBr	94–96,128,334	SmSCl	94,96
GdSI	94,96,128	GdSBr	94–96,128,334	GdSCl	94
TbSI	96,128	TbSBr	95,96,128,334		
YSI	94,96,128	β-YSBr	94–96	YSCl	94,96
DySI	94,96,128	β-DySBr	94–96,128	DySCl	128
		β-HoSBr	95,96,128	HoSCl	128
ErSI	94,96,128	ErSBr	94–96,128	ErSCl	85,94,96,128,355
				TmSCl	128
YbSI	94,96,128	YbSBr	94–96,128	YbSCl	128
LuSI	94,96,128	LuSBr	85,94–96,128	LuSCl	128

mides and iodides by Dagron and Thevet (96). Three types of ampoule were used (see Fig. 16): (i) an open, quartz-glass ampoule containing very fine sheets of the metal (L); (ii) an ampoule (containing sulfur) that had to be sealed for the sulfide bromides and chlorides (S); and (iii) an ampoule containing iodine, or a sealed ampoule for bromine or chlorine (I, Br, Cl).

The ampoules are introduced into a Pyrex-glass tube which is sealed under vacuum. The first step is the reaction of the metal with the halogen. For this, iodine is heated to a temperature of 110°C, whereas, after opening of the ampoule, bromine and iodine, are kept at room temperature and −70°C, respectively,. The ampoule containing the metal is slowly heated to ~300°C, yielding grains of the metal covered with the respective halides. In the second step, the reaction product is reacted with the sulfur vapor at a temperature lying between 400 and 500°C, the other part of the ampoule being at 230°C. The sulfide halide forms, and the halogen thereby liberated attacks the unreacted metal. In this way, extremely microcrystalline, homogenous sulfide halide is slowly formed. The quartz-glass ampoule is removed from the tube, and sealed under vacuum, and the crude product is recrystallized for some weeks at 450 to 500°C.

For single crystals, the same method is applied, using an excess of metal and of halogen. After reaction, the mixture of the sulfide halide and the halide is heated to a temperature slightly above the melting point of the respective halide. Perfect, small crystals for X-ray determination are formed. The excess of the trihalide is removed by treatment with anhydrous alcohol (92, 93, 96). CeSI may also be prepared from the sulfides (68, 93); CeSCl is formed by reaction 4 (92).

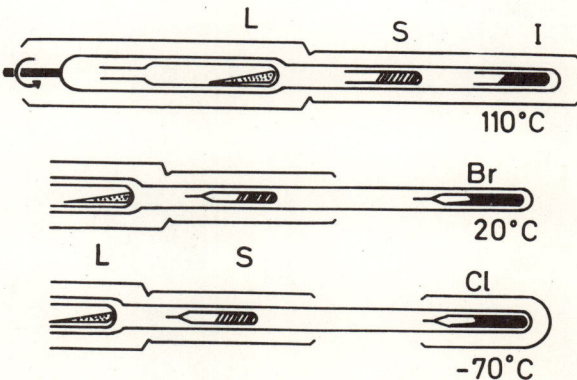

FIG. 16. Preparation of lanthanide chalcogenide halides: experimental arrangement.

$$2\ \text{CeSI} + \text{PbCl}_2 \xrightarrow{500°C} 2\ \text{CeSCl} + \text{PbI}_2 \qquad (4)$$

2. Chemical Properties

Besides a few colorless compounds, such as YSBr, most of these compounds are slightly colored, e.g., yellow (YbSBr), green (PrSI), rose (ErSI), and orange (YbSI). They are very sensitive to humidity, and must be kept under vacuum or dry nitrogen.

3. Crystallographic Data

Whereas the structures of the sulfide iodides and sulfide bromides seem well established, this is not the case for the sulfide chlorides. As the sulfide and chloride ions possess the same number of electrons, they cannot be distinguished by X-ray diffraction, and neutron diffraction would have to be applied in order to obtain an unambiguous picture. Five different structure types have been observed so far (see Table VIII), based on the structure determinations shown in Table IX.

TABLE VIII

CHALCOGENIDE HALIDES OF GROUP IIIB AND LANTHANIDES

(1)[a]	LaSI		LaSBr	LaSCl
	α-CeSI		CeSBr	CeSCl
	β-CeSI			
	PrSI		α-PrSBr	PrSCl[b]
(2)			β-PrSBr	
	NdSI		NdSBr	NdSCl
	SmSI	(4)	SmSBr	SmSCl
	GdSI		GdSBr	GdSCl
	TbSI		TbSBr	
	YSI		β-YSBr	YSCl
(3)	DySI		β-DySBr	DySCl[b,c]
		(5)	β-HoSBr	HoSCl[b,c]
	ErSI		ErSBr	ErSCl
				TmSCl[b,c]
	YbSI		YbSBr	YbSCl[b,c]
	LuSI		LuSBr	LuSCl[b,c]

[a] Structure types (1) α-CeSI; (2) SmSI; (3) hexagonal, structure unknown; (4) NdSBr; (5) FeOCl. [b] Structure type doubtful. [c] Existence doubtful.

TABLE IX
Chalcogenide Halides of Lanthanides: Structure Determinations

Structure type[a]	Compound	Ref.	a (Å)	b (Å)	c (Å)	Degrees	Crystallographic system
(1)	α-CeSI	115	7.33	14.35	7.05		Orthorhombic
(2)	SmSI	335	11.21			23.37 (α)	Rhombohedral
(3)[b]	GdSI	—	10.73	4.24			Hexagonal
(4)	NdSBr	334	6.94	6.91	7.05	99.28 (β)	Monoclinic
(5)	LuSBr	85	5.274	3.995	8.085		Orthorhombic
	ErSCl	355	5.31	3.96	7.45		

[a] See footnote a, Table VIII. [b] Structure unsolved.

For the still-unsolved, hexagonal structure type of GdSI, only the cell dimensions are presented. The structures were described in detail by Dagron and Thevet (96). Their common building-principle is a layer structure formed by planar layers $[LS]_n$ (L = metal), separated by a double layer of bromine or iodine.[3] The plane layers $[LS]_n$ are formed by the juxtaposition of metal tetrahedra enclosing sulfur, $[L_4S]$. These layers exhibit two different symmetries.

(i) This is tetragonal. Fig. 17 shows schematically the sequence of the tetrahedra, which is such that each of the four apexes of a tetrahedron belongs simultaneously to four neighbors, and two edges of the tetrahedra are common to two neighbors. This arrangement can support deformations leading to the orthorhombic symmetry of the CeSI and FeOCl type, respectively, and to the monoclinic NdSBr type.

(ii) This is hexagonal. Here, in the SmSI type, the apexes of the tetrahedra $[L_4S]$ are common to four neighbors as well (see Fig. 18). Three tetrahedra are arranged at an angle of 120° around a common apex, so that one of the faces lies in the plane of the base. The fourth tetrahedron has an inverse arrangement with respect to the other three, and has three of its edges in common with the others. This arrangement leads to octrahedral holes, forming channels that cross the layers of tetrahedra, and these holes contain the iodide ions.

In the CeSI (115) and NdSBr (334) type of structure, bromine and iodine are coordinated to five metal ions (four of the same layer, and one of the opposite layer) and four halogen ions of the double layer. In the SmSI type (335), iodine is coordinated to three metal ions of a $[LS]_n$ layer and three other iodine ions of the double layer. In the FeOCl type of compound, such as ErSCl (355) and LuSBr (85), the halogen is surrounded by a polyhedron formed by six sulfur and four halogenide ions.

---

[3] In the case of the sulfide chlorides, S and Cl are indistinguishable.

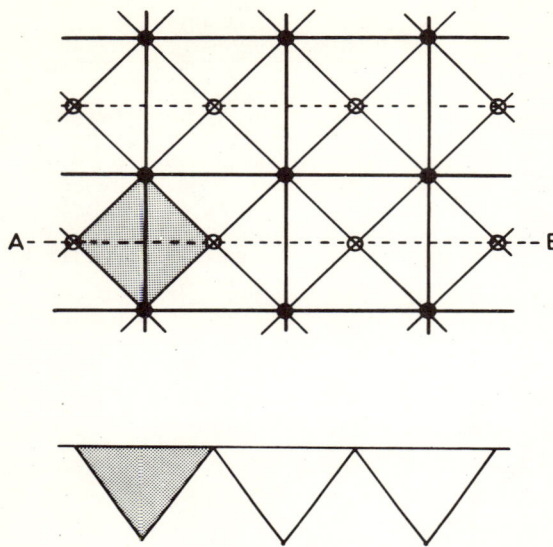

FIG. 17. Tetragonal symmetry: projective view (above) and cut A–B (below) of a [LS]$_n$-layer formed by a sequence of [L$_4$S] tetrahedrons (ideally: PbFCl-type; distorted: CeSI-, FeOCl-, and NdSBr-types). (Redrawn from C. Dagron and F. Thevet, *Ann. Chim.* **6,** 67 (1971), Fig. 6, p. 77.)

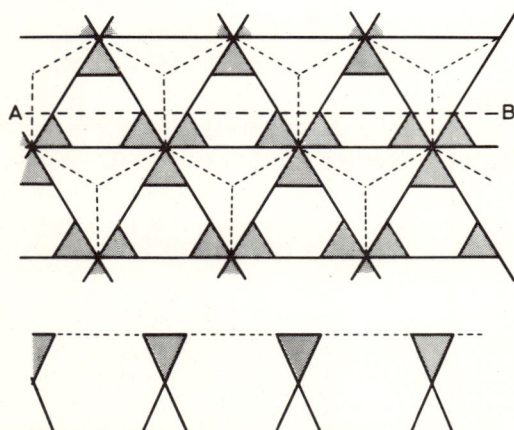

FIG. 18. Hexagonal symmetry: projective view (above) and cut A–B (below) of a [LS]$_n$-layer formed by a sequence of [L$_4$S] tetrahedrons (SmSI-type). (Redrawn from C. Dagron and F. Thevet, *Ann. Chim.* **6,** 67 (1971), Fig. 7, p. 77.)

4. The Lanthanide Contraction

In the sequence of structures from the large to the small rare-earth elements, the lanthanide contraction is manifested as shown in Figs. 19a and 19b. Within a structure, the cell volume diminishes linearly with the atomic number. If a certain, limiting value is reached, there is

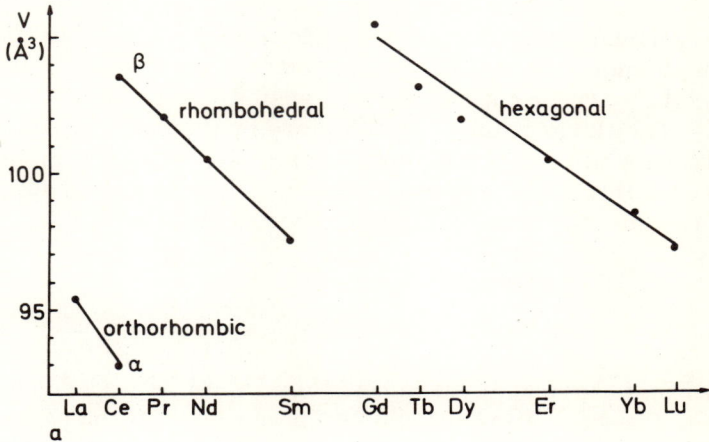

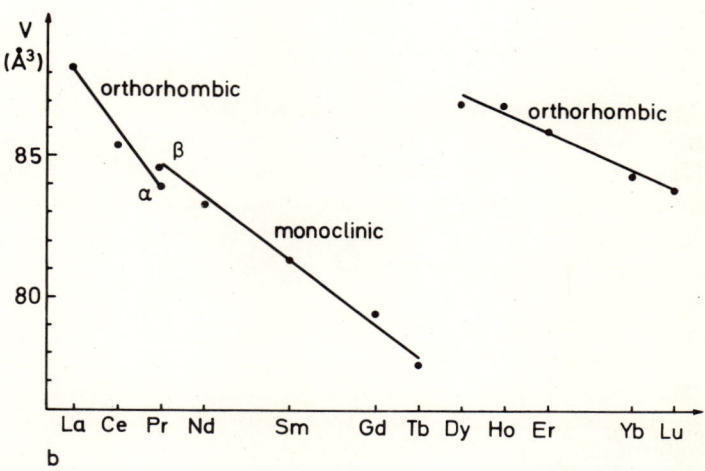

FIG. 19. Molecular-volume variation in relation to the atomic number: a, sulfide iodides; b, sulfide bromides. (From C. Dagron and F. Thevet, *Ann. Chim.* **6**, 67 (1971), Figs. 2 and 3, p. 72.)

a jump to another structure more adapted to the respective ionic radii. This change in structure may be accompanied by a large jump in the cell volume.

V. Group IVB

Very little is known about chalcogenide halides of Group IVB elements. Although the existence of sulfide chlorides (*45, 274, 329, 365*) and of a selenide chloride (*329*) of titanium was claimed in early publications, their true composition, and even their existence, remains doubtful. They have usually been obtained by the reaction of titanium chlorides with sulfur and selenium, respectively, or with hydrogen sulfide. The synthesis of a pure compound, $TiSCl_2$, was published in 1959 (*113*). It is an intermediate of the reaction of $TiCl_4$ with H_2S.

$$TiCl_4 + H_2S \xrightarrow{<0°} TiCl_4 \cdot H_2S \xrightarrow{>0°} TiSCl_2 \xrightarrow{>136°} TiS_2 \quad (5)$$

An optimal reaction-temperature of 65°C was claimed. The blackish brown $TiSCl_2$ is very sensitive to oxygen and humidity. As no X-ray measurements have been made, additional work is needed. Titanium sulfide halides have also been claimed as catalysts for the polymerization of propylene (*363*).

Nothing has thus far been reported concerning chalcogenide halides of other elements of Group IVB.

VI. Group VB

Although not yet published in a journal, according to thesis work, the vanadium compounds VSCl (*11*), VSBr, and VSI (*208*) seem to exist. Numerous niobium chalcogenide halides have been reported, and among these are the best characterized examples of Group VB (see Table X). Only two tantalum compounds, TaS_2Cl_2 (*361*) and $TaSCl_3$ (*13*) have thus far been described in the literature.

1. Preparative Methods

Compounds of the type NbY_2X_2 (Y = S, Se; X = Cl, Br, I) were prepared by the chemical-transport technique (*336*, see Section VI,5), using a slight excess of the halogen and a stoichiometric metal: chalcogen ratio. The starting materials, elements and, in some cases, S_2Cl_2 or $NbCl_5$, are heated in sealed glass or quartz-glass tubes in a temperature gradient in the temperature region between 300 and 500°C. Ex-

TABLE X

CHALCOGENIDE HALIDES OF NIOBIUM: BIBLIOGRAPHY

Compound	Preparation	Crystallographic data	Optical properties	Magnetic data	Other physical data
NbS_2Cl_2	271,322,339	322,339,347	271,322	322,339	322
NbS_2Br_2	271,322,339	322,339	271	322,339	
NbS_2I_2	271,322,339	322,339	271	322,339	
$NbSe_2Cl_2$	322,339	322,339		322,339	
$NbSe_2Br_2$	322,339	322,339		322,339	
$NbSe_2I_2$	322,339,406	322,339		322,339,406	322,406
$Nb_{1.1}Se_2I_2$	322	322			
NbS_2Cl	322,361	322			
NbS_2I	322	322			
$Nb_5Se_8Cl_{10}$	322	322			
$NbSe_2Br$	322				
$NbSe_3Cl$	322	322			
$NbSe_4I_{0.33}$	223	223			
$NbSCl_3$	13,131		13,131		
$NbSBr_3$	131		131		
$NbSeCl_3$	131		131		
$NbSeBr_3$	131		131		

perimental details were given (*322, 339*), and additional information has been published about $NbSe_2I_2$ (*406*). Besides NbY_2X_2 and niobium chalcogenides, other chalcogenide halides are also formed by these reactions: NbS_2Cl, NbS_2I, $NbSe_2Br$, $NbSe_3Cl$, $Nb_5Se_8Cl_{10}$, and $Nb_{1.10}Se_2I_{1.96}$ (*322*). Single crystals of $NbSe_4I_{0.33}$ (together with NbS_2) have been obtained in this way by transport of $NbSe_3$ in iodine vapor (*223*). This compound may also be obtained by direct reaction of a mixture of $Nb + 4 Se + I$ in a sealed tube at 700°C. The excess of iodine is removed by extraction with CCl_4 (*223*). $NbSCl_3$ and $TaSCl_3$ were prepared by the reaction of stoichiometric amounts of the respective pentachlorides and B_2S_3 in a sealed tube at 90°C (Nb) or 80°C (Ta) (*13*).

$$6\ NbCl_5 + 2\ B_2S_3 \rightarrow 6\ NbSCl_3 + 4\ BCl_3 \qquad (6)$$

The reaction starts at room temperature. The BCl_3 formed is condensed in a side arm cooled by liquid nitrogen.

In a similar way, the compounds $NbSCl_3$, $NbSeCl_3$, and $NbSeBr_3$, as well as $TaSCl_3$, $TaSBr_3$, and $TaSeBr_3$, are obtained by reacting the respective pentahalide with Sb_2S_3 and Sb_2Se_3, respectively, in the presence of CS_2 as the solvent at room temperature for 1 to 3 days. In the case of $TaSeBr_3$, the reaction mixture is heated to 50°C. VSCl has been prepared by the direct reaction of VCl_5 and Sb_2S_3 in a sealed tube (*11*).

By mixing solutions of sulfur and the pentachlorides of niobium and tantalum in benzene in a dry, inert atmosphere at room temperature, fine crystalline powders are formed having the compositions NbS_2Cl and TaS_2Cl_2, respectively (*361*). By the direct interaction of sulfur with the pentachlorides in a sealed tube at 240–300°C, a reaction product sublimes; it has the composition $M(S_2)_mCl_n$, where M is Nb or Ta, and $m = 1–3$, $n = 1–4$, depending on the initial ratio of MCl_5 to S.

2. Chemical Properties

The platelike crystals of the compounds NbY_2X_2 are stable in air, and dissolve in hot HNO_3 with precipitation of Nb_2O_5, and in 20% $KOH–H_2O_2$, affording a clear solution. At 10^{-4} torr, they completely decompose as follows.

$$3\ NbS_2Cl_2 \rightarrow NbS_2 + S_2,g + 2\ NbSCl_3,g \qquad (7)$$

(also, for NbS_2Br_2, $NbSe_2Cl_2$, and $NbSe_2Br_2$).

$$NbS_2I_2 \rightarrow NbS_2 + I_2,g \qquad (8)$$

(also for $NbSe_2I_2$) (*339*).

At atmospheric pressure, $NbSe_2I_2$ is thermally stable up to 380°C (*406*).

The compounds $NbSCl_3$ and $TaSCl_3$ (*13*), as well as NbS_2Cl and TaS_2Cl (*361*), are reported to be sensitive to humidity. They are insoluble in organic solvents, but $NbSCl_3$ and $TaSCl_3$ dissolve in acetonitrile, forming $MSCl_3\cdot 2\ CH_3CN$ (*13*).

3. Crystallographic Data

The compounds NbY_2X_2 were investigated by X-ray powder diffraction, and by the single-crystal, Weissenberg technique. They crystallize in two different structures, monoclinic and triclinic. Some of them occur in both crystallographic forms (see Table XI). Both the monoclinic and the triclinic NbY_2X_2 compounds have layer structures. The layers are parallel to the a,b plane for the monoclinic form, and parallel to the a,c plane for the triclinic form (*322*). For NbS_2Cl_2, a complete structure-determination has been performed (*347*). A layer structure comparable to that of $MoCl_3$ or $AlCl_3$ if one Cl is replaced by an S_2 group was found. Besides the S_2 groups, it is the occurrence of metal–metal bonds that is most characteristic for this compound, and that is typical for a great number of transition-metal compounds in lower oxidation states (*340*) (see also, Section VII,B). Here, NbS_2Cl_2 exhibits the specific properties of almost all Nb(IV) compounds: the connection of

TABLE XI
Crystallographic Data for Niobium Chalcogenide Halides

Compound	a (Å)	b (Å)	c (Å)	α (°)	β (°)	γ (°)	Symmetry
NbS$_2$Cl$_2$	6.27	11.09	6.68		111.07		
NbS$_2$Br$_2$	6.54	11.32	6.91		110.54		
NbSe$_2$Cl$_2$	6.65	11.44	6.96		109.02		Monoclinic
NbSe$_2$Br$_2$	6.76	11.53	7.20		113.90		C2/m
NbSe$_2$I$_2$	6.89	12.34–12.46	7.51		112.26		
NbS$_2$Br$_2$	6.589	7.254	6.528	112.63	120.05	67.72	
NbS$_2$I$_2$	6.80	7.23	6.77	102.3	117.4	73.8	
NbSe$_2$Cl$_2$	6.538	7.261	6.350	11.35	119.01	66.93	Triclinic
NbSe$_2$I$_2$	7.207	7.757	7.060	113.20	121.10	67.59	
NbS$_2$Cl	8.615	24.754	6.719	98.77	106.58	128.88	
NbS$_2$I	7.032	24.892	6.818	94.66	118.75	104.07	
NbSe$_3$Cl	11.790	18.847	7.196		95.41		Monoclinic
Nb$_5$Se$_8$Cl$_{10}$	9.365	12.673	7.579		124.78		P2$_1$ or P2$_1$/m
NbSe$_4$I$_{0.33}$	9.489		19.13				Tetragonal P4/mnc
Nb$_{1.1}$Se$_2$I$_2$	12.874	25.829	7.815		99.98		Monoclinic P2$_1$/c

two niobium by a metal–metal bond. The coordination is shown in Fig. 20, and the situation is best described by the formula $^2_\infty[\text{Nb}_2(\text{S}_2)\text{Cl}_{8/2}]$. Every Nb$_2$ group is coordinated to two S$_2$ groups, and two such Nb$_2$(S$_2$)$_2$ units are linked by 4 Cl.

NbSe$_4$I$_{0.33}$ is tetragonal, and has the space group P4/mnc. The structure is built up of chains of rectangular, NbSe$_8$ antiprisms. Iodine atoms are situated between these chains. With respect to the crystallographic data, a great similarity to the β-phase of "NbSe$_4$" exists. The structure is better expressed by the formula Nb$_{12}$Se$_{48}$I$_4$, instead of NbSe$_4$I$_{0.33}$ (223). Crystallographic data for these and other niobium compounds are summarized in Table XI.

X-Ray powder diffraction photographs suggest that NbSCl$_3$ and TaSCl$_3$ are isostructural (13).

4. Physical Properties

IR and Raman spectra of the NbS$_2$X$_2$ compounds have been measured, and assigned to the different modes (271). Especially in the absence of X-ray data, such measurements serve to prove the presence of S–S and Nb–S modes (see Table X). Magnetic data are available for

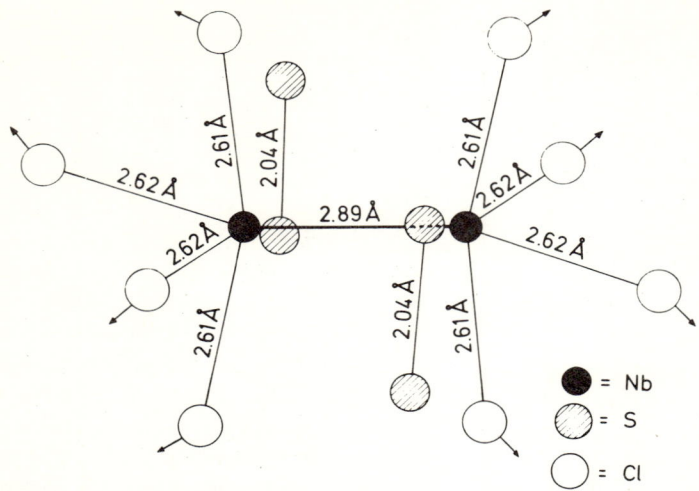

FIG. 20. Coordination polyhedron of NbS_2Cl_2. (Redrawn from H. G. v. Schnering and W. Beckmann, Z. Anorg. Allg. Chem. **347**, 231 (1966), Fig. 3, p. 238.)

the NbY_2X_2 compounds (*322, 339*). They exhibit a temperature- and field-strength-independent diamagnetism; this proves the absence of isolated, Nb^{4+} ions, in accordance with the result of the structure determination (*339*). Conductivity and photoelectric effects have been measured on $NbSe_2I_2$ single crystals. An activation energy of 0.26 eV was found for the high-temperature branch of the conductivity curve (*406*). The absorption spectrum of NbS_2Cl_2 single crystals is caused by indirect band-to-band transitions, the indirect band-gap observed being 1.89 eV at ambient temperature (*322*). NbY_2X_2 compounds are diamagnetic semiconductors having high electrical resistivities.

5. Chemical-Transport Reactions

A "chemical transport reaction" may be defined as the transference of a condensed phase through a gaseous phase by means of a chemical reaction in which gases $B_{(g)}$, $C_{(g)}$, $D_{(g)} \cdots$ are involved.

$$A_{(s)} + B_{(g)} \rightleftharpoons C_{(g)} \cdots ; \Delta H^0_{react.} \qquad (9)$$

This definition of the process shows that it differs essentially from sublimation and distillation. A chemical-transport reaction is necessarily reversible; a concentration gradient is induced, e.g., by means of a temperature gradient that reverses the reaction and causes substance A to precipitate out of the gas phase. The direction of transport of a solid in a temperature gradient ($T_1 < T_2$) is given by the sign of $\Delta H^0_{react.}$.

Endothermic reaction: $T_2 \to T_1$

Exothermic reaction: $T_1 \to T_2$

In preparative work, it may be advantageous to superimpose a transport reaction on the synthesis. In this way, the reaction product is prevented from covering the surface of the starting material, a process that would slow, or completely stop, the reaction. As the vapor pressure of the condensed phases may be negligibly small in the range of temperature within which transport takes place, lower temperatures than would otherwise be feasible may be employed. The desired substance is transported in a temperature gradient, and separated from the starting material, and is, in this way, at the same time also purified (see Fig. 21). In the case of niobium sulfide halides, the chemical transport can be described by the following equations.

$NbS_2Cl_{2,(s)} + NbCl_{4,(g)} \to 2\ NbSCl_{3,(g)}$; endothermic
(analogous NbS_2Br_2, $NbSe_2Cl_2$, $NbSe_2Br_2$) (10)

$NbS_2I_{2,(s)} + I_{2,(g)} \to NbI_{4,(g)} + S_{2,(g)}$; endothermic (analogous $NbSe_2I_2$) (11)

Generally, it is not important in which form the components are introduced into the transport tube, because the transport reaction transfers the whole system into a reversible state. In the preparation of NbS_2Cl_2, for example, Nb and S_2Cl_2, as well as Nb, S, and $NbCl_5$, have been used as starting materials (*339*). The principles of chemical transport reactions are treated in refs. (*241, 336, 337, 338*).

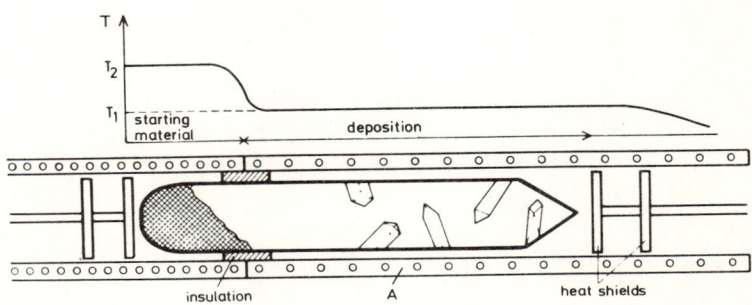

Fig. 21. Reaction scheme of crystal growth by chemical transport: A, transport furnace having two consecutive heating coils.

VII. Group VIB

A. Chromium

The work on chromium compounds is based on two publications only (*17, 181*). By the reaction between Cr_2S_3 and $CrBr_3$ or CrY (Y = S, Se, Te) and I_2 in sealed, quartz or glass tubes, the compounds shown in Table XII have been obtained.

Preliminary X-ray work on CrSBr crystals, probably formed via a chemical-transport reaction, shows rhombic symmetry (*181*). Powder patterns of the iodides can be indexed with respect to a hexagonal lattice, and are similar to those of CrI_3 (*17, 181*).

The ideal composition CrYI does not seem to be attained in all cases (see Table XII). Whether this is due to the existence of extended phases (e.g., $CrSI_{1-x}$) or represents an individual phase is not known. The compounds are described as black, and sensitive to air. Additional work should be done for the chromium system.

B. Molybdenum

Just as, in Group VB, niobium, so, in this Group, molybdenum provides most of the examples of the chalcogenide halides. The occurrence and preparation of such compounds are described in numerous publications. In most cases, they have been obtained as powders, with the composition based on chemical analyses only. The presence of defined, homogeneous phases is, therefore, in many cases doubtful. In addition, some published results are contradictory. A decision is possible where a complete structure analysis has been made. As will be shown later, the formation of metal–metal bonds (so-called clusters), as in the case of niobium, is the most characteristic building-principle. Such clusters

TABLE XII
CHALCOGENIDE HALIDES OF CHROMIUM

Phase	Reaction temperature (°C)	Density (g/cm³)	Ref.
CrSBr	870	3.99	*181*
$CrSI_{0.83}$	420	4.51	*181*
CrSeI	400	5.41	*181*
$CrTe_{0.73}I$	315	5.77	*181*
CrTeI	190	5.50	*17*

may or may not be maintained in a chemical reaction, and they therefore account for some divergent observations.

1. Preparative Methods

In this section, a review is given of preparative work on chalcogenide halides. Table XIII summarizes the proposed compositions thus far described in the literature.

The first report dates from 1894: Mo and S_2Cl_2 heated to red heat should yield $Mo_5S_8Cl_9$ (362). No further work was published until the 1950s. In 1959, a compound having the composition of $Mo_5S_8Cl_9$ was considered as an intermediate step in the chlorination of MoS_2 (138). $MoCl_5$ reacts at 150°C with S_2Cl_2 to give MoS_2Cl_3 (129). A compound having this composition was also obtained, as black needles, by the reaction of S with $MoCl_3$ in S_2Cl_2 at 450°C under pressure. The composition was confirmed by structure analysis (215, 217). $MoTe_2$ reacts with Br at room temperature to afford $MoTe_2Br_{10}$, which loses Br_2 dur-

TABLE XIII
Chalcogenide Halides of Molybdenum: Bibliography

	Cl	Br	I
S		$Mo_6S_6Br_2$ (354)	$Mo_6S_6I_2$ (354)
		Mo_6SBr_{10} (272)	
	$Mo_5S_8Cl_9$ (362)		
	$Mo_3S_7Cl_4{}^a$ (101,215,216,217,222,261)	$Mo_3S_7Br_4{}^a$ (101,215,261)	$Mo_3S_7I_4{}^a$ (101,186)
	$Mo_2S_5Cl_3$ (271,311)	$Mo_2S_5Br_3$ (311)	
	$Mo_2S_4Cl_5$ (311)		
	$MoS_2Cl_3{}^b$ (129,215,216,217)		
	MoS_2Cl_2 (58,311,356)	MoS_2Br_2 (311)	
	$MoSCl_3$ (12,13,57,58,131)	$MoSBr_3$ (131)	
	$MoSCl_2$ (58,138)		
	$MoSCl$ (269)	$MoSBr$ (269,270)	$MoSI$ (269)
Se		$Mo_6Se_5Br_3{}^c$ (273,354)	$Mo_6Se_5I_3{}^c$ (273,354)
		Mo_6SeBr_{10} (272)	Mo_6SeI_{10} (272)
	$Mo_3Se_7Cl_4{}^a$ (101,222,261)	$Mo_3Se_7Br_4{}^a$ (101,265)	$Mo_3Se_7I_4{}^a$ (101,186)
	$MoSeCl_3$ (58,131)	$MoSeBr_3$ (131)	
	$MoSeCl_2$ (58)		
Te		$Mo_6Te_5Br_3{}^c$ (273,354)	$Mo_6Te_5I_3{}^c$ (273,354)
		Mo_6TeBr_{10} (272)	Mo_6TeI_{10} (272)
		$MoTe_2Br_{10}$ (260)	
		$MoTe_2Br_8$ (260)	

[a] Mentioned in (101, 186) as $Mo_6Y_{14}X_8$. [b] Mentioned in (129) as $Mo_2S_4Cl_6$. [c] $Mo_6Y_5X_3$ composition with the highest X content in the solid solutions $Mo_6Y_{8-x}X_x$.

ing heating by forming $MoTe_2Br_8$, which is stable up to 300°C (260). Rannou and Sergent (311) reported in 1967 the preparation of five sulfide halides: from MoS_3 and S_2X_2 (X = Cl, Br), they obtained between 350 and 400°C $Mo_2S_4Cl_5$ and between 420 and 480°C $Mo_2S_5X_3$ (X = Cl, Br) (311). The reaction of elemental Mo and S_2Cl_2 and S_2Br_2 at 500°C, however, resulted in MoS_2Cl_2 and MoS_2Br_2, respectively (311). Formation of the compound MoS_2Cl_2 was also observed on reaction of $MoCl_4$ with H_2S in benzene (356) or CS_2 (58).

Compounds having the composition $Mo_3Y_7X_4$ were reported by Opalovskii et al. (222, 261). $Mo_3S_7Cl_4$ and $Mo_3Se_7Cl_4$ which are formed at 400 and 450°C by the reaction of $MoCl_3$ with S and Se, respectively. The same workers also prepared $Mo_3S_7Br_4$ and $Mo_3Se_7Br_4$ by replacing $MoCl_3$ with $MoBr_3$ and $MoBr_2$ in the temperature region between 250 and 300°C. These phases are, however, amorphous (265). Red, single crystals of $Mo_3S_7Cl_4$ and of orange $Mo_3S_7Br_4$ have been obtained by reaction of S with $MoCl_3$ and $MoBr_3$ in S_2Cl_2 and S_2Br_2 at 400 to 500°C under pressure. The compositions were confirmed independently by structure investigations (215). The d-values were not in agreement with those published by Opalovskii et al. (261) for $Mo_3S_7Cl_4$, but agreed with those published for $Mo_2S_5Cl_3$ (311). According to later data (101, 186, 222), all of the compounds having the composition $Mo_3Y_7X_4$ (Y = S, Se; X = Cl, Br) have now been obtained. $MoYCl_2$ compounds (Y = S, Se) were prepared by the introduction of Sb_2S_3 and Sb_2Se_3 into the syntheses by Britnell et al.; the reaction of Sb_2Y_3 with $MoCl_5$, $MoCl_4$, and $MoBr_5$ to form these compounds takes place between 100 and 300°C (57, 58, 131). $MoSCl_3$ may also be obtained from B_2S_3 and $MoCl_5$ (12, 13). Compounds of the type MoSX (X = Cl, Br, I) have been obtained from MoX_2, Mo, and S in stoichiometric amounts at 1000°C (269). The compounds are isotypic; the composition was confirmed by structure analysis of MoSBr single crystals (270). In a similar way, Mo_6YX_{10} compounds have been prepared, with Y = S, Se, or Te, and X = Cl or Br, as well as Mo_6YI_{10}, with Y = S or Se (272, 273). In this case, the structure analysis on Mo_6SeCl_{10} (273) also confirmed the composition proposed.

A group of compounds interesting because of their physical properties has been found by Sergent et al. (272, 354). They are derived formally and structurally, by substitution of chalcogen in Mo_6Y_8 (Mo_3Y_4) (Y = S, Se, Te) corresponding to $Mo_6Y_{8-x}X_x$ (X = Cl, Br, I). The preparation again starts from Mo, Y, and MoX_2. In the case of the selenides, solid solutions are formed, with $0 \leq x \leq 3$. The maximum halogen content corresponds to $Mo_6Y_5X_3$. Sulfur compounds exist only for Br and I. Their composition is $Mo_6S_6X_2$ (354), with a deficiency of halogen.

2. Chemical Properties

Little has been reported concerning the chemical properties of the numerous compounds mentioned in Table XIII. Also, the behavior of fine powders may differ from that of large, single crystals. Most of the compounds seem to be stable in air: MoSX (X = Cl, Br); $Mo_3Y_7Cl_4$ (Y = S, Se), $MoTe_2Br_8$, $Mo_2S_4Cl_5$, $Mo_2S_5X_3$ (X = Cl, Br), $Mo_5S_8Cl_9$, MoS_2Cl_3, and $Mo_3S_7Cl_4$. Others, such as $MoYCl_3$ (Y = S, Se) and $MoYCl_2$ (Y = S, Se), are sensitive to humidity, and decompose more or less rapidly in air. $MoTe_2Br_8$ is hydrolyzed in water, but is soluble in such polar, organic solvents as ethanol and acetone. The colorless compound can be sublimed (260). $MoSCl_3$ seems to form adducts with polar, organic solvents (58).

3. Crystallographic Data

The discussion will be restricted to cases for which a complete structure-determination exists. Again, as with the niobium compounds, but even more pronouncedly, the occurrence of metal–metal bonds (340) (see Table XIV) is most significant.

MoSBr: The cubic structure (F$\bar{4}$3m) consists of Mo_4S_4 cubes that contain Mo_4 clusters with a metal–metal distance of 2.80 Å (see Fig. 22). The clusters are connected by Br atoms in such a way that Mo is in the center of an octahedron ($MoBr_3S_3$). A molybdenum atom of a cluster is, therefore, connected to three other clusters by three Br bridges, and the compound is best described by the formula [(Mo_4S_4) $Br_{12/3}$]. The motive Mo_4S_4 is the first example in the chemistry of Mo(III) having the tetrahedral, Mo_4 cluster (269, 270).

MoS_2Cl_3, monoclinic ($P2_1/c$). Figure 23 shows the translational unit, with atom designations. The sulfur atoms form pairs having a distance of 1.98 Å. The molybdenum atoms also occur in pairs enclosed by two bridging S_2 groups. With two chlorine atoms in terminal and

TABLE XIV

METAL–METAL BONDS IN MOLYBDENUM CHALCOGENIDE HALIDES

Compound	Coordination	Mo–Mo distance Å	Ref.
MoS_2Cl_3	Pairs	2.83	215
$Mo_3S_7Cl_4$	Triangle	2.74–2.75	215
MoSBr	Tetrahedron	2.80	270
$Mo_6Cl_{10}Se$	Octahedron	2.60–2.63	354
Metal	Cube	2.72	

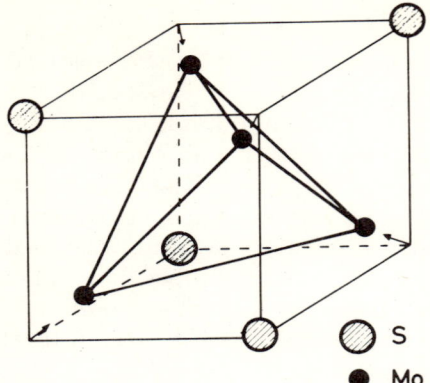

FIG. 22. Tetrahedral Mo$_4$-cluster in MoSBr. (Redrawn from Ch. Perrin, R. Chevrel, and M. Sergent, *C. R. Acad. Sci., Ser. C* **280**, 949 (1975), p. 951.)

two more in bridging positions, the coordination number of molybdenum is nine. The metal coordination in MoS$_2$Cl$_3$ is very similar to that in NbS$_2$Cl$_2$ (*347*), the difference in stoichiometry being adjusted by a different bridging mode of the chlorine atoms of $^1_\infty$[Mo$_2$(S$_2$)$_2$Cl$_4$Cl$_{4/2}$] and $^1_\infty$[Nb$_2$(S$_2$)$_2$Cl$_{8/2}$], respectively (*215–217*).

Mo$_3$S$_7$X$_4$ (X = Cl, Br), monoclinic (P2$_1$/c). Figure 24 shows the asymmetrical unit of the crystal structure. The three, independent molybdenum atoms form an almost equilateral triangle. Six of the seven sulfur atoms occur in three S$_2$ groups, each one bridging one Mo–Mo

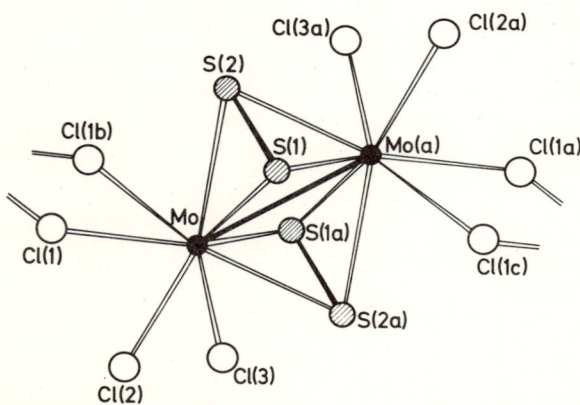

FIG. 23. Two asymmetric units forming the translational element along the *a* axis of MoS$_2$Cl$_3$. (Redrawn from J. D. Marcoll, A. Rabenau, D. Mootz, and M. Wunderlich, *Rev. Chim. Miner.* **11**, 607 (1974), Fig. 1, p. 611.)

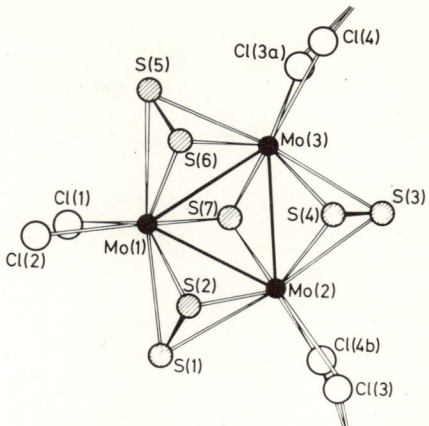

FIG. 24. Asymmetric unit of $Mo_3S_7Cl_4$, including bridging chlorine atoms. (Redrawn from J. D. Marcoll, A. Rabenau, D. Mootz, and M. Wunderlich, *Rev. Chim. Miner.* **11**, 607 (1974), Fig. 4, p. 613.)

bond on the same side of the triangular plane. The seventh sulfur is bonded from the opposite side to all three molybdenum atoms. Hence, two chlorine atoms at each molybdenum atom complete the coordination at one Mo in terminal, at the other two in bridging, positions, forming an unlimited, zigzag chain $^1_\infty[Mo_3S(S_2)_3Cl_2Cl_{4/2}]$ (*215–217* Cl; *215* Br).

$Mo_6Cl_{10}Se$ has an orthorhombic structure, space group Pccn (*273*), and is isostructural with Nb_6I_{11} (*357*). Structure units $Mo_6X'_8$, where a Mo_6 cluster in the form of a strongly deformed octahedron is surrounded by a cube of 8 X' (X' = ⅞ I + ⅛ Se), are linked by bridging I atoms (see Fig. 25). This results in the formula $(Mo_6Cl_7Se)Cl_{6/2}$. The chalcogenide chlorides $Mo_6Cl_{10}Y$ (Y = S, Se, Te) are isotypic (*273*).

4. Physical Properties

For reasons mentioned in the previous paragraph, only properties of well-defined phases will be reported.

The thiohalides MoSX are diamagnetic (*269*). In $Mo_4S_4Br_4$, molybdenum transfers 12 electrons to the anions, and 12 electrons remain for the cluster Mo_4, forming a doublet for each edge of the Mo_4 tetrahedron (*269*). The diamagnetism of the compounds of the type $Mo_6Cl_{10}Se$ can be interpreted in a similar way (*273*) from structure considerations. These compounds have been found to be dielectric (*273*).

The compounds of type $Mo_6Y_{8-x}X_x$ are conductors, and exhibit a

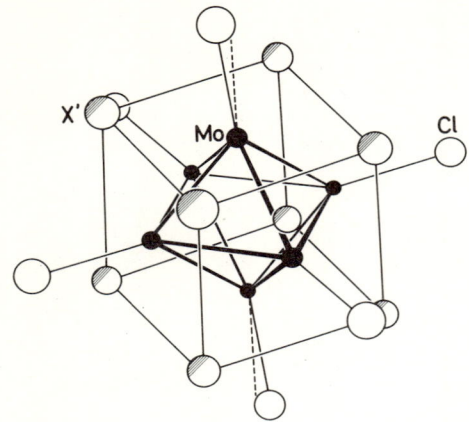

FIG. 25. The group $(Mo_6X'_8)Cl_{6/2}$ (X' = 7/8 Cl, 1/8 Se) in Mo_6SeCl_{10}. (Redrawn from A. Simon, H. G. v. Schnering, and H. Schäfer, *Z. Anorg. Allg. Chem.* **355**, 295 (1967), Fig. 7, p. 308.)

small paramagnetism (*354*). They belong to the superconducting, Chevrel phases.

5. *Chevrel Phases*

Superconductors having extremely high critical fields have been found in ternary molybdenum chalcogenides (*127*). This class of compounds was first described by Chevrel et al. (*83*). The structure may be regarded as being built up by units of Mo_6Y_8, similar to the $Mo_6X'_8$ unit in $Mo_6Cl_{10}Se$ (see the preceding, and Fig. 25), but linked together directly without bridging halogens. In these high-field superconductors, metal elements M are arranged in channels formed by the Mo_6Y_8 units, the ideal formula being $M_xMo_6Y_8$, where x may lie between 0 and 2, depending on M. The superconducting properties of these compounds were discussed on the basis of the number of valence electrons (*354*). The highest critical temperatures T_c have been observed with $PbMo_6S_8$ and $SnMo_6S_8$. Of the 36 electrons of the Mo_6 cluster, 16 are transferred to the 8 S of the Mo_6S_8 unit. As Pb and Sn are supposedly divalent in MMo_6S_8, giving two p electrons to the Mo_6 4d bands, there will be 22 electrons on the Mo_6S_8 cluster in stoichiometric MMo_6S_8.

With the chacogenide halides of molybdenum of the type $Mo_6Y_{8-x}X_x$ (see the preceding), this effect of varying the number of valence electrons is achieved by replacement of the chalcogen of valency 2^- by halogen of valency 1^- of the Mo_6Y_8 unit. In $Mo_6S_6Br_2$ and $Mo_6S_6I_2$, there are, again, 22 electrons per Mo_6 cluster, exactly as in $PbMo_6S_8$. It is not surprising, therefore, that these chalcogenide halides exhibit critical temperatures T_c as high as in $PbMo_6S_8$ (*354*).

C. Tungsten

Only two compounds, $W_2S_7Cl_8$ (*362*) and $W_4S_9Cl_6$ (*97*) are mentioned in the older literature, their true nature being uncertain. The existence of the other compounds in Table XV seems to be well established. All of them were reported by the same group, and, with few exceptions, it remains the only work (*57, 58, 131*). This example illustrates that the lack of information on chalcogenide halides, especially of transition elements, has its main origin in the lack of systematic investigations.

1. *Preparative Methods*

General methods of preparation are summarized in (*58*), and the specific conditions for a particular compound may be found in the references cited next. (*i*) Reaction of a metal halide with sulfur: $WSCl_4$ (*57, 58, 130*). (*ii*) Reaction of metal halides with either Sb_2S_3 or Sb_2Se_3: $WSCl_4$ (*57, 58, 130*); $WSBr_4$ (*57, 58*), $WSeCl_3$ (*58*), $WSeBr_4$ (*58*), $WSCl_3$ (*58, 130*), $WSeBr_3$ (*58*), and $WSBr_2$ (*58*). A variation is the preparation of $WSCl_4$ from WCl_6 and B_2S_3 (*13*) and of WS_2Cl_2 from $WSCl_4$ and Sb_2S_3 in CS_2 (*131*). (*iii*) Chlorination of tungsten sulfides: $WSCl_4$ (*58*). (*iv*) Reaction of metal halides with dry hydrogen sulfide in CS_2: $WSCl_4$, WS_2Cl_2 (*58*). Similarly, WS_2Cl_2 from $WOCl_4$ + H_2S in benzene (*356*). (*v*) Thermal decomposition of tungsten(V) halide sulfides and selenides: $WSCl_2$, $WSeBr_2$ (*58*).

TABLE XV

Chalcogenide Halides of Tungsten: Bibliography

Compound	Ref.	Compound	Ref.
$W_2S_3Cl_4$	*362*		
$W_4S_9Cl_6$	*97*		
$WSCl_2$	*58*	$WSBr_2$	*58*
$WSCl_3$	*57,58*		
$WSCl_4$	*13,57,58,109,130,359*	$WSBr_4$	*57,58,109*
WS_2Cl_2	*58,131,356*		
$WSeCl_2$	*58*	$WSeBr_2$	*58*
$WSeCl_3$	*58*	$WSeBr_3$	*58*
$WSeCl_4$	*56,58*	$WSeBr_4$	*58*

2. Chemical Properties

(i) Tungsten(VI) compounds. The compounds $WSCl_4$, $WSBr_4$, $WSeCl_4$, and $WSeBr_4$ are very sensitive to moisture. They can be sublimed *in vacuo*, and are soluble in benzene, carbon tetrachloride, and carbon disulfide (58). (ii) Tungsten(V) compounds. $WSCl_3$ and $WSeBr_3$ are much less reactive than the hexavalent compounds, and are hydrolyzed only slowly in moist air. They are insoluble in all nonpolar solvents. $WSeBr_3$ disproportionates readily during heating, forming $WSeBr_2$ and $WSeBr_4$. Disproportionation may, therefore, be the reason for lack of success in the preparation of $WSBr_3$ (58). (iii) Tungsten(IV) compounds. $WSCl_2$ and $WSBr_2$ are insoluble in organic solvents. They are slowly hydrolyzed by moisture (58).

3. Crystallographic Data

The structures of the tungsten(VI) sulfide halides $WSCl_4$ and $WSBr_4$ have been solved (109). Although the two compounds crystallize in different space groups, $P\bar{1}$ and $P2_1/c$ for $WSCl_4$ and $WSBr_4$, respectively, their molecular structures are very similar. As tungsten is in the highest valence state, no metal–metal bonds are formed. In the dimeric structure, each tungsten is 6-coordinate, being bonded to four halogen atoms in a plane with a terminal W=S at right angles to this plane (see Fig. 26). The difference between the structures lies in the arrange-

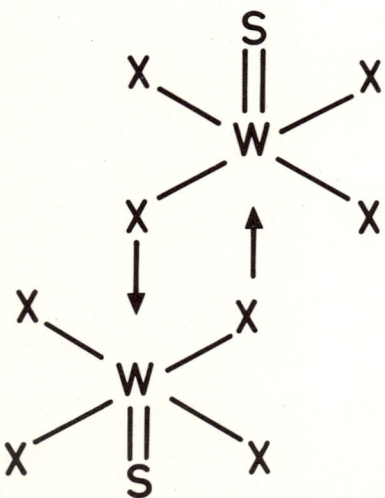

FIG. 26. Structure of WSX_4: dimeric unit of two WSX_4.

ment of the dimeric units. WSeCl$_4$ and WSeBr$_4$ are isomorphous with WSCl$_4$ and WSBr$_4$, respectively (58, 109).

4. *Miscellaneous*

IR spectra have been reported for WSCl$_4$ (13, 57), and WSBr$_4$ and WSCl$_3$ (57). WSCl$_4$ was also the subject of a mass-spectrometric study of the vaporization (359). The reaction of WSeCl$_4$ with 1,2-dimethoxyethane has been investigated (56).

VIII. Group VIIB

A. MANGANESE

Only two papers have been published about manganese chalcogenide halides, both by Batsanov and Gorogotskaya (18, 19). They reacted pink γ-MnS (wurtzite type, tetrahedral coordination) and green α-MnS (rocksalt type, octahedral coordination) with the respective halogens at moderate temperatures (19), and obtained two series of sulfide halogenides: γ-MnSCl$_2$, γ-MnSBr, and γ-MnSI; and α-MnSCl$_2$, α-MnSBr, and α-MnSI, respectively. The reaction products were identified as new compounds by their manganese contents and their X-ray powder patterns. The compounds are stable in air, and practically insoluble in water or the common organic solvents. Densities, refractive indices, and molecular reflections were given (19).

The reactions of MnSe with the halogens at room temperature or 150–170°C (for iodine) yielded the following selenide halides: MnSeCl, MnSeBr$_2$, and MnSeI$_2$ (19). The comparably stable MnSeI$_2$ occurs in two modifications, of which one is cubic, with a = 6.54 Å. MnSeCl and MnSeBr$_2$ decompose in water and organic solvents, and are thermally unstable. Heating to 70–100°C leads to formation of Mn$_2$Se$_2$Cl and Mn$_2$SeBr$_2$. The elimination of "SeBr$_2$" from MnSeBr$_2$ on heating indicates that the compounds MnSeX$_2$ (X = Br, I) are addition products of selenium halides, rather than Mn(IV) compounds. Densities, refractive indices, and X-ray powder data were reported for the more stable compounds (19).

B. RHENIUM

The known rhenium chalcogenides are of the composition ReYX$_2$ and Re$_3$Se$_2$X$_5$ (Y = S, Se; X = Cl, Br), including Re$_3$Te$_2$Br$_5$ (140, 263, 264, 353, 393). Additionally, the compounds Re$_2$S$_3$Cl$_4$ (140) and ReS$_2$Br

(353) are documented in the literature. Chalcogenide iodides of rhenium are not known.

1. Preparative Methods

There are three reasonable routes for preparing rhenium chalcogenide halides. Reactions of rhenium chalcogenides (ReS_2, $ReSe_2$, Re_2S_7, and Re_2Te_5) with halogens (Cl_2, Br_2) at elevated temperatures in sealed tubes yield the compounds $ReYCl_2$ (Y = S, Se), $Re_3Y_2Br_5$ (Y = Se, Te), and $Re_2S_3Cl_4$ (*140, 263, 264, 393*). The compounds $ReYX_2$ (Y = S, Se; X = Cl, Br) may also be obtained by the action of chalcogens on rhenium trihalides (*353*). An intermediate in the reaction of $ReBr_3$ with sulfur is ReS_2Br. Fusion of rhenium diselenide with rhenium trihalides at 700–720°C yields the compounds $Re_3Se_2X_5$ (X = Cl, Br) (*264*). The detailed conditions for all of these preparations are summarized in Table XVI.

2. Chemical Properties

All of the known rhenium chalcogenide halides are stable in air. With the exception of $Re_2S_3Cl_4$, they are insoluble in water, acids, and the common organic solvents. They dissolve readily in hot, 50% KOH (*263, 264*). $Re_2S_3Cl_4$ is soluble in water, and ethanol, but insoluble in nonpolar organic solvents. With acids, alkalis, or hot water, hydrolytic decomposition takes place. Alkaline solutions can be oxidized to produce perrhenate compounds.

TABLE XVI

PREPARATIVE ROUTES AND EXPERIMENTAL CONDITIONS FOR THE FORMATION OF RHENIUM CHALCOGENIDE HALIDES

Educts	Temperature range (°C)	Specific conditions	Products	Ref.
ReS_2, Cl_2	400–450		$ReSCl_2$, S_2Cl_2	*140,393*
Re_2S_7, Cl_2	100–120	Cl_2 diluted with CO_2	$Re_2S_3Cl_4$, S_2Cl_2	*140*
$ReSe_2$, Cl_2	480–500		$ReSeCl_2$	*264*
$ReSe_2$, Br_2	640–650	28–30 h	$Re_3Se_2Br_5$	*263*
Re_2Te_5, Br_2	60–70		$Re_3Te_2Br_5$, $TeBr_4$	*264*
$ReCl_3$, S			$ReSCl_2$	*353*
$ReCl_3$, Se			$ReSeCl_2$	*353*
$ReBr_3$, S			$ReSBr_2$	*353*
$ReBr_3$, Se			$ReSeBr_2$	*353*
$ReSe_2$, $ReCl_3$	700–720	Sealed ampoule, 10–15 h	$Re_3Se_2Cl_5$	*264*
$ReSe_2$, $ReBr_3$	700–720	Sealed ampoule, 10–15 h	$Re_3Se_2Br_5$	*264*

Treatment of the rhenium sulfide chlorides with an excess of chlorine at temperatures above 450°C results in formation of ReCl$_5$ (140). Reduction between 350 and 500°C gives binary rhenium chalcogenides (139, 262). Action of water vapor on rhenium sulfide chlorides at 350–500°C produces oxysulfides (139).

3. Crystallographic Data

The structures of the rhenium chalcogenide halides have not been studied. X-Ray powder data were collected, in order to prove the homogeneity of the compounds (140, 263, 264, 353).

IX. Group VIIIB

Only chalcogenide halides of palladium and platinum are mentioned in the literature.

Thiele and co-workers (389) prepared the only known palladium chalcogenide halides, PdTeI and Pd$_2$SeI$_3$, by hydrothermal synthesis in HI (see Section II,D,2) at 300°C, starting with the elements. Crystalline Pd$_2$SeI$_3$ is better obtained by reaction of PdI$_2$ with Se and an excess of iodine in a closed ampoule at 250°C (reaction time, 2 days).

The brass-colored PdTeI consists of Jahn–Teller distorted PdTe$_{2/2}$I$_{4/4}$ octahedra, which are interconnected by common edges and corners to afford a loose, spatial network. The compound is considered to be ionic, containing Pd^{3+} and Te^{2-}, although the observed diamagnetism, electronic conduction, and color suggest some metallic character.

In the black, lustrous Pd$_2$SeI$_3$, four quadratic PdSe$_{1/2}$I$_{3/2}$ entities are connected by a common Se$_2^{2-}$ to a Pd$_4$Se$_2$I$_4$I$_{4/2}$ group (see Fig. 27). These groups combine via common iodines to form corrugated layers. The formal oxidation state of Pd is 2+.

No platinum chalcogenide halides of invariable, exact, stoichiometric compositions are known.

Batsanov et al. (23) reacted sulfur with PtCl$_2$ and PtBr$_2$ by heating mixtures of the reactants in evacuated, sealed ampoules. At 100–200°C after 12–24 h, sulfide chlorides PtCl$_x$S$_y$ (1.70 ≤ x ≤ 2; 0.6 ≤ y ≤ 3.35) and sulfide bromides PtBr$_x$S$_y$ (1.87 ≤ x ≤ 2.06; 0.84 ≤ y ≤ 1.80) were formed. The compositions depended on the initial PtX$_2$:S ratio, and the temperature. At 320–350°C, loss of chlorine led to the compounds PtClS$_y$ (1.7 ≤ y ≤ 1.9). According to their X-ray powder patterns, all of these products retained the main structural features of the original platinum halides. From considerations of molar volumes, the authors deduced the presence of polysulfide anions.

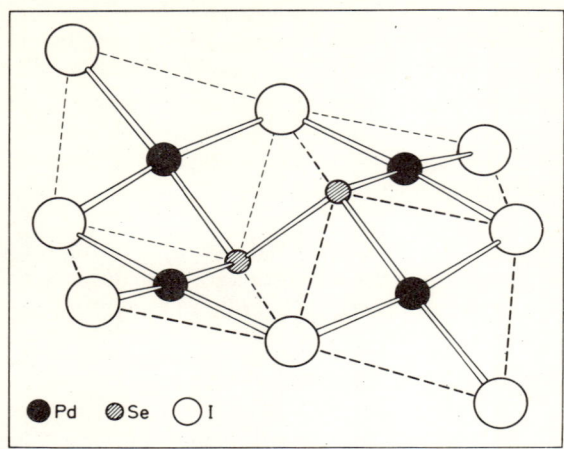

Fig. 27. Pd$_4$Se$_2$I$_4$I$_{4/2}$ unit of the structure of Pd$_2$SeI$_3$. (Redrawn from G. Thiele, M. Köhler-Degner, K. Wittmann, and G. Zoubek, *Angew. Chem.* **90,** 897 (1978), Fig. 2, p. 898.)

Thiele and co-workers, who tried to prepare platinum chalcogenide halides, could neither isolate nor identify any pure, homogeneous compound (*389*).

X. Group IIIA

With few exceptions, the compositions of the chalcogenide halides of the Group IIIA elements correspond to the general formula MYX (M: IIIA metal, Y: chalcogen, X: halogen). But, despite this uniformity, the structures and properties vary considerably with the metal atom, reflecting the general trends of chemical behavior within this group. The semimetal boron forms covalent, organiclike BYX (Y = S, Se; X = Cl, Br, I) and B$_2$S$_3$X$_2$ (X = Cl, Br, I) compounds, which are liquids or volatile solids having molecular structures. With decreasing electronegativity of the metal atom, the ionic character of the compounds increases, along with a gain in thermal and chemical stability, which reach a maximum for indium. The low stability of the TlYX compounds (Y = S, Se; X = Cl, Br, I) may be ascribed to the inert, electron-pair effect, which favors the oxidation state +1 for thallium. Therefore, thallium also forms chalcogenide halides of Tl(I). TlTeX compounds are not known.

A. Boron

As already mentioned, the boron chalcogenide halides are covalent, organiclike compounds. The sulfide halides $B_3S_3X_3$ (X = Cl, Br, I) (*345, 409–411*) and $B_2S_3X_2$ (X = Cl, Br, I) (*343–345*), as well as $B_2Se_3I_2$ (*346*), form five- or six-membered rings. They are liquids or low-melting solids that readily dissolve in many organic solvents. The boron selenide halides BSeX (X = Cl, Br, I) (*91, 152*) are solids that do not dissolve in organic solvents, and decompose without melting. Their degree of polymerization therefore seems to exceed the value of three. Because of their organiclike nature, the boron chalcogenide halides will not be considered in detail.

B. Aluminum

Aluminum forms a complete series of AlYX compounds (Y = S, Se, Te; X = Cl, Br, I). Furthermore, a number of compounds with Se(IV) and Te(IV) are known, such as $TeCl_3^+$ $AlCl_4^-$ (*197*) and $SeCl_3^+$ $AlCl_4^-$ (*364*), which are not considered here. A borderline case consists in the reduced phases found in the systems $(TeCl_4 + 4AlCl_3)$–Te and $(SeCl_4 + 4AlCl_3)$–Se (*88*), e.g., Te_4^{2+} $(AlCl_4^-)_2$ (*89*), Te_6^{2+} $(Al_2Cl_7^-)_2$ (*89*), Te_6^{2+} $(AlCl_4^-)_2$ (*88*), Se_4^{2+} $(AlCl_4^-)_2$ (*88*), and Se_8^{2+} $(AlCl_4^-)_2$ (*87*), which contain cyclic polytellurium and polyselenium cations. For a detailed review of homopolyatomic ions of the posttransition elements, see (*86*).

1. Preparative Methods

The compounds AlYX are best prepared by direct reaction between the respective aluminum halide and chalcogenide in a sealed ampoule at 350°C. The reaction is complete after 2 weeks. In the case of the iodides, a mixture of Al and I_2 (molar ratio 3:10) is used instead of AlI_3. Other preparative methods, such as the reaction of an aluminum halide with Zn or Cd chalcogenide, or with the chalcogen itself, are applicable to the bromide and chloride only, and give poor yields (15–20%) (*158, 159, 266, 327, 328*).

Growth of single crystals. Crystals of the aluminum selenide halides (needles, maximum length 15 mm) were grown by vapor transport in sealed ampoules between two temperatures (380 and 320°C for AlSeCl, and 350 and 300°C for AlSeBr and AlSeI) over a period of two months. A large excess of the halogenide was used (*266*).

2. Chemical Properties

The colorless, hygroscopic compounds AlYX are insoluble in common organic solvents. They decompose into the corresponding halide and chalcogenide at temperatures between 280 (AlYI) and 400°C (AlYCl). With water vapor, hydrolysis to Al(OH)$_3$ occurs. In the case of the sulfide halides, the intermediate oxide halide can be isolated. With oxygen γ-Al$_2$O$_3$ is formed at temperatures between 25 (AlTeI) and 200°C (AlSCl). Again, the sulfide halides, which are more inert against oxygen than the other compounds, react via an intermediate oxide halide. Reaction with chlorine leads to the formation of AlCl$_3$. With NH$_3$, chalcogenide amides of the general formula AlY(NH$_2$)·NH$_3$ are obtained (*158, 159, 266, 327, 328*).

3. Crystallographic Data

AlSCl has an orthorhombic structure, with the lattice constants $a = 8.09$, $b = 10.52$, $c = 3.86$ Å, and $Z = 4$. It is probably isotypic with SbSCl and BiSCl, crystallizing in a layer type of lattice (*157*) (see Section XII,C,5). The selenide halides are monoclinic, with the probable space-group P2$_1$/m. The lattice constants are given in Table XVII. The constancy of the b parameters for all three compounds suggests the general presence of an Al–Se chain extending in that direction (*266*).

Structural data, except powder patterns, for the other compounds are not known.

C. GALLIUM

Like aluminum, gallium forms a series of GaYX compounds (Y = S,Se,Te; X = Cl,Br,I) (*160*). Hardy and Cottreau (*165*) noticed that at least GaSCl and GaSBr do not correspond exactly to this stoichiometry. These phases are halogen-rich, according to the formula Ga$_9$S$_8$X$_{11}$ (X = Cl,Br).

TABLE XVII

CRYSTALLOGRAPHIC DATA FOR ALUMINUM CHALCOGENIDE HALIDES

Compound	Symmetry	a (Å)	b (Å)	c (Å)	β (degrees)	Z	d_{ex} (g/cm^3)	Ref.
AlSCl	Orthorhombic	8.09	10.52	3.86		4	1.87	*157*
AlSeCl	Monoclinic	16.33	12.87	18.172	94.0	4	2.43	*266*
AlSeBr	Monoclinic	16.97	12.87	18.731	94.0	4	2.98	*266*
AlSeI	Monoclinic	18.11	12.87	19.973	94.0	4	3.30	*266*

1. Preparative Methods

The chalcogenide bromides and chlorides may be prepared by the reaction of the halide with the respective chalcogenide in a sealed ampoule. A mixture of gallium metal and chalcogen may be used, instead of the chalcogenide. The chalcogenide iodides are synthesized directly from the elements. The exact preparative conditions are listed in Table XVIII (160, 165).

Growth of single crystals. $Ga_9S_8Cl_{11}$ and $Ga_9S_8Br_{11}$ have been obtained as needle-shaped, single crystals, 40–50 mm long. They were prepared by the reaction of Ga, S, and an excess of the halogenide at 400 ($Ga_9S_8Cl_{11}$) or 450°C ($Ga_9S_8Br_{11}$) (165).

2. Chemical Properties

The gallium chalcogenide halides are hygroscopic compounds that decompose (without melting) at temperatures between 240 and 380°C to a mixture of chalcogenide and halide (see Table XVIII). The telluride halides are yellow, and the other compounds are colorless (160, 165).

TABLE XVIII

PREPARATION AND PROPERTIES OF GALLIUM CHALCOGENIDE HALIDES

Compound	Preparation[a]	Temperature (°C)	Time (days)	Temperature of decomposition (°C)	d_{ex} (g/cm³)	Ref.
GaSCl	A,B	260	14	250–260	2.96	160
$Ga_9S_8Cl_{11}$	A	250	0.1	400	2.52	165
	B		14–21			
GaSBr	A,B	305–345	14	360–370	3.37	160
$Ga_9S_8Br_{11}$	A	330	0.1	450	3.20	165
	B		14–21			
GaSI	C	300	10–15	370–380	3.74	160
GaSeCl	A,B	260	8–12	240–250	3.54	160
GaSeBr	A,B	300–340	12–15	280–290	3.96	160
GaSeI	C	300	14–21	320–330	4.23	134,160
GaTeCl	A,B	260–270	10–15	230–240	4.17	160
GaTeBr	A,B	260–270	10–15	250–260	4.83	160
GaTeI	C	260–280	5–10	240–250	4.75	160
				800		200

[a] A: $GaX_3 + Ga_2Y_3$; B: $GaX_3 + 2\,Ga + 3Y$; C: from the elements.

3. Crystallographic Data

$Ga_9S_8Cl_{11}$ and $Ga_9S_8Br_{11}$ are monoclinic; probable space-group P2/m, Z = 20. The lattice constants are a = 17.74 and 18.15, b = 58.9 and 59.65, c = 18.92 and 19.88 Å, and β = 122.2 and 121.5° for the chloride and bromide, respectively (165).

D. INDIUM

Again, the complete series of InYX compounds (Y = S,Se,Te; X = Cl,Br,I) exists. Of the Group IIIA chalcogenide halides, the indium compounds have been the most extensively studied.

1. Preparative Methods

Indium chalcogenide chlorides and bromides are obtained by heating mixtures of the chalcogenide and the halide in sealed ampoules to temperatures between 200 and 420°C (see Table XIX). For the preparation of the chalcogenide iodides, In and I_2 are used instead of InI_3. InTeI may be synthesized directly from the elements (161).

The preparation conditions are listed in Table XIX. InTeCl and InTeI may also be obtained by fusing stoichiometric mixtures of Te and InCl, or InI (80, 331).

Growth of single crystals. Single crystals of the selenide and telluride halides can be grown by recrystallization at 360–390°C, or sublimation at 420 (InSeCl) and 370°C (InTeI), respectively (162).

TABLE XIX

PREPARATION AND PROPERTIES OF INDIUM CHALCOGENIDE HALIDES

Compound	Preparation[a]	Temperature (°C)	Time (days)	Color	d_{ex} (g/cm³)	Ref.
InSCl	A	200–225	35	Light grey	3.68	*161,162*
InSBr	A	200–225	35	Light grey	4.32	*161,162*
InSI	B	170–190	14	Yellow	4.11	*161,162*
InSeCl	A	400–420	14	Dark yellow	4.52	*161,162*
InSeBr	A	380–440	14–20	Dark brown	5.11	*161,162*
InSeI	B	380–400	7–14	Light yellow	4.69	*161,162*
InTeCl	A	400–410	14	Brown	4.52	*81,161,162*
InTeBr	A	380–420	14–20	Yellow–brown	4.40	*161,162*
InTeI	C	350–400	8–14	Red–brown	5.07	*80,161,162*

[a] A: $InX_3 + In_2Y_3$; B: $In_2Y_3 + In + 3/2\ I_2$; C: from the elements.

2. Chemical Properties

The indium chalcogenide halides are more stable than their aluminum and gallium analogs. Thus, only the sulfide halides are moisture-sensitive, and must be kept and handled under an inert-gas atmosphere. Generally, the chemical stability increases in the orders S < Se < Te and I < Br < Cl. The indium chalcogenide halides are only slowly, if at all, attacked by water. With strong acids, decomposition, usually accompanied by evolution of H_2Y gas, takes place. The attack of hot bases leads to the formation of $In(OH)_3$. Only the sulfide halides decompose before melting. The thermal behavior is described in (80, 81, 161). At elevated temperatures, all of the compounds become air-sensitive. InTeCl, the most stable one, is oxidized in air at 360°C, but is stable up to 680°C under argon (81).

3. Phase Diagrams

Safonov et al. (331) determined the liquidus surface of the ternary In–Te–Cl system by DTA, X-ray diffraction, and crystal optical methods. Only one ternary compound, InTeCl, exists. The crystallization field of InTeCl occupies 6% of the diagram, which demonstrates the considerable thermodynamic stability of this compound. InTeCl melts congruently at 453°C. It forms part of the two pseudobinary systems In_2Te_3–$InCl_3$ and InCl–Te (81). Whereas the first consists of the two eutectic parts In_2Te_3–InTeCl and InTeCl–$InCl_3$, the latter is more complicated. It is composed of the monotectic system InTeCl–InCl and the eutectic system Te–InTeCl, where tellurium forms a solid solution with InTeCl containing from 100 to 82 atom% of Te at the eutectic temperature (81).

Safonov et al. (330) also investigated the In–Te–I system by DTA and X-ray diffraction. The system contains three, congruently melting, ternary compounds: InTeI, $InI_3 \cdot 2TeI_4$ and $InI_3 \cdot 3TeI_4$, of which the latter two are out of the scope of this review. InTeI melts congruently at 475°C. It forms part of the quasibinary cuts In_2Te_3–InI_3 and InI–Te, where the sections InTeI–In_2Te_3, InTeI–InI_3, InTeI–InI, and InTeI–Te are all of the eutectic type (80).

4. Crystallographic Data

InSCl, InSBr, InSeCl, and InSeBr are isotypic, and crystallize in the hexagonal $CdCl_2$ lattice type. The halide and chalcogenide ions are statistically distributed among the Cl sites. As in $CdCl_2$, the bonding within the $InY_{3/3}X_{3/3}$ octahedra should be predominantly ionic (162).

The symmetry of the isotypic compounds InSI and InSeI is tetragonal (*162*).

The telluride halides crystallize in monoclinic lattices, but only InTeBr and InTeI are isotypic (*162*). InTeCl forms a layer type of structure, as do InSCl and its analogs, but, owing to the size of the Te atom and the enhanced covalency of the In–Te bond, only a coordination number of 4 for indium is realized. The structure is built up of strongly distorted, $InTe_{3/3}Cl_{1/1}$ tetrahedra that share the corners and edges occupied by Te atoms. The Cl atoms are coordinated to one tetrahedron each, and do not take part in the layer formation (*324, 325*).

The crystallographic data are summarized in Table XX.

E. THALLIUM

Three groups of thallium chalcogenide halides have been reported, corresponding to the formulas TlYX (Y = S,Se;X = Cl,Br,I) (*22*), Tl_4YCl_4 (Y = S,Se) (*321*), and Tl_2YX_4 (Y = S,Se;X = Br,I) (*21*). Thallium telluride halides are not known.

1. Preparative Methods

The sulfide halides TlSX are prepared by heating a stoichiometric mixture of the thallium halogenide and sulfur in a sealed ampoule at 180°C for 30 h. The mixture is then slowly cooled to room temperature. The compounds TlSeX are obtained by reaction between thallium metal and selenide halide at 280°C during 40 h (*22*). On heating TlYCl to 500°C *in vacuo,* the compounds Tl_4YCl_4 result (*321*). Dissolution of

TABLE XX

CRYSTALLOGRAPHIC DATA FOR INDIUM CHALCOGENIDE HALIDES

Compound	Symmetry	a (Å)	b (Å)	c (Å)	β (degrees)	Z	Ref.
InSCl	Hexagonal	3.728		17.78		3	*162*
InSBr	Hexagonal	3.820		18.59		3	*162*
InSI	Tetragonal	18.28		9.96		32	*162*
InSeCl	Hexagonal	3.860		18.58		3	*162*
InSeBr	Hexagonal	3.935		19.13		3	*162*
InSeI	Tetragonal	18.70		10.15		32	*162*
InTeCl	Monoclinic	7.42	14.07	7.13	92.1	8	*162*
		7.42	14.06	7.07	92.1	8	*324,325*
InTeBr	Monoclinic	7.350	7.577	8.343	117.61	4	*434*
InTeI	Monoclinic	8.42	7.73	7.44	94.3	4	*162*

Tl$_2$S in an alcoholic Br$_2$ solution, and evaporation of the solvent, yields Tl$_2$SBr$_4$. The same procedure with iodine gives Tl$_2$SI$_4$. Tl$_2$SeBr$_4$ is prepared by the action of Br$_2$ on powdered Tl$_2$Se, whereas I$_2$ and Tl$_2$Se, heated together in a sealed tube at 140°C for 40 h, form Tl$_2$SeI$_4$ (*21*).

2. *Miscellaneous*

The compounds TlYX are insoluble in water and organic solvents. Decomposition with oxidizing acids and bases leads to segregation of the chalcogen. On heating, disintegration into Tl(I) halide and chalcogen takes place between 96 and 132°C (*22*). The decomposition is incomplete for the chlorides, which form Tl$_4$YCl$_4$ (*321*). The electric conductivity of Tl$_2$S–TlCl melts has been measured (*404*).

Tl$_4$SCl$_4$ and Tl$_4$SeCl$_4$ melt at 440 and 442°C, respectively. They can be distilled between 650 and 700°C without decomposition. They are insoluble in H$_2$O and organic solvents, but soluble in aqueous alkaline solutions. With conc. acids, decomposition takes place. The electric conductivity has been determined to be $1.4 \cdot 10^{-7}$ and $2.1 \cdot 10^{-7}$ Ω^{-1} cm^{-1} for Tl$_4$SCl$_4$ and Tl$_4$SeCl$_4$, respectively. The probable structural formula is Tl$_3$(TlCl$_4$Y). The compounds thus, presumably, consist of Tl$_{1/4}$Cl$_{4/4}$Y$_{2/8}$ octahedra that are interconnected by the chalcogen atoms to linear chains (*321*).

The Tl$_2$YX$_4$ compounds are all hygroscopic. The greenish-yellow Tl$_2$SBr$_4$ reacts with water, or HCl, to give TlBr$_2$. The black-violet iodine derivatives are insoluble in these solvents. Organic solvents tend to wash the halogen away (*21*).

XI. Group IVA

A. Carbon, Silicon, and Germanium

For these three elements, for the first time in this Chapter, combinations of a nonmetal with chalcogens and halogens have to be discussed. As regards their structure and properties, most of these do not belong to the field of this review as described in the introductory Chapter. For those interested in the ternary carbon compounds that have the composition CSX$_2$ (X = Cl, Br, I), information may be found in an issue of "Gmelin" (*142*).

Only sulfide chlorides of silicon have thus far been described (e.g., *143*). Among these, a nonvolatile compound, Si$_2$S$_3$Cl$_2$, was mentioned (*351*), but no details were given. Areas of vitrification have been found in the systems Si–S–Br (*185*), Si–S–I (*185*), Si–Se–Br (*185*), and Si–

Se–I (*100, 185*). These chalcogenide halide glasses belong to the so-called "semiconducting glasses" that have attracted much attention in solid-state physics, because of their physical properties (see Section XII,B); they are, however, not treated in this review.

The tendency to glass formation is even more pronounced in the systems containing germanium, such as Ge–S–Br (*185*), Ge–S–I (*77, 99, 185*), Ge–Se–Br (*185*), Ge–Se–I (*100, 185*), and Ge–Te–I (*118*). In these systems, such phases as $GeSI_2$ (*99, 185, 278*), $GeSeI_2$ (*100, 185*), $Ge_2Se_3I_2$ (*100*), and Ge_2SeI_6 (*100*) have been investigated, and some of them can be recrystallized, to yield individual compounds. A germanium sulfide bromide, $Ge_4S_6Br_4$, and, probably, the homologous iodide were discovered by Pohl (*275*). The almost colorless bromide, which decomposes at 305°C, has a molecular structure. According to their phase diagrams (*213, 254*), GeS and GeI_2 do not form ternary compounds.

B. Tin

The first chalcogenide halides of tin were reported in 1963 (*24*). Although numerous publications since then have been devoted to this subject, the existence and true composition of some the compounds described here still seem questionable. It is, therefore, advisable to start with a discussion of the systems $SnY-SnX_2$ (Y = S, Se, Te; X = Cl, Br, I).

1. Phase Diagrams

a. $SnS-SnX_2$. The systems $SnS-SnX_2$ are presented in Fig. 28 (*48, 144*); however, the discussion that follows should be noted. SnS and $SnCl_2$ form a simple eutectic system without any ternary phase (*48, 231*). In the eutectic system $SnS-SnBr_2$, an incongruently melting compound having the approximate composition $Sn_9S_2Br_{14}$ was reported by Blachnik and Kasper (*48*). For this compound, however, Thevet et al. (*386*), gave the composition Sn_2SBr_2 and a hexagonal structure (*385*). The composition $Sn_7S_2Br_{10}$ has since been claimed, from a single-crystal measurement for the ternary compound, and it was claimed to be isostructural with $Pb_7S_2Br_{10}$ (*104*). The situation as regards the eutectic system SnS–SnI is even more complicated. Novoselova et al. (*253*) claimed the formation of the compounds Sn_3SI_4 and $Sn_7S_3I_8$, melting incongruently at 330 and 410°C, respectively, the first of which exhibited an additional, thermal effect at 266°C. Their phase diagram is included in Fig. 28 (*48*). By contrast, the phase diagram published by Thevet et al. (*386, 388*) differed with respect to the composition, not only of the eutectic but also of the ternary phase. The latter compound,

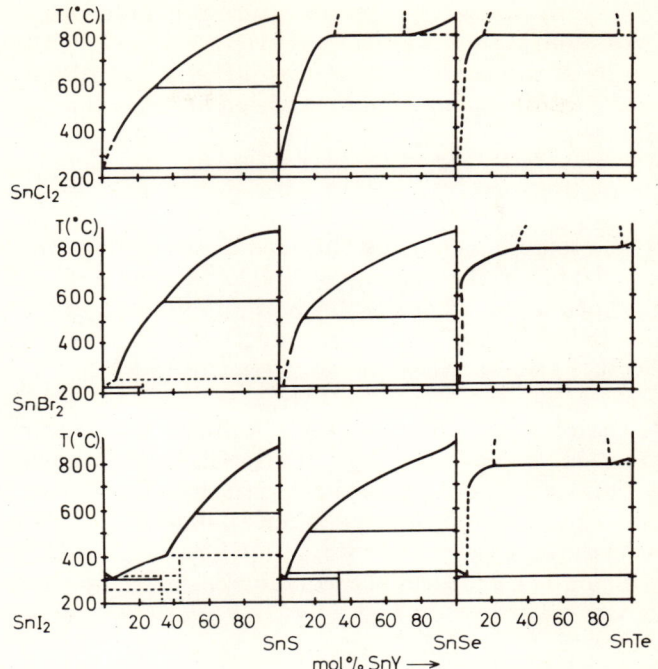

FIG. 28. Systems SnY–SnX$_2$ (Y = S, Se, or Te; X = Cl, Br, or I). (Redrawn from R. Blachnik and F. W. Kasper, *Z. Naturforsch.* **29B**, 159 (1974), Fig. 2, p. 160; and R. Blachnik and G. Kudermann, *Z. Naturforsch.* **28B**, 1 (1973), Fig. 1, p. 2.)

Sn$_2$SI$_2$, was said (*385*) to exist in an α and a β form. The crystal structures of both forms have been elucidated (*119, 239, 240, 388*). Apart from questions that still remain concerning the true relationship of α- and β-Sn$_2$SI$_2$, a publication by Fenner (*121*) on the synthesis and structure of a new ternary phase, Sn$_4$SI$_6$, contradicts both of the phase diagrams already mentioned.

b. *SnSe–SnX$_2$*. All systems belong to the eutectic class, the system SnSe–SnCl$_2$ being of the monotectic type (see Fig. 28). The compound Sn$_4$SeBr$_6$ is mentioned in the literature (*104*), and Blachnik and Kasper (*48*) reported the peritectic compound Sn$_3$SeI$_4$, the composition of which has been confirmed by Thevet *et al.* (*387*), correcting an earlier (wrong) composition of Sn$_2$SeI$_2$ (*385*).

c. *SnTe–SnX$_2$*. All three systems are of the monotectic type (see Fig. 28) and show no ternary. A ternary compound of the composition Sn$_4$TeBr$_6$ has, however, been reported (*104*).

d. Sn(IV) Compounds. In the literature the, following compounds with the oxidation state +4 for tin were reported, and identified as single phases by X-ray patterns (*24*): $Sn_2Cl_2S_3$ (*16, 24*), $Sn_3Se_5Cl_2$, and $Sn_4Te_7Cl_2$ (*16, 24*). No compounds have been found in the systems SnS_2–SnI_4 (*383*).

2. Preparative Methods

All tin(IV) compounds may be obtained by annealing stoichiometric mixtures of the binary tin compounds in closed ampoules. α-Sn_2SI_2 can only be obtained free from the β-phase by applying an excess of SnI_2 at a temperature below 350°C (*388*). An excess of SnI_2 cannot be removed from α-Sn_2SI_2 by sublimation without transforming the α into the β form (*388*).

Tin(IV) chalcogenide chlorides are obtained by oxidation of $SnCl_2$ with the respective chalcogen (*24*). For $Sn_2Cl_2S_3$, the chloride and sulfur, in the ratio of 1:1, or 1:2, are heated in sealed tubes at 140–150°C for 8 hours. The volatile by-products $SnCl_4$ and SCl_2 are removed in a vacuum desiccator. In a similar way, $Sn_2Cl_2Se_5$ and $Sn_4Cl_2Te_7$ are obtained by fusing 1:1 mixtures of $SnCl_2$ and Se, or Te, at 240 and 480°C, respectively.

3. Crystallographic Data

Table XXI shows the crystallographic data as far as they are known. The structures of the three sulfide iodides have been determined by single-crystal studies. Powder patterns of the Sn(IV) compounds not given here have been reported elsewhere (*24*).

a. SnS_2I_2. The structures of the α and β modification are similar, and contain common, structure elements. A detailed description is therefore restricted to β-SnS_2I_2. This orthorhombic structure (see Table XXI), published independently by two groups (*119, 240, 388*), is built up of $(Sn_4S_2I_4)_\infty$ ribbons extending in the direction of the shortest axis (the b-axis). The main part of such an arrangement is a folded ribbon formed of

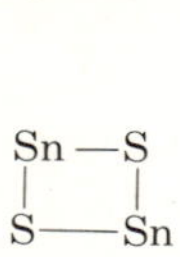

$$\begin{array}{cc} Sn & -S \\ | & | \\ S & -Sn \end{array}$$

groups with outward-pointing SnI_2 (see Fig. 29). Tin forms five, sulfur two, and iodine two or four, bonds, respectively. The coordination of tin may be considered to be a deformed octahedron, the sixth coordinate

TABLE XXI

Crystallographic Data for Tin(II) Chalcogenide Halides

Compound	Symmetry	Space group	a (Å)	b (Å)	c (Å)	β (degrees)	Z	Ref.
α-Sn_2SI_2	Monoclinic	B2/m	14.305	17.281	4.435	110.47	6	239,388
β-Sn_2SI_2	Orthorhombic	Pnma	17.475	4.412	25.390		12	119,240,388
Sn_4SI_6	Monoclinic	C2/m	14.129	4.425	25.15	93.42	4	121
Sn_2SBr_2	Hexagonal	P4$_2$/nmc	12.239		4.390		4	385
Sn_3SeI_4	Tetragonal	P6$_3$/m or P6$_3$	8.455		15.87		4	387
$Sn_7S_2Br_{10}$	Hexagonal	Pnma or Pn2$_1$a	11.35		4.40		1	104
Sn_4SeBr_6	Orthorhombic	Pnma or Pn2$_1$a	8.74	4.15	10.50			104
Sn_4TeBr_6	Orthorhombic		8.95	4.22	10.65			104

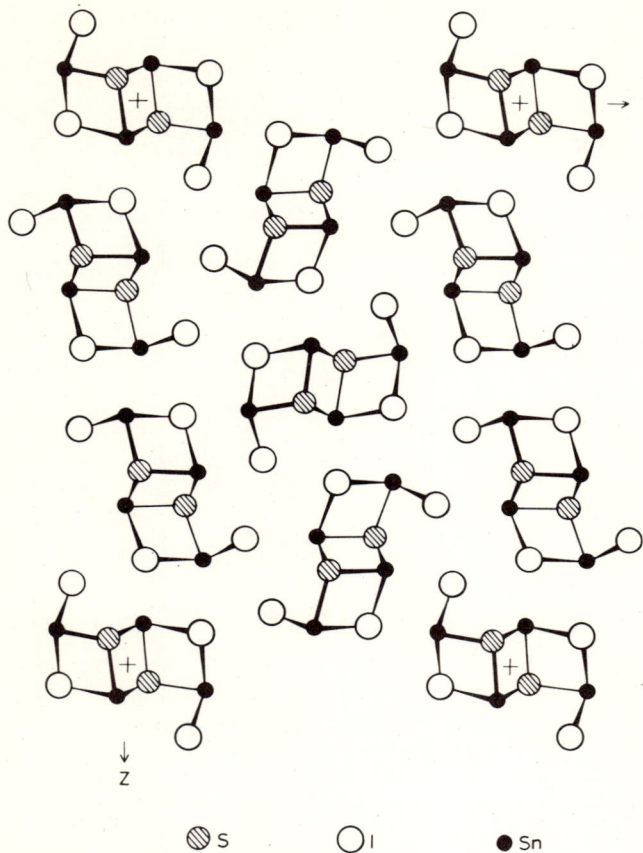

FIG. 29. Structure of β-Sn_2SI_2. (Redrawn from J. Fenner, *Naturwissenschaften* **63**, 244 (1976), Fig. 1, p. 244.)

being represented by a relatively distant iodine. If the longest Sn–I distance of this octahedron that exceeds the common bonding-length is neglected, a truncated quadratic pyramid is obtained that has tin below the basal plane (see Fig. 30). Two types of such pryamids exist, those in which (A) the base consists of four I atoms and the apex of an S atom, and (B) the base consists of two I atoms and two S atoms, the apex being a third S atom. The pyramids of type A are connected with those of type B in two ways: by a face consisting of S and I atoms, and by a vertical edge of S atoms. Whereas the pyramids of type B are interconnected by a face consisting of S atoms, the pyramids of type A do not share common atoms. In the structure of β-Sn_2SI_2, hexagonal

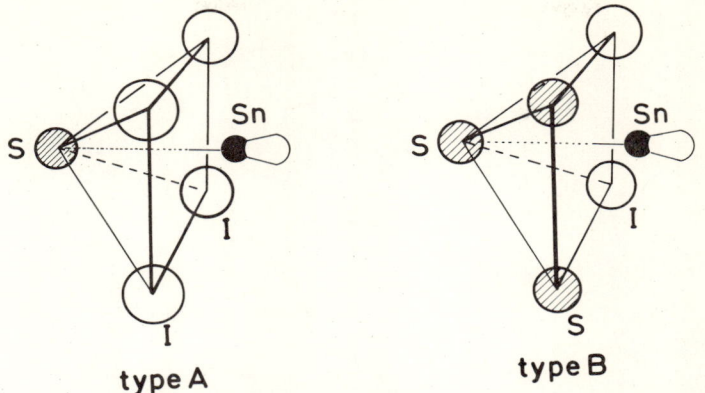

Fig. 30. Pyramids of type A and B, including the lone-electron pair. (Redrawn from Nguyen-Huy-Dung and F. Thevet, *Acta Cryst.* **B32**, 1112 (1976), Fig. 3, p. 1114.)

channels of iodine exist that contain tin and sulfur in an ordered arrangement. The structure elements just described are also found in the monoclinic α-Sn_2SI_2 *(239, 388)*. The main difference consists in the occupation of the vast, hexagonal, iodine channels which, in the α modification, should contain tin and sulfur statistically distributed on the same crystallographic sites. Although the authors used this argument for explaining the possibility of a deviation from stoichiometry *(239)*, this unusual result requires further discussion. With the exception of the tin atoms in the hexagonal channels, each tin has the same pyramidal environment as in SnS and SnI_2, the binary constituents of the phase diagram.

b. Sn_4SI_6. In addition to structure elements of β-Sn_2SI_2, the third ternary compound, Sn_4SI_6, contains domains of pure SnI_2 (Sn_6I_{12}), as shown in Fig. 31 *(121)*. The structure may be considered to be a variety of the SnI_2 structure, with layers perpendicular to the c axis. Sulfur is built into these layers in such a way that structure elements of β-Sn_2SI_2 are formed *(121)*.

4. Physical Properties

The ^{119}Sn Mössbauer data for $Sn_7S_2Br_{10}$, Sn_4SeBr_6, and Sn_4TeBr_6 have been measured, and compared with those of the parent halides and chalcogenides. The data are consistent with a random distribution of halide and chalcogenide anions *(104)*. Thus far, the reflection spectra of a few tin(IV) compounds assumed to be semiconductors have

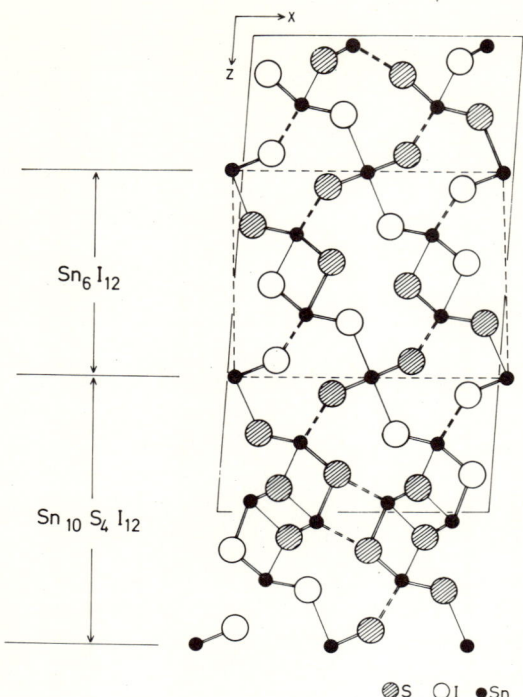

Fig. 31. Structure of Sn_4SI_6. (Redrawn from J. Fenner, Z. Naturforsch. **33B**, 479 (1978), Fig. 1, p. 480.)

been measured. The following, optical band-gaps, ΔE, have been reported: $Sn_2S_3Cl_2$, 2.7 eV; $Sn_3Se_5Cl_2$, 1.5 eV; and $Sn_4Te_7Cl_2$, 1.1 eV (16).

C. Lead

Although the phase diagrams of all of the systems $PbY-PbX_2$ (Y = S, Se, Te; X = Cl, Br, I) have been published (see Fig. 32), contradictory data are to be found in the literature with respect to the existence and composition of ternary phases. Only four of them could be identified by structure determinations. They are summarized in Table XXII.

1. Phase Diagrams (Fig. 32)

a. Lead Sulfide Halides. The phase diagrams $PbS-PbBr_2$ (247, 249, 302) and $PbS-PbI_2$ (247, 248, 302) are of the peritectic type, with

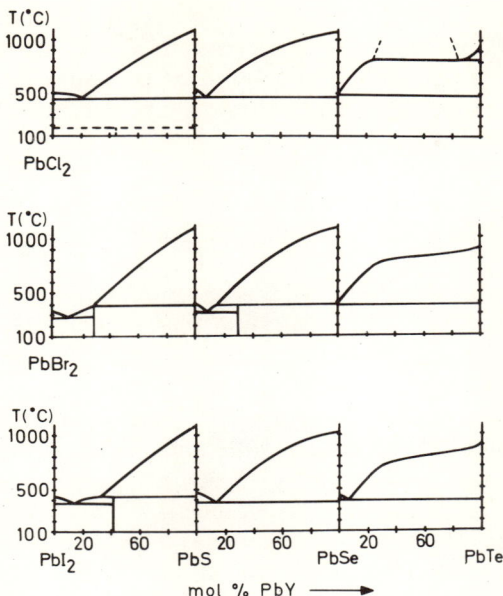

Fig. 32. Phase diagrams of the systems $PbY-PbX_2$ (Y = S, Se, or Te; X = Cl, Br, or I). (Redrawn from A. V. Novoselova, I. N. Odin, and B. A. Popovkin, *Russ. J. Inorg. Chem.* **14**, 1402 (1969), Fig. 1, p. 1403.)

$Pb_7S_2Br_{10}$ and $Pb_5S_2I_6$ as the only thermodynamically stable lead sulfide halides. They decompose peritectically at 381°C ($Pb_7S_2Br_{10}$) and 418°C ($Pb_5S_2I_6$), and their composition has been confirmed by structure analyses. No compound exists in the eutectic system $PbS-PbCl_2$ (*26, 302, 394, 401, 412*). A phase having the composition Pb_4SCl_6, which is isotypic with Pb_4SeBr_6 (*196, 302*), can, however, be obtained by quenching a melt of this composition, and it may be stable at tempera-

TABLE XXII

CRYSTALLOGRAPHIC DATA FOR LEAD CHALCOGENIDE HALIDES

Compound	Symmetry	a (Å)	b (Å)	c (Å)	β (degrees)	Z	Space group	Ref.
Pb_4SCl_6[a]	Orthorhombic	4.23	14.96	9.22		2	Imm2	*196,302*
$Pb_7S_2Br_{10}$	Hexagonal	12.273		4.319		1	$P6_3/m$	*196*
$Pb_5S_2I_6$	Monoclinic	14.338	4.437	14.552	98.04	2	C2/m	*196*
Pb_4SeBr_6	Orthorhombic	4.361	15.780	9.720		2	Imm	*196*

[a] Structure not determined; isotypic with Pb_4SeBr_6.

tures below 180°C (247, 250, 302). All other lead sulfide halides mentioned in the literature seem to be metastable with respect to the binary phases. Typically, they are found as precipitates obtained from lead salts in solution. Thus far, only some of them have been identified as single phases by their X-ray patterns, as, for instance, $Pb_3S_2Br_2$ [$2PbS \cdot PbBr_2$, (302)], Pb_2SI_2, and Pb_2SBr_2 (164). Such "precipitation reactions" were discussed in (203) and (302). The numerous reports that appeared before 1964 are cited in (145).

b. Lead Selenide Halides. No ternary compound exists in the eutectic systems $PbSe-PbCl_2$ (247, 250) and $PbSe-PbI_2$ (246, 247). The only stable lead selenide halide has been found in the peritectic system $PbSe-PbBr_2$ (247, 249), but, in contrast to the data in these studies, the true composition of the ternary is Pb_4SeBr_6, as has been shown by the synthesis of single crystals (302) and a structure determination (196).

c. Lead Telluride Halides. No ternary compounds have been observed in the systems that PbTe forms with $PbCl_2$ [monotectic (247, 252)], $PbBr_2$ [eutectic (247, 249)], and PbI_2 [eutectic (43, 247, 251)]. The same holds for the systems PbI_2-TeI_4 [eutectic (212, 332)], PbI_2-Te (256), and PbTe–I [not quasibinary (211)]. In the eutectic system $PbTe-TeI_4$, an incongruently melting compound having the composition $Pb_3Te_5I_8$ has been observed, and confirmed as a single phase by X-ray diffraction (42).

2. Preparative Methods

The stable compounds $Pb_7S_2Br_{10}$, $Pb_5S_2I_6$, Pb_4SeBr_6, and $Pb_3Te_5I_8$ can be obtained in a polycrystalline form by annealing stoichiometric amounts of the binary phases just below the peritectic temperatures of 381, 418, 383, and 385°C, respectively. Pure, and well-recrystallized, Pb_4SCl_6 is formed by quenching a melt of the composition $PbS \cdot 3PbCl_2$ after annealing at 600°C (302). Single crystals of the sulfide and selenide halides can be prepared by hydrothermal syntheses in the respective hydrohalic acids (300, 302; see also, Section II,D,2). The crystallization conditions of $Pb_5S_2I_6$ in the system $Pb-S-I-R-H_2O$, R being a solvent, have been investigated (226, 429). So far, few of the metastable sulfide halides have been prepared in pure form by precipitation from solution. A phase of the composition $2PbS \cdot PbBr_2$ has been isolated by dilution of a solution prepared from PbS and hydrobromic acid (302). Another example is the phase Pb_2SI_2, formed by the reaction of H_2S with a solution of $KPbI_3$ in acetone (164).

3. Crystallographic Data

The crystal structures of Pb_4SeBr_6, $Pb_5S_2I_6$, and $Pb_7S_2Br_{10}$ have been determined from single-crystal, X-ray analyses (*196*). The unit-cell data are given in Table XXII. The compounds have common structural features with respect to the pure halides of lead. In Pb_4SeBr_6, all of the Pb atoms are coordinated by a trigonal prism of Br (Se) atoms. Additional neighbors above the prism faces complete the coordination number to 7, 8, or 9. In $Pb_5S_2I_6$, some of the Pb atoms are surrounded by 6 I + 1 S, or 5 I + 3 S, in the same, extended, trigonal, prismatic arrangement, and others occupy the centers of PbI_6 octahedra. In $Pb_7S_2Br_{10}$, which is isostructural with Th_7S_{13} (*417*), the Pb atoms are coordinated by 8, or 9, nonmetal atoms in a trigonal-prismatic 6 + 2, or 6 + 3, coordination. Part of the metal and part of the nonmetal positions show a statistical occupancy. The types of coordination in the

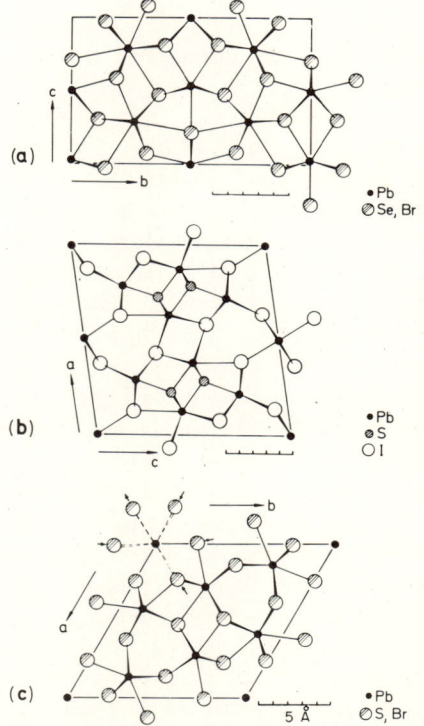

FIG. 33. Structure relationships and coordinations of lead chalcogenide halides: a, Pb_4SeBr_6; b, $Pb_5S_2I_6$; c, $Pb_7S_2Br_{10}$. (Redrawn from B. Krebs, *Z. Anorg. Allg. Chem.* **396**, 137 (1973), Figs. 1, 2, and 3, pp. 141, 147, and 148.)

three structures are given in Fig. 33, a–c. The compound Pb_4SCl_6 is isotypic with Pb_4SeBr_6. In the case of the bromides $Pb_7S_2Br_{10}$ and Pb_4SeBr_6, unusually high temperature-factors for the lead atoms have been observed (196) that indicate an incomplete occupancy of the metal atoms. For both compounds, a homogeneity range has been reported in the literature (247, 249).

4. Physical Properties

Reflection spectra of some sulfide and selenide halides have been measured (302, 303). The compounds are assumed to be semiconductors. Their optical band-gaps are found to be Pb_4SeBr_6, 1.6 eV; $Pb_5S_2I_6$, ~2 eV; $Pb_7S_2Br_{10}$, ~2 eV; and Pb_4SCl_6, 2.5 eV. The reflection spectra of Pb_2SI_2 and Pb_2SBr_2 were reported to correspond to those of $Pb_5S_2I_6$ and $Pb_7S_2Br_{10}$ (164).

XII. Group VA

A. Nitrogen and Phosphorus

As in the case of Group IVA, combinations with these typical non-metals will not be treated in detail insofar as nitrogen compounds are concerned. Of interest with respect to this review are the highly conducting compounds that are obtained by the room-temperature reaction of tetrasulfur tetranitride with halogens, e.g., $(SNBr_{0.4})_x$ (1, 366, 423).

Phosphorus thiohalides are reviewed in (146), and, with more relevance to this review, in (407). Four structural types exist. These are as follows. (i) PSX_3 [X = Cl (203), Br]: tetrahedral molecules $S=PX_3$; (ii) $P_2S_2I_4$ (90) and $P_2Se_2I_4$ (25), probably $I_2YP-PYI_2$; (iii) $P_2S_5Br_4$ and $P_2S_4Br_2$. The structure of the latter, formed by the action of Br_2 on P_4S_6, is known. It consists of a P_2S_4 ring in a skew conformation, each P being attached to two atoms, either Br or S (114). (iv) $P_4S_3I_2$ and $P_4Se_3I_2$. The former exists in two forms. α-$P_4S_3I_2$, prepared in CS_2 solution from the elements, consists of one six- and two five-membered rings in different relative arrangements of the P and S atoms as compared with P_4S_3 (413). β-$P_4S_3I_2$, which forms by the reaction of P_4S_3 with I_2, has a structure closely related to that of P_4S_3 (268), and is isostructural with $P_4Se_3I_2$ (267).

B. Arsenic

Although arsenic is a nonmetal, the chalcogenide halides of this element bear a strong relationship to the antimony and bismuth systems, and will therefore be treated separately.

Numerous compounds are mentioned in the literature, but only in a few cases, however, was their existence based on X-ray powder patterns or the preparation of single crystals: AsSI and AsSeI (74, 76, 79, 98); $As_4Te_5I_2$ (75, 77, 79, 98) and $As_8Te_7I_5$ (77). The preparation of these compounds from the elements or binary arsenides requires a special annealing treatment in order to overcome the glassy state. Single crystals of AsSI, AsSeI, and $As_4Te_5I_2$ have been obtained by sublimation of the polycrystalline material at temperatures below 200°C in a vertical-gradient furnace (79); AsSI has also been obtained by hydrothermal synthesis (282). With the exception of $As_4Te_5I_2$, which is said to be cubic (77), no crystallographic data are available for the arsenic chalcogenide halides. Phase diagrams of these systems, as far as they have been published, are included in Tables XXV–XXIX. In contrast to the antimony and bismuth compounds, the crystalline material has not attracted any attention with regard to its physical properties.

Vitreous Semiconductors

The main interest in these systems has focused on the glass-forming regions, because of the search for vitreous semiconductors. This category of glasses is sometimes referred to as "chalcogenide glasses" (232), distinguished from the oxide type of glasses by their high electronic conductivity and absence of ionic conductivity. They constitute an extensive group of materials whose optical, electrical, and other properties may vary with composition within a fairly wide range. Many of them are typical semiconductors whose distinctive features are the absence of long-range order, their homogeneity, the independence of their properties on small amounts of impurities, and small mobility of charge carriers. For further study of this subject, a chapter on glasses in a book by Krebs (195), and, especially, a review article by Kolomiets (190), should be of interest.

Most of the contributions on arsenide systems have been made by Russian scientists, and their publications are not available in English. An X-ray diffraction study (277), viscosity and structure investigation (78) of the system $As_2Y_3–AsI_3$ (Y = S,Se), and optical and magneto-optical properties of the systems As–S–I and As–S–Br (60) appeared in the translation "Inorganic Materials." A good insight into the situation is provided by the extensive work on As–Te–I glasses by Quinn and Johnson (295), where many references may be found.

C. ANTIMONY AND BISMUTH

Chalcogenide halides of antimony and bismuth have been known since the end of the last century, mainly through contributions by Ouvrard and Schneider. With the exception of BiSeCl, only thiohalides were reported. The early work is summarized in (147) (Sb, before 1948) and (148) (Bi, before 1927), and it will only occasionally be dealt with. Beyond that, these materials had not found much interest.

However, the situation changed dramatically with the discovery of the photoconductivity of SbSI in 1960 (243) and its ferroelectric properties (117). Innumerable publications on the chalcogenide halides of antimony and bismuth then appeared, and are still appearing, with the result that these compounds, and, especially, SbSI, now belong to the most well known chalcogenide halides. As it is impossible to cover the whole field of these investigations in this review, especially where the physical properties, glassy systems, and films are concerned, these topics will only be mentioned. The bismuth and antimony compounds will be treated together, as they possess much similarity. Tables XXIII and XXIV list the chalcogenide halides of antimony and bismuth, the existence of which has been established.

In the early 1950s, in a systematic study, Dönges (106–108) discovered most of the chalcogenide halides of antimony and bismuth that are known today, and then solved their structures.

1. Preparative Methods

The chalcogenide halides in Tables XXIII and XXIV having a reference in the column "solid-state reactions" can be obtained in a polycrystalline form by annealing an intimate mixture of the respective chalcogenides M_2Y_3 and halides MX_3 in stoichiometric ratios in closed ampoules. Other ways of preparation that partially date back to the early work of Ouvrard and Schneider (especially for the thiochalcogenides), and that may still be of interest in special cases, are summarized in (147) (Sb) and (148) (Bi). Preparative work has since been devoted, however, to the preparation of single crystals for X-ray structure determinations and physical measurements. Small crystals, sufficiently large for X-ray work, can be obtained by the aforementioned method, provided that prolonged annealing-times are applied. The methods mentioned in Tables XXIII and XXIV are of greater importance.

TABLE XXIII

Chalcogenide Halides of Antimony: Preparation, Bibliography

References

Compound	Color	Melt growth	Vapor growth	Hydrothermal synthesis	Solid-state reactions
Sb$_4$S$_5$Cl$_2$	Red–brown				106
SbSCl				279,282,283	
SbSBr	Orange	117,244	187	279,282,283,285,312	106
SbSI	Red	38,117,163, 220,235,391,421	112,163,166,183, 236,255,281,418,425	279,280,282,283,285,312	106
SbSeCl				282	
SbSeBr	Dark red	117,244	171,187	279,282,283,285	106
SbSeI	Black	117,244	166,237,238	279,282–285	106,238
SbTeBr				285	
SbTeI	Black	117	166,403	283,285,286	108

TABLE XXIV
Chalcogenide Halides of Bismuth: Preparation, Bibliography

Compound	Color	Melt growth	Vapor growth	Hydrothermal synthesis	Solid-state reactions
BiSCl	Red	117,244		279,282,283	106
$Bi_{19}S_{27}Cl_3$					193
$Bi_4S_5Cl_2$	Dark grey				194
BiSBr	Dark red	117,244	191	279,282,283,285	106
$Bi_{19}S_{27}Br_3$	Black		192	219	192
BiSI	Black	117,244		279,282,283,285	106
$Bi_{19}S_{27}I_3$			67,225		
BiSeCl	Dark red	117		279,283	107
BiSeBr	Black	117,244	171	279,283,285	107
BiSeI	Black	117,244,398	172	279,283,285	107
BiTeBr	Black		259,395	283,285	108,259
BiTeI	Black	170,428	71,170,259	283,285	71,108,170,259

Under these headers: References

a. Melt growth. The method was applied by Nitsche et al. (243, 244). Single crystals in the form of needles up to 2 cm long have been obtained by slow cooling of the molten compounds at a rate of 5°C/h. The crystals are washed with dilute hydrochloric aicd to remove surplus trihalide, rinsed with alcohol, and dried (243). In the Bridgman–Stockbarger technique (55), the melt is homogenized in a pointed-growth ampoule and then lowered in a temperature gradient in a two-zone furnace (1 mm/h). The cylindrical ingot consists of densely packed, single-crystal needles whose axes, corresponding to the crystallographic c axis, are parallel to the ingot axis (38, 170, 244, 391, 398). The equipment used for BiTeI (170) is shown in Fig. 34. The growth of SbSI under pressure has been reported (220). The melt need not necessarily have the stoichiometric composition. In some cases, an excess of the halogenide proved to be useful [flux technique (55), e.g., (170, 235, 391, 398, 421)].

b. Vapor growth. The common feature of this method is simple sublimation in a closed tube. The tubes are placed, vertically or horizontally, in a two-zone furnace, with the starting material at the higher temperature. The method, first reported by Kern for SbSI (183), has been applied for a number of substances mainly by Horak and co-workers (71, 166, 170–172, 187, 191, 237, 238, 395; see also, Tables XXIII and XXIV, where experimental details may be found). The arrangement is similar to that depicted in Fig. 21 (Section VI,B,5).

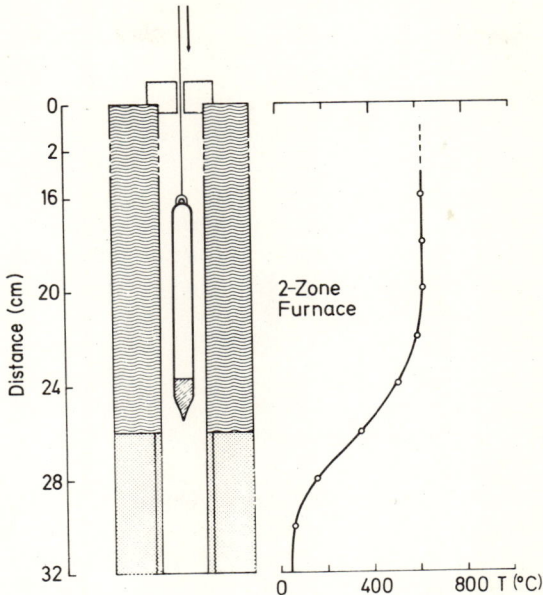

FIG. 34. Bridgman–Stockbarger technique for BiTeI. (Redrawn from J. Horak and H. Rodot, *C. R. Acad. Sci., Ser. B.* **267**, 363 (1968), Fig. 2, p. 364.)

BiTeI has been grown by chemical transport from Bi_2Te_3 and BiI_3, using Br_2 as the transport agent (218).

c. *Hydrothermal growth.* The applicability of hydrothermal synthesis in acid solution (see Section II,D,2) to the growth of Group VA chalcogenide halides was first observed by Rau and Rabenau for SbSI and SbSBr (312). In succeeding years, the method was extensively studied by Popolitov and co-workers for a great number of compounds (see Tables XXIII and XXIV, and 279–287). They used standard autoclaves lined with Teflon for low temperatures (≤300°C) and quartz glass, titanium, and platinum for higher temperatures (282, 283). In order to evaluate optimal growth-conditions, they constructed Eh–pH diagrams for different A^V–B^{VI}–C^{VII}–H_2O systems (282, 285), and determined crystallization regions in such systems (284–286). Although, for practical purposes, the experimental parameters lie in the range of 200–400°C and pressures of 200–530 atm (283), the influence of high pressure has been investigated for the systems Sb–Se–I–R–H_2O (284), Sb–Te–I–R–H_2O (286), and Bi–Se–I–R–H_2O (287), R being a solvent such as HCl, HI, or C_2H_5OH, at pressures up to 2000 atm.

It was found that the region of formation of the chalcogenide halides depends on the pH, the solvent concentration, and the ratios of the initial components in the charge. Temperature and pressure have practically no influence on the phase formation in these systems (285). The use of bromine (283) and $SeBr_2$ as the solvent leads to a different mechanism, having different kinetics of formation and different growth-forms of the crystals (285).

The main problem in the crystal growth of antimony and bismuth chalcogenide halides is the growth of large, isometric crystals. In most of the experiments, a needle morphology predominated, with the growth rate estimated to be more than 50 times larger parallel the c direction than perpendicular to it (227). This growth-rate anisotropy seems to be an inherent characteristic of the material (235). For the growth of SbSI from the vapor phase, the relationship between the morphology of the crystals and the condition of the crystallization process was investigated (112). It has been found (418) that isometric crystals up to $5 \times 7 \times 10$ mm^3 may be grown from the vapor on a seed by applying crystallization by temperature-gradient reversal in a Scholz ampoule (348–350) geometry. Principally, the Scholz apparatus is an axially symmetrical, vertical ampoule, the crystallization taking place in a zone along its axis, whereas the source material is placed on the periphery, concentric to this zone. The needle morphology of large SbSI crystals has also been successfully controlled by a modified, flux technique using an excess of SbI_3 as the solvent. By preventing additional, disturbing nucleation during growth, the crystals are forced to grow in thickness,· yielding crystals up to 1 cm in diameter. The method seems also to be applicable to other chalcogenide halides (235).

2. Chemical Properties

The chalcogenide halides of antimony and bismuth are stable in air, and do not dissolve in H_2O or diluted acids. Their colors, mainly referring to single-crystal needles, are given in Tables XXIII and XXIV.

3. Phase Diagrams

Knowledge of phase diagrams is not only a prerequisite for efficient crystal-growth, but also provides information on the formation of solid solutions, in which, for example, physical properties may change continuously. The numerous publications concerning Group VA systems are summarized in Tables XXV–XXVII, together with the respective references and the most important information. Abbreviations used

TABLE XXV

System Investigations: Cuts Me_2Y_3–MeX_3

System	Ref.	Remarks
As_2S_3–AsI_3	76	T – x; AsSI
As_2Se_3–$AsBr_3$	73	T – x; no compound
As_2Se_3–AsI_3	76	T – x; AsSeI
As_2Te_3–AsI_3	75	T – x; $As_4Te_3I_2$
Sb_2S_3–SbI_3	38,255,391	T – x; SbSI
Sb_2S_3–$SbSI$	2	SbSI – $(Sb_2S_3)_x$, x = 0.03–0.12, p-T
Sb_2Se_3–SbI_3	37, 103	T – x; SbSeI
Sb_2Te_3–SbI_3	3,27,75,397,403	p – T; SbTeI (3,27); T – x, SbTeI (397,403)
Bi_2S_3–$BiBr_3$	202	BiSBr, thermodynamics
Bi_2S_3–BiI_3	202	BiSI, thermodynamics
Bi_2Se_3–$BiBr_3$	201	Vapor pressure
Bi_2Se_3–BiI_3	36,398,399	BiSeI; reaction (399), T – x; BiSeI, $9Bi_2Se_3 \cdot BiI_3$
Bi_2Te_3–$BiBr_3$	201	Vapor pressure
Bi_2Te_3–BiI_3	27,170,390,402	T – x; $Bi_{1.00}Te_{0.98}I_{1.02}$ (390). BiTeI, T – x; p – T; BiTeI (402), T – x; BiTeI (170)

are T – x = temperature–concentration diagram; p – T = pressure–temperature diagram; s.s. = solid solution.

The composition and pressure of the gas phase over SbSI and Sb_3S_3 (392) and SbSeI and BiSeI (102) have been measured, leading to the thermodynamic values: $-\Delta H_f^0$ = 25.4 ± 2.4, 22.3 ± 2.7, and 27.2 ± 2.4 kcal/mol, and $\Delta S°$ = 26.9 ± 1.3, 31.1 ± 1.7, and 33.4 ± 2.5 cal/(mol·deg) for SbSI, SbSeI, and BiSeI, respectively.

TABLE XXVI

System Investigations: MeYX – Me'Y'X'

System	Ref.	Remarks
AsSI–AsSeI	74	T – x; s.s.
AsSI–SbSI	74,333	T – x; eutectic (74), peritectic finite s.s. (333)
SbSBr–SbSI	61,136,151,396	s.s. (396), T – p – x (61)
SbSI–SbSeI	31,32,294	T – x; s.s. (32)
SbSeI–SbTeI	33,245	T – x; s.s. (33)
SbSI–BiSI	30,137,245	T – x; s.s. (30)
SbTeI–BiTeI	31,34	T – x; finite s.s. (34)
SbSeI–BiSeI	29,35,214	T – x; s.s. (35)

TABLE XXVII

System Investigations: Miscellaneous

System	Ref.	Remarks
SbI_3–Se	82	$T - x$, eutectic
SbSeI–Se	82	$T - x$, eutectic
Sb_2Se_3–I	82	$T - x$
Bi_2Te_3–I	28	$Bi_2Te_3 \cdot I_2$, $Bi_2Te_3 \cdot 3I_2$(?)
As–Se–I	199	Vapor pressures
$As_4Te_5I_2$–As	77	$T - x$
Sb_2S_3–I	38	$T - x$, $Sb_2S_3 \cdot I_2$, $Sb_2S_3 \cdot 3I_2$
BiTeI–Te	116	$T - x$, eutectic
BiI_3–Te	116	$T - x$, eutectic

4. Crystallographic Data

The crystallographic data are summarized in Table XXVIII. The best known structure is that of the compounds belonging to the so-called SbSI type (see Table XXVIII), which was proposed by Dönges (*106, 107*), and confirmed for SbSBr (*84*), SbSI (*153, 174, 184, 258*), and BiSI (*153*).

Characteristic of this SbSI type is the formation of infinite, $(Sb_2S_2I_2)_\infty$ double-chains parallel to the *b* axis, as shown in Fig. 35 (notation: International Tables for X-ray Crystallography), which explains the needle-like shape of these crystals. The relationship to the needle-like sulfides Sb_2S_3 and Bi_2S_3, which crystallize in the same space-group, has been pointed out (*106, 153*): one S of the sulfide is replaced by two halogens in such a way that the $(Me_4S_6)_\infty$ chains of the sulfides split into two $(Me_2S_2X_2)_\infty$ chains. The relationship to the sesquisulfide is even stronger for the $Bi_{19}S_{27}X_3$ [X = Cl (*193*), Br (*192, 219*), I (*225*)] compounds, which consist of $(Bi_4S_6)_\infty$ chains connected by the halogens and one extra Bi atom, leading to the formulation $Bi(Bi_2S_3)_9X_3$.

The phase transiton from a paraelectric to a ferroelectric state, most characteristic for the SbSI type compounds, has been extensively studied for SbSI, because of its importance with respect to the physical properties of this compound (e.g., *153, 173–177, 184, 257*). The first-order transition is accompanied by a small shift of the atomic parameters and loss of the center of symmetry, and is most probably of a displacement nature. The true structure of $Sb_4S_5Cl_2$ (*106*), $Bi_4S_5Cl_2$ (*194*), and SbTeI (*108, 403*) is still unknown. In contrast to the sulfides and selenides of bismuth, BiTeBr (*108*) and BiTeI (*108, 390*) exhibit a layer structure similar to that of the CdI_2 structure, if the difference between Te, Br, and I (see Fig. 36) is ignored.

TABLE XXVIII

CRYSTALLOGRAPHIC DATA FOR ANTIMONY AND BISMUTH COMPOUNDS

Compound	Symmetry	a (Å)	b (Å)	c (Å)	β (degrees)	Z	Space group	Ref.
$Sb_4S_5Cl_2$	Orthorhombic	10.53	9.38	11.08		4	Pnma (?)	367
SbSBr		8.20	3.95	9.70				
SbSI		8.49	4.16	10.10				
SbSeBr		8.30	3.95	10.20				
SbSeI		8.65	4.12	10.38				
BiSCl	Orthorhombic	7.70	4.00	9.87		4	Pnma	367
BiSBr		8.02	4.01	9.70				
BiSI		8.46	4.14	10.15				
BiSeBr		8.18	4.11	10.47				
BiSeI		8.71	4.19	10.45				
BiSeCl	Orthorhombic	12.37	18.10	4.08		12	Pmmm	367
SbTeI	Orthorhombic	9.18	10.8	4.23		4	Pmmm, Pnmm, Pmmn	108
SbTeI	Monoclinic	14.55	4.23	13.72	81.12	8	C2/c, Cc	403
BiTeBr	Hexagonal	4.23		6.47		1	P$\bar{3}$m1	108
BiTeI		4.31		6.83				
$Bi_{19}S_{27}Cl_3$		15.40		4.02				193
$Bi_{19}S_{27}Br_3$	Hexagonal	15.55		4.02		2/3	P6$_3$	219
$Bi_{19}S_{27}I_3$		15.63		4.02				225
$Bi_4S_5Cl_2$	Rhombohedral	19.80		12.4		15	R32, R3m, R$\bar{3}$m	194

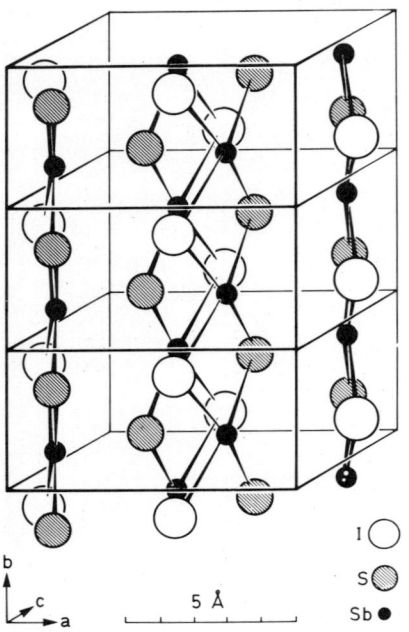

FIG. 35. [Sb$_2$S$_2$I$_2$]$_\infty$ double chain in the structure of SbSI. (Redrawn from E. Dönges, Z. Anorg. Allg. Chem. **263,** 112 (1950), Fig. 11, p. 129.)

5. Physical Properties

The semiconducting properties of the compounds of the SbSI type (see Table XXVIII) were predicted by Mooser and Pearson in 1958 (228). They were first confirmed for SbSI, for which photoconductivity was found in 1960 (243). The breakthrough was the observation of ferroelectricity in this material (117) and other SbSI type compounds (244; see Table XXIX), in addition to phase transitions (184), nonlinear optical behavior (156), piezoelectric behavior (44), and electromechanical (183) and other properties. These photoconductors exhibit abnormally large temperature-coefficients for their band gaps; they are strongly piezoelectric. Some are ferroelectric (see Table XXIX). They have anomalous electrooptic and optomechanical properties, namely, elongation or contraction under illumination. As already mentioned, these fields cannot be treated in any detail in this review; for those interested in ferroelectricity, review articles (224, 352) are mentioned. The heat capacity of SbSI has been measured from −180 to +40°C and, from these data, the excess entropy of the ferro-paraelectric transition

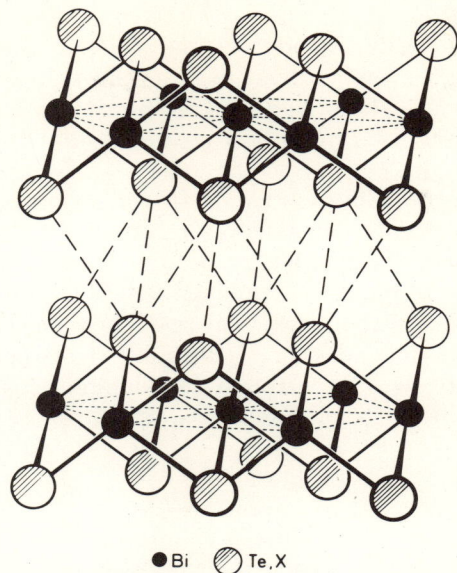

● Bi ◯ Te, X

Fig. 36. Structure of BiTeBr and BiTeI.

has been calculated to be $\Delta S = 53$ mcal/(mol·deg), plausible for a displacement transition (384; see also, *Crystallographic Data*).

^{121}Sb Mössbauer spectra of SbSBr, SbSI, SbSeBr, SbSeI, and SbTeI have been obtained at 4.2 K. The quadrupole coupling-constant of the iodides decreases in the order SbSI > SbSeI > SbTeI, becoming nega-

TABLE XXIX

Some Physical Properties of Antimony and Bismuth Chalcogenide Halides

Compound	Band gap (eV)	Ref.	Magnetic susceptibility ref.	Curie point (°C)	Ref.
SbSBr	2.22	6	221	−180	244
SbSI	2.12	6	41,221	22	117
SbSeBr	1.84	169			
SbSeI	1.61	6	221		
BiSCl	1.89	39	221		
BiSBr	1.88	169		−170	244
BiSI			221	−160	244
BiSeBr			221		
BiSeI	1.32	40	41,221		
BiTeBr	0.54	293			
BiTeI	0.38	293	41,420		

tive for the last (105). A selection of band gaps of MXY compounds is included in Table XXIX. Although somewhat differing values are reported for some of the compounds, and the uncertainty is ~0.2 eV, the sequence of decreasing band gaps shows that it is mainly the decrease in electronegativity differences that influences the band gap (149; see also, Section II,A,5).

With respect to the physical properties mentioned, band-structure calculations have attracted considerable interest, e.g., for SbSBr, SbSI, and SbSeI (234). For the compounds having reference 221 in column 4 of Table XXIX, a temperature-independent diamagnetism has been found, with values of about 10^{-7} cm^{-3}g^{-1} between 77 and 340 K. A small temperature-dependence is exhibited by BiTeI, a narrow-gap semiconductor (41). The anisotropy of the magnetic susceptibility has been studied for SbSI, BiSeI, and BiTeI (41, 420).

Appendix

After submitting the manuscript some interesting new results were published, the most important of which are presented here. Less relevant contributions are not specifically mentioned. They are all listed in the Appendix References (420–433) and referred to by number herein.

II. Copper: The first two structures of copper selenide halides have now been solved. $CuSe_2Cl$ turned out to be a true isotype of $CuTe_2Cl$, containing infinite, pseudofourfold, selenium helices (433). The crystals are systematically twinned, which is the reason for the wrong lattice constants given earlier (307). $CuSe_3Br$ is unique by the occurrence of chair-type Se_6-rings (424). As in $CuSe_2Cl$, the copper atoms are coordinated by two selenium and two halogen atoms in a distorted tetrahedral arrangement.

II. Silver: According to DTA data and conductivity data, Ag_3SBr and Ag_3SI undergo phase transitions at 120 and 163 K, respectively. In Ag_3SI the transition to the low-temperature phase is accompanied by the appearance of additional x-ray reflections (427).

VI. Niobium: Two new niobium selenide halides, $Nb_3Se_5Cl_7$ and $Nb_3Se_5Br_7$, were prepared by Rijnsdorp and Jellinek (431). $Nb_3Se_5Cl_7$ was obtained by annealing $NbSe_2Cl_2$ and $NbCl_4$ in a 2:1 ratio at 530°C for 2–3 weeks. The isotypic bromine compound was formed by thermal decomposition of $NbSe_2Br_2$ in the presence of $NbSeBr_3$. The compounds are monoclinic with $a = 7.599$, $b = 12.675$, $c = 8.051$ Å, $\beta = 106.27°$ (space group $P2_1/m$) and $a = 7.621$, $b = 12.833$, $c = 8.069$ Å, $\beta =$

106.21°, respectively. The crystal structure and XPS spectra show $Nb_3Se_5Cl_7$ to be a mixed valence compound which can be formulated as $[Nb_2^{4+}Nb^{5+}(Se_2)_2^{2-}Se^{2-}Cl_7^-]$. $[Nb_2Se_4]$-cages with Nb–Nb pairs (Nb–Nb = 2.94 Å), as in $NbSe_2Cl_2$ (347, see Fig. 20), are interconnected by faces of three bridging chlorines to form linear chains of $[Nb_2^{4+}(Se_2)_2^{2-}Cl_5^-]$, to which side chains $[Nb^{5+}Se^{2-}Cl_2^-]$ are attached. In accordance with these results, the compound is a diamagnetic semiconductor with a band gap (1.49 eV at 300 K) very close to that of $NbSe_2Cl_2$.

Bullett has calculated band structures and densities of states for $NbSe_2Cl_2$, $NbSe_2Cl_2$, $Nb_3Se_5Cl_7$, and $NbSe_4I_{0.33}$ on the basis of the geometries of the Nb_2Y_4-clusters. Metal–metal bonding results in very narrow bands largely composed of niobium d-orbitals at the band edges (422).

VII. Molybdenum: A recent review of molybdenum chalcogenide halides is contained in a paper of Tsigdinos (432).

XII. Antimony: Of various new papers on the crystal growth of SbSI, the one by Ishikawa *et al.* (425) is worth mentioning, since it describes a new device especially designed to avoid formation of hollows in vapor-grown crystals.

XII. Bismuth: As shown by a recent single-crystal structure determination by Krämer (426), $Bi_4S_5Cl_2$ crystallizes in the rhombohedral space group $R\bar{3}$. The Cl^- and S^{2-} ions display a light-packed framework of statistically distributed anions. Unusually high coordination numbers of eight and nine for bismuth are attained by extended planar–pyramidal and trigonal–prismatic coordinations, i.e., three additional ions are located either below the basal planes of the pyramids or above the rectangular faces of the prisms with cation–anion distances ranging from 2.62 to 3.66 Å.

References

1. Akhtar, M., Kleppinger, J., MacDiarmid, A. G., Milliken, J. A., Moran, M. J., Chiang, C. K., Cohen, M. J., Heeger, A. J., and Peebles, D. L., *J. Chem. Soc., Chem. Commun.* p. 473 (1977).
2. Alekhina, N. S., Grekov, A. A., Kachalov, N. P., Palandov, I. N., Syrkin, L. N., and Chekhunova, N. P., *Poluprovodn. Segnetoelektr.* p. 90 (1973).
3. Aleshin, V. A., Valitova, N. R., Popovkin, B. A., and Novoselova, A. V., *Russ. J. Phys. Chem.* **48**, 1421 (1974).
4. v. Alpen, U., Fenner, J., Marcoll, J. D., and Rabenau, A., *Electrochim. Acta* **22**, 801 (1977).
5. v. Alpen, U., Fenner, J. Predel, B., Rabenau, A., and Schluckebier, G., *Z. Anorg. Allg. Chem.* **438**, 5 (1978).

6. Alward, J. F., Fong, C. Y., El Batanouny, M., and Wooten, F., *Solid State Commun.* **25,** 307 (1978).
7. Andreev, Yu. V., and Loginova, M. V., *Russ. J. Inorg. Chem.* **15,** 1761 (1970).
8. Atherton, M. J., and Holloway, J. H., *Adv. Inorg. Radiochem.* **22,** 171 (1979).
9. Aurivillius, K., *Acta Chem. Scand.* **4,** 1413 (1950).
10. Aurivillius, K., *Ark. Kemi* **26,** 497 (1967).
11. Baghlaf, A. O., M.Sc. Thesis, University of Manchester, England (1972).
12. Baghlaf, A. O., Ph.D. Thesis, University of Manchester, England (1975).
13. Baghlaf, A. O., and Thompson, A., *J. Less-Common Met.* **53,** 291 (1977).
14. Baroni, A., *Atti Accad. Naz. Lincei Cl. Sci. Fis. Mat. Nat. Rend.* **29,** 76 (1939).
15. Batsanov, S. S., and Abaulina, L. I., *Izv. Sib. Otd. Akad. Nauk SSSR* p. 67 (1961).
16. Batsanov, S. S., Derbeneva, S. S., and Shestakova, N. A., *Inorg. Mater.* **2,** 1945 (1966).
17. Batsanov, S. S., and Doronina, L. M., *Inorg. Mater.* **2,** 423 (1966).
18. Batsanov, S. S., and Gorogotskaya, L. I., *Izv. Sib. Otd. Akad. Nauk SSSR* p. 42 (1959).
19. Batsanov, S. S., and Gorogotskaya, L. I., *Russ. J. Inorg. Chem.* **4,** 24 (1959).
20. Batsanov, S. S., Kolomiichuk, V. N., Derbeneva, S. S., and Erenburg, R. S., *Inorg. Mater.* **9,** 979 (1973).
21. Batsanov, S. S., and Petrova, I. Kh., *Izv. Vyssh. Uchebn. Zaved. Khim. Khim. Tekhnol.* **4,** 349 (1961).
22. Batsanov, S. S., and Rigin, V. I., *Dokl. Chem.* **158,** 1094 (1964).
23. Batsanov, S. S., Ruchkin, E. D., and Khripin, L. A., *Inorg. Mater.* **2,** 88 (1966).
24. Batsanov, S. S., Shestakova, N. A., and Khripin, L. A., *Dokl. Chem.* **152,** 729 (1963).
25. Baudler, M., Voiland, B., and Valpertz, H. W., *Chem. Ber.* **106,** 1049 (1973).
26. Bell, M. C., and Flengas, S. N., *J. Electrochem. Soc.* **113,** 27 (1966).
27. Belotskii, D. P., Dodik, S. M., Antipov, I. N., and Nefedova, Z. I., *Ukr. Khim. Zh.* **36,** 897 (1970).
28. Belotskii, D. P., Dodik, S. M., and Demkiv, O. A., *Ukr. Khim. Zh.* **38,** 766 (1972).
29. Belotskii, D. P., Gavrilenko, N. V., and Lapshin, V. F., *Khal'kogenidy* p. 88 (1974).
30. Belotskii, D. P., Gavrilenko, N. V., Zhornovyi, I. V., Kulikovskaya, S. M., Kormish, M. E., Noval'kovskii, N. P., and Onipko, A. F., *Inorg. Mater.* **14,** 159 (1978).
31. Belotskii, D. P., Gorchinskii, N. I., Dodik, S. M., Boichuk, R. F., Noval'kovskii, N. P., and Smirnov, A. V., *Khal'kogenidy* p. 83 (1974).
32. Belotskii, D. P., Gorchinskii, N. I., and Noval'kovskii, N. P., *Dopov. Akad. Nauk Ukr. RSR, Ser. B* **34,** 1079 (1972).
33. Belotskii, D. P., Gavrilenko, N. V., Noval'kovskii, N. P., Zhornovyi, I. V., Katerinyuk, D. M., Levenets, M. D., and Gorchinskii, N. I., *Inorg. Mater.* **10,** 25 (1974).
34. Belotskii, D. P., Kulikovskaya, S. M., Gavrilenko, N. V., Noval'kovskii, N. P., and Legeta, L. V., *Inorg. Mater.* **13,** 529 (1977).
35. Belotskii, D. I., and Lapshin, V. F., *Inorg. Mater.* **7,** 1727 (1971).
36. Belotskii, D. P., Lapshin, V. F., and Boichuk, R. F., *Inorg. Mater.* **7,** 1724 (1971).
37. Belotskii, D. P., Lapshin, V. F., Boichuk, R. F., and Noval'kovskii, N. P., *Inorg. Mater.* **8,** 499 (1972).
38. Belyaev, L. M., Lyakhovitskaya, V. A., Netesov, G. B., Mokhosoev, M. V., and Aleikina, S. M., *Inorg. Mater.* **1,** 1969 (1965).
39. Belyaev, A. D., Slivka, V. Yu., and Chepur, D. V., *Ukr. Fiz. Zh.* **13,** 162 (1968).
40. Bercha, D. M., and Zayachkovskii, M. P., *Sov. Phys. Solid State* **14,** 766 (1972).
41. Bercha, D. M., Zayachkovskii, M. P., Zayachkovskaya, N. F., and Maslyuk, V. T., *Sov. Phys. Solid State* **17,** 1393 (1975).

42. Berg, L. G., and Malkova, T. I., *Izv. Vyssh. Uchebn. Zaved. Khim. Khim. Tekhnol.* **12,** 691 (1969).
43. Berg, L. G., Malkova, T. I., and Pavlova, A. K., *Russ. J. Inorg. Chem.* **12,** 107 (1967).
44. Berlincourt, D., Jaffe, H., Merz, W. J., and Nitsche, R., *Appl. Phys. Lett.* **4,** 61 (1964).
45. Biltz, W., and Ehrlich, P., *Z. Anorg. Allg. Chem.* **234,** 97 (1937).
46. Blachnik, R., and Alberts, J. E., *Z. Naturforsch., Teil B* **31,** 163 (1976).
47. Blachnik, R., and Hoppe, A., *J. Chem. Thermodyn.* **8,** 631 (1976).
48. Blachnik, R., and Kasper, F. W., *Z. Naturforsch., Teil B* **29,** 159 (1974).
49. Blachnik, R., and Kudermann, G., *Z. Naturforsch., Teil B* **28,** 1 (1973).
50. Bode, H., and Hettwer, E., *Z. Anal. Chem.* **173,** 285 (1960).
51. Bodroux, F., *C. R. Acad. Sci.* **130,** 1398 (1900); *Bull. Soc. Chim. Fr.* **23,** 502 (1900).
52. Boncheva-Mladenova, Z., and Aramov, N., *Proc. Int. Conf. Therm. Anal., 4th, 1974,* p. 349, 1975.
53. Boncheva-Mladenova, Z., Aramov, N., and Raikowa, D., *Z. Anorg. Allg. Chem.* **401,** 306 (1973).
54. Brebrick, R. F., *in* "Progress in Solid State Chemistry" (H. Reiss, ed.), Vol. 3, pp. 213–264. Pergamon, Oxford, 1967.
55. Brice, J., "The Growth of Crystals from Liquids", pp. 4–5. North-Holland Publ., Amsterdam, 1973.
56. Britnell, D., Drew, M. G. B., Fowles, G. W. A., and Rice, D. A., *Inorg. Nucl. Chem. Lett.* **9,** 415 (1973).
57. Britnell, D., Fowles, G. W. A., and Mandyczewsky, R., *J. Chem. Soc., Chem. Commun.* p. 608 (1970).
58. Britnell, D., Fowles, G. W. A., and Rice, D. A., *J. Chem. Soc., Dalton Trans.* p. 2191 (1974).
59. Brukl, A., *Monatsh. Chem.* **45,** 471 (1924).
60. Burkov, V. I., Dembovskii, S. A., Kizel', V. A., Kozlova, N. L., Krasilov, Yu. I., and Chernov, A. P., *Inorg. Mater.* **9,** 333 (1973).
61. Buturlakin, A., Gerzanich, E. I., Chepur, D. V., and Groshik, I. I., *Sov. Phys. J.* **19,** 1333 (1976).
62. Capitaine, H., *J. Prakt. Chem.* **18,** 422 (1839).
63. Carkner, P. M., Thesis, University of New Hampshire, 1975.
64. Carkner, P. M., and Haendler, H. M., *J. Solid State Chem.* **18,** 183 (1976).
65. Carkner, P. M., and Haendler, H. M., *J. Cryst. Growth* **33,** 196 (1976).
66. Carlson, E. H., *J. Cryst. Growth* **1,** 271 (1967).
67. Carlson, E. H., *J. Cryst. Growth* **12,** 162 (1972).
68. Carter, F. L., *Metallurg. Soc. Conf., 15th,* p. 245, 1961.
69. Cava, R. J., Reidinger, F., and Wuensch, B. J., *Solid State Commun.* **24,** 411 (1977).
70. Chandrasekhar, H. R., and Genzel, L., *Solid State Commun.* **25,** 73 (1978).
71. Chepur, D. V., Gorak, Ya. A., Kovach, D. Sh., Turyanitsa, I. D., Borets, A. N., and Yatskovich, I. I., *Inorg. Mater.* **6,** 336 (1970).
72. Cherin, P., and Unger, P., *Acta Crystallogr.* **23,** 670 (1967).
73. Chernov, A. P., Borisenkova, A. F., and Dembovskii, S. A., *Inorg. Mater.* **11,** 822 (1975).
74. Chernov, A. P., Dembovskii, S. A., Borisenkova, A. F., and Reis, I. A., *Inorg. Mater.* **10,** 1323 (1974).
75. Chernov, A. P., Dembovskii, S. A., and Chubirka, L. A., *Inorg. Mater.* **6,** 411 (1970).
76. Chernov, A. P., Dembovskii, S. A., and Kirilenko, I. A., *Inorg. Mater.* **6,** 228 (1970).

77. Chernov, A. P., Dembovskii, S. A., and Luzhnaya, N. P., *Russ. J. Inorg. Chem.* **20,** 1208 (1975).
78. Chernov, A. P., Dembovskii, S. A., and Makhova, V. I., *Inorg. Mater.* **6,** 722 (1970).
79. Chernov, A. P., Kanishcheva, A. S., and Dembovskii, S. A., *Inorg. Mater.* **5,** 320 (1969).
80. Chernykh, S. M., and Safonov, V. V., *Russ. J. Inorg. Chem.* **23,** 465 (1978).
81. Chernykh, S. M., Safonov, V. V., Ksenzenko, V. I., Korshunov, B. G., and Fedorov, P. I., *Russ. J. Inorg. Chem.* **21,** 1590 (1976).
82. Chervenyuk, G. I., Niiger, F., Belotskii, D. P., and Noval'kovskii, N. P., *Inorg. Mater.* **13,** 806 (1977).
83. Chevrel, R., Sergent, M., and Prigent, J., *J. Solid State Chem.* **3,** 515 (1971).
84. Christofferson, G. D., and McCullough, J. D., *Acta Crystallogr.* **12,** 14 (1959).
85. Collin, G., Dagron, C., and Thevet, F., *Bull. Soc. Chim. Fr.* p. 418 (1974).
86. Corbett, J. D., *Prog. Inorg. Chem.* **21,** 129 (1976).
87. Corbett, J. D., McMullan, R. K., and Prince, D. J., *Inorg. Chem.* **10,** 1749 (1971).
88. Corbett, J. D., and Prince, D. J., *Inorg. Chem.* **9,** 2731 (1970).
89. Couch, T. W., Lokken, D. A., and Corbett, J. D., *Inorg. Chem.* **11,** 357 (1972).
90. Cowley, A. H., and Cohen, S. T., *Inorg. Chem.* **3,** 780 (1964).
91. Cueilleron, J., and Hillel, R., *Bull. Soc. Chim. Fr.* p. 3635 (1968).
92. Dagron, C., *C. R. Acad. Sci., Ser. C* **262,** 1575 (1966).
93. Dagron, C., *C. R. Acad. Sci.* **260,** 1422 (1965).
94. Dagron, C., Etienne, J., and Laruelle, P., *Int. Conf. Solid Comp. Trans. Elements, 2nd, 1967.*
95. Dagron, C., and Thevet, F., *C. R. Acad. Sci., Ser. C* **271,** 677 (1977).
96. Dagron, C., and Thevet, F., *Ann. Chim.* **6,** 67 (1971).
97. Defagr, E. D., *Ann. Chim. Phys.* **7,** 266 (1901).
98. Dembovskii, S. A., and Chernov, A. P., *Inorg. Mater.* **4,** 1079 (1968).
99. Dembovskii, S. A., Kirilenko, V. V., and Buslaev, Yu. A., *Inorg. Mater.* **7,** 290 (1971).
100. Dembovskii, S. A., and Popova, N. P., *Inorg. Mater.* **6,** 116 (1970).
101. Dmitrievich, D. S., and Vasil'evich, M. A., *Mater. Vses. Nauchn. Stud. Konf. Khim., 13th* p. 39, 1975.
102. Dolgikh, V. A., Popovkin, B. A., Ivanova, G. I., and Novoselova, A. V., *Inorg. Mater.* **11,** 548 (1975).
103. Dolgikh, V. A., Popovkin, B. A., Odin, I. N., and Novoselova, A. V., *Inorg. Mater.* **9,** 823 (1979).
104. Donaldson, J. D., Laughlin, D. R., and Silver, J., *J. Chem. Soc.* p. 996 (1977).
105. Donaldson, J. D., Kjekshus, A., Nicholson, D. G., and Southern, J. T., *Acta Chem. Scand., Ser. A* **29,** 220 (1975).
106. Dönges, E., *Z. Anorg. Allg. Chem.* **263,** 112 (1950).
107. Dönges, E., *Z. Anorg. Allg. Chem.* **263,** 280 (1950).
108. Dönges, E., *Z. Anorg. Allg. Chem.* **265,** 56 (1951).
109. Drew, M. G. B., and Mandyczewsky, R., *J. Chem. Soc., A* p. 2815 (1970).
110. Durovic, S., *Acta Crystallogr.* **24,** 1661 (1968).
111. Durovic, S., *Chem. Zvesti* **22,** 858 (1968).
112. Dziuba, Z., *J. Cryst. Growth* **35,** 340 (1976).
113. Ehrlich, P., and Siebert, W., *Z. Anorg. Allg. Chem.* **301,** 288 (1959).
114. Einstein, F. W. B., Penfold, B. R., and Tapsell, Q. T., *Inorg. Chem.* **4,** 186 (1965).
115. Etienne, J., *Bull. Soc. Fr. Mineral. Cristallogr.* **92,** 134 (1969).

116. Evdokimenko, L. T., and Tsypin, M. I., *Inorg. Mater.* **7,** 1172 (1971).
117. Fatuzzo, E., Harbeke, G., Merz, W. J., Nitsche, R., Roetschi, H., and Ruppel, W., *Phys. Rev.* **127,** 2034 (1962).
118. Feltz, A., Buettner, H. J., Lippmann, F. J., and Maul, W., *J. Non-Cryst. Solids* **8–10,** 64 (1972).
119. Fenner, J., *Naturwissenschaften* **63,** 244 (1976).
120. Fenner, J., *Acta Crystallogr., Sect. B* **32,** 3084 (1976).
121. Fenner, J., *Z. Naturforsch., Teil B* **33,** 479 (1978).
122. Fenner, J., unpublished results
123. Fenner, J., to be published
124. Fenner, J., and Mootz, D., *J. Solid State Chem.* **24,** 367 (1978).
125. Fenner, J., and Rabenau, A., *Z. Anorg. Allg. Chem.* **426,** 7 (1976).
126. Fenner, J., and Schulz, H., *Acta Crystallogr., Sect. B* **35,** 307 (1979).
127. Fischer, O., *Proc. Int. Conf. Low Temp. Phys.*, *14th* pp. 172–191, Elsevier, New York, 1975.
128. Flahaut, J., Laruelle, P., Dagron, C., Adolphe, C., Etienne, J., Ghemard, G., Loye, O., Rysanek, N., Savigny, N., and Thevet, F., *Proc. Rare Earth Res. Conf., 11th* pp. 947–953, 1974.
129. Fortunatov, N. S., and Timostschenko, N. I., *Ukr. Khim. Zh.* **31,** 1078 (1965).
130. Fortunatov, N. S., and Timostschenko, N. I., *Ukr. Khim. Zh.* **35,** 1207 (1969).
131. Fowles, G. W. A., Hobson, R. J., Rice, D. A., and Shanton, K. J., *J. Chem. Soc., Chem. Commun.* p. 552 (1976).
132. Frueh, A. J., and Gray, N., *Acta Crystallogr., Sect. B* **24,** 156 (1968).
133. Funke, K., *Prog. Solid State Chem.* **11,** 345 (1975).
134. Gadzhiev, S. M., Bakhyshov, R. G., Suleimanov, D. M., and Kuliev, A. A., *Russ. J. Phys. Chem.* **45,** 1518 (1971).
135. Geller, S., in "Solid Electrolytes" (S. Geller, ed.), pp. 51–65. Springer-Verlag, Berlin and New York, 1977.
136. Gerzanich, E. I., *Inorg. Mater.* **6,** 1403 (1970).
137. Gerzanich, E. I., Buturlakin, A. P., and Chepur, D. V., *Izv. Akad. Nauk SSSR, Ser. Fiz.* **39,** 774 (1975).
138. Glukhov, I. A., *Izv. Otd. Estestven. Nauk Akad. Nauk Tadzhik, SSR* **24,** 21 (1957).
139. Glukhov, I. A., Davidyants, S. B., El'manova, N. A., and Yunusov, M. A., *Russ. J. Inorg. Chem.* **8,** 47 (1963).
140. Glukhov, I. A., Davidyants, S. B., Yunusov, M. A., and El'manova, N. A., *Russ. J. Inorg. Chem.* **6,** 649 (1961).
141. "Gmelins Handbuch der Anorganischen Chemie," Syst. No. 34, "Quecksilber," 8th ed., Vol. B3, pp. 1034–1042, 1104–1109, 1169–1172. Verlag Chemie, Weinheim, 1968.
142. "Gmelins Handbuch der Anorganischen Chemie," Syst. No. 14, "Kohlenstoff," 8th ed., Vol. D6, pp. 143–161, 163, 165–171. Springer-Verlag, Berlin and New York, 1978.
143. "Gmelins Handbuch der Anorganischen Chemie," Syst. No. 15, "Silicium," 8th ed., Vol. B, pp. 753–754. Verlag Chemie, Weinheim, 1959.
144. "Gmelins Handbuch der Anorganischen Chemie," Syst. No. 46, "Zinn," 8th ed., Vol. C2, pp. 74–75, 78–79, 123–124, 209–211. Springer-Verlag, Berlin and New York, 1975.
145. "Gmelins Handbuch der Anorganischen Chemie," Syst. No. 47, "Blei," 8th ed., Vol. C2, pp. 605–607, Verlag Chemie, Weinheim, 1969.

146. "Gmelins Handbuch der Anorganischen Chemie," Syst. No. 16, "Phosphor," 8th ed., Vol. C, pp. 585–588, 590–596, 599–601, 603–604, 609. Verlag Chemie, Weinheim, 1965.
147. "Gmelins Handbuch der Anorganischen Chemie," Syst. No. 18, "Antimon," 8th ed., Vol. B3, pp. 548–550, 555. Gmelin Verlag, Clausthal-Zellerfeld, 1949.
148. "Gmelins Handbuch der Anorganischen Chemie," Syst. No. 19, "Wismut," 8th ed., Vol. B1, pp. 168–169, 175. Verlag Chemie, Weinheim, 1927.
149. Goodman, C. H. L., *J. Phys. Chem. Solids* **6**, 305 (1958).
150. van Gool, W., *Annu. Rev. Mater. Sci.* **4**, 311 (1974).
151. Grekov, A. A., Malitskaya, M. A., and Fridkin, V. M., *Sov. Phys. Crystallogr.* **17**, 504 (1972).
152. Gurrieri, S., *Boll. Sedute Accad. Gioenia Sci. Nat. Catania* **72**, 667 (1960).
153. Haase-Wessel, W., Thesis, Göttingen, 1973.
154. Haase-Wessel, W., *Naturwissenschaften* **60**, 474 (1973).
155. Haendler, H. M.; Mootz, D., Rabenau, A., and Rosenstein, G., *J. Solid State Chem.* **10**, 175 (1974).
156. Häfele, H. G., Wachernig, H., Irslinger, C., Grisar, R., and Nitsche, R., *Phys. Status Solidi* **42**, 531 (1970).
157. Hagenmüller, P., and Rouxel, J., *C. R. Acad. Sci.* **250**, 1859 (1960).
158. Hagenmüller, P., Rouxel, J., David, J., and Colin, A., *C. R. Acad. Sci.* **253**, 667 (1961).
159. Hagenmüller, P., Rouxel, J., David, J., Colin, A., and LeNeindre, B., *Z. Anorg. Allg. Chem.* **323**, 1 (1963).
160. Hahn, H., and Katscher, H., *Z. Anorg. Allg. Chem.* **321**, 85 (1963).
161. Hahn, H., and Nickels, W., *Z. Anorg. Allg. Chem.* **304**, 100 (1960).
162. Hahn, H., and Nickels, W., *Z. Anorg. Allg. Chem.* **314**, 307 (1962).
163. Hamano, K., Nakamura, T., Ishibashi, Y., and Ooyane, T., *J. Phys. Soc. Jpn.* **20**, 1886 (1965).
164. Hardt, H. D., and Scheepker, H., *Naturwissenschaften* **57**, 39 (1970).
165. Hardy, A., and Cottreau, D., *C. R. Acad. Sci., Ser. C* **262**, 739 (1966).
166. Havrankova, V., and Horak, J., *Collect. Czech. Chem. Commun.* **31**, 1256 (1966).
167. Holzäpfel, G., and Rickert, H., *in* "Advances in Solid State Physics" (H. J. Queisser, ed.), Vol. XV, pp. 317–349. Pergamon/Vieweg, Braunschweig, 1975.
168. Holzäpfel, G., and Rickert, H., *Naturwisschenschaften* **64**, 53 (1977).
169. Horak, J., Kozakova, M., and Klazar, J., *Collect. Czech. Czech. Chem. Commun.* **37**, 2309 (1972).
170. Horak, J., and Rodot, H., *C. R. Acad. Sci., Ser. B* **267**, 363 (1968).
171. Horak, J., Turjanica, I. D., Klazar, J., and Kozakova, M., *Krist. Tech.* **3**, 241 (1968).
172. Horak, J., Turjanica, I. D., and Nejezchleb, K., *Krist. Tech.* **3**, 231 (1968).
173. Itoh, K., Matsunaga, H., and Nakamura, E., *J. Phys. Soc. Jpn.* **41**, 1679 (1976).
174. Itoh, K., Ogusu, K., Shiozaki, Y., and Toyoda, K., *Ferroelectrics* **7**, 79 (1974).
175. Iwata, Y., Fukui, S., Koyano, N., and Shibuya, I., *J. Phys. Soc. Jpn.* **21**, 1846 (1966).
176. Iwata, Y., Koyano, N., and Shibuya, I., *J. Phys. Soc. Jpn.* **20**, 875 (1965).
177. Iwata, Y., Koyano, N., and Shibuya, I., *Annu. Rep. Res. React. Inst., Kyoto Univ.* **9**, 9 (1976).
178. Jander, G., and Brodersen, K., *Z. Anorg. Allg. Chem.* **264**, 57 (1951).
179. Jander, G., and Brodersen, K., *Z. Anorg. Allg. Chem.* **264**, 76 (1951).
180. Karbanov, S., Boncheva-Mladenova, Z., and Aramov, N., *Monatsh. Chem.* **103**, 1496 (1972).
181. Katscher, H., and Hahn, H., *Naturwissenschaften* **53**, 361 (1966).

182. Keppert, D. L., "The Early Transition Metals," Academic Press, New York, 1972.
183. Kern, R., *J. Phys. Chem. Solids* **23**, 249 (1962).
184. Kikuchi, A., Oka, Y., and Sawaguchi, E., *J. Phys. Soc. Jpn.* **23**, 337 (1967).
185. Kirilenko, V. V., and Dembovskii, S. A., *Fiz. Khim. Stekla* **1**, 225 (1975).
186. Kirillovich, E. V., *Mater. Vses. Nauchn. Stud. Konf. Khim., 13th* p. 27 (1975).
187. Klazar, J., and Horak, J., *Collect. Czech. Chem. Commun.* **33**, 973 (1968).
188. Kniep, R., Mootz, D., and Rabenau, A., *Z. Anorg. Allg. Chem.* **422**, 17 (1976).
189. Köhler, K., and Breitinger, D., *Naturwissenschaften* **61**, 684 (1974).
190. Kolomiets, B. T., *Phys. Status Solidi* **7**, 359, 713 (1964).
191. Kozakova, M., Horak, J., and Klikorka, J., *Z. Chem.* **6**, 431 (1966).
192. Krämer, V., *J. Appl. Crystallogr.* **6**, 499 (1973).
193. Krämer, V., *Z. Naturforsch., Teil B* **29**, 688 (1974).
194. Krämer, V., *Z. Naturforsch., Teil B* **31**, 1542 (1976).
195. Krebs, H., "Fundamentals of Inorganic Crystal Chemistry," McGraw-Hill, London, 1968.
196. Krebs, B., *Z. Anorg. Allg. Chem.* **396**, 137 (1973).
197. Krebs, B., Buss, B., and Altena, D., *Z. Anorg. Allg. Chem.* **386**, 257 (1971).
198. Kulakov, M. P., and Sokolovskaya, Zh.D., *Inorg. Mater.* **7**, 1282 (1971).
199. Kuliev, A. A., Gadzhiev, S. M., Bakhyshov, R. G., and Khudiev, Kh. G., *Izv. Akad. Nauk Azerb. SSR, Ser. Fiz.-Tekh. Mat. Nauk* p. 126 (1970).
200. Kuliev, A. A., Gadzhiev, S. M., Filatova, S. I., Suleimanova, D. M., and Zaidova, G. A., *Izv. Vyssh. Uchebn. Zaved. Khim. Khim. Tekhnol.* **14**, 1619 (1971).
201. Kulieva, S. A., and Kuliev, A. A., *Uch. Zap. Azerb. Gos. Univ. Ser. Khim. Nauk* p. 38 (1974).
202. Kulieva, S. A., Kuliev, A. A., and Gadzhiev, S. M., *Azerb. Khim. Zh.* p. 131 (1974).
203. Lenher, V., *J. Am. Chem. Soc.* **23**, 680 (1902).
204. Leonova, T. M., and Sviridov, V. V., *Vestsi Akad. Navuk B. SSR, Ser. Khim. Navuk* p. 40 (1966).
205. Leonova, T. M., and Sviridov, V. V., *Inorg. Mater.* **5**, 1016 (1969).
206. Leonova, T. M., and Sviridov, V. V., *Vestn. Beloruss. Univ.* **2**, 23 (1970).
207. Leonova, T. M., and Sviridov, V. V., *Fotokhim. Radiats.-Khim. Protsessy Vodn. Rastvorakh. Tverd. Telakh.* p. 46 (1970).
208. Levayer-Cauquais, C., Thesis, University of Nantes, *Nantes Impr. Fac. Sci.* (s.d.) 1971.
209. Liang, C. C., *in* "Fast Ion Transport in Solids" (W. van Gool, ed.), pp. 19–31. North-Holland Publ., Amsterdam, 1973.
210. Loginova, M. V., and Andreev, Yu. V., *Inorg. Mater.* **6**, 1661 (1970).
211. Malkova, T. I., and Latypov, Z. M., *Inorg. Mater.* **10**, 1057 (1974).
212. Malkova, T. I., and Latypov, Z. M., *Izv. Vyssh. Uchebn. Zaved. Khim. Khim. Tekhnol.* **17**, 933 (1974).
213. Maneglier-Lacordaire, S., Rivet, J., and Flahaut, J., *Ann. Chim. (Paris)* **10**, 291 (1975).
214. Marchenko, V. I., Gavrilenko, N. V., Moik, N. B., and Zhornovyi, I. V., *Pishch. Promst. Kaz. Mezhved. Resp. Nauchno Tekh. Sb.* p. 53 (1974).
215. Marcoll, J. D., Thesis, Stuttgart, 1975.
216. Marcoll, J. D., Mootz, D., Rabenau, A., and Wunderlich, H., *Eur. Crystallogr. Meet. 2nd* p. 247 (1974).
217. Marcoll, J. D., Rabenau, A., Mootz, D., and Wunderlich, H., *Rev. Chim. Miner.* **11**, 607 (1974).
218. Marinkovic, V., Pejovski, S., and Vene, N., *J. Cryst. Growth* **44**, 615 (1978).

219. Mariolacos, K., *Acta Crystallogr., Sect. B* **32,** 1947 (1976).
220. Masuda, Y., Sakata, K., Hasegawa, S., Ohara, G., Wada, M., and Nishizawa, M., *Jpn. J. Appl. Phys.* **8,** 692 (1969).
221. Matyas, M., and Horak, J., *Phys. Status Solidi A* **36,** к137 (1976).
222. Mazhara, A. P., Opalovskii, A. A., Fedorov, V. E., and Kirik, S. D., *Russ. J. Inorg. Chem.* **22,** 991 (1977).
223. Meerschaut, A., Palvadeau, P., and Rouxel, J., *J. Solid State Chem.* **20,** 21 (1977).
224. Merz, W. J., *Festkörperprobleme* **4,** 101 (1965).
225. Miehe, G., and Kupcik, V., *Naturwissenschaften* **58,** 219 (1971).
226. Mininzon, Yu. M., Popolitov, V. I., and Lobachev, A. N., *Sov. Phys. Crystallogr.* **22,** 717 (1977).
227. Molnar, B., Johannes, R., and Haas, W., *Bull. Am. Phys. Soc.* **10,** 109 (1965).
228. Mooser, E., and Pearson, W. B., *J. Phys. Chem. Solids* **7,** 65 (1958).
229. Mootz, D., Rabenau, A., Wunderlich, H., and Rosenstein, G., *J. Solid State Chem.* **6,** 583 (1973).
230. Moritani, T., Kuchitsu, K., and Morino, Y., *Inorg. Chem.* **10,** 344 (1971).
231. Morozov, I. S., and Li Ch'ih-fa, *Russ. J. Inorg. Chem.* **8,** 878 (1963).
232. Mott, N. F., *in* "Amorphous and Liquid Semiconductors" (W. E. Spear, ed.), pp. 497 –503. University of Edinburgh, Edinburgh, 1977.
233. Nakamura, S., Takei, K., and Mizuno, K., *Radioisotopes* **18,** 423 (1969).
234. Nako, K., and Balkanski, M., *Phys. Rev. B* **8,** 5759 (1973).
235. Nassau, K., Shiever, J. W., and Kowalchik, M., *J. Cryst. Growth* **7,** 237 (1970).
236. Neels, H., and Schmitz, W., *Krist. Tech.* **3,** к85 (1968).
237. Nejezehleb, K., and Horak, J., *Czech. J. Phys.* **18,** 138 (1968).
238. Nejezehleb, K., Turjanica, I. D., and Horak, J., *Collect. Czech. Chem. Commun.* **33,** 674 (1968).
239. Nguyen-Huy-Dung, and Thevet, F., *Acta Crystallogr., Sect. B* **32,** 1108 (1976).
240. Nguyen-Huy-Dung, and Thevet, F., *Acta Crystallogr., Sect. B* **32,** 1112 (1976).
241. Nitsche, R., *Fortschr. Mineral.* **44,** 231 (1967).
242. Nitsche, R., *Mater. Res. Bull.* **7,** 679 (1972).
243. Nitsche, R., and Merz, W. J., *J. Phys. Chem. Solids* **13,** 154 (1960).
244. Nitsche, R., Roetschi, H., and Wild, P., *Appl. Phys. Lett.* **4,** 210 (1964).
245. Noval'kovskii, N. P., Belotskii, D. P., Gavrilenko, N. V., and Chervenyuk, G. I., *Tezisy Dokl. Vses. Konf. Kristallokhim. Intermet. Soedin., 2nd* (R. M. Rykhal, ed.), p. 173. Lvov, USSR, 1974.
246. Novoselova, A. V., Odin, I. N., and Popovkin, B. A., *Inorg. Mater.* **2,** 1193 (1966).
247. Novoselova, A. V., Odin, I. N., and Popovkin, B. A., *Russ. J. Inorg. Chem.* **14,** 1402 (1969).
248. Novoselova, A. V., Odin, I. N., and Popovkin, B. A., *Inorg. Mater.* **6,** 113 (1970).
249. Novoselova, A. V., Odin, I. N., and Popovkin, B. A., *Inorg. Mater.* **6,** 224 (1970).
250. Novoselova, A. V., Odin, I. N., and Popovkin, B. A., *Inorg. Mater.* **6,** 332 (1970).
251. Novoselova, A. V., Odin, I. N., Trifonov, V. A., and Popovkin, B. A., *Inorg. Mater.* **3,** 1827 (1967).
252. Novoselova, A. V., Odin, I. N., Valitova, N. R., and Popovkin, B. A., *Inorg. Mater.* **4,** 680 (1968).
253. Novoselova, A. V., Todriya, M. K., Odin, I. N., and Popovkin, B. A., *Inorg. Mater.* **7,** 437 (1971).
254. Novoselova, A. V., Todriya, M. K., Odin, I. N., and Popovkin, B. A., *Inorg. Mater.* **7,** 1125 (1971).

255. Novoselova, A. V., Zlomanov, V. P., Popovkin, B. A., and Tananaeva, O. I., *Eur. Conf. Cryst. Growth, 1st, Abstract Book,* Post Deadline Paper, Zürich, 1976.
256. Odin, I. N., Popovkin, B. A., and Novoselova, A. V., *Inorg. Mater.* **6,** 424 (1970).
257. Oka, Y., Kikuchi, A., Mori, T., and Sawaguchi, E., *J. Phys. Soc. Jpn.* **21,** 405 (1966).
258. Omote, O., and Ito, K., *Shinku Kagaku* **13,** 110 (1965).
259. Onopko, L. V., Onopko, V. V., Chepur, D. V., Dovgoshei, N. I., Turyanitsa, I. D., and Zayachkovskaya, N. F., *Inorg. Mater.* **10,** 975 (1974).
260. Opalovskii, A. A., and Fedorov, V. E., *Dokl. Chem.* **176,** 810 (1967).
261. Opalovskii, A. A., Fedorov, V. E., and Khaldoyanidi, K. A., *Dokl. Chem.* **182,** 907 (1968).
262. Opalovskii, A. A., Fedorov, V. E., and Lobkov, E. U., *Izv. Sib. Otd. Akad. Nauk SSSR, Ser. Khim. Nauk* p. 144 (1971).
263. Opalovskii, A. A., Fedorov, V. E., and Lobkov, E. U., *Russ. J. Inorg. Chem.* **16,** 790 (1971).
264. Opalovskii, A. A., Fedorov, V. E., Lobkov, E. U., and Erenburg, B. G., *Russ. J. Inorg. Chem.* **16,** 1685 (1971).
265. Opalovskii, A. A., Fedorov, V. E., Mazhara, A. P., and Cheremisina, I. M., *Russ. J. Inorg. Chem.* **17,** 1510 (1972).
266. Palvadeau, P., and Rouxel, J., *Bull. Soc. Chim. Fr.* p. 2698 (1967).
267. Penney, G. J., and Sheldrick, G. M., *Acta Crystallogr., Sect. B* **26,** 2092 (1970).
268. Penney, G. J., and Sheldrick, G. M., *J. Chem. Soc. A* p. 1100 (1971).
269. Perrin, C., Chevrel, R., and Sergent, M., *C. R. Acad. Sci., Ser. C* **280,** 949 (1975).
270. Perrin, C., Chevrel, R., and Sergent, M., *C. R. Acad. Sci., Ser. C* **281,** 23 (1975).
271. Perrin, C., Perrin, A., and Prigent, J., *Bull. Soc. Chim. Fr.* p. 3086 (1972).
272. Perrin, C., Sergent, M., and Prigent, J., *C. R. Acad. Sci., Ser. C* **277,** 465 (1973).
273. Perrin, C., Sergent, M., Le Traon, F., and Le Traon, A., *J. Solid State Chem.* **25,** 197 (1978).
274. von der Pfordten, O., *Justus Liebigs Ann. Chem.* **234,** 257 (1886).
275. Pohl, S., *Angew. Chem. Int. Ed. Engl.* **15,** 162 (1976).
276. Poleck, T., and Goercki, C., *Ber. Dtsch. Chem. Ges.* **21,** 2412 (1888).
277. Poltavtsev, Yu. G., *Inorg. Mater.* **11,** 1492 (1975).
278. Poltavtsev, Yu. G., and Pozdnyakova, V. M., *Inorg. Mater.* **9,** 766 (1973).
279. Popolitov, V. I., *Sov. Phys. Crystallogr.* **14,** 312 (1969).
280. Popolitov, V. I., and Litvin, B. N., *Sov. Phys. Crystallogr.* **13,** 483 (1968).
281. Popolitov, V. I., and Litvin, B. N., *Sov. Phys. Crystallogr.* **15,** 1116 (1970).
282. Popolitov, V. I., and Litvin, B. N., *in* "Studies in Soviet Science: Crystallization Processes under Hydrothermal Conditions" (A. N. Lobachev, ed.), pp. 57–72. Consultants Bureau, New York, 1973.
283. Popolitov, V. I., Litvin, B. N., and Lobachev, A. N., *Phys. Status Solidi A* **3,** к1 (1970).
284. Popolitov, V. I., and Lobachev, A. N., *Inorg. Mater.* **8,** 1082 (1972).
285. Popolitov, V. I., and Lobachev, A. N., *Inorg. Mater.* **8,** 1389 (1972).
286. Popolitov, V. I., and Lobachev, A. N., *Inorg. Mater.* **9,** 191 (1973).
287. Popolitov, V. I., Zver'kova, O. N., and Lobachev, A. N., *J. Appl. Chem. USSR* **47,** 1533 (1974).
288. Puff, H., *Angew. Chem.* **75,** 681 (1963).
289. Puff, H., Harpain, A., and Hoop, K. P., *Naturwissenschaften* **53,** 274 (1966).
290. Puff, H., and Kohlschmidt, R., *Naturwissenschaften* **49,** 299 (1962).
291. Puff, H., and Küster, J., *Naturwissenschaften* **49,** 299 (1962).

292. Puff, H., and Küster, J., *Naturwissenschaften* **49**, 464 (1962).
293. Puga, G. D., Kovach, D. Sh., Turyanitsa, I. D., Borets, A. N., and Chepur, D. V., *Ukr. Fiz. Zh.* (*Russ. Ed.*) **16**, 276 (1971).
294. Puga, G. D., Puga, P. P., Maksimets, V. V., Borets, A. N., Groshik, I. I., Bercha, D. M., and Chepur, D. V., *Ferroelectrics* **6**, 111 (1974).
295. Quinn, R. K., and Johnson, R. T., *J. Non-Cryst. Solids* **7**, 53 (1972).
296. "Problems of Nonstoichiometry" (A. Rabenau, ed.). North-Holland Publ., Amsterdam, 1970.
297. Rabenau, A., *in* "Crystal Growth: an Introduction" (P. Hartmann, ed.), pp. 198–209. North-Holland Publ., Amsterdam, 1973.
298. Rabenau, A., *in* "Die feste Materie" (L. Genzel, ed.), pp. 223–246. Umschau Verlag, Frankfurt/Main, 1973.
299. Rabenau, A., *Rost Krist.* **13**, (1979), in press.
300. Rabenau, A., and Rau, H., *Inorg. Synth.* **14**, 160 (1973).
301. Rabenau, A., and Rau, H., *Phillips Techn. Rev.* **30**, 89 (1969).
302. Rabenau, A., and Rau, H., *Z. Anorg. Allg. Chem.* **369**, 295 (1969).
303. Rabenau, A., Rau, H., and Rosenstein, G., *Naturwissenschaften* **55**, 82 (1968).
304. Rabenau, A., Rau, H., and Rosenstein, G., *Naturwissenschaften* **56**, 137 (1969).
305. Rabenau, A., Rau, H., and Rosenstein, G., *Angew. Chem. Int. Ed. Engl.* **8**, 145 (1969).
306. Rabenau, A., Rau, H., and Rosenstein, G., *Solid State Commun.* **7**, 1281 (1969).
307. Rabenau, A., Rau, H., and Rosenstein, G., *Z. Anorg. Allg. Chem.* **374**, 43 (1970).
308. Rabenau, A., Rau, H., and Rosenstein, G., *J. Less-Common Met.* **21**, 395 (1970).
309. Rabenau, A., Rau, H., and Rosenstein, G., *Monatsh. Chem.* **102**, 1425 (1971).
310. Raghava Rao, B. S. V., and Watson, H. E., *J. Indian Inst. Sci., Sect. A* **12**, 17 (1929); *J. Phys. Chem.* **32**, 1354 (1928).
311. Rannou, J. P., and Sergent, M., *C. R. Acad. Sci., Ser. C* **265**, 734 (1967).
312. Rau, H., and Rabenau, A., *Solid State Commun.* **5**, 331 (1967).
313. Ray, P. C., *J. Chem. Soc.* **111**, 101 (1917).
314. Reisman, A., and Berkenblit, M., *J. Electrochem. Soc.* **109**, 1111 (1962).
315. Reuter, B., and Hardel, K., *Angew. Chem.* **72**, 138 (1960).
316. Reuter, B., and Hardel, K., *Naturwissenschaften* **48**, 161 (1961).
317. Reuter, B., and Hardel, K., *Z. Anorg. Allg. Chem.* **340**, 158 (1965).
318. Reuter, B., and Hardel, K., *Z. Anorg. Allg. Chem.* **340**, 168 (1965).
319. Reuter, B., and Hardel, K., *Ber. Bunsenges. Phys. Chem.* **70**, 82 (1966).
320. Rice, D. A., *Coord. Chem. Rev.* **25**, 199 (1978).
321. Rigin, V. I., and Batsanov, S. S., *Russ. J. Inorg. Chem.* **10**, 950 (1965).
322. Rijnsdorp, J., *Int. Conf. Solid Comp. Trans. Elements, 5th, Extended Abstracts,* p. 45. Upplands Grafiska AB, Uppsala, 1976.
323. Rodionov, Yu. I., Klokman, V. R., and Myakishev, K. G., *Russ. J. Inorg. Chem.* **17**, 440 (1972).
324. Roos, G., Eulenberger, G., and Hahn, H., *Naturwissenschaften* **59**, 363 (1972).
325. Roos, G., Eulenberger, G., and Hahn, H., *Z. Anorg. Allg. Chem.* **396**, 284 (1973).
326. Rose, H., *Ann. Phys.* **13**, 59 (1828).
327. Rouxel, J., *Ann. Chim.* (*Paris*) **7**, 49 (1962).
328. Rouxel, J., and Palvadeau, P., *Bull. Soc. Chim. Fr.* p. 2044 (1966).
329. Ruff, O., and Neumann, F., *Z. Anorg. Allg. Chem.* **128**, 81 (1923).
330. Safonov, V. V., Chernykh, S. M., and Korshunov, B. G., *Russ. J. Inorg. Chem.* **23**, 271 (1978).

331. Safonov, V. V., Chernykh, S. M., Korshunov, B. G., and Ksenzenko, V. I., *Russ. J. Inorg. Chem.* **22,** 438 (1977).
332. Safonov, V. V., Vasilishcheva, I. V., and Korshunov, B. G., *Russ. J. Inorg. Chem.* **16,** 1232 (1971).
333. Savchenko, N. D., Dovgoshei, N. I., Turyanitsa, I. D., Golovei, M. I., Chepur, D. V., and Semrad, E. E., *Fiz. Khim. Tverd. Tela* **6,** 129 (1975).
334. Savigny, N., Adolphe, C., Zalkin, A., and Templeton, D. H., *Acta Crystallogr., Sect. B* **29,** 1532 (1973).
335. Savigny, N., Laruelle, P., and Flahaut, J., *Acta Crystallogr., Sect. B* **29,** 345 (1973).
336. Schäfer, H., "Chemical Transport Reactions," Academic Press, New York, 1964.
337. Schäfer, H., *J. Cryst. Growth* **9,** 17 (1971).
338. Schäfer, H., in "Crystal Growth: an Introduction" (P. Hartman, ed.), pp. 143–151. North-Holland Publ., Amsterdam, 1973.
339. Schäfer, H., and Beckmann, W., *Z. Anorg. Allg. Chem.* **347,** 225 (1966).
340. Schäfer, H., and von Schnering, H. G., *Angew. Chem.* **76,** 833 (1964).
341. Schmidt, C., and Gmelin, E., *Solid State Commun.* **21,** 987 (1977).
342. Schmidt, C., Gmelin, E., and van Alpen, U., *Nuovo Cimento Soc. Ital. Fis. B* **38,** 206 (1977).
343. Schmidt, M., and Siebert, W., *Angew. Chem. Int. Ed. Engl.* **3,** 637 (1964).
344. Schmidt, M., and Siebert, W., *Z. Anorg. Allg. Chem.* **345,** 87 (1966).
345. Schmidt, M., and Siebert, W., *Angew. Chem. Int. Ed. Engl.* **5,** 597 (1966).
346. Schmidt, M., Siebert, W., and Gast, E., *Z. Naturforsch., Teil B* **22,** 557 (1967).
347. von Schnering, H. G., and Beckmann, W., *Z. Anorg. Allg. Chem.* **347,** 231 (1966).
348. Scholz, H., *Phillips Techn. Rev.* **28,** 316 (1967).
349. Scholz, H., *Acta Electron.* **17,** 69 (1974).
350. Scholz, H., and Kluckow, R., in "Crystal Growth" (H. S. Peiser, ed.), pp. 475–482. Pergamon, Oxford, 1967.
351. Schumb, W. C., and Bernard, W. J., *J. Am. Chem. Soc.* **77,** 862 (1955).
352. Scott, J. F., *Rev. Mod. Phys.* **46,** 83 (1974).
353. Sergeevna, G. E., *Mater. Vses. Nauchn. Stud. Konf.: Khim., 13th.* p. 26 (1975).
354. Sergent, M., Fischer, O., Decroux, M., Perrin, C., and Chevrel, R., *J. Solid State Chem.* **22,** 87 (1977).
355. Sfez, G., and Adolphe, C., *Bull. Soc. Fr. Mineral. Cristallogr.* **96,** 37 (1973).
356. Sharma, K. M., Anand, S. K., Multani, R. K., and Jain, B. D., *Chem. Ind. (London)* p. 1556 (1969).
357. Simon, A., von Schnering, H. G., and Schäfer, H., *Z. Anorg. Allg. Chem.* **355,** 295 (1967).
358. Simon, G., and Zeller, G. R., *J. Phys. Chem. Solids* **35,** 187 (1974).
359. Singleton, D. L., and Stafford, F. E., *Inorg. Chem.* **11,** 1208 (1972).
360. Sinha, K. P., and Biswas, A. B., *J. Chem. Phys.* **23,** 404 (1955).
361. Sinitsyna, S. M., Khlebodarov, V. G., and Bukhtereva, N. A., *Russ. J. Inorg. Chem.* **20,** 1267 (1975).
362. Smith, E. F., and Oberholtzer, V., *Z. Anorg. Allg. Chem.* **5,** 63 (1894).
363. Solvay & Cie., Netherlands Patent 288,255 (May 11, 1964; Appl. Jan. 28, 1963).
364. Stork-Blaisse, B. A., and Romers, C., *Acta Crystallogr., Sect. B* **27,** 386 (1971).
365. Strähler, A., and Bachran, F., *Ber. Dtsch. Chem. Ges.* **44,** 2906 (1911).
366. Street, G. B., Bingham, R. L., Crowley, J. I., and Kuyper, J., *J. Chem. Soc., Chem. Commun.* p. 464 (1977).
367. A. J. C. Wilson, ed., "Structure Reports," Vol. 13, p. 206. Oosthoek, Utrecht, 1954.

368. Tai, H., and Hori, S., *Nippon Kinzoku Gakkaishi* **40,** 722 (1976).
369. Takahashi, T., Kuwabara, K., Yamamoto, O., and Watanabe, S., *Denki Kagaku* **37,** 717 (1969).
370. Takahashi, T., and Yamamoto, O., *Denki Kagaku* **32,** 610 (1964).
371. Takahashi, T., and Yamamoto, O., *Denki Kagaku* **32,** 664 (1964).
372. Takahashi, T., and Yamamoto, O., *Denki Kagaku* **33,** 346 (1965).
373. Takahashi, T., Yamamoto, O., and Mori, H., *Denki Kagaku* **35,** 181 (1967).
374. Takahashi, T., and Yamamoto, O., U.S. Patent 3,558,357 (Jan. 26, 1971); Jpn. Appl. Mar. 30, 1968.
375. Takei, K., *Bull. Chem. Soc. Jpn.* **28,** 403 (1955).
376. Takei, K., *Bull. Chem. Soc. Jpn.* **28,** 406 (1955).
377. Takei, K., *Bull. Chem. Soc. Jpn.* **28,** 408 (1955).
378. Takei, K., *Nippon Kagaku Zasshi* **77,** 830 (1956).
379. Takei, K., *Nippon Kagaku Zasshi* **77,** 965 (1956).
380. Takei, K., and Hagiwara, H., *Radioisotopes* **24,** 715 (1975).
381. Takei, K., and Hagiwara, H., *Bull. Chem. Soc. Jpn.* **49,** 1425 (1976).
382. Takei, K., Hagiwara, H., and Tanaka, H., *Bull. Chem. Soc. Jpn.* **50,** 1341 (1977).
383. Tananaeva, O. I., Novoselova, A. V., and Kul'bachevska, E. V., *Inorg. Mater.* **13,** 434 (1977).
384. Taraskin, S. A., Lyakhovitskaya, V. A., and Ivanov-Shits, A. K., *Sov. Phys. Crystallogr.* **17,** 597 (1972).
385. Thevet, F., Nguyen, H. D., and Dagron, C., *C. R. Acad. Sci., Ser. C* **275,** 1279 (1972).
386. Thevet, F., Nguyen, H. D., and Dagron, C., *C. R. Acad. Sci. Ser. C* **276,** 1787 (1973).
387. Thevet, F., Nguyen, H. D., and Dagron, C., *C. R. Acad. Sci. Ser. C* **281,** 865 (1975).
388. Thevet, F., Nguyen, H. D., Dagron, C., and Flahaut, J., *J. Solid State Chem.* **18,** 175 (1976).
389. Thiele, G., Köhler-Degner, M., Wittmann, K., and Zoubek, G., *Angew. Chem. Int. Ed. Engl.* **17,** 852 (1978).
390. Tomokiyo, A., Okada, T., and Kawano, S., *Jpn. J. Appl. Phys.* **16,** 291 (1977).
391. Tomura, H., and Mori, T., *J. Phys. Soc. Jpn.* **19,** 1247 (1964).
392. Trifonov, V. A., Dernovskii, V. I., Popovkin, B. A., Lyakhovitskaya, V. A., Belousov, V. I., and Novoselova, A. V., *Russ. J. Inorg. Chem.* **48,** 458 (1974).
393. Tronev, V. G., Bekhtle, G. A., and Davidyants, S. B., *Tr. Akad. Nauk Tadzh. SSR* **84,** 105 (1958).
394. Truthe, W., *Z. Anorg. Chem.* **76,** 161 (1912).
395. Turjanica, I. D., Horak, J., and Kozakova, M., *Collect. Czech. Chem. Commun.* **33,** 300 (1968).
396. Turjanica, I. D., Koperles, V. M., and Chepur, D. V., *Fiz. Elektron. (Lvov)* **8,** 30 (1974).
397. Turjanica, I. D., Olekseyuk, I. D., and Kozmanko, I. I., *Inorg. Mater.* **9,** 1275 (1973).
398. Turjanica, I. D., Zayachkovskii, N. F., and Kozmanko, I. I., *Inorg. Mater.* **10,** 1617 (1974).
399. Turyanitsa, I. D., Zhdankin, A. P., Dovgoshei, N. I., Gryadil, I. A., and Chepur, D. V., *Poluprovodn. Tekh. Mikroelektron.* p. 82 (1974).
400. Urazov, G. G., and Celidse, L. A., *Izv. Akad. Nauk SSSR, Sekt. Fiz. Khim. Anal.* **13,** 263 (1940).
401. Urazov, G. G., and Sokolova, M. A., *Izv. Akad. Nauk SSSR, Sekt. Fiz. Khim. Anal.* **14,** 317 (1941).
402. Valitova, N. R., Aleshin, V. A., Popovkin, B. A., and Novoselova, A. V., *Inorg. Mater.* **12,** 194 (1976).

403. Valitova, N. R., Popovkin, B. A., Novoselova, A. V., and Aslanov, L. A., *Inorg. Mater.* **9**, 1960 (1973).
404. Velikanov, A. A., and Zinchenko, V. F., *Sov. Electrochem.* **11**, 1733 (1975).
405. Walton, R. A., *in* "Progress in Inorganic Chemistry" (S. J. Lippard, ed.), Vol. 16, pp. 1–226. Wiley (Interscience), New York, 1972.
406. Weissenstein, J., and Horak, J., *Czech. J. Phys.* **24**, 235 (1974).
407. Wells, A. F., "Structural Inorganic Chemistry," London and New York, 1975. 4th ed., Oxford Univ. Press (Clarendon).
408. Whitmore, D. H., *J. Cryst. Growth* **39**, 160 (1977).
409. Wiberg, E., and Sturm, W., *Z. Naturforsch., Teil B* **8**, 529 (1953).
410. Wiberg, E., and Sturm, W., *Z. Naturforsch., Teil B* **10**, 112 (1955).
411. Wiberg, E., and Sturm, W., *Angew. Chem.* **67**, 483 (1955).
412. Winterhager, H., and Kammel, R., *Z. Erzbergbau Metallhuettenwes.* **9**, 97 (1956).
413. Wright, D. A., and Penfold, B. R., *Acta Crystallogr.* **12**, 455 (1959).
414. Yamamoto, O., and Takahashi, T., *Denki Kagaku* **34**, 833 (1966).
415. Yushina, L. D., Karpachev, S. V., and Ovchinnikov, Yu. M., *Sov. Electrochem.* **6**, 1344 (1970).
416. Yushina, L. D., Karpachev, S. V., and Ovchinnikov, Yu. M., *Sov. Electrochem.* **6**, 1379 (1970).
417. Zachariasen, W. H., *Acta Crystallogr.* **2**, 288 (1949).
418. Zadarozhnaya, L. A., Lyachovitskaya, V. A., Givargizov, E. I., and Belyaev, L. M., *J. Cryst. Growth* **41**, 61 (1977).
419. Zaidova, G. A., Kuliev, A. A., and Gadzhiev, S. M., *Uch. Zap. Azerb. Gos. Univ. Ser. Khim. Nauk* p. 51 (1973).

APPENDIX REFERENCES

420. Bercha, D. M., Zayachkovskii, M. P., and Zayachkovskaya, N. F., *Sov. Phys. Solid State* **20**, 1834 (1978).
421. Bhalla, A. S., Spear, K. E., and Cross, L. E., *Mater. Res. Bull.* **14**, 423 (1979).
422. Bullett, D. W., *J. Phys. C*, in press.
423. Greene, R. L., Kwak, J. F., and Fuller, W. W., *J. Phys., Colloq.* (Orsay) **6**, 1401 (1978).
424. Haendler, H. M., and Carkner, P. M., *J. Solid State Chem.* **29**, 35 (1979).
425. Ishikawa, K., Tomoda, W., and Toyoda, K., *Shizuoka Daigaku Denshi Kogaku Kenkyosho Kenkyu Hokoku* **13**, 17 (1978).
426. Krämer, V., *Acta Crystallogr., Sect. B* **35**, 139 (1979).
427. Magistris, A., Chiodelli, G., and Schiraldi, A., *Z. Phys. Chem. Neue Folge* **112**, 251 (1978).
428. Nguyen Tat Dih, Lostak, P., and Horak, J., *Czech. J. Phys., Sect. B* **28**, 1297 (1978).
429. Popolitov, V. I., Lobachev, A. N., Peskin, V. F., and Mininzon, Yu. M., *Ferroelectrics* **21**, 421 (1978).
430. Razzini, G., Lazzari, M., and Scrosati, B., *Electrochim. Acta* **23**, 805 (1978).
431. Rijnsdorp, J., and Jellinek, F., *J. Solid State Chem.* **28**, 149 (1979).
432. Tsigdinos, G. A., *In* "Topics in Current Chemistry" (F. L. Boschke, ed.), Vol. 76, pp. 65–105. Springer-Verlag, New York, 1978.
433. Wichelhaus, W., to be published.
434. Kniep, R., and Wilms, A., *Mater. Res. Bull.* **15**, in press.

Subject Index

A

Aluminum
  chalcogenide halides, 383–384
  reaction of atoms with acetylene, 155
Ammonia synthesis, catalyzed by graphite intercalation compounds, 318
Antimony
  chalcogenide halides, 402–413
    band structure calculations, 412
    Mössbauer studies, 411
    optomechanical properties, 410
    phase diagrams, 406–408
    as semiconductors, 410
    structural data, 408–410
    synthesis, 402–406, 413
  trifluoromethyl compounds, 180
Arene complexes, metal atom synthesis, 145–149
Arenediazo complexes, from oxomolybdenum complexes and hydrazines, 225–226
Arsenic
  chalcogenide halides, 401–402
    phase diagrams, 401, 407–408
    as vitreous semiconductors, 401–402
  trifluoromethyl compounds, 179–180, 189

B

Band structure, of graphite–alkali metal compounds, 287
Biological activity, of organotin compounds, 41–48
Bismuth
  chalcogenide halides, 402, 404–413
    magnetic properties, 412
    phase diagrams, 406–408
    structural data, 408–409, 411, 413
    synthesis, 402–406
  trifluoromethyl compounds, 181, 187
Boron
  chalcogenide halides, 382–383
  difluoride radicals, 207

C

Carbon disulfide
  insertion reactions
    with metal amides, 216, 220–222
    with metal–carbon bonds, 233
    with metal hydrides, 233, 247
    with nickel–aziridine bond, 254
    with platinum–fluorine bond, 261–262
  reactions
    with nickel atoms, 163
    with nucleophiles, 211–215
Carbonyls, matrix isolation studies, 115–118, 130–138
Catalysis, by graphite intercalation compounds, 308, 314, 316
Chalcogenide halide compounds, *see also* individual elements, 329–425
  Chevrel phases, 331, 376–377
  electrical conductivity, 331, 337–339, 342, 346–349
  photoelectric effects, 368, 410
  magnetic properties, 375–376
    temperature-independent paramagnetism, 336, 347, 368
  phototropism, 356–357
  as semiconductors, 368, 395–396, 400, 410–412
    glasses, 390, 401–402
  single crystal preparation, 353–354, 359, 383, 385–386, 404–406
  as superconductors, 331, 375–377
  synthesis
    via Bridgman–Stockbarger method, 331–332, 404–405
    via chemical transport reactions, 330–332, 364, 368–369
    via hydrothermal methods, 332, 350–351, 405
Chemical transport reactions, and synthesis of chalcogenide halide compounds, 330–332, 364, 368–369
Chemisorption
  atom clusters as models, 115
  of carbon monoxide on metals, 116–118
  of copper surface, 117
  of ethylene on metal surfaces, 124, 129
  of nitric oxide on iron, 143
  of nitrogen on iron, 130

Chemisorption (cont'd.)
  of oxygen on rhodium, 120
Chevrel phases, in molybdenum chalcogenide halides, 331, 376–377
Chromium
  chalcogenide halides, 370
  cryochemistry
    atom photoaggregation, 108, 112, 114–115
    dimetal species, 86–87, 97–99
    dinitrogen complexes, 142
    and organic rearrangements, 160–163
    in organometallic synthesis, 145–148, 155–156
  1,1-dithiolato complexes, 221–224
    dithiocarbamates, 221–224
    dithiocarboxylates, 223
    xanthates, 222–224
Cobalt
  cryochemistry
    atom clustering, 86–90
    ethylene complexes, 124–128
    in organometallic synthesis, 146–148, 155, 164–166
    and structure of octacarbonyl, 133–134
  1,1-dithiolato complexes, 248–253
    dithiocarbamates, 248–253
    dithiocarbimate, 252
    1,1-ethylene dithiolates, 249–250
    stereochemical nonrigidity, 253
    xanthates, 251–252
Copper
  chalcogenide halides, 332–339, 412
    free energy of formation, 335
    ionic conductivity, 336–339, 348
    reflectance spectra, 336–338
    structural data, 335–336, 412
    synthesis, 332
  cryochemistry
    acetylene complexes, 153–154
    atom clustering, 92
    carbonyl clusters, 117
    dioxygen compounds and oxides, 139–140
    ethylene complexes, 121–123, 152–153
    phosphine compounds, 167
    selective aggregation of atoms, 103–104

1,1-dithiolato complexes, 254–255, 265–268
  dithiocarbamates, 254, 265, 268
  dithiolenes, 267

D

Dinitrogen complexes
  in cryochemistry, 140–143, 167
  rhenium dithiocarbamate, 235
Dioxygen complexes, in cryochemistry, 118–120, 137–140
1,1-Dithiolato metal complexes, see also individual metals, 211–280
  ligands, 211–215
  stability and metal hardness, 215–216, 218

E

Electrical conductivity
  of chalcogenide halide compounds, 331
    of Group IB, 337–339, 342, 346–349
    photoelectric effects, 368, 410
    semiconductors, 368, 390, 395–396, 400–402, 410–412
    superconductors, 375–377
  of graphite intercalation compounds, 290, 294, 309–310, 312, 317–318
Electrochemistry
  of 1,1-dithiolato complexes, 220, 228–229, 232–233, 245, 247, 261–262
  silver chalcogenide halide batteries, 342, 348–349
  synthesis of graphite acid salts, 289
  use of graphite intercalation compounds
    in batteries, 316–317
    as reversible electrodes, 317
ESCA spectroscopy
  and amino acid complexation of organotin compound, 43
  dithiocarbamate complexes, 237, 262
ESR spectroscopy
  dithiocarbamate complexes, 218–220, 225, 235, 241, 265–268
  matrix isolated species, 81, 95
    calcium silver molecule, 100–101
    chromium cyclophane, 161

copper acetylene complex, 154
manganese hydrides, 167
manganese oxides, 138
silver carbonyl, 134
silver ethylene complex, 152
organotin radicals, 22, 26
and phototropism in mercury chalcogenide halides, 356
and plasma reactions, 191
Ethylene complexes, cryochemistry, 120–130

F

F-Centers, and mercury chalcogenide halides, 356–357
Fischer–Tropsch catalysis, by graphite intercalation compounds, 318
Fluorination of metal alkyls, 197–203
experimental method, 197–198

G

Gallium, chalcogenide halides, 384–386
Germanium
chalcogenide halides, 390
trifluoromethyl compounds, 184, 186–187, 191, 193–194, 197
divalent, 186
via fluorination of alkyls, 198–201
ligand redistribution reactions, 184, 192
reactivity of germanium–carbon bond, 196
tetrakis complex, 178, 181, 184
Gold
chalcogenide halides, 342–348
magnetic properties, 347
as metallic conductors, 346–347
phase diagrams, 334
structural data, 344–348
synthesis, 342–343
cryochemistry
anion formation, 100–101
carbon dioxide compounds, 137–138
carbonyls, 135–138
dioxygen complexes and oxides, 139–140
ethylene complexes, 152–153
lithium compound, 97
phosphine clusters, 164
1,1-dithiolato complexes, 268

Graphite
anomalous diamagnetism, 291
changes on nitration, 290
Raman spectrum, 287
structure, 282
Graphite, intercalation compounds, 281–327
acid salts, 289–290, 316
bonding, 289–290
nitrates, 290
with alkali metals, 285–289
band structure, 287
electronic properties, 287
optical spectra, 287–288
reaction with water, 288
as reductants, 285, 288–289, 316
structures, 286
synthesis, 285
with bromine, 291–294
intercalation mechanism, 293–294
carbon monofluoride, 284–285
as lubricant, 285
structure, 284–285
as catalysts, 308–309, 314, 316, 318
as chemical reagents, 288–290, 315–316
halogenation 299, 315–316
with chlorine, 291–292
with chromium trioxide, 314
covalent, 282–285
electrical conductivity, 290, 294, 309–310, 312, 317–318
electrochemical application, 290, 308, 316–317
with Group IIA metals, 281
with Group V pentahalides, 309–313
antimony, 309–310
arsenic, 310–312
with interhalogens, 295–296
fluorides, 295–296
with krypton difluoride, 300
lamellar, 282–283, 285–314
with lanthanides, 281
with metal halides, 300–314
bonding, 307
bromides, 304
chlorides, 302–304
ferric chloride, 301, 303–308
fluorides, 305, 312–314
from nonaqueous solvents, 301, 306
reduction properties, 308

Graphite, intercalation compounds (cont'd.)
  role of chlorine, 303–304, 307
  structures, 305–306
  synthesis, 301
Mössbauer studies, 299, 307–310
^{19}F-NMR studies, 295, 297–300, 309–310, 313
oxide, 283
Raman studies, 287, 294
residue compounds, 283, 314–315
staging, 282–283
with xenon fluorides, 296–300

H

Hafnium complexes
  arenes via metal atom synthesis, 167
  dithiocarbamates, 216
Hall effect, in graphite intercalation compounds, 290, 294, 307
Hammett parameters in organotin chemistry, 10, 15, 26
Hexafluoroethane
  C—C and C—F bond strengths, 181–182
  and plasma generation of trifluoromethyl radicals, 180–192
Hydrogenation catalysis, by matrix isolated metal clusters, 91

I

Indium, chalcogenide halides, 386–388
  phase diagrams, 387
  structural data, 387–388
  synthesis, 386
Infrared spectroscopy, of matrix isolated species, 81
  arene compounds, 145–147
  ethylene complexes, 122–123, 127–128
  goldlithium molecule, 97
  metal carbonyls, 116–118, 130–136, 141
  metal hydrides, 144–145
  nickel carbon disulfide complex, 163
  plutonium oxide, 140
  rhodium dioxygen clusters, 118–120, 122
Interhalogens, graphite intercalation, 295–296
Iridium
  atoms in matrices, 88–90
  dithiocarbamates, 253–254
Iron
  cryochemistry
    arene complexes, 145–147, 167
    dinitrogen complexes, 142–143
    heteronuclear diatomics, 99–100
    nitrosyls, 143
    oxides, 138
    phosphine complexes, 143–144
    polyene complexes, 155–157, 167–168
  1,1-dithiolato complexes, 236–246
    dithiocarbamates, 236–241, 244–246
    dithiocarboxylates, 243
    1,1-ethene dithiolates, 243–244
    spin equilibria, 237–241
    stereochemical nonrigidity, 253
    xanthates, 241–243

J

Jahn–Teller distortions
  in metal carbonyls, 131, 141
  in palladium telluride iodide, 381
  in titanium dinitrogen complex, 141

K

Kinetics
  of dimerization of silver carbonyl, 118
  of oxidation of nitrosyl ligand, 250
  of oxygen transfer from oxomolybdenum complex, 228
  of photonucleation in matrices, 107

L

Lanthanides
  chalcogenide halides, 357–364
    crystallographic data, 360–364
    structure and lanthanide contraction, 363–364
    synthesis, 357–360
  cryochemistry in organometallic synthesis, 157–158
  graphite intercalation compounds, 281
Lead, chalcogenide halides, 396–400
  phase diagrams, 396–398
  reflectance spectra, 400
  structural data, 399–400
  synthesis, 398

M

Magnetic circular dichroism (MCD) spectra, of matrix isolated species, 131–132, 138
Magnetism
  of chalcogenide halide compounds, 336, 347, 368, 375–376, 412
  of iron(III) dithiocarbamates, 237–241
Manganese
  chalcogenide halides, 379
  cryochemistry
    hydrides, 167
    nitrosylcarbonyl, 132–133
    oxides, 138
  1,1-dithiolato complexes, 232–233
Markownikoff orientation, and hydrostannation, 7
Mercury
  chalcogenide halides, 351–357
    phototropism, 356–357
    single crystal preparation, 353–354
    structural data, 355–356
    synthesis, 351–353
  trifluoromethyl derivatives, 178–180
    bis complex in synthesis, 192–197
  trifluorosilyl derivative, 207
Metal–metal bonding, in chalcogenide halides, 330, 373
Metal vapor synthesis
  complexes, 130–145
    clusters, 114–130
  metal clusters, 81–114
  organometallics, 145–166
    methyls, 204–206
    trifluoromethyls, 203–205, 207–208
Mössbauer spectra
  antimony chalcogenide halides, 411–412
  graphite intercalation compounds
    of antimony pentafluoride, 310
    of ferric halides, 307–309
    of xenon fluorides, 299
  of matrix isolated species, 81
    dimetallic iron species, 99–100, 167
    gold phosphine clusters, 164
    iron dinitrogen complexes, 130, 142–143
  of tin compounds, 2, 23
    chalcogenide halides, 395
    organo derivatives, 28, 40, 43

Molecular orbital (MO) calculations
  nickel ethylene complexes, 123–124
  and optical spectra of matrix isolated species
    copper ethylene complexes, 121–122
    dimetal molecules, 83, 86–90, 98–99, 104
    metal carbonyls, 131
  PN molecule, 143
  structure of triatomics
    nickel, 116
    silver, 96
Molybdenum
  chalcogenide halides, 370–377, 413
    Chevrel phases, 376–377
    metal–metal bonding, 330, 373
    structural data, 373–376
    as superconductors, 376
    synthesis, 371–372
  cryochemistry
    arene complexes, 147
    dimetal species, 85–87, 97–99
    photoaggregation of atoms with chromium, 108–114
  dithiocarbamate complexes, 244–231
    activation of molecular oxygen, 228
    industrial use, 225
    as oxidation catalysts, 228
    oxo derivatives, 224–229
    sulfur derivatives, 225–227, 230
    xanthates, 224–225

N

Nickel
  cryochemistry
    acetylene compounds, 153–154
    arene complexes, 145–147
    aryl compound, 149
    atom clustering, 89, 91–92
    carbon disulfide complexes, 163
    carbonyl clusters, 116–117
    and catalysis of hydrogenation, 91
    ethylene complexes, 81, 122–124
    formation of reactive metal slurries, 91–92
    hexafluoro-2-butyne compounds, 153–154
    olefin complexes, 149–152
    and organic rearrangements, 168
    phosphine complexes, 167

Nickel (cont'd.)
  1,1-dithiolato complexes, 254–261
    dithiocarbamates, 254–256, 260–261
    dithiocarbonates, 260
    dithiocarboxylates, 257–259, 265
    dithiophosphates, 260
    phenylation, 259
    trithiocarbonates, 260
    xanthates, 256–257, 261
Niobium
  chalcogenide halides, 364–369, 412–413
    as diamagnetic semiconductors, 368
    infrared and Raman spectra, 367
    magnetic properties, 367–368
    structural data, 366–368, 412–413
    synthesis, 364–366, 412
    XPS spectra, 413
  cryochemistry
    arene complex, 148
    dimetallic species, 85–87
    dinitrogen complex, 142
  dithiocarbamates, 219–221
Nitrogen, chalcogenide halides, 400
Nitrosyl complexes, with 1,1-dithiolato ligands, 218, 223–224, 231, 233, 235–237, 250, 253–254, 260
^{13}C-Nuclear magnetic resonance (NMR) spectra, of fluxional cyclopentadienyl molybdenum compounds, 231
^{19}F-NMR spectra
  of graphite intercalation compounds, 295, 297–300, 309–310, 313
  of polyfluorotetramethyl germanes, 199
  and reaction of CF_3 radicals with mercuric halides, 184–186
  of substituted silanes, 202
^{1}H-NMR spectra
  of nickel xanthate adducts, 257
  of polyfluorotetramethyl germanes, 200
  and stereochemical nonrigidity
    of chromium(III) dithiocarbamates, 222, 224
    of cyclopentadienyl molybdenum compounds, 231
  of substituted silanes, 201

O

Optical inversion, of tris(dithiocarbamate) complexes, 222, 253

Optical purity, of active organotin compounds, 8–9
Optical spectra
  of graphite–alkali metal compounds, 287–288
  of matrix isolated species
    acetylene complexes, 154
    carbonyls, 131, 133, 135, 137
    cobalt atom clusters, 86–89
    copper atom clusters, 103–104, 167
    copper group oxides, 139
    dimetal molecules, 83–87, 91, 97–99
    dinitrogen complexes, 131, 133, 142
    dioxygen complexes, 138–139
    gold anion, 100–101
    heteronuclear bimetallics, 97–99, 109–112, 115
    iridium atoms in argon, 90, 149–154
    olefin complexes, 121–123, 127–128, 149–154
    photodissociation of Group VI hexacarbonyls, 91
    silver atom clusters, 193–95, 102, 104–112, 167
Organotin compounds, 1–77
  acetates, 16
  alkoxides, 17–18
  allyls, 13–14
  amino derivatives, 18
  angle strain in stannacycloalkanes, 12
  biological activity, 41–48
    and alkyl chain length, 41, 45–47
    amino acid interactions, 42–43
    diorganotins, 45–47
    enzyme interactions, 45–46
    inhibition of oxidative phosphorylation, 41–42
    as insecticides, 41
    monoorganotins, 47
    nature of alkyl group, 42
    protein interactions, 43–44
    triorganotins, 41–45
  carbodiimides, 19
  cyclostannazanes, 20
  disproportionation, 4, 6, 13, 15–16, 19–21
  enamines, 18–19
  environmental degradation, 48–51
    biomethylation, 50–51
    via carbon hydroxylation, 48–49
    metabolism in mammals, 48–49

photochemical, 49
fluxionality, 14–15
functional substitution, 24–25
hydrides, 15–16
hydrostannation, 7, 16
industrial use, 51–61
  agrochemicals, 52–53
  as catalysts for polyurethane foaming, 60–61
  disinfectants, 53
  fungicides, 53–55
  in glass coating, 61
  in marine antifouling paints, 55–58
  pesticides, 51–53
  as PVC stabilizers, 9, 58–60
  silicone vulcanization catalysts, 60
  water soluble biocides, 55
Mössbauer spectra, 28, 40, 43
  and protein binding, 43
optical activity, 8–9
radicals, 22, 25–26
  addition to double bonds, 26
  ESR spectra, 22, 26
reaction with sulfur dioxide, 11
stannoxanes, 20
stannylenes, 26–28
  insertion reactions, 28
  oxidative addition, 27
structures of tin(IV) compounds, 28–39
  five-coordinate, 30–32
  four-coordinate, 29
  seven-coordinate, 35
  six-coordinate, 33–35
  table of data, 35–39
synthesis, 2–28
  chlorides, 16
  tin–nitrogen bonds, 18–20
  tin–oxygen bonds, 16–18
  tin–sulfur bonds, 19–21
thiolates, 19
tin–carbon bond cleavage, 10–15
  acidolysis, 10–11, 14–15
  mechanism, 10–12, 14
  by mercuric halides, 10–11
  by radicals, 12–13
tin–carbon bond formation, 2–10, 13–14
  via arylcopper(I) compounds, 6
  catalysis, 3–4
  from elemental tin, 3–4
  via Grignards, 4–7
  via organoaluminums, 4
  via organolithiums, 4–7
  via tin–Group IA compounds, 9–10
  via ylides, 6
  via zinc alkyls, 6
  with tin–metal bonds, 22–24
    Group IIB metals, 23–24
    magnesium, 22–23
  with tin–tin bonds, 21–22
    cleavage, 21–22
Osmium, 1,1-dithiolato complexes, 247–248
Oxidative addition reactions, in cryochemistry, 158–160

P

Palladium
  chalcogenide halides, 381–382
    structure, 382
  cryochemistry
    atom clustering, 89, 91
    dinitrogen complexes, 91
    dioxygen complexes, 138–139
    hexafluoro-2-butyne carbonyls, 153–154
    olefin complexes, 149, 151
    oxidative addition reactions, 158–160
  1,1-dithiolato complexes
    dithiocarbamates, 254, 260–262
    sulfur-rich dithiocarboxylates, 258–259
    xanthates, 256, 262
Phase transitions, in chalcogenide halide compounds, 332, 408, 412
Phosphorus
  chalcogenide halides, 400
  trifluoromethyl compounds, 179–180
Plasma generation, of trifluoromethyl radicals, 180–192
  experimental method, 181–183
  spectra of intermediates, 190–191
Platinum
  chalcogenide halides, 381–382
  cryochemistry, 89, 91
  1,1-dithiolato complexes, 258, 260, 262–265
    trithiocarbonates, 264
Plutonium, cryochemistry, 140, 142

R

Raman spectra
  of graphite, 287
    alkali metal compounds, 287
    bromine intercalation compound, 294
  of matrix isolated species, 81, 96
  of niobium sulfide halides, 367
Rhenium
  chalcogenide halides, 379–381
  1,1-dithiolato complexes, 233–236
Rhodium
  cryochemistry
    carbonyls, 117–118
    dioxygen complexes, 118–120, 122, 138–139
  1,1-dithiolato complexes, 253
Ruthenium
  1,1-dithiolato complexes, 246–248
    structure, 246–247
  organometallics via metal atom synthesis, 168

S

Scandium, matrix isolated diatomic, 83
Secondary ion mass spectrometry (SIMS), and matrix isolation studies, 167
Selenium, trifluoromethyl compounds, 180
Silicon
  chalcogen halides, 389–390
    as semiconducting glasses, 389–390
  trifluoromethyl compounds, 180
    via alkyl fluorination, 198–203
Silver
  chalcogen halides, 338–343, 412
    ionic conductivity, 341, 348
    phase diagrams, 341–343, 412
    in solid-electrolyte cells, 342, 348–349
    structural data, 340–341
    synthesis, 338–339
  cryochemistry
    atom clustering, 92–96, 101–102, 104–108
    carbonyls, 118–119, 134–135, 140
    ethylene complexes, 152–153
    heteronuclear diatomics, 100–101, 112–115
    optical spectra, 93–95, 102, 104–106, 108–112, 167
    oxides, 139
    PN compounds, 143
    synthesis of silver particles, 93–95
  1,1-dithiolato complexes, 268
  structure of hexametal cluster in zeolite, 95–96
Spin state changes, in iron(III) dithiocarbamates, 237–241
Stereochemical nonrigidity, of tris(dithiocarbamate) complexes, 253
Sulfur, trifluoromethyl compounds, 188–189

T

Tantalum
  chalcogenide halides, 364–369
    structural data, 366–367
    synthesis, 365–366
  dithiocarbamate complexes, 221
Technetium, carbonyl dithiocarbamate complex, 236
Tellurium, trifluoromethyl compounds, 181, 187–188, 191
Thallium, chalcogenide halides, 382
  synthesis, 388–389
Thorium, dinitrogen complex and nitride in matrix, 167
Tin, *see also* Organotin compounds
  chalcogenide halides, 390–396
    Mössbauer spectra, 395–396
    phase diagrams, 390–392
    reflectance spectra, 395–396
    structural data, 392–395
    synthesis, 392
  trifluoromethyl compounds, 186–187, 192, 194–195, 197
    via alkyl fluorination, 199–203
    reactivity of tin–carbon bond, 196
Titanium
  chalcogenide halides, 364
  cryochemistry
    dimetal species, 83
    dinitrogen complexes, 140–141
    hexacarbonyl, 140–141
    organometallic synthesis, 147, 155–156
    oxygen abstraction reactions, 162

dithiocarbamates, 216–218
  cyclopentadienyl compounds, 217–218
  xanthates, 217
Transition metal vapor cryochemistry,
    see also individual elements, 79–175
  in alkane matrices, 84–85
  alkene complexes, 149–152
    clusters, 120–130
  alkyne complexes, 152–155
  arene complexes, 145–149
  carbonyls, 130–138
    clusters, 115–118
  carboranes, 164–166
  cryophotoaggregation, 93, 101–107
    cluster distribution, 107–108
    mixed-metal species, 108–114
    selectivity, 103–106
  dimetallic species, 96–101
    via photoselective aggregation, 108–114
  dinitrogen complexes, 140–143
  dioxygen complexes, 138–140
    clusters, 118–120
  metal anion formation, 100–101
  nitrides, 140–143
  organic reactions, 160–163
    rearrangements, 168
  oxidative addition, 158–160
  oxides, 138–140
  polyatomic metal species, 81–96
  polyolefin complexes, 155–158
  siloxanes, 163–164
Trifluoromethyl compounds, *see also* individual elements, 178–210
  comparison with methyl analogs, 179
  synthesis, 179–208
    by direct fluorination, 179–180, 197–203
    from mercury compound, 192–197
    via metal atom synthesis, 179–181, 203–208
Trifluoromethyl radicals, plasma generation, 180–192

reactions, 183–192
  with halides, 183–187
  with metal atom vapors, 204–208
  with sulfur vapor, 188–189
Trifluorosilyl radicals, 206
  reactions with mercury, 207
Tungsten
  chalcogenide halides, 377–379
    structural data, 378–379
    synthesis, 377
  dithiocarbamate complexes, 229, 231
  metal atom synthesis of arene complexes, 148

V

Vanadium
  chalcogenide halides, 364–365
  cryochemistry
    carbonyls, 130–133
    dimetal species, 83–87
    dinitrogen complexes, 141
    organometallics, 147, 155
  dithiocarbamates, 218–220
  dithiocarboxylates 220–221
  xanthates, 220

X

Xenon fluorides, intercalation with graphite, 297–300

Y

Ytterbium hydrides, metal atom synthesis, 144

Z

Zinc, sulfur-rich dithiocarboxylate, 258–259
Zirconium
  cryochemistry, 167
  dithiocarbamate, 216

CONTENTS OF PREVIOUS VOLUMES

VOLUME 1

Mechanisms of Redox Reactions of Simple Chemistry
H. Taube

Compounds of Aromatic Ring Systems and Metals
E. O. Fischer and H. P. Fritz

Recent Studies of the Boron Hydrides
William N. Lipscomb

Lattice Energies and Their Significance in Inorganic Chemistry
T. C. Waddington

Graphite Intercalation Compounds
W. Rüdorff

The Szilard-Chambers Reactions in Solids
Garman Harbottle and Norman Sutin

Activation Analysis
D. N. F. Atkins and A. A. Smales

The Phosphonitrilic Halides and Their Derivatives
N. L. Paddock and H. T. Searle

The Sulfuric Acid Solvent System
R. J. Gillespie and E. A. Robinson

AUTHOR INDEX—SUBJECT INDEX

VOLUME 2

Stereochemistry of Ionic Solids
J. D. Dunitz and L. E. Orgel

Organometallic Compounds
John Eisch and Henry Gilman

Fluorine-Containing Compounds of Sulfur
George H. Cady

Amides and Imides of the Oxyacids of Sulfur
Margot Becke-Goehring

Halides of the Actinide Elements
Joseph J. Katz and Irving Sheft

Structure of Compounds Containing Chains of Sulfur Atoms
Olav Foss

Chemical Reactivity of the Boron Hydrides and Related Compounds
F. G. A. Stone

Mass Spectrometry in Nuclear Chemistry
H. G. Thode, C. C. McMullen, and K. Fritze

AUTHOR INDEX—SUBJECT INDEX

VOLUME 3

Mechanisms of Substitution Reactions of Metal Complexes
Fred Basolo and Ralph G. Pearson

Molecular Complexes of Halogens
L. J. Andrews and R. M. Keefer

Structure of Interhalogen Compounds and Polyhalides
E. H. Wiebenga, E. E. Havinga, and K. H. Boswijk

Kinetic Behavior of the Radiolysis Products of Water
Christiane Ferradini

The General, Selective, and Specific Formation of Complexes by Metallic Cations
G. Schwarzenbach

Atmosphere Activities and Dating Procedures
A. G. Maddock and E. H. Willis

Polyfluoroalkyl Derivatives of Metalloids and Nonmetals
R. E. Banks and R. N. Haszeldine

AUTHOR INDEX—SUBJECT INDEX

VOLUME 4

Condensed Phosphates and Arsenates
Erich Thilo

Olefin, Acetylene, and π-Allylic Complexes of Transition Metals
R. G. Guy and B. L. Shaw

Recent Advances in the Stereochemistry of Nickel, Palladium, and Platinum
J. R. Miller

The Chemistry of Polonium
  K. W. Bagnall
The Use of Nuclear Magnetic Resonance in Inorganic Chemistry
  E. L. Muetterties and W. D. Phillips
Oxide Melts
  J. D. Mackenzie
AUTHOR INDEX—SUBJECT INDEX

VOLUME 5

The Stabilization of Oxidation States of the Transition Metals
  R. S. Nyholm and M. L. Tobe
Oxides and Oxyfluorides of the Halogens
  M. Schmeisser and K. Brandle
The Chemistry of Gallium
  N. N. Greenwood
Chemical Effects of Nuclear Activation in Gases and Liquids
  I. G. Campbell
Gaseous Hydroxides
  O. Glenser and H. G. Wendlandt
The Borazines
  E. K. Mellon, Jr., and J. J. Lagowski
Decaborane-14 and Its Derivatives
  M. Frederick Hawthorne
The Structure and Reactivity of Organophosphorus Compounds
  R. F. Hudson
AUTHOR INDEX—SUBJECT INDEX

VOLUME 6

Complexes of the Transition Metals with Phosphines, Arsines, and Stibines
  G. Booth
Anhydrous Metal Nitrates
  C. C. Addison and N. Logan
Chemical Reactions in Electric Discharges
  Adli S. Kana'an and John L. Margrave
The Chemistry of Astatine
  A. H. W. Aten, Jr.

The Chemistry of Silicon–Nitrogen Compounds
  U. Wannagat
Peroxy Compounds of Transition Metals
  J. A. Connor and E. A. V. Ebsworth
The Direct Synthesis of Organosilicon Compounds
  J. J. Zuckerman
The Mössbauer Effect and Its Application in Chemistry
  E. Fluck
AUTHOR INDEX—SUBJECT INDEX

VOLUME 7

Halides of Phosphorus, Arsenic, Antimony, and Bismuth
  L. Kolditz
The Phthalocyanines
  A. B. P. Lever
Hydride Complexes of the Transition Metals
  M. L. H. Green and D. L. Jones
Reactions of Chelated Organic Ligands
  Quintus Fernando
Organoaluminum Compounds
  Roland Köster and Paul Binger
Carbosilanes
  G. Fritz, J. Grobe, and D. Kummer
AUTHOR INDEX—SUBJECT INDEX

VOLUME 8

Substitution Products of the Group VIB Metal Carbonyls
  Gerard R. Dobson, Ingo W. Stolz, and Raymond K. Sheline
Transition Metal Cyanides and Their Complexes
  B. M. Chadwick and A. G. Sharpe
Perchloric Acid
  G. S. Pearson
Neutron Diffraction and Its Application in Inorganic Chemistry
  G. E. Bacon

Nuclear Quadrupole Resonance and Its
    Application in Inorganic Chemistry
    Masaji Kubo and Daiyu Nakamura

The Chemistry of Complex
    Aluminohydrides
    E. C. Ashby

AUTHOR INDEX—SUBJECT INDEX

VOLUME 9

Liquid-Liquid Extraction of Metal Ions
    D. F. Peppard

Nitrides of Metals of the First Transition
    Series
    R. Juza

Pseudohalides of Group IIIB and IVB
    Elements
    M. F. Lappert and H. Pyszora

Stereoselectivity in Coordination
    Compounds
    J. H. Dunlop and R. D. Gillard

Heterocations
    A. A. Woolf

The Inorganic Chemistry of Tungsten
    R. V. Parish

AUTHOR INDEX—SUBJECT INDEX

VOLUME 10

The Halides of Boron
    A. G. Massey

Further Advances in the Study of
    Mechanisms of Redox Reactions
    A. G. Sykes

Mixed Valence Chemistry—A Survey and
    Classification
    Melvin B. Robin and Peter Day

AUTHOR INDEX—SUBJECT INDEX—
    VOLUMES 1–10

VOLUME 11

Technetium
    *K. V. Kotegov, O. N. Pavlov, and
    V. P. Shvedov*

Transition Metal Complexes with Group
    IVB Elements
    J. F. Young

Metal Carbides
    William A. Frad

Silicon Hydrides and Their Derivatives
    B. J. Aylett

Some General Aspects of Mercury
    Chemistry
    H. L. Roberts

Alkyl Derivatives of the Group II Metals
    B. J. Wakefield

AUTHOR INDEX—SUBJECT INDEX

VOLUME 12

Some Recent Preparative Chemistry of
    Protactinium
    D. Brown

Vibrational Spectra of Transition Metal
    Carbonyl Complexes
    Linda M. Haines and M. H. Stiddard

The Chemistry of Complexes Containing
    2,2′-Bipyridyl, 1,10-Phenanthroline,
    or 2,2′,6′,2″-Terpyridyl as Ligands
    W. R. McWhinnie and J. D. Miller

Olefin Complexes of the Transition
    Metals
    H. W. Quinn and J. H. Tsai

Cis and Trans Effects in Cobalt(III)
    Complexes
    J. M. Pratt and R. G. Thorp

AUTHOR INDEX—SUBJECT INDEX

VOLUME 13

Zirconium and Hafnium Chemistry
    E. M. Larsen

Electron Spin Resonance of Transition
    Metal Complexes
    B. A. Goodman and J. B. Raynor

Recent Progress in the Chemistry of
    Fluorophosphines
    John F. Nixon

Transition Metal Cluster with π-Acid
  Ligands
  R. D. Johnston
AUTHOR INDEX—SUBJECT INDEX

VOLUME 14

The Phosphazotrihalides
  M. Bermann
Low Temperature Condensation of High
  Temperature Species as a Synthetic
  Method
  P. L. Timms
Transition Metal Complexes Containing
  Bidentate Phosphine Ligands
  W. Levason and C. A. McAuliffe
Beryllium Halides and Pseudohalides
  N. A. Bell
Sulfur–Nitrogen–Fluorine Compounds
  O. Glemser and R. Mews
AUTHOR INDEX—SUBJECT INDEX

VOLUME 15

Secondary Bonding to Nonmetallic
  Elements
  N. W. Alcock
Mössbauer Spectra of Inorganic
  Compounds: Bonding and Structure
  G. M. Bancroft and R. H. Platt
Metal Alkoxides and Dialkylamides
  D. C. Bradley
Fluoroalicyclic Derivatives of Metals and
  Metalloids
  W. R. Cullen
The Sulfur Nitrides
  H. G. Heal
AUTHOR INDEX—SUBJECT INDEX

VOLUME 16

The Chemistry of Bis(trifluoromethyl)-
  amino Compounds
  H. G. Ang and Y. C. Syn

Vacuum Ultraviolet Photoelectron
  Spectroscopy of Inorganic Molecules
  R. L. DeKock and D. R. Lloyd
Fluorinated Peroxides
  Ronald A. De Marco and Jean'ne
  M. Shreeve
Fluorosulfuric Acid, Its Salts, and
  Derivatives
  Albert W. Jache
The Reaction Chemistry of Diborane
  L. H. Long
Lower Sulfur Fluorides
  F. Seel
AUTHOR INDEX—SUBJECT INDEX

VOLUME 17

Inorganic Compounds Containing the
  Trifluoroacetate Group
  C. D. Garner and B. Hughes
Homopolyatomic Cations of the Elements
  R. J. Gillespie and J. Passmore
Use of Radio-Frequency Plasma in
  Chemical Synthesis
  S. M. L. Hamblyn and B. G. Reuben
Copper(I) Complexes
  F. H. Jardine
Complexes of Open-Chain Tetradenate
  Ligands Containing Heavy Donor
  Atoms
  C. A. McAuliffe
The Functional Approach to Ionization
  Phenomena in Solutions
  U. Mayer and V. Gutmann
Coordination Chemistry of the Cyanate,
  Thiocyanate, and Selenocyanate Ions
  A. H. Norbury
SUBJECT INDEX

VOLUME 18

Structural and Bonding Patterns in
  Cluster Chemistry
  K. Wade

Coordination Number Pattern Recognition Theory of Carborane Structures
  Robert E. Williams

Preparation and Reactions of Perfluorohalogenoorganosulfenyl Halides
  A. Haas and U. Niemann

Correlations in Nuclear Magnetic Shielding. Part I
  Joan Mason

Some Applications of Mass Spectroscopy in Inorganic and Organometallic Chemistry
  Jack M. Miller and Gary L. Wilson

The Structures of Elemental Sulfur
  Beat Meyer

Chlorine Oxyfluorides
  K. O. Christe and C. J. Schack

SUBJECT INDEX

VOLUME 19

Recent Chemistry and Structure Investigation of Nitrogen Triiodide, Tribromide, Trichloride, and Related Compounds
  Jochen Jander

Aspects of Organo-Transition-Metal Photochemistry and Their Biological Implications
  Ernst A. Koerner von Gustorf, Luc H. G. Leenders, Ingrid Fischler, and Robin N. Perutz

Nitrogen–Sulfur-Fluorine Ions
  R. Mews

Isopolymolybdates and Isopolytungstates
  Karl-Heinz Tytko and Oskar Glemser

SUBJECT INDEX

VOLUME 20

Recent Advances in the Chemistry of the Less-Common Oxidation States of the Lanthanide Elements
  D. A. Johnson

Ferrimagnetic Fluorides
  Alain Tressaud and Jean Michel Dance

Hydride Complexes of Ruthenium, Rhodium, and Iridium
  G. L. Geoffroy and J. R. Lehman

Structures and Physical Properties of Polynuclear Carboxylates
  Janet Catterick and Peter Thornton

SUBJECT INDEX

VOLUME 21

Template Reactions
  Maria De Sousa Healy and Anthony J. Rest

Cyclophosphazenes
  S. S. Krishnamurthy, A. C. Sau, and M. Woods

A New Look at Structure and Bonding in Transition Metal Complexes
  Jeremy K. Burdett

Adducts of the Mixed Trihalides of Boron
  J. Stephen Hartman and Jack M. Miller

Reorganization Energies of Optical Electron Transfer Processes
  R. D. Cannon

Vibrational Spectra of the Binary Fluorides of the Main Group Elements
  N. R. Smyrl and Gleb Mamantov

The Mossbauer Effect in Supported Microcrystallites
  Frank J. Berry

SUBJECT INDEX

VOLUME 22

Lattice Energies and Thermochemistry of Hexahalometallate(IV) Complexes, A_2MX_6, which Possess the Antifluorite Structure
  H. Donald B. Jenkins and Kenneth F. Pratt

Reaction Mechanisms of Inorganic Nitrogen Compounds
  G. Stedman

Thio-, Seleno-, and Tellurohalides of the Transition Metals
M. J. Atherton and J. H. Holloway

Correlations in Nuclear Magnetic Shielding, Part II
Joan Mason

Cyclic Sulfur–Nitrogen Compounds
H. W. Roesky

1,2-Dithiolene Complexes of Transition Metals
R. P. Burns and C. A. McAuliffe

Some Aspects of the Bioinorganic Chemistry of Zinc
Reg H. Prince

SUBJECT INDEX